Dietmar Findeisen

System Dynamics and Mechanical Vibrations

Springer
Berlin
Heidelberg
New York
Barcelona
Hong Kong
London
Milan
Paris
Singapore
Tokyo

Dietmar Findeisen

System Dynamics and Mechanical Vibrations

An Introduction

With 89 Figures

Springer

Professor Dr.-Ing. Dietmar Findeisen
Institute of Machine Design
Department 11 Mechanical Engineering and Production Technology
Technical University Berlin
Straße des 17. Juni 135
10623 Berlin
Germany

Division S.3 Scientific Equipment Design
Department S Interdisciplinary Scientific and Technological Operations
Federal Institute for Materials Research and Testing
Unter den Eichen 87
12205 Berlin
Germany

Library of Congress Cataloging-in-Publication Data
Findeisen, Dietmar.
System dynamics and mechanical vibrations / Dietmar Findeisen.
Includes bibliographical references and index.
ISBN 3-540-67144-7
1. Vibration. 2. System analysis. 3. Linear systems. 4. Mechatronics. I. Title.
TA 355.F47 2000
621.8'11--dc21 00-032969

ISBN 3-540-67144-7 Springer-Verlag Berlin Heidelberg New York

Springer-Verlag Berlin Heidelberg New York
a member of BertelsmannSpringer Science+Business Media GmbH

Typesetting: Camera-ready copy from author, layout by Marianne Schillinger-Dietrich, Berlin
Cover-Design: de'blik, Berlin

Printed on acid-free paper SPIN: 10573437 68/3020 – 5 4 3 2 1 0 –

Dedicated to Hanne-Marie

Preface

The Aim of the Book. This book is concerned with the subjects of vibrations and system dynamics on an integrated basis.

Design engineers find themselves confronted with demands made on machinery, structures and dynamic systems which are increasing at such a rate that dynamic performance requirements are always rising. Hence, advances in analysis and design techniques have to keep pace with recent developments in strong lightweight materials, more extensive knowledge of materials properties and structural loading. Whereas the excitation applied to structures is always increasing, the machine mass and damping is reduced. Consequently, unwanted vibrations can have very serious effects on dynamic systems. It is, therefore, essential to carry out vibration analysis as an inherent part of machine design.

The problems arising either from the observed or predicted dynamic behaviour of systems are of particular interest in control theory. Vibration theory places emphasis on analysis, which implies determining the response to given excitations, and any design amounts to changing the system parameters so as to bring about a satisfactory response. The improvement in performance achieved by changing solely the parameters of the mechanical system is very limited. However, a new approach to system design has proved to be more successful. It consists of designing forces that, when exerted on the system, produce a satisfactory response. This approach, known as control, has become a ubiquitous part of the engineering curriculum, completing the conventional mechanical disciplines.

It was L. Meirovitch who anticipated from a philosophical point of view that the three subjects rigid-body dynamics, vibrations and control really belong together. With this most persuasive argument he prepared the ground for an integrated approach.

Accordingly, C. F. Beards has pointed out, from a pedagogical point of view, that an integrated approach also leads to greater efficiency in the teaching of vibrations and control, with the duplication of teaching material being eliminated. The understanding of the individual subjects is likely to be enhanced because the basic equations governing the behaviour of vibratory and control systems are the same.

The reader interested in modern analysis and control techniques using state space approach and progressive matrix methods should refer to the many excellent advanced specialized texts. The aim of this book is to give practising designers, engineers and students of mechanical engineering a thorough understanding of the fundamentals of system dynamics. This more general area, covering vibrations and control as essential parts, is linked with advanced texts, thus providing a theoretical basis appropriate to further studies. Methods associated with "classical" control theory are still widely used in dynamic system analysis. Most applications can be handled with relatively simple models. In fact, tried-and-tested classical design methods have been significantly enhanced by modern computational techniques. The graphical tools of classical design can now be more easily used with computer graphics, and the effects of nonlinearities and model approximations evaluated by computer simulation. Today's engineers should be familiar with both classical methods and new computational improvements.

This book, which has been conceived as a professional reference book, should also prove suitable as a textbook for courses ranging from the junior level to the senior level. To help with the tailoring of the material to a given course, a chapter-by-chapter review of the material follows.

Contents. *Chapter 1* introduces the subject of systems modelling, and presents a general classification of physical quantities. The quantities are classified in proven categories following A. G. J. MacFarlane.

In *Chapter 2* the interaction of dynamic system variables is visualized through diagrams. Significant types of systematic diagram which apply to both electrical circuits and dynamics of control are presented to demonstrate their use for creating mechanical model systems. Though the key point of the representation of mechanical systems is the network diagram (mechanical circuit), a comprehensive overview of useful types of diagram is given with respect to the mixed domain system structure of present-day systems predominating in real engineering situations.

Chapter 3 provides mathematical relations between the system variables of interacting mechanical elements (subsystems). System responses to relevant specific excitations are evaluated by the classical method, including phasor response analysis and subsequently Fourier series analysis. Passing to the Fourier integral the benefits of the Fourier transform method are demonstrated. Considering modern frequency concepts in data reduction (spectral analysis) as well as response calculations in shock and vibrations, special emphasis is placed on non-periodic forcing functions (pulse-type excitations) as well as on stochastic force time histories (random excitations). Subsequently, the Laplace transform method, which is of great significance both for control and vibrations, is presented. Its suitability for response calculations in shock and vibrations is exemplified with regard to suddenly applied external forces (step-type excitations). Finally, the special features of Fourier and Laplace transform methods are contrasted.

In *Chapter 4* frequency-response analysis comes into focus. Integral transform methods introduced in Chapter 3 provide the link to transform models (ω-domain models) covering sinusoidal steady-state analysis as well as frequency concepts. An effective method for lumped-system analysis is gained by combining fundamental laws of general networks (force and motion interconnective requirements) with frequency-response characteristics. The mechanical mobility has proved a useful concept in vibration data analysis. The dynamic compliance (receptance) is a characteristic related to mobility which turns out to be the more convenient concept for machinery considering deformations of structures as the result of vibratory effects. By comparison, the principles underlying both concepts are presented alongside the historical reciprocals known as mechanical impedance and dynamic stiffness. According to mechanical circuit theorems, even for complex structures simplifications are possible in order to provide the equivalent model system. As pointed out, the network diagram reduction (repeated structures in parallel and series) results in a direct viewing procedure for gaining the overall dynamic characteristic of structures.

Chapter 5 treats the behaviour of energetic systems already taken as the basis for H. M. Paynter's dynamic systems approach. The theory of vibrations is favourably covered in the following two problems, namely that of measuring instruments and that of system analysis, both of which are designed to reduce unwanted vibrations. Looking at desired vibrations being applied to a variety of industrial processes or testing procedures (vibration method), the problem of system synthesis comes into focus. In order to meet the demand for rating and optimizing the driving power flow in machinery, this book has made two attempts at an energetic system approach. The transmission problem of power is dealt with using either the mechanical 2-port or the mechanical circuit as significant ω-domain models and expressing the stationary flow of energy in terms of an algebraic function, the complex power. This relationship is graphically interpreted by vector representation modified to power relations. Spectral decomposition of power and its polar representation are efficient tools for specifying the significant power parameters of vibrating systems over a frequency range of interest. Finally, the concepts of phasor power and dynamic compliance are combined to utilize this relationship for an integrated system design based on a dynamic and an energetic approach.

The book concludes with *Appendices* giving an illustrative survey of normalized temporal and frequency responses and visualizing response specifications with the aid of graph (scale) papers. The polar plots of power originally introduced in this text follow the rectangular spectral decomposition of power on coordinate (squared) paper to facilitate the extraction of the relevant performance criteria.

Acknowledgements. This book has been evolved from a set of classnotes that I have prepared at the Technical University Berlin for senior grade students of mechanical engineering. I have attempted to develop the view of the field in my own way, keeping up with technological advances as well as experience gained by teaching and practice over several years. I wish to acknowledge the helpful comments and suggestions offered by my students. I am also indebted to the Federal Institute for Materials Research and Testing, Berlin, which supported my research into the complex dynamical system of the fatigue testing machine during my activity in the design of scientific testing equipment.

I would like to express my appreciation to my former instructor and colleague Professor K. Federn who contributed indirectly to this book through his early publications and his excellent lecture on vibration machines. From this senior-level course I received the decisive impulse to deal with the subject of mechanical vibrations by placing the focal point on vibration generator systems.

I wish to acknowledge gratefully the contribution of Dr. P. Schmiechen who translated the classnotes of my course on vibration machines which form part of this graduate text. Furthermore, Peter Craven provided his editorial assistance.

Special thanks are due to Christel Blaß for her excellent job in typing the manuscript, which has been completed with endeavour by Renate Landgraf. I would also like to thank Simone Nickel for her efficient help in producing the figures, in parts supplemented by Gudrun Blamberg.

The editorial and production staff of Springer-Verlag deserve my thanks for their cooperation and thorough professional work in producing this book.

Finally, I thank my wife for her patience and encouragement during the period of preparation of this book.

Berlin, April 2000 *Dietmar Findeisen*

Contents

Index of Formula Symbols XI

1 Theory of Dynamic Systems 1
1.1 Definitions and Overview of Systems Modelling 1
1.2 General Classification of Dynamic System Variables 4
1.2.1 Classification in Terms of Spatial Relationships 4
1.2.2 Classification in Terms of Local Energy State 5

2 System Representation by Diagrams (Model System) 9
2.1 Block-diagram and Signal-flow-diagram Representation 9
2.1.1 Transfer Function Block Diagram 10
2.1.2 Control System Structure 12
2.1.3 Control System Design. *Fundamental Aspects* 13
2.1.4 Signal Flow Graphs. *Reduction of the Diagram* 18
2.2 Two-port-diagram Representation 20
2.2.1 Generic Two Port. *Two-terminal-pair Network* 20
2.2.2 Connection of Two Ports. *Fundamental Configurations* 22
2.2.3 Mechanical Two Ports 25
2.2.4 Fundamental Mechanical Elements. *Fundamental Configurations* 28
2.2.5 Supplementary Mechanical Elements. *Couplers and Sources* 32
2.2.6 Connections with Block Diagrams. *Transfer Function Block Diagrams* 39
2.2.7 Analysis of Complex System Structures. *Fluid System* 46
2.3 Network-diagram Representation (Circuit Diagram) 48
2.3.1 Direct Representation of Simple Systems by Networks 49
2.3.2 Fundamental and Supplementary Mechanical Elements 50
2.3.3 Construction of Mechanical Network Diagram. *Mechanical Circuit* 53
2.3.4 Derivation of Mechanical Network Equations. *Equation of Motion* 54
2.3.5 Connections with Signal-flow Diagrams. *Oriented Linear Graphs* 56
2.4 Combined-flow-diagram Representation 58
2.4.1 Symbolic Measurement of Dynamic System Variables 59
2.4.2 Symbolic Regulation of Supplementary Elements. *Couplers and Controlled Sources* 59

2.4.3 Connections between Network and Block Diagrams. *Mixed Domain System Structures* 60
2.5 Bond-graph Representation (Multiports) 63
2.5.1 Classification of Multiports 65
2.5.2 Conventions for Interconnected Multiports. *Augmented Bond Graphs* 66
2.5.3 Fundamental Interconnective Relationships. *Generalized Kirchhoff's Laws* 66
2.5.4 Construction of Mechanical Bond Graph. *Mechanical Multiport* 68
2.6 Comparison of Diagram Representations (References to Applications) 69
2.6.1 Schematic Diagrams. *Visually Descriptive Diagrams* 70
2.6.2 Systematic Diagrams. *Interconnection Diagrams* 71

3 System Representation by Equations (Mathematical Model) 74
3.1 Representation of Mechanical Systems by Differential Equations of Motion (Classical Method) 74
3.1.1 System Specifications by Normalization of the Differential Equation. *Time-response Analysis* 75
3.1.2 Free and Forced Response of Damped Second-order Systems 77
3.1.3 Forced Response of a Single Degree-of-freedom System to Complex Excitation. *Phasor-response Analysis* 89
3.2 Representation of Mechanical Systems by Integral-transformed Models (Transform Methods) 95
3.2.1 Periodic Vibration. *Fourier Series Analysis* 96
3.2.2 Non-periodic Vibration. *The Fourier Integral* 101
3.2.3 Fourier Transform Method. *Frequency-response Analysis* .. 106
3.2.4 System Response to Transient Excitation. *Pulse-type Functions* 116
3.2.5 Random Vibration. *Data Processing* 125
3.2.6 System Response to Random Excitation. *White Noise* 133
3.2.7 Transient Vibration. *The Laplace Integral* 139
3.2.8 Laplace Transform Method. *Transfer-function Analysis* 143
3.2.9 System Response to Transient Excitation. *Step-type Functions* 154
3.2.10 The Graphical Interpretation of the Transfer Function. *Conformal Mapping* 166
3.2.11 The Graphical Interpretation of the Frequency-response Function. *Frequency Response Plots* 176
3.3 Comparison of Fourier and Laplace Transform Methods (References to Applications) 194
3.3.1 Fourier Transform Method. *Advantages and Disadvantages* 194
3.3.2 Laplace Transform Method. *Advantages and Disadvantages* 196

4 Transform Analysis Methods of Vibratory Systems (Frequency-response Characteristics) 198
4.1 Formulation of Dynamic Equations (Equations of Motion) 198
4.1.1 Analytical Dynamics. *Mathematical System by Analytical Methods* 199
4.1.2 Synthetical Dynamics. *Mathematical System by Synthetical Methods* 200
4.2 Frequency-response Characteristics (Concepts of Mobility and Dynamic Compliance) 207
4.2.1 Equivalent Definitions of Frequency-response Function.... 207
4.2.2 Dynamic Characteristics of Mechanical Elements. *Component Mobilities and Dynamic Compliances*........... 212
4.2.3 Dynamic Characteristics of Composite Systems. *Overall Mobility and Dynamic Compliance* 217
4.2.4 General Transform Analysis Principles. *Mechanical Circuit Theorems* 231
4.2.5 Graphical Methods to Mechanical System Design. *Selecting Vibratory Specifications by Polar Diagrams* 237
4.2.6 Some Exercises in Transform Analysis Methods. *Applying Dynamic Compliance Techniques* 249

5 The Flow of Power and Energy in Systems (Energy Transactions)... 257
5.1 Power Transmission through Linear Two Ports (Generalized Transport Process)... 262
5.1.1 The Transmission Problem of Two-port Networks. *Unrestricted Terminal Conditions*............................ 263
5.1.2 The Transmission Problem related to Complex Power. *Generalized Quadratic Forms* 265
5.1.3 The Power Transmission Factor. *Generalized Transmission Ratio* 270
5.2 Power Transmission through Mechanical Networks (Generalized Impedance) 281
5.2.1 The Transmission Problem of One-port Networks. *Functional Relationships*...................................... 281
5.2.2 The Transmission Problem related to Complex Power. *Phasor Power* 283
5.2.3 Connections with Frequency-response Characteristics. *Combining Dynamic Compliance and Phasor Power Concepts*... 297

Appendix A
Time-history Curves
(Displacement Response and Force Excitation) 317

Appendix B
Frequency Response Plots
(Normalized Dynamic Compliance and related Characteristics) 343

Appendix C
Frequency Response Plots of Power (Normalized Complex Power).... 361

References 369

Index 375

Index of Formula Symbols

Symbol	Quantity	Symbol for Unit
a_n	Fourier coefficient	–
$A(\omega)$	amplitude (frequency-) response	–
$A(\omega_f)$	displacement response factor (amplitude ratio or gain assigned to a single forcing frequency)	m/N
A_P	active-power transmission factor	1
A_Q	real-power transmission factor	1
$\underline{A}_s$	transmisson factor of complex power	–
b_n	Fourier coefficient	–
c	(viscous) damping coefficient	Ns/m
c_c	critical damping coefficient	Ns/m
c_n	Fourier coefficient	–
$\hat{c}_n$	Fourier coefficient amplitude	–
$\underline{c}_n$	complex Fourier coefficient, complex amplitude spectrum	–
$\underline{C}\,(j\omega)$	dynamic compliance (receptance)	m/N
$\underline{C}_{ii}\,(j\omega)$	direct (driving-point) dynamic compliance	m/N
$\underline{C}_{ij}\,(j\omega)$	transfer dynamic compliance	m/N
C_c	compliance of the damper	m/N
C_k	compliance of the spring	m/N
C_m	compliance of the mass	m/N
$e\,(t)$	effort (variable)	–
E	error variable (transform)	–
E_p	potential energy (T-storage element state)	J
E_k	kinetic energy (P-storage element state)	J
$f\,(t)$	flow (variable)	–
f_d	damped natural frequency (cyclic)	s^{-1}

f_n	harmonic frequency (cyclic)	s^{-1}
f_0	natural frequency (cyclic)	s^{-1}
f_1	fundamental frequency (cyclic)	s^{-1}
$F(t)$	force (excitation, driving force)	N
$\underline{F}(t)$	complex excitation (of force)	N
$\underline{F}(\omega)$	(Fourier) excitation transform, spectral density (of force)	Ns
$\underline{F}(p)$	(Laplace) excitation transform (driving transform)	Ns
$\hat{F}$	force (excitation-) amplitude	N
$\underline{\hat{F}}$, $\underline{\tilde{F}}$	force phasor, (r.m.s.) force phasor	N
F_d	damping (or damper) force	N
F_g	(local) force of gravity	N
F_m	inertial (or mass) force	N
F_s	elastic (or spring) force	N
F_R	rectangular pulse (of force)	N
F_0	maximum height	N
g	acceleration of gravity	m/s^2
$g(t)$	unit pulse response (weighting function)	–
$G(\omega)$, $G(j\omega)$	frequency response function, frequency transfer function	–
$G(p)$, $(G(s))$	transfer function	–
$h(t)$	unit step response	–
$H(\omega)$	system function	–
I	current, (r.m.s.) value	A
I_R	impulse (pulse area) of rectangular pulse (of force)	Ns
J	moment of inertia	$kg \cdot m^2$
k	gear ratio	1
k	elastic (spring) constant (stiffness)	N/m
k	torsional stiffness	Nm/rad
$\underline{K}(j\omega)$	dynamic stiffness	N/m
$\underline{K}_{ii}(j\omega)$	direct (driving-point) dynamic stiffness	N/m
$\underline{K}_{ij}(j\omega)$	transfer dynamic stiffness	N/m
$\underline{K}^{*\prime}(j\omega)$	converted dynamic stiffness (conjugate)	N/(ms)
K_c	stiffness of the damper	N/m
K_k	stiffness of the spring	N/m
K_m	stiffness of the mass	N/m
K_0	proportional action coefficient	–
m	mass (parameter)	kg
p, (s)	complex frequency (angular),	

	complex pulsatance (Laplace domain variable)	
p	momentum	kg · m/s
$p(s)$	probability density function (of displacement magnitudes)	m^{-1}
$P(s)$	cumulative probability distribution function (of displacement magnitudes)	1
$P(t)$, $S(t)$	instantaneous power, actual power (rate of energy transfer)	V·A
P, P_{-}, S_{-}	active power (average power)	W
Q	reactive power	V·A
Q	quality factor (Q factor)	1
$\mathrm{r}\,(t)$	ramp response	–
R	feed back signal (transform)	–
R_{d}	(displacement) magnification factor	1
$R\,(\tau)$	correlation function	–
R_{FF}	autocorrelation function (at force)	N^2
R_{Fs}, R_{sF}	cross-correlation function (between force and displacement, v.v.)	N·m
R_{ss}	autocorrelation function (at displacement)	m^2
$s\,(t)$	displacement (response)	m
$\underline{s}\,(t)$	complex response (of displacement)	m
$\underline{s}\,(\omega)$	(Fourier) response transform, spectral density (of displacement)	m·s
$\underline{s}\,(p)$	(Laplace) response transform	m·s
$\hat{s}$	displacement (-response) amplitude	m
$\underline{\hat{s}}$	displacement phasor	m
s_{Init}	primary (initial) response	m
s_{Res}	residual response	m
s_{stat}	static deflection	m
$S\,(\omega)$	power spectral density	–
S_{FF}	auto-spectral density (auto-spectrum) (at force)	N^2·s
S_{Fs}, S_{sF}	cross-spectral density (cross-spectrum) (between force and displacement, v.v)	N·m·s
S_{ss}	auto-spectral density (at displacement)	m^2 s
S_0	white noise auto-spectrum (at force)	N^2·s
$S\,(t)$	instantaneous power (actual scalar quantity or energy flow)	V·A
$\underline{S}\,(t)$	phasor of the instantaneous power	V·A
$\underline{S}\,(\mathrm{j}\omega)$	complex power (phasor power)	V·A
S	apparent power	V·A

t	time (variable),	s
	time of observation (running parameter)	s
T, M	torque	N·m
$\hat{T}$	torque (-excitation) amplitude	N·m
$\underline{\hat{T}}$	torque phasor	N·m
T, T^{F}, T^{v}	transmissibility (force, velocity)	1
T_d	damped natural period	s
T_f	forcing period	s
T_r, τ	time constant (relaxation time)	s
T_0	natural period	s
T_1, T_2	time constant	s
u	input, excitation	–
$u(t)$	excitation displacement	m
$\hat{u}$	amplitude of excitation displacement	m
$\underline{\hat{u}}$	excitation displacement phasor	m
$u_0(t)$	unit step (excitation), Heaviside function	1
$u_1(t)$	unit ramp (excitation)	s
U	input (transform)	–
U	voltage, (r.m.s.)value	V
υ	output, response	–
$\upsilon(t)$	velocity	m/s
$\hat{\upsilon}$	amplitude of velocity	m/s
$\underline{\hat{\upsilon}}$, $\underline{\tilde{\upsilon}}$	velocity phasor, (r.m.s.) velocity phasor	m/s
V	output (transform)	–
W	reference variable (transform)	–
W, (A)	work	J
X	controlled variable (transform)	–
$\underline{X}_k(\mathrm{j}\eta)$	normalized dynamic stiffness (excitation external or via spring)	1
$\underline{X}_k^{*'}(\mathrm{j}\eta)$	normalized phasor power (conjugate)	1
$\underline{\tilde{X}}_k(\mathrm{j}\eta)$	normalized mechanical impedance	1
$\underline{X}_m(\mathrm{j}\eta)$	normalized dynamic stiffness (excitation via unbalanced rotating mass)	1
X_{kc}	normalized stiffness of the damper	1
X_{kk}	normalized stiffness of the spring	1
X_{km}	normalized stiffness of the mass	1
Y	manipulated variable (transform)	–
$\underline{Y}(\mathrm{j}\omega)$	(mechanical) mobility (mechanical admittance)	m/(N · s)
$\underline{Y}_{ii}(\mathrm{j}\omega)$	direct (driving-point) (mechanical) mobility	m/(N · s)

$\underline{Y}_{ij}(j\omega)$	transfer (mechanical) mobility	$m/(N \cdot s)$
Y_c	mobility of the damper	$m/(N \cdot s)$
Y_k	mobility of the spring	$m/(N \cdot s)$
Y_m	mobility of the mass	$m/(N \cdot s)$
$\underline{Y}_k(j\eta)$	normalized dynamic compliance (excitation external or via spring)	1
$\underline{Y}_m(j\eta)$	normalized dynamic compliance (excitation via unbalanced rotating mass)	1
Y_{kc}	normalized compliance of the damper	1
Y_{kk}	normalized compliance of the spring	1
Y_{km}	normalized compliance of the mass	1
z	complex variable	–
$\underline{Z}(j\omega)$	mechanical impedance	Ns/m
$\underline{Z}_{ii}(j\omega)$	direct (driving-point) mechanical impedance	Ns/m
$\underline{Z}_{ij}(j\omega)$	transfer mechanical impedance	Ns/m
Z_c	impedance of the damper	Ns/m
Z_k	impedance of the spring	Ns/m
Z_m	impedance of the mass	Ns/m
α	T-variable rate (across power variable, effort variable)	–
γ	imaginary part of complex variable z	–
δ	damping coefficient	s^{-1}
$\delta(t)$	unit pulse (excitation) (Dirac or delta function, δ functional)	s^{-1}
ζ, ϑ	damping ratio (fraction of critical damping)	1
η	frequency ratio (ratio of forcing frequency to undamped natural frequency)	1
η	efficiency (ratio of an output power to an input power)	1
η_r	resonance frequency ratio	1
η_1	forcing frequency ratio (assigned to a single forcing frequency)	1
ϑ	duty cycle (pulse control factor)	1
Θ	angular position	rad
λ	T-variable state (across energy variable)	–

λ	power factor	1
Λ, (δ)	logarithmic decrement	Np
σ	P-variable state (through energy variable)	–
σ	real part of complex variable *p* (or *s*)	s
σ_F	standard deviation (of force magnitudes)	N
σ_s	standard deviation (of displacement magnitudes)	m
σ^2_s	variance (of displacement magnitudes)	m^2
τ	P-variable rate (through power variable, flow variable)	–
τ	non-dimensional time (variable) (natural time)	1
τ	dummy variable in time (variable of integration)	s
τ_d	normalized natural period	1
τ_r	normalized time constant (non-dimensional relaxation time)	1
τ_0	pulse duration	s
τ_1	normalized forcing period	1
φ	angular displacement	rad
$\hat{\varphi}$	angular displacement amplitude	rad
$\underline{\hat{\varphi}}$	angular displacement phasor	rad
φ_0	phase angle	rad
φ_{0F}	initial phase of (excitation) force	rad
φ_{0s}	initial phase of displacement (response)	rad
φ_1	impedance angle	rad
ψ	phase difference (phase shift)	rad
$\psi(\omega)$	phase (frequency-) response	–
ψ_1	phase difference (assigned to a single forcing frequency)	rad
ω	(real) angular frequency, pulsatance (frequency domain variable)	s^{-1}
ω_{cf}	break point (corner) frequency	s^{-1}
ω_d	damped natural frequency (angular)	s^{-1}
ω_f	forcing angular frequency	s^{-1}

ω_r	(displacement) resonance frequency (angular)	s^{-1}
ω_0	natural frequency (angular)	s^{-1}
ω_1	fundamental frequency (angular)	s^{-1}
ω_n	harmonic frequency (angular)	s^{-1}

Symbol for Operator	**Operator**
d	differential
F	Fourier transformation
F^{-1}	inverse Fourier transformation
Im	imaginary part of
L	Laplace transformation
L^{-1}	inverse Laplace transformation
Prob	probability
Re	real part of
Res, R_k	residue, residue of $f(p)$ at p_k
R_S	modified Rayleigh quotient
T	operator, system operator
δ	change of state
Δ	discriminant
Π	product of
Σ	sum of
Φ	static functional operator
Ψ	general functional operator

Remarks. Symbols for quantities being used are in the main conforming to ISO 2041:1990 (E/F) (Vibration and shock-Vocabulary), and IEC 50 (101):1977 (International Electrotechnical Vocabulary; Chapter 101: Mathematics).

Furthermore the following standards are of use concerning *Quantities and units*: ISO 31 (Part 1: Space and time; Part 2: Periodic and related phenomena; Part 3: Mechanics; Part 11: Mathematical signs and symbols for use in the physical sciences and technology), and *Vibration and shock-Experimental determination of mechanical mobility*: ISO 7626/1:1986 (E) (Part 1: Basic definitions and transducers), International Electrotechnical Vocabulary: IEC 50 (131):1978 (Chapter 131: Electric and magnetic circuits).

The sympol p is given for the complex frequency (complex pulsatance) though the symbol s is in use as Laplace variable in mathematics and electrotechnical science. In mechanical science the symbol s is recommended for displacement as the prime motion variable quantity. Consequently it will be justified to give the "reserve symbol" p for the complex quantity or Laplace domain variable.

1 Theory of Dynamic Systems

The theory of dynamic systems was treated in a fundamental way by *K. Küpfmüller*, [1], to derive the general relationships between input and output quantities in telecommunications. Later it was extended to automatic controller design and evolved, along with information theory and the theory of automata, into mathematical kybernetics, with applications not restricted to a particular area.

Physical systems or real-world arrangements are studied to understand and predict their behaviour or to gain an insight into their mechanism. Real arrangements can be studied either directly by observation (experiment) or indirectly by studying models of the physical system. The choice of study – direct or indirect – depends on several factors including the existence and availability of the physical system, its complexity, and the associated cost and time.

Some terms needed to distinguish between the various models are detailed in the following paragraphs:

Experimental modelling treats selected mathematical relationships through "induction" concerning an already existing system by fitting its observed input-output data.

Scale models are of similar shape to the physical object, which can be used directly for measurements. Wind-tunnel facilities are examples of this type.

Other types of physical models are the *prototype* and the *pilot model*. The first possesses an almost one-to-one correspondence with the system under consideration; the latter may be a scaled-down physical representation of the real arrangement. Physical models are usually costly and time consuming in addition to presenting a modest level of flexibility in terms of modification.

Mathematical modelling of systems is a process which can be separated into three partial problems: the initial identification and idealization of a system's elements (subsystems); secondly, of their interaction; and finally, the systematic application of basic (physical, biological, economic, etc.) laws. This process involves "deduction" and offers distinct methods for reducing the disadvantages of experimental modelling which have already been discussed.

The modern theory of dynamic systems includes modelling, analysis and control of systems. It offers mathematical methods for handling, reducing and analysing data from abstract objects by using digital computers, [2] to [4].

1.1 Definitions and Overview of Systems Modelling

The purpose of *systems modelling* is to predict the behaviour of an engineering device consisting of a known collection of physical objects. This ordered arrangement defines the *physical system*. The *model system* is a collection of abstract objects with properties determined from experiments with the physical system. Its components interact with one another and with their environment.

If they interact in such a way that a certain input results in a certain output, the arrangement of components is called a *control system*; otherwise a system is uncontrolled. At this stage, the internal interactions are not considered ("black box") so that the control system may be realized either as an open-loop or a closed-loop arrangement.

A *deterministic control system* is one in which the input-output relationship is predictable and repeatable at any time; otherwise, a system is stochastic. For example, an electromechanical system such as the direct current motor (DC motor) is a deterministic control system that contains several electrical and mechanical components. They interact in such a way that a specific value of the "input voltage" will result in a specific "output velocity" at any time.

A *dynamic system* is one where the output depends not only on the input at the present time, but also on its previous behaviour. In contrast to a static system, a system depending on past behaviour is said to have internal dynamics. This is another way of saying that it has internal energy storage elements. Taking up the example of the DC motor, its "output velocity" can only be predicted as long as its entire time history is known and not just the instantaneous value of the "input voltage".

A *linear dynamic system* is a system that has an input-output relationship where the output to two inputs applied together (simultaneously) is simply the sum of the individual outputs; otherwise, a system is nonlinear. An important property of linear systems concerns the transfer of combined signals. If the system excitation can be represented as a linear combination of some independent input signals, the system response will also be a linear combination of some independent output signals, each corresponding to the output of the individual input. Direct consequences of linearity are that the input-output relationship is scalable and that zero excitation provides zero response.

A *time-invariant dynamic system* is a system where the characteristics of its internal dynamics do not change with time; otherwise, a system is time-variant. An important property of time-invariant systems concerns the transfer of time-shifted signals. The system response to an input signal applied at a later time is identical to that to an input signal applied at an earlier time. The only difference between the two system responses is that they are shifted in time by the excitation time shift.

For *continuous-time systems* the data about its internal dynamics are known as continuous functions of time. If they are known at discrete instants of time only, the systems are called discrete-time systems.

System analysis is continued by observing physical systems, in particular by performing experiments. The only reason for this is that systems modelling has to accept limitations in representing physical phenomena. The *mathematical model* based on a set of abstract objects only imperfectly describes the arrangement of interacting physical objects. The model response will exhibit the phenomena observed on the physical system only if the model contains all relevant features of the physical system. To ensure the correspondence between model and physical system, current terms accepted in computer simulation should be defined.

The *computerized model* is an operational computer program that implements a system's model. A record of predicted behaviour of the system is obtained from computer run(s). Measurements, on the other hand, make it possible to obtain a record (table, graph) of observations of the physical system behaviour.

The *model verification* is defined as the substantiation that a computerized model represents the system's model within specified limits of accuracy.

The *model validation* implies in its essence the level of agreement between observed and predicted behaviour. Validating a model requires comparing its behaviour (simulation results) with that of the physical system (measured or observed data). This assumes that the verification step has been performed to avoid confusing faults of the program with faults in the model.

Experimentation is a way of enhancing the understanding of physical phenomena and fine-tuning model systems by expediently adapting mathematical methods. It is obvious that the building of a credible model is an iterative process; in other words, a model is modified so as to reduce the differences between model and system behaviour.

This book is concerned with deterministic, linear, time-invariant, continuous-time, dynamic systems for which the relationships between excitation and system response are derived from physical laws. The model behaviour can be represented as an ordinary differential equation with constant coefficients.

Physical laws relating to mechanical vibrating systems include Newton's second law (for discrete systems) and other basic relations describing the behaviour of mechanical elements. Excitations treated in the following are continuous-time signals of various deterministic or stochastic types as they are applied to engineering devices.

In system analysis terminology, systems are often referred to as *plants*, or *processes*. Moreover, the excitation is known as *input signal*, or simply input, and response as *output signal*, or simply output. It is convenient to represent the relationship between input and output schematically in terms of a block diagram as shown in Fig. 1.1.

First step of modelling. To define the functions or operations of a physical system three entities must be identified: *inputs*, *system functional block diagram*, and *outputs*. Inputs are the stimuli that cause the system to produce responses or outputs. There are two types of input variables acting on the system. Those inputs which can be controlled or changed are considered as desirable, and those which occur out of control are undesirable. The latter type is called a disturbance input.

Identified relations are marked in the block diagram by noting the cause and the response in terms of physical variables and noting nominally the functional rela-

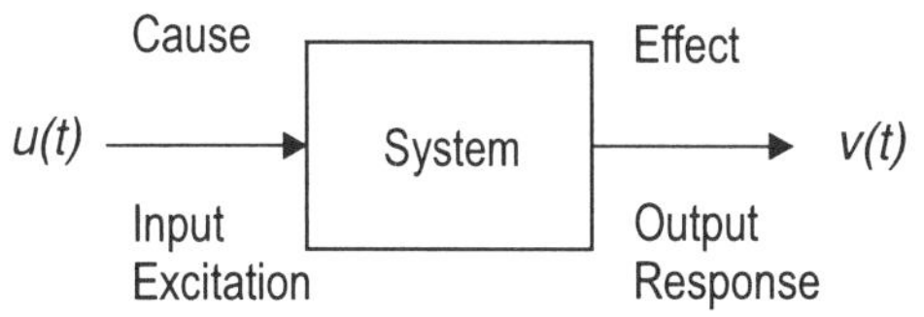

Fig. 1.1. Graphical symbol of the basic system (system functional block diagram)

tionship between input and output. In this way, the functional block diagram for a specific system will be formed.

For example, if "input voltage" and "motor shaft speed" are noted at the action line entering and leaving the box respectively, and furthermore "speed control drive system" is noted on the inside of the box, Fig. 1.1 may represent the system functional block of the referred DC motor.

1.2 General Classification of Dynamic System Variables

The interaction between a system functional block and its environment or between functional subsystems happens by related signals. They will be identified as input and output signals by measurements at terminal points. Applied to systems or their components these quantities represent the system variables.

Any general discussion of system analysis procedure requires the adoption of names for the different types of variables and for the general types of elements. A first systematic classification of physical quantities was introduced by *A. Sommerfeld,* [8]. This fundamental statements were taken up and developed by *A. G. J. MacFarlane*, [9], in his definition of dynamic system variables. The physical quantities are found to be expediently classified in terms of both spatial relationships and local energy state.

1.2.1 Classification in Terms of Spatial Relationships

The physical variables which determine the flow of energy in the dynamic system differ in the way they are related in physical space.

P-Variables **(*1-point* or *through variables*).** The fundamental physical quantities of force, momentum, charge, current, and entropy are said to be *1-point variables* since their specification (or ideal measurement) at a given point in space involves only that single point in space. To identify this type of spatially intensive variables, they will be termed pervariables, abbreviated *P-variables.* Spatial-intensive variables are often termed *through variables* since certain variables are propagated through the measuring instruments normally used for their measurement. For example, force is propagated through a spring balance, current through an amperemeter.

T-Variables **(*2-point* or *across variables*).** Variables such as displacement, velocity, temperature, and voltage are said to be *2-point variables* since their specification (or ideal measurement) involves two points in space. In most cases one of the two points involved is a reference point. To identify this type of spatially extensive variables, they will be termed transvariables, abbreviated *T-variables.* Spatially extensive variables are often termed *across variables* since they are propagated through an ideal measuring instrument across the two points in space. For example, velocity measurement is performed on two specific points between physical object and inertial reference system.

Classification of Dynamic Model Components

Each component of a dynamic system model may be considered as representing a relationship between pairs of measurements performed on the physical system. In all cases one of the measurements is of a T-variable and the other is of a P-variable. Physical reasoning shows that all the objects of which dynamic system models are composed involve a measured or defined relationship between one T- and one P-variable. For example, a spring is defined in terms of a relationship between the pervariable force and the transvariable displacement. This relationship is called the spring characteristic.

The objects of dynamic system models are not usually directly related to measurements on distinct physical objects, as may be the case for simple physical systems, e. g., arrangements consisting of springs and rigid bodies. However, the overall dynamic behaviour of the complex physical object, satisfactorily represented by the complete dynamic system model, is based on concepts ultimately derived from measurements. They will be performed on distinct physical objects at pairs of points in space.

The types of dynamic variables to be used for systems modelling also allow storage elements to be classified as T- and P-storage elements.

Translational mechanical system variables. Mechanical systems involve two distinct types of state. Isolated physical objects remain in a state of uniform translational motion. This tendency is attributed to a property of matter termed inertia.

Physical objects return to a given configuration after being deformed by an acting force. This tendency is attributed to a property of materials termed elasticity.

The quantitative measure of the inertia of an object is its mass, that of the elasticity of a given object configuration is its stiffness. The first translational state will be termed the kinetic type, the second one represents the static type.

Change of mechanical state. Force can be defined in terms of measurements of changes in translational state. The force exerted on an isolated body is proportional to the resultant acceleration of the body. The constant relating force and acceleration is the mass of the body. The force on a constrained body is a function of the resultant deformation. In general, the relationship between applied force and resultant displacement for a constrained body must be determined, by measurement or calculation, for each relevant configuration. Such a relationship will be termed a stiffness characteristic.

Relationships between mechanical system variables will be derived from a change in translational mechanical state. A change of state includes the incremental work done as well as the instantaneous power.

The symmetrical set of relationships between the four translational mechanical system variables is shown in Fig. 1.2.

1.2.2 Classification in Terms of Local Energy State

For systems of all types the description of dynamic behaviour requires the use of four distinct types of physical variable. Both the terms for these types and the general types of storage element have to be defined. For the translational mechanical

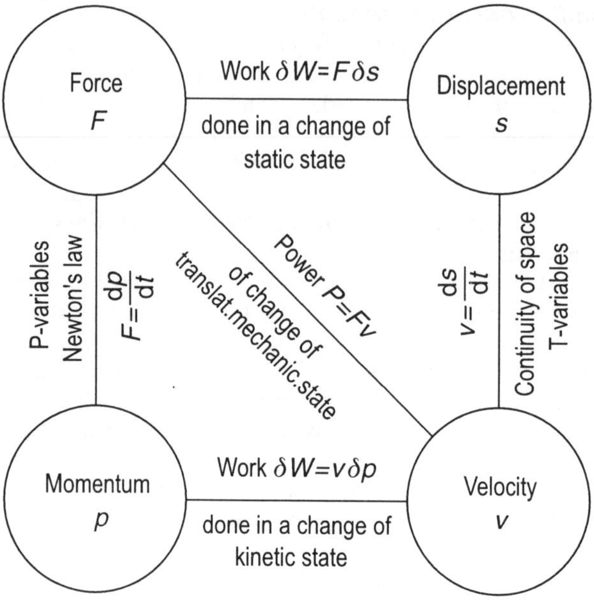

Fig. 1.2. Relationships between translational mechanical systems variables by MacFarlane [9]

system the four variables are:

– displacement, momentum, force and velocity.

The spatially related distinction of dynamic variables between P- and T-variables will be completed by classification of variables related to energy storage and flow. This may be provided by introducing the terms quantity, [8], or state [9], and intensity, [8], or rate, [9].

***State Variables* (*quantity* or *energy variables*).** Variables such as displacement and momentum in the translational mechanical case are termed *quantity* or *state variables.* State variables are direct measures or quantities of stored system energy. Alternatively the term *energy variable* is convenient to system analysis.

***Rate Variables* (*intensity* or *power variables*).** Variables such as force and velocity in the translational mechanical case are termed *intensity* or *rate variables.* Their product gives the rate at which work is done on a dynamic object or implies the intensity of energy flowing. Alternatively the term *power variable* is convenient to system analysis.

Rate variables for storage elements are the time rate of change of the storage element state variable:

$$\frac{\mathrm{d}}{\mathrm{d}t}\,[\text{state (or energy) variable}] = \text{rate (or power) variable.}$$

This provides the corresponding relationships between mechanical quantities, such as momentum and force or displacement and velocity:

$$\frac{\mathrm{d}p}{\mathrm{d}t} = F \qquad \text{and} \qquad \frac{\mathrm{d}s}{\mathrm{d}t} = v$$

General Classification of Dynamic System Variables

For a given type of system, the physical kind of objects and signals suffices to classify the variables as rate and state variables. Since there is always one P-variable and one T-variable in the pair of state and also of rate variables, the fundamental types of physical variable used in the analysis of dynamic systems may be classified and marked by general symbols as:

T-variable state:	λ
T-variable rate:	α
P-variable state:	σ
P-variable rate:	τ.

To describe the interaction of components by their intensity of energy flowing the preferably used terms for spatially corresponding rates or power variables are:

effort variable (or across power variable): α
flow variable (or through power variable): τ.

In the case of translational mechanical variables the variables are classified and termed as:

T-variable state (or across energy variable):	displacement s
T-variable rate (or *effort variable*):	*velocity* υ
P-variable state (or through energy variable):	momentum p
P-variable rate (or *flow variable*):	*force* F.

A *dynamic model system storage element* for which a T-variable state is a direct measure of stored energy will be termed *T-storage element*, and a dynamic model system storage element for which a P-variable state is a direct measure of stored energy will be termed a *P-storage element*.

With reference to the two different states of the physical system and the distinct types of energy, the storage elements for the mechanical case are termed:

T-storage element (or *potential energy storage element*):	*spring*
P-storage element (or *kinetic energy storage element*):	*mass*

In general, the incremental work done by change of state as well as the resulting instantaneous power are

$$\delta(\text{work}) = (\text{P-variable rate}) \times \delta(\text{T-variable state})$$

and

$$\text{power} = (\text{P-variable rate}) \times (\text{T-variable rate}) = \text{flow variable} \times \text{effort variable}.$$

This provides the corresponding relationships between mechanical quantities as:

$$\delta W = F\delta s \quad ; \quad \delta W = \upsilon\delta p$$

and

$$P = F\upsilon.$$

The relationships between the general dynamic system variables are summarized in Fig. 1.3.

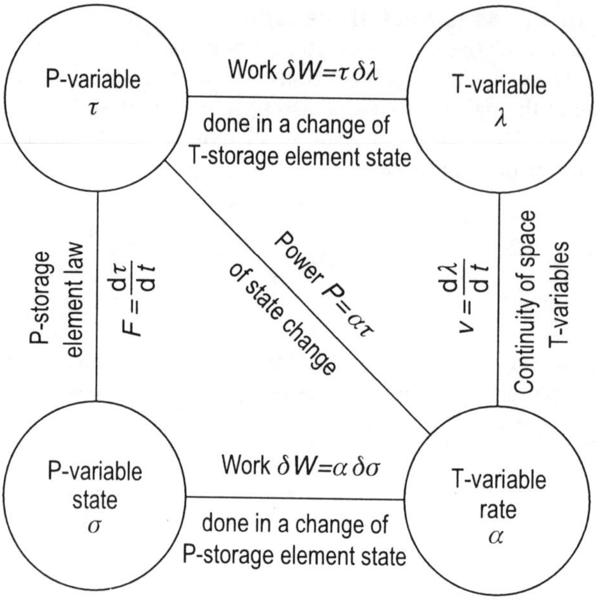

Fig. 1.3. Relationships between general dynamic system variables by MacFarlane [9]

2 System Representation by Diagrams (Model System)

Besides the abstract mathematical description of the input-output relations, the relations between the system variables can be visualized by diagrams.

Figurative proceeding is more convenient to solve problems in *device engineering*. However, simple design representations – called *schematic diagrams* – are helpful only for depicting methods of operation of the physical system. The description of dynamic system behaviour is based on a model system which requires specific types of diagram. Models for a structured system, especially for the system components are restricted to defined types of element representing the abstract objects of the model system. Contrary to schematic diagrams whose purpose is solely descriptive systematically constructed diagrams are used for the analysis procedure in *systems engineering*. An important part of systematic approach to the design of complex systems is to express the system specifications in a form suitable for the following mathematical parts of the problem.

Significant types of *systematic diagram* preferably used in electrical engineering are treated in the following sections with regard to mechanical model systems.

2.1 Block-diagram and Signal-flow-diagram Representation

In portraying systems block diagrams have proven to be an effective tool which is not only useful in visualizing the structure of the model but also helpful in the communication between builders and users of models.

Block Diagram Definitions

This type of systematic diagram represents the functional relationships in terms of which an observer may describe the physical system behaviour. That equals the representation of the system behaviour in terms of the spatial flow of signals.

The signals are indicated by orientation arrows on the line segments, and the interconnections are displayed by blocks or boxes, in the case of unspecified or unknown relations by empty or "black" boxes, [5] to [7].

Known relations are marked either by denoting the functional relationship between input and output quantities or by depicting a block symbol or a picture in the box, here called the functional block, Fig. 2.1.

Key element of the theory of dynamic systems is the *transmission system* that relates uniquely the output to the input. *Single-variable systems* relate one output signal to one input signal (one-input, one-output systems).

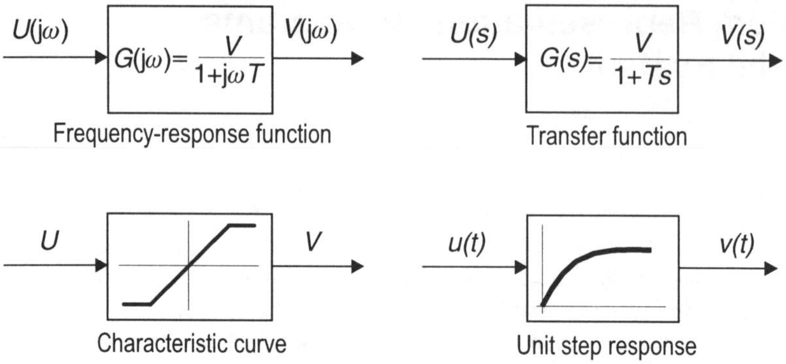

Fig. 2.1. Basic system with notation of the transmission behaviour (functional block)

A block may represent a linear or a nonlinear element. Obviously the block diagram is an oriented diagram which defines the cause-and-effect relation between variables in its causal sense. It consists of 4 basic elements: (functional) block, line segment, pickoff point and summing node.

The *functional block diagram* of a system is adequate for the identification of quantities associated with subsystems. The interconnection of components will be found by following the paths of signal flow along the connecting lines. However, for the representation of a system in terms of interconnected subsystems, more detailed developments of block diagrams are required. In addition, the mathematical operations relating the inputs and outputs of the subsystems must be indicated on the figurations of systematic diagrams.

The *simulation block diagram* represents operations pictorially by notations in the blocks. Upgraded to a computerized model it is convenient for elementary operations on time domain signals, such as multiplication by a constant or by a time-varying coefficient, differentiation, and integration.

The *transfer function block diagram* offers an efficient means for using diagrams in system analysis. This computerized model makes use of the transforms of input and output signals on the corresponding functional block. The transfer function is indicated inside the block, here called transfer function block.

2.1.1 Transfer Function Block Diagram

One of the advantages of the transfer function representation is the simplicity of the algebraic relations between the subsystem or component transfer functions. Thus, confining systems modelling to linear time-invariant behaviour the overall system transfer function can be easily obtained.

The resulting algebraic relationships between (the transforms of) the variables, especially at a summing node and at a pickoff point, are illustrated in Fig. 2.2.

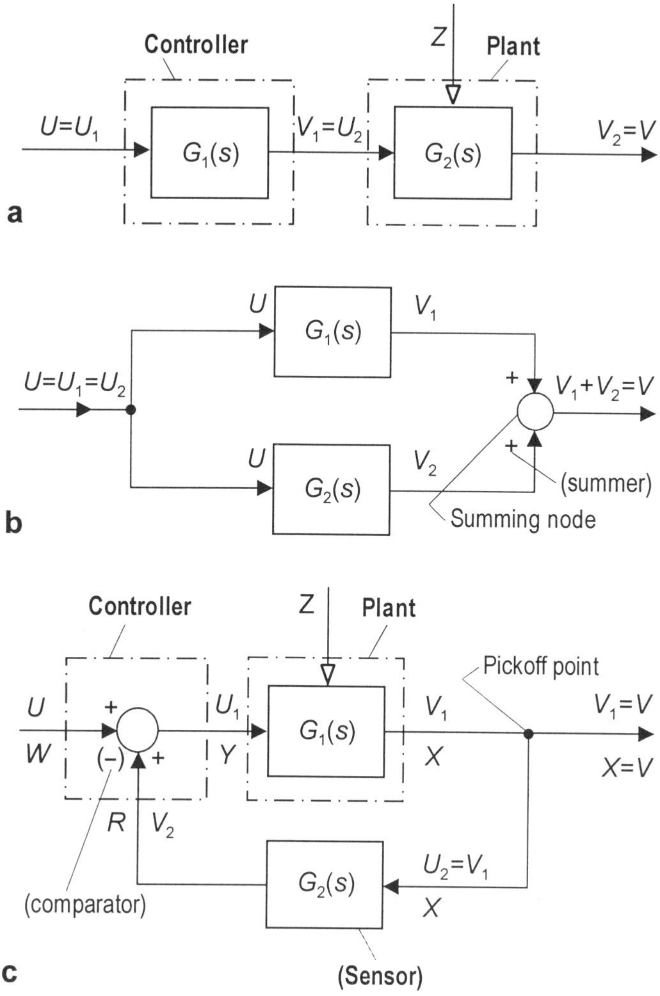

Fig. 2.2. Block diagrams of fundamental configurations (basic structures of control systems). **a** Cascade; **b** parallel; **c** feedback loop

Overall system transfer functions $G(s)$

$$G(s) = G_1(s) \cdot G_2(s) \tag{2.1}$$

$$G(s) = G_1(s) + G_2(s) \tag{2.2}$$

$$G(s) = \frac{G_1(s)}{1 \mp G_1(s)\, G_2(s)} \tag{2.3}$$

For two components which are connected in *cascade* or *tandem*, Fig. 2.2a, the output V_1 of the first block is the input U_2 to the second block. The transfer function of the cascade is equal to the product of the individual transfer functions, Eq. (2.1).

For two components in *parallel* whose inputs U_1, U_2 branching out at pickoff point are the same and whose outputs V_1, V_2 joining at summing node are added, Fig. 2.2b, the transfer function of the combined system is equal to the sum of the individual transfer functions, Eq. (2.2).

For two components connected in a *circuit* or a *feedback loop*, Fig. 2.2c, the loop transfer function can be derived from cascade and parallel operations.

Using a *summer* to join the inputs U, V_2 the input line segments are indicated by a plus sign. Therefore the product of the individual transfer functions is subtracted from 1 in the denominator of the system transfer function, Eq. (2.3).

2.1.2 Control System Structure

The fundamental configurations of a block diagram even serve as *basic structures of control systems*. As mentioned before, a control system is arranged to regulate or adjust the signal flow in some desired manner. To distinguish a system that acts without outside intervention from a manually controlled system sometimes the term *automatic control system* is used that can be either an open or a closed loop.

Open-loop System. The type of diagram depicted in Fig. 2.2a represents the basic structure of an *open-loop control system*. The desired output U (reference variable) acts as input U_1 on the controller and the output V_1 of the controller acts as input U_2 on the plant. The latter component defines the object to be controlled.

Using feedforward blocks in series, the output does not affect the input. Therefore the transfer function of the cascade corresponds to that one related to an open-loop control system, Eq. (2.1).

The desired output of the plant U remains the same irrespective of the actual output V, so that open-loop controllers are limited to situations where events are quite predictable. However, this type of controller is not satisfactory for interaction in processes with occurring disturbances. This occurrence is marked by adding a line segment for the disturbance input Z on the plant block.

Closed-loop System. The diagram modified to a general single-loop system in Fig. 2.2c represents the basic structure of a *closed-loop* or *feedback control system*. This type of control system uses measurements of the output to modify the system's action in order to achieve the intended goal. Therefore a sensor forming the feedback block is added to the controller which detects the error between the actual output V (controlled variable X) and the desired output U. This control operation implies the comparison of two converted signals of the same type, i.e., the transduced actual output (feedback signal R) must be subtracted from the desired output U (reference variable W; here $W = U$), to form the actuating (error) signal (E) being subsequently transduced to the input of the plant U_1 (manipulated variable Y). Thus, by error detection the input on the plant is manipulated so that the error caused by the disturbance input Z is reduced. To perform the connection variables by subtraction instead of addition the summer has to be replaced by a node called comparator which represents a closed-loop controller in its simplest form. Sometimes the output transducer (sensor) is not of interest. By eliminating the feedback block the control system may be simplified to a single-loop system with unity feedback.

Using a *comparator* to join the inputs W, R one of the input line segments (feedback input) is indicated by a minus sign. The addition of feedback input R and forward input W would create a positive feedback system which acts by increasing the deviation between desired and actual output (regenerative feedback). Control operation requires diminishing of the output deviation (degenerative feedback), hence control design presupposes a system with negative feedback. Therefore the product of the individual transfer functions is added to 1 in the denominator of the system transfer function related to closed-loop control system (negative nonunity feedback system), Eq. (2.3).

2.1.3 Control System Design. *Fundamental Aspects*

The closed-loop system is much superior to the open-loop system in that it responds satisfactorily to changes in commands and maintains system performance in the presence of disturbances.

Performance Requirements. The *robustness* and the *linearity* of the original system are improved because feedback (or parallel) compensation reduces the parameter sensitivity and extends the linear range of nonlinear element characteristics.

On the other hand the introduction of a feedback loop around a stable system has the disadvantage that the compensated system can potentially become unstable.

To analyse the performance of control systems with higher specifications, more detailed models and analytical techniques for determining the *stability* must be introduced.

The general function of the controller is to keep the controlled variable near its desired value whenever a change in command or a disturbance is caused. To produce the control signal by acting on the error signal control logic elements must be arranged for following up a specifically implemented algorithm. By this algorithm describing the *control law* specifications are layed down to govern the process. Specific control objectives, such as

- minimize the rise time and/or the settling time,
- minimize the transient error and/or the steady-state error

are explicit statements for control design to evaluate the system performance in terms of *speed* and *accuracy*.

To realize specific control laws in automatic control systems, it will be necessary either to generate the reference variable W by transducing the command input U as well as to form the manipulated variable Y by affecting the error variable E. Thus, the design of a controller generally requires that the error detector will be completed by input elements and control logic elements ("brain" of the system).

Hence the controller before acting on the plant interacts with the actuator ("muscle" of the system) which finally produces the driving signal Y. Both the actuator containing final control elements and the output transducer consisting of feedback elements are essential implements at the interface coupling the control equipment and the process, Fig. 2.3.

Block diagram algebra can be used for the evaluation of control system's performance by deriving the error variable E and its interacting variables from standard diagram, Fig. 2.3.
Assuming the simplifications

$$A_c(s) = G_{c1}(s) = 1 \quad ; \qquad U = W, \quad E = Y_R$$
$$G_c(s) = G_{c2}(s)$$

the input-output relationships of standard types of control systems are stated

Open-loop system $H(s) \equiv 0$

If there is noise (or a disturbance) applied to the plant the plant output will be

$$X = G_p(s)(Y + Z) = G_p(s)G_c(s)W + G_p(s)Z \tag{2.4}$$

such that the output deviation (actuating difference or error) is

$$\Delta X = G_p(s)Z \tag{2.5}$$

i.e., noise affects directly the output.

Closed-loop system $H(s) > 0$

Introducing a feedback loop plant output will be

$$X = G_p(s)(Y + Z) = G_p(s)\left[G_c(s)(W - H(s)X) + Z\right] \tag{2.6}$$

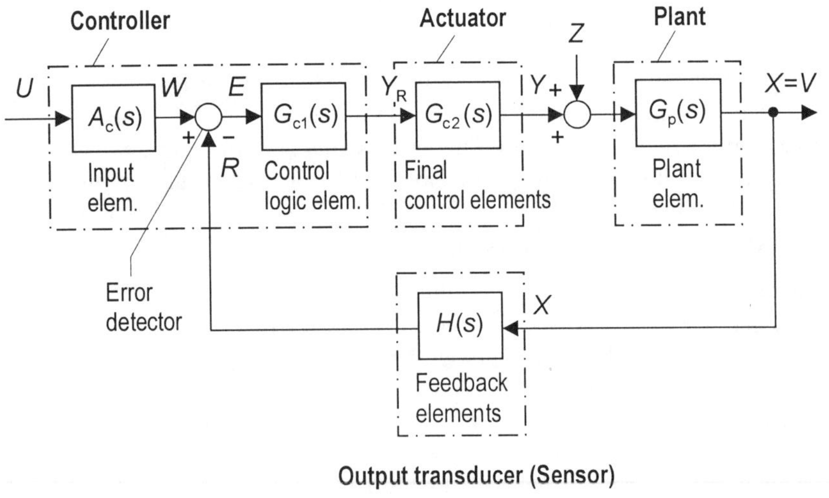

Fig. 2.3. Block diagram of control system configuration (standard diagram of feedback control systems)

so that

$$X = \frac{G_c(s)G_p(s)}{1 + H(s)G_c(s)G_p(s)} W + \frac{G_p(s)}{1 + H(s)G_c(s)G_p(s)} Z \tag{2.7}$$

The effect of change in command and of occuring disturbance on controlling is indicated either by
the *primary* or *reference transfer function* $G_r(s)$

$$\frac{X}{W} = \frac{G_c(s)G_p(s)}{1 + H(s)G_c(s)G_p(s)} = G_r(s) \tag{2.8}$$

or by the *disturbance transfer function* $G_d(s)$

$$\frac{X}{Z} = \frac{G_p(s)}{1 + H(s)G_c(s)G_p(s)} = G_d(s) \tag{2.9}$$

The effect on the plant output due to noise is described by the disturbance transfer function $G_d(s)$, Eq. (2.9). Hence it follows the remaining output deviation

$$\Delta X = G_d(s)Z \approx \frac{1}{H(s)G_c(s)} Z \qquad for \quad H_s(s)G_c(s)G_p(s) \gg 1 \tag{2.10}$$

which can be made arbitrarily small by increasing the overall gain of the feedback system $H_s(s)G_c(s)$, Eq. (2.10).

To study the effect of parameter variations the *sensivity* S will be considered which is defined as the ratio between change in control system parameters ΔG to change in plant parameters ΔG_p, in each case related to the overall system transfer function G

$$S = \frac{\Delta G / G}{\Delta G_p / G_p} \approx \frac{dG / G}{dG_p / G_p} = \frac{dG}{dG_p} \frac{G_p}{G} \tag{2.11}$$

For the open-loop system the transfer behaviour is given by feedforward blocks in series $G(s) = G_c(s)G_p(s)$, Eq. (2.1); therefore

$$S = \frac{G_c dG_p}{dG_p} \frac{G_p}{G_c G_p} \equiv 1 \tag{2.12}$$

However, for the closed-loop system the transfer behaviour is determined by a feedback loop as described in the reference transfer function $G_r(s)$, Eq. (2.8), so that

$$S = \frac{G_c dG_p}{(1 + HG_c G_p)^2 \, dG_p} \frac{G_p (1 + HG_c G_p)}{G_c G_p} = \frac{1}{1 + HG_c G_p} \tag{2.13}$$

Similar to the susceptibility to noise, in a closed-loop system, the sensivity to system parameter changes can be made arbitrarily small by increasing the overall gain of the control system HG_c.

Performance of Position Control. If a load with inertia J is to be positioned at some desired angle Θ_r by means of the mentioned dc-motor the taken up angle position Θ equals not necessarily the desired value Θ_r (command input U). This may result from the disturbance (disturbance input Z) of different origin, acting as a disturbance torque T_d on the load. For this reason the feedback loop will be closed by a feedback potentiometer (feedback element) measuring the actual position (controlled variable X) and transducing it in a proportional wiper voltage U_a. A command potentiometer (input element) generates a voltage U_r proportional to the desired angle Θ_r. The measured voltage U_a (feedback signal R) and the generated voltage U_r (reference variable W) are compared by a differential amplifier resulting in the error voltage U_e (error variable E) being a nonzero voltage when Θ does not equal Θ_r. The differential amplifier (error detector) will be completed by a power amplifier (control logic element).

Command potentiometer and amplifier together are forming the operating unity (controller) which produces the control signal U_c (controller output variable Y_R) affecting the DC motor (actuator). Its response is the motor torque T which represents the actuating signal (manipulated variable Y) to drive the load (plant elements). The positioning drive acts by closed loop in the sense to diminish the error of angle position.

Thus, noting the related physical variables at the line segments, Fig. 2.3 may represent the block diagram for a specific system, e.g., for an electromechanical servomechanism acting as a *position control system*.

To describe the overall behaviour of the electromechanical system an adapted control logic element must be supposed with an implemented algorithm referring to a specific control law. As a basis of many control systems the *proportional control* may be considered. Block diagrams for controllers are drawn in terms of the deviations from a zero-error equilibrium condition. Applying this convention to the general terminology in Fig. 2.3, the proportional control is described by

$$Y_R = K_p E \qquad (2.14).$$

This relation denotes an algorithm where the change in the control signal Y_R is proportional to the error signal E, K_p is the proportional gain.

Assuming the differential and the power amplifier linear their gains are combined into one, denoted K_p. The system is thus seen to have proportional control in which the motor voltage U_c is proportional to the difference between the command voltage U_r and the feedback voltage U_a of the potentiometers

$$U_c = K_p(U_r - U_a) = K_p U_e \qquad (2.15).$$

According to the block diagram, Fig. 2.3, the interconnection of physical variables involves the relationships along the controller

$$U_r / \Theta_r = A_c(s) = K_1\,; \quad U_a / \Theta = H(s) = K_2$$

with the amplifier transfer function

$$U_c / U_e = G_c(s) = K_p$$

and the motor transfer function

$$T / U_c = G_m(s) = K_T / R.$$

The relationship around the plant is given by the plant transfer function

$$\frac{\Theta}{T + T_d} = G_p(s) = \frac{1}{s(Js + c)} \qquad (2.16)$$

which identifies the controlled load as a neutrally stable second-order plant containing a moment of inertia J and a viscous damping c as plant elements. For a meaningful error signal U_e to be generated the simplifications are introduced:

$$K_1 = K_2 \,; \quad K = K_1 K_p K_T / R.$$

Hence it follows by Eqs. (2.8), (2.9) that the specified reference transfer function

$$\frac{\Theta}{\Theta_r} = \frac{K}{Js^2 + cs + K} = G_r(s) \qquad (2.17)$$

and the disturbance transfer function

$$\frac{\Theta}{T_d} = \frac{1}{Js^2 + cs + K} = G_d(s) \qquad (2.18)$$

describe the overall behaviour of the electromechanical system reacting upon a change in command or disturbance input.

The performance of the *proportional control of a second-order system* can be summarized as follows:

The closed-loop system is stable if J, c and K are positive. For no damping ($c = 0$), the closed-loop system is neutrally stable.

A change in desired position can be simulated by a unit step for the command input Θ_r. Using the corresponding transform of unit-step function $\Theta_r(s) = 1/s$, and applying the final value theorem to the transform of actual position, the steady-state output is

$$\Theta_{ss,r} = \lim_{s \to 0} s\, G_r(s) \frac{1}{s} = \frac{K}{K} = 1 \qquad (2.19).$$

The steady-state error signal, being the difference between original unit-step input and steady-state output, is thus zero if the system is stable ($c > 0$, $K > 0$).

A sudden change in load torque also can be modelled by a unit step for the disturbance input T_d. Using the corresponding transform $T_d(s) = 1/s$ the steady state error signal due to a unit-step disturbance is

$$\Theta_{ss,d} = \lim_{s \to 0} s\, G_d(s) \frac{1}{s} = \frac{1}{K} \qquad (2.20).$$

This deviation can be reduced by choosing the overall gain K large.

Nevertheless, the steady-state error does not generally run up to zero for second-order systems.

If the plant's transfer function were instead

$$G_p(s) = \frac{1}{Js^2 + cs + k} \qquad (2.21),$$

then a steady-state error of $k/(k + K)$ would remain.

The transient behaviour is characterized by the damping ratio

$$\zeta = \frac{c}{2\sqrt{JK}} \qquad (2.22).$$

For slight damping, the response to a step input will be distinctly oscillatory and the overshoot large, hence the transient error signal is running the risk to exceed the specified tolerance.

Thus, control system design implies *alternative control laws* for higher specifications to improve the system performance in terms of *speed* and *accuracy* – possibly being conflicting objectives –, without having to change the existent plant.

Advanced Control Methods. The performance requirements and the expense of modern systems can be quite high, so that an analytical approach is usual necessary to design a control system with sufficient accuracy, speed, and stability characteristics. For this problem the stated fundamental aspects concerning classical control methods are rather incomplete, because there are many more topics that should be covered to be complete. A lot of them are simply powerful techniques for quantifying and visualizing performance and stability of closed-loop control systems as a function of feedback controller gains. It would go far beyond the scope of this review chapter to reiterate them here. Instead, the reader being interested in control theory is referred to more specialized texts, [2] to [4].

By example of an electromechanical system the efficiency of feedback compensation could be demonstrated by reducing system sensitivity and enhancing system performance. However, the main purpose dating from the stated fundamental aspects is to show the role of modelling for a successful design of control systems by using diagrams, especially to point out the efficiency of transfer function block diagrams in system analysis.

2.1.4 Signal Flow Graphs. *Reduction of the Diagram*

Transfer function block diagrams possibly representing an extensive system structure may be reduced by systematically applying interconnection conventions of variables. Some intermediate variables may disappear in the simplification process.

Block Diagram Reduction

An elementary example of simplification is given by the block diagrams of fundamental configurations, Fig. 2.2. The two blocks representing differently connected components may be replaced by only one equivalent block. The original system structure is noted by depicting the overall system transfer function in the block reflecting the related interconnection of subsystems. Thus, Eqs. (2.1) to (2.3), may be interpreted as common reduction formulas for cascaded, parallel elements, and for feedback loops, in case to be completed by formulas for relocating a summer or a takeoff point. A summary of aiding tools is given in literature by *block-diagram-reduction (transformation) tables* contrasting original and reduced (equivalent) configurations [3], [4].

Basic reduction rules simply require that the relationships between the transformed variables are maintained. Hence, any two diagram arrangements are equivalent if they correctly express the algebraic relations defined by component transfer functions and have the same input and output variables.

Signal-flow-graph (SFG) Models

That type of systematic diagram represents an alternative method for determining the relationship between system variables. It is the matter of a graphical representation in which variables are represented as nodes (or vertices) noted by dots and the operations on these variables are represented by branches noted by directed line segments between the dots. Each branch is an unidirectional path of signals and is labelled with the input-output-relationship. This contrasts with the block diagram's representation of variables as lines and operations as blocks. Because of this difference, that outlined type of diagram is a simplified notation for block diagrams. Signal flow graphs are attractive since block-diagram-reduction procedure may be difficult for complex systems with many loops.

An important convention to note by using SFG is that a node is summing the signals of all incoming branches and transmits this sum (the value of the variable) to all outgoing branches. The oriented line segments show the relationship between the variables whose nodes are connected by the branch, with the arrow indicating the direction of causality. Subtraction is indicated by a negative sign with the operation denoted by the proper line segment. Corresponding to the block diagrams of Fig. 2.2, representing fundamental configurations, the equivalent signal flow graphs are shown in Fig. 2.4.

Finally it has to be mentioned that Mason's Circuit Rule for SFG provides the necessary relations between system variables without any required manipulation or reduction. By applying some special definitions this circuit rule must give the correct value for the relationship between nodes. It is invariant under the removal of an intermediate node, that is it gives the same answer for the transfer function when applied to the original signal flow graph and to the reduced graph obtained by removal of an intermediate node. Thus, the required overall system transfer function may be written down by inspection even when the number of loops or forward paths in diagram is large, [3], [4].

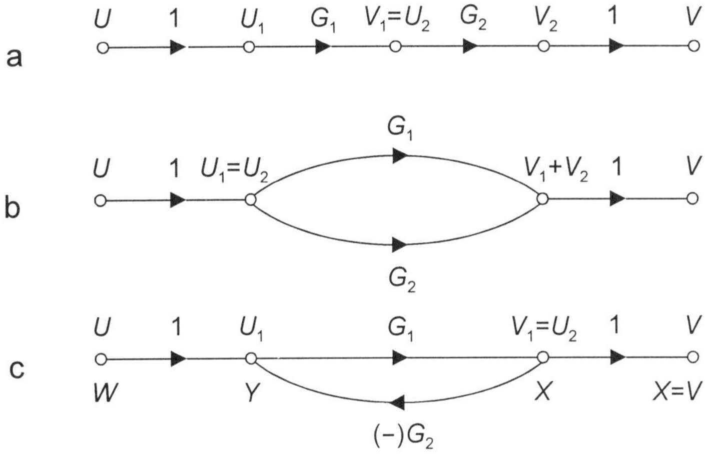

Fig. 2.4. Signal flow graphs (SFG) of fundamental configurations. **a** Cascade; **b** parallel; **c** feedback loop

2.2
Two-port-diagram Representation

As a particular line of general network theory the *theory of linear n-port networks*, originally confined to *four-pole theory*, is an approved method for analysing electrical circuits [10].

The conception of four-pole or four-terminal networks bases on describing the interconnections between the terminals of a single entity ("box") encompassing a rather complicated arrangement of components. Overall relationships are defined by input and output variables obtainable by measurements at the four accessible terminals of the "box" (terminal voltages and currents) without being acquainted with the details of component circuits inside the "box". By combining two single terminals to a terminal pair the interconnection between four variables implies the statement that four poles are networks for *transmission of power through* from an input-terminal pair to an output-terminal pair. Therefore, it is convenient to prefer the term *two port* instead of four pole characterizing by this the essential attribute of a network being connected to external circuits by two pairs of terminals forming two ports.

The results of two-port theory can be used for systems modelling of different engineering devices transmitting energy, (power circuits), e.g., electromechanical or mechanical systems, if their objects are defined by an input port and an output port.

2.2.1
Generic Two Port. *Two-terminal-pair Network*

In the beginning the variables at single terminal pairs must be considered for determining the characteristics of a two port.

A generic two-port network is supposed to be passive, hence it does not dispose of effort or flow variable sources. First direction arrows of currents and voltages are chosen unsymmetrically at input and output port. These conventional reference directions (chain arrow directions) are transmitted subsequently to opposite terminals in generalized variable notation, Fig. 2.5, according to 1.2.2.

A two-terminal-pair network consists of an interconnected set of terminal pairs, each of which represents a known relationship between *a pair of rate variables* (intensity or power variables). *T*- and *P*-variable rates, in preferably used terms

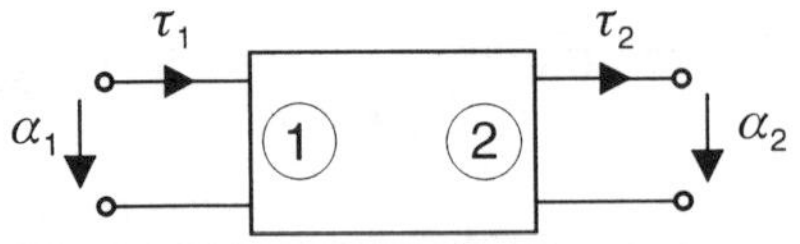

Fig. 2.5. Graphical symbol of the basic 2-port with subsequent reference directions (generic two port)

effort variables α and *flow variables* τ quantify the (transforms of the) received or supplied performance at opposite terminals, namely the input power $\alpha_1\tau_1$ and the output power $\alpha_2\tau_2$. The variable relationship of the general two port is described by a pair of linear equations, where input quantities are represented as functions of the output quantities.
4-pole equations in transmission form (chain form)

$$\begin{aligned}\alpha_1 &= A_{11}\alpha_2 + A_{12}\tau_2\\ \tau_1 &= A_{21}\alpha_2 + A_{22}\tau_2\end{aligned} \tag{2.23a}$$

or, in matrix notation

$$\begin{bmatrix}\alpha_1\\ \tau_1\end{bmatrix} = \underbrace{\begin{bmatrix}A_{11} & A_{12}\\ A_{21} & A_{22}\end{bmatrix}}_{=A}\begin{bmatrix}\alpha_2\\ \tau_2\end{bmatrix} \tag{2.23b}$$

where $\boldsymbol{A}$ is the *transmission* or *chain matrix* of the two-terminal-pair with the 2-port parameters A_{ik} (i,k = 1,2) as its 4 elements

$$\boldsymbol{A} = \left[A_{ik}\right] = \begin{bmatrix}A_{11} & A_{12}\\ A_{21} & A_{22}\end{bmatrix} \tag{2.24}$$

These 2-port parameters (transmission or chain parameters) can be determined by physical equations or measurements at the terminal pairs on the following *conditions of constraint*, Fig. 2.6.
If, for example, $\alpha = U$, $\tau = I$ (electrical system) the transmission parameters will be interpreted physically as:

$A_{11} = (U_1/U_2)_{I_2=0}$: the open-circuit voltage ratio
$A_{21} = (I_1/U_2)_{I_2=0}$: the open-circuit transfer admittance
$A_{12} = (U_1/I_2)_{U_2=0}$: the short-circuit transfer impedance
$A_{22} = (I_1/I_2)_{U_2=0}$: the short-circuit current ratio

Since the elements of the two port are linear and since only the zero-state responses are considered (all initial conditions zero), the superposition theorem

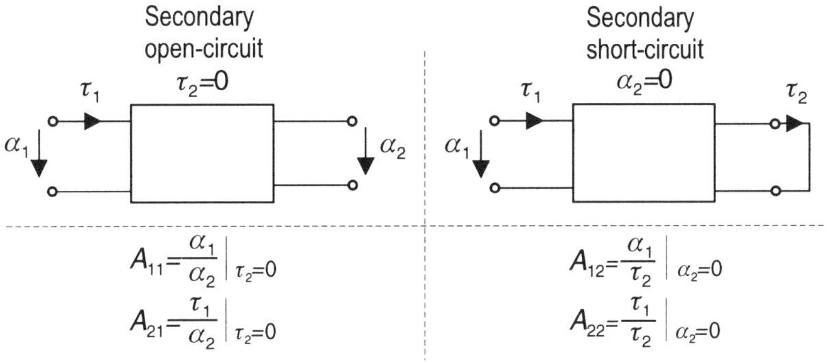

Fig. 2.6. 2-port under output-terminal constraints

guarantees that each input variable is the sum of contributions due to the output variables independently acting alone.
Transmission (chain) matrix of the general 4-pole with the 2-port parameters determined from restrained-state operations:

$$A = \begin{bmatrix} \left.\dfrac{\alpha_1}{\alpha_2}\right|_{\tau_2=0} & \left.\dfrac{\alpha_1}{\tau_2}\right|_{\alpha_2=0} \\[2ex] \left.\dfrac{\tau_1}{\alpha_2}\right|_{\tau_2=0} & \left.\dfrac{\tau_1}{\tau_2}\right|_{\alpha_2=0} \end{bmatrix} \tag{2.25}$$

Output quantities as functions of the input quantities
4-pole equations of the inverse form in matrix notation

$$\begin{bmatrix} \alpha_2 \\ \tau_2 \end{bmatrix} = \underbrace{\begin{bmatrix} B_{11} & B_{12} \\ B_{21} & B_{22} \end{bmatrix}}_{=\boldsymbol{B}} \begin{bmatrix} \alpha_1 \\ \tau_1 \end{bmatrix} \tag{2.26}$$

Computation of *inverse transmission matrix* $\boldsymbol{B}$ by inversion of $\boldsymbol{A}$

$$\boldsymbol{B} = \left[B_{\ell m}\right] = \boldsymbol{A}^{-1} = \frac{1}{\det \boldsymbol{A}} \begin{bmatrix} A_{22} & -A_{12} \\ -A_{21} & A_{11} \end{bmatrix} \tag{2.27}$$

2.2.2 Connection of Two Ports. *Fundamental Configurations*

The competent significancy of two-port theory derives from its facilities to connect several two ports. It is useful in analysis problems to be able to express the parameters of the resultant two-terminal-pair network in terms of those characterizing the individual or component two-terminal pairs.

One of the advantages of the two-port representation is the simplicity of the algebraic relations between the individual 4-pole equations by using matrix notation. Thus, confining systems modelling to linear two-port networks the overall 4-pole matrix equations can easily be obtained.

The resulting algebraic relationships between (the transforms of) the variables, especially the *connection rules* at the junctions, are illustrated in Fig. 2.7.

Cascade Connection. For several component 4-poles which are connected in *cascade* or *tandem*, Fig. 2.7a, the output from any network is precisely the input of the next one.

$$\tau_1 = \tau_1' \; ; \; \tau_2' = \tau_1'' = \tau_2 \; ; \cdots \; ; \; \tau_2^{(n)} = \tau_{n+1}$$
$$\alpha_1 = \alpha_1' \; ; \; \alpha_2' = \alpha_1'' = \alpha_2 \; ; \cdots \; ; \; \alpha_2^{(n)} = \alpha_{n+1}$$

4-pole equations of the transmission form in matrix notation

$$\begin{bmatrix} \alpha_1 \\ \tau_1 \end{bmatrix} = \underbrace{\begin{bmatrix} A_{11}^{(1)} & A_{12}^{(1)} \\ A_{21}^{(1)} & A_{22}^{(1)} \end{bmatrix} \begin{bmatrix} A_{11}^{(2)} & A_{12}^{(2)} \\ A_{21}^{(2)} & A_{22}^{(2)} \end{bmatrix} \cdots}_{=\boldsymbol{A}} \begin{bmatrix} \alpha_{n+1} \\ \tau_{n+1} \end{bmatrix} \tag{2.28}$$

where $\boldsymbol{A}$ is the *transmission* or *chain matrix* of the composite 2-port.

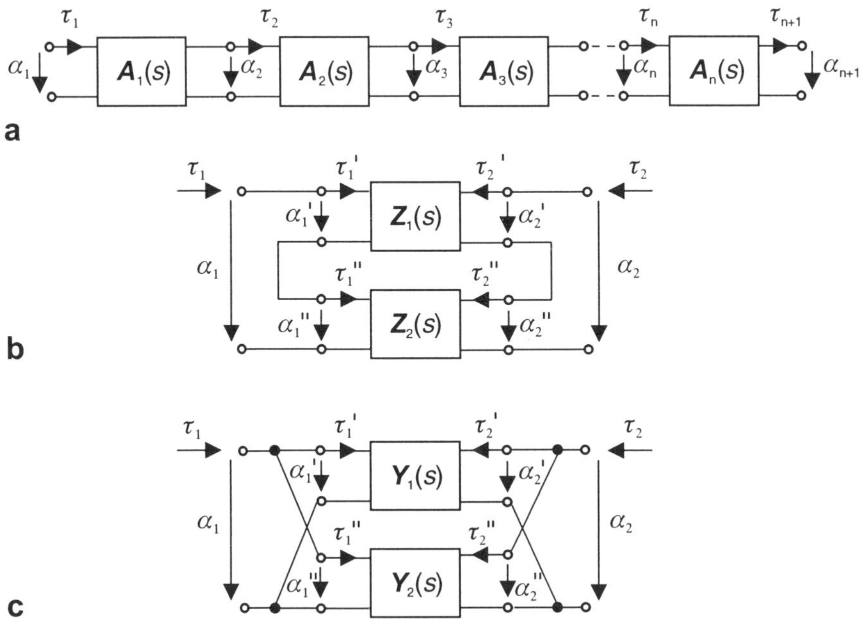

Fig. 2.7. 2-ports of fundamental connections (basic connection types of 4-poles). **a** Cascade; **b** series; **c** parallel

The overall matrix of a cascade is equal to the multiple product of all transmission matrices of the components arranged in the same sequence as the respective networks to which they pertain, Eq. (2.29).

$$A = A_1 \cdot A_2 \cdots A_h \cdots A_n = \prod_{h=1}^{n} A_h \tag{2.29}$$

Because most of circuits in electrical networks or communication theory are composed in sequential structure the 4-pole cascade connection is the most important of all interconnections of component networks. Due to the convention of subsequent reference directions instead of symmetrical ones the cascade connection involves that the output flow variable at the preceding two port corresponds in amount and direction to the input flow variable at the adjoining two port.

Series Connection. For two component 4-poles which are connected in *series*, Fig. 2.7b, the parameters for the resultant 2-port network can be found by adding the corresponding equations for the separate components.
Flow variables are identical

$$\tau_1 = \tau_1' = \tau_1''$$

$$\tau_2 = \tau_2' = \tau_2''$$

Effort variables add

$$\alpha_1 = \alpha_1' + \alpha_1''$$

$$\alpha_2 = \alpha_2' + \alpha_2''$$

4-pole equations in impedance form

$$\begin{bmatrix} \alpha_1 \\ \alpha_2 \end{bmatrix} = \underbrace{\begin{bmatrix} Z_{11} & Z_{12} \\ Z_{21} & Z_{22} \end{bmatrix}}_{=\boldsymbol{Z}} \begin{bmatrix} \tau_1 \\ \tau_2 \end{bmatrix} \tag{2.30}$$

where $\boldsymbol{Z}$ is the *impedance* or *Z-matrix* of the composite 2-port.
The component impedance (or resistance) matrices of individual 4-poles add to the overall impedance matrix:

$$\boldsymbol{Z} = \boldsymbol{Z}_1 + \boldsymbol{Z}_2 \tag{2.31}$$

Parallel Connection. For two component 4-poles which are connected in *parallel*, Fig. 2.7c, the parameters for the resultant 2-port network can also be found by addition.
Flow variables add

$$\tau_1 = \tau_1' + \tau_1''$$

$$\tau_2 = \tau_2' + \tau_2''$$

Effort variables are identical

$$\alpha_1 = \alpha_1' = \alpha_1''$$

$$\alpha_2 = \alpha_2' = \alpha_2''$$

4-pole equations in admittance form

$$\begin{bmatrix} \tau_1 \\ \tau_2 \end{bmatrix} = \underbrace{\begin{bmatrix} Y_{11} & Y_{12} \\ Y_{21} & Y_{22} \end{bmatrix}}_{=\boldsymbol{Y}} \begin{bmatrix} \alpha_1 \\ \alpha_2 \end{bmatrix} \tag{2.32}$$

where $\boldsymbol{Y}$ is the *admittance* or *Y-matrix* of the composite 2-port.

The component admittance (or conductance) matrices of individual 4-poles add to the overall admittance matrix:

$$\boldsymbol{Y} = \boldsymbol{Y}_1 + \boldsymbol{Y}_2 \tag{2.33}$$

Mixed Connections. Further interconnections of two component 4-poles are given by combining the input terminals in series whereas the output terminals are joined in parallel, at last by the inverted combination, Fig. 2.8.

First one is the *series-parallel connection*, Fig. 2.8a, with the 4-pole equation in matrix notation

$$\begin{bmatrix} \alpha_1 \\ \tau_2 \end{bmatrix} = \underbrace{\begin{bmatrix} H_{11} & H_{12} \\ H_{21} & H_{22} \end{bmatrix}}_{=\boldsymbol{H}} \begin{bmatrix} \tau_1 \\ \alpha_2 \end{bmatrix} \tag{2.34}$$

where the series-parallel matrices (component hybrid matrices) of the individual 4-poles add to the *hybrid* or *H-matrix* of the composite 2-port

$$\boldsymbol{H} = \boldsymbol{H}_1 + \boldsymbol{H}_2 \tag{2.35}$$

Latter one is the *parallel-series connection*, Fig. 2.8b, with the 4-pole equation in matrix notation

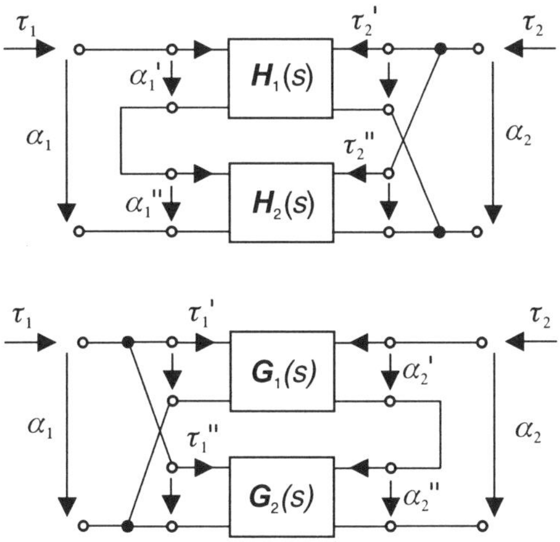

a

b

Fig. 2.8. Two ports of fundamental connections. **a** Series - parallel; **b** parallel-series

$$\begin{bmatrix} \tau_1 \\ \alpha_2 \end{bmatrix} = \underbrace{\begin{bmatrix} G_{11} \; G_{12} \\ G_{21} \; G_{22} \end{bmatrix}}_{=\boldsymbol{G}} \begin{bmatrix} \alpha_1 \\ \tau_2 \end{bmatrix} \tag{2.36}$$

where the parallel-series matrices (component inverse hybrid matrices) of the individual 4-poles add to the *inverse hybrid* or *G-matrix* of the composite 2-port

$$\boldsymbol{G} = \boldsymbol{G}_1 + \boldsymbol{G}_2 = \boldsymbol{H}^{-1} \tag{2.37}$$

2.2.3 Mechanical Two Ports

The 2-port parameter method of analysing vibration problems embodies the use of a pair of linear equations to relate the mechanical system variables of the output to the input of a general linear elastic system, [11], Fig. 2.9.

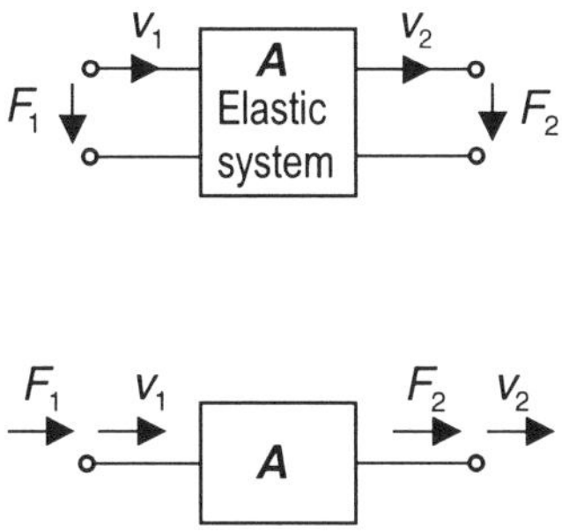

a

b

Fig. 2.9. Basic 2-port of a general linear elastic system. **a** Generic mechanical two port; **b** simplified diagram by Molloy [11]

The elastic system may be a combination of linear, lumped mechanical elements such as masses, springs and dampers. It also can be a combination of linear, distributed mechanical components, such as beams, plates, diaphragms, etc. The elastic system must have two identifiable connection points (1) and (2) which are called the input and output points, Fig. 2.9b.

Electromechanical Analogies. Supposing a translational mechanical system the instantaneous power of state change under transmitting through the single entity "2-port" is described by the appropriate rate (or power) variables force F and velocity υ, as shown in Fig. 1.2.

Nevertheless, by denoting the terminal pairs as shown in Fig. 2.9a, the definition of T- and P-variables is inverse to the before presented general classification of dynamic system variables. Provided that the *reverse relationship*

effort variable $\alpha = F$ *force*
flow variable $\tau = \upsilon$ *velocity*

is applied to define the pair of rate variables for the translational mechanical case a two-port-network modelling is pointed out which prefers the "classical" type of the two possible *electromechanical analogies.*

By this type called *dual* or *force-to-voltage analogy* and being historically founded on mass-inductance dualogue (or *impedance analogue*) the physical variables are related in the "linear" correspondence

force $\leftrightarrow$ voltage ; mass $\leftrightarrow$ inductance
velocity $\leftrightarrow$ current ; spring $\leftrightarrow$ capacitance.

Though, this relationship ranks as the naturally perceived one it provides a difference in the topology of related mechanical and electrical networks. To remove this disadvantage alternatively the "new" type called *true-connected* or *force-to-current analogy* was introduced (Hähnle [12], Firestone [13]), and definitively treated (Trent [14]). By that type basing on mass-capacitance analogue (or *mobility analogue*) the physical variables are related in the "reciprocal" correspondence

force $\leftrightarrow$ current ; mass $\leftrightarrow$ capacitance
velocity $\leftrightarrow$ voltage ; spring $\leftrightarrow$ inductance.

While it is pointless to discuss which analogy is "correct", as both are equally valid when they exist, the mass-capacitance analogy has considerable advantages. These chiefly stem from the fact that while the mass-capacitance analogue of a nonplanar electrical or mechanical system always can be constructed, the mass-inductance analogue of such a system does not exist [15].

The input force and velocity are produced by connection of point (1) to that portion of the complete mechanical system which precedes it. At the output point (2) there exists a force (F_2) and a velocity (υ_2) which result from the application of (F_1) and (υ_1) at input point (1) and the reaction of the portion of the mechanical system following the 2-port. The positive direction of forces and velocities are chosen to coincide with the direction of energy flow from the vibrating source.

Phasor Performance Equations. 4-pole equations in transmission form (chain form) representing the performance equations of the translational mechanical system are given by

$$\underline{F_1} = A_{11}\underline{F_2} + A_{12}\underline{\upsilon_2}$$
$$\underline{\upsilon_1} = A_{21}\underline{F_2} + A_{22}\underline{\upsilon_2} \tag{2.38a}$$

or, in matrix notation

$$\begin{bmatrix} \underline{F_1} \\ \underline{\upsilon_1} \end{bmatrix} = \underbrace{\begin{bmatrix} A_{11} & A_{12} \\ A_{21} & A_{22} \end{bmatrix}}_{=\boldsymbol{A}} \begin{bmatrix} \underline{F_2} \\ \underline{\upsilon_2} \end{bmatrix} \tag{2.38b}$$

where $\boldsymbol{A}$ is the *transmission* or *chain matrix* of the *mechanical* two-terminal pair with the 2-port parameters A_{ik} (i,k = 1,2) as its 4 elements.

Confining oneself to harmonic excitations and responses and adopting the conventional complex number representation the components of the force-velocity-vectors and the 2-port parameters are pointed out in terms of complex quantities, so that $\underline{F}_j$, $\underline{\upsilon}_i$ are *phasors*, whereas A_{ik} in general are complex system parameters formed by the ratios of two phasors according to the indicated terminals. The physical interpretation of these ratios, called *complexors*, depends upon the types of related physical variables in conformity with the chosen force-to-voltage or impedance analogy. Thus, the phasor ratios can be determined by inspection, Fig. 2.10.

Transposing the 4-pole equations, Eq. (2.38a), it is readily seen, that A_{11} and A_{22} are a non-dimensional ratio either of forces or of velocities, defining a *transmissibility*. A_{12} is the ratio of the phasors force to velocity which defines a *mechanical*

	Output port termination constraints	Physical interpretation of complex system parameters (phasor ratios)
Mechanical 2-port parameters (transmission or chain parameters)	blocked: $A_{11} = \left.\frac{\underline{F_1}}{\underline{F_2}}\right\vert_{\underline{V_2}=0} = \underline{T}_{12}^{F}$	Force transmissibility
	free: $A_{12} = \left.\frac{\underline{F_1}}{\underline{V_2}}\right\vert_{\underline{F_2}=0} = \underline{Z}_{12}$	Transfer (mechanical) impedance (free impedance)
	blocked: $A_{21} = \left.\frac{\underline{V_1}}{\underline{F_2}}\right\vert_{\underline{V_2}=0} = \underline{Y}_{12}$	Transfer (mechanical) mobility (blocked mobility)
	free: $A_{22} = \left.\frac{\underline{V_1}}{\underline{V_2}}\right\vert_{\underline{F_2}=0} = \underline{T}_{12}^{V}$	Velocity transmissibility

Fig. 2.10. Mechanical 2-port parameters from output-terminal constraints

impedance $\underline{Z}_{12}$. A_{21} is expressed in terms of the reciprocal of the mentioned dynamic characteristic and defines a *(mechanical) mobility* $\underline{Y}_{12}$.

The general procedure for determining 2-port parameters of a system requires several analytical steps, such as setting up the performance equations, subjecting them to boundary conditions, solving them and casting the solutions in Eq. (2.38). The 2-port parameters of a structure also can be measured experimentally by applying output-terminal constraints for disconnecting definably the supply of mechanical energy.

Let the measurement point 2 on the structure be blocked, i.e. constrained to have zero velocity, $\underline{\upsilon}_2 = 0$. Then, solving Eq. (2.38) for this boundary condition A_{11} is defined in terms of the *force transmissibility* $\underline{T}^{\mathrm{F}}_{12}$ and A_{21} is determined by the *transfer (mechanical) mobility* $\underline{Y}_{12}$, both between point 1 and point 2 being blocked.

Let the measurement point 2 on the structure be allowed to respond freely without any (motion) constraint, $\underline{F}_2 = 0$. Then, solving Eq. (2.38) for that boundary condition A_{12} is equal to the *transfer (mechanical) impedance* $\underline{Z}_{12}$ and A_{22} is defined in terms of the *velocity transmissibility* $\underline{T}^{\upsilon}_{12}$, both between point 1 and point 2 being free, Fig. 2.9.

2.2.4 Fundamental Mechanical Elements. *Fundamental Configurations*

Instead of implementing the general procedure for determining 2-port parameters of a complete two-port network model representing complex mechanical objects (structures) it is more convenient to consider it as a two-terminal-pair network resulting from the connection of simpler component two-terminals pairs. Thus, it is useful to determine the 2-port parameters of mechanical network components, such as storage elements and dissipators (masses, springs and dampers) defining linear element characteristics by fundamental physical laws. Suchlike deduced *elementary* or *degenerate 2-ports* are composed of parameters, in particular reduced to real or imaginary system parameters. Subsequently the 2-port parameters of composite 2-ports (structures) are computed with the aid of basic connection types of 4-poles by applying connection rules at the input and output points of linear elements.

Mechanical Elementary 2-ports

Confining oneself to translational mechanical systems the two-terminal pairs of translational mechanical components are defined as follows, Fig. 2.11.

Spring. Potential energy storage element: *linear time-invariant spring*, Fig. 2.11a.
$\mathrm{j}\omega$: differential operator, $1/\mathrm{j}\omega$: integral operator

Force requirement (dynamic equilibrium)

$$\underline{F_1} = \underline{F_2} = \underline{F_s}$$

T-storage element law (*Hooke's law*)

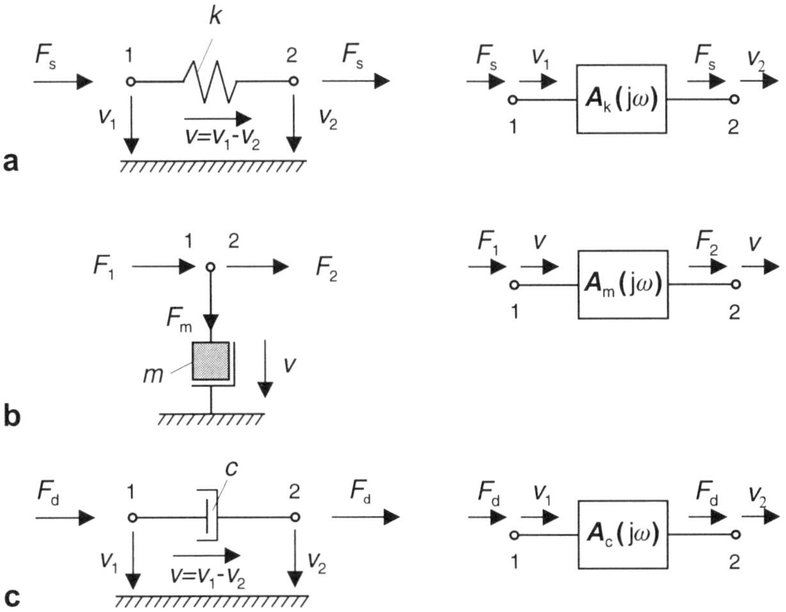

Fig. 2.11. Basic 2-ports of linear time-invariant mechanical elements. **a** Spring; **b** mass; **c** damper

force-motion performance equation

$$\underline{F}_s = k\left(\frac{\upsilon_1}{j\omega} - \frac{\upsilon_2}{j\omega}\right) = \frac{k}{j\omega}\underline{\upsilon}$$

4-pole equations

$$\underline{F_1} = 1 \cdot \underline{F_2} + 0 \cdot \underline{\upsilon_2}$$
$$\underline{\upsilon_1} = \frac{j\omega}{k}\underline{F_2} + 1 \cdot \underline{\upsilon_2} \tag{2.39a}$$

or, in matrix notation

$$\begin{bmatrix} \underline{F_1} \\ \underline{\upsilon_1} \end{bmatrix} = \underbrace{\begin{bmatrix} 1 & 0 \\ \frac{j\omega}{k} & 1 \end{bmatrix}}_{=A_k(j\omega)} \begin{bmatrix} \underline{F_2} \\ \underline{\upsilon_2} \end{bmatrix} \tag{2.39b}$$

where $\boldsymbol{A}_k$ is the *transmission matrix of the spring.*
Basic 2-port determined by its *impedance-analogous parameter*:

A_{21} is the inverse of the mechanical impedance of the spring Z_k, i.e., the *mobility of the spring*, Y_k

$$A_{21} = \frac{j\omega}{k} = \frac{\underline{\upsilon}}{\underline{F}_s} = 1/Z_k = Y_k \tag{2.40}$$

with the *elastic (spring) constant* or *stiffness k*.

Mass. Kinetic energy storage element: *time-invariant mass*, Fig. 2.11b.
Motion requirement (geometric constraints)

$$\underline{\upsilon_1} = \underline{\upsilon_2} = \underline{\upsilon}$$

P-Storage element law (*Newton's second law*)
force-motion performance equation

$$\underline{F_\text{m}} = \underline{F_1} - \underline{F_2} = m\text{j}\omega\underline{\upsilon_1} = m\text{j}\omega\underline{\upsilon}$$

4-pole equations in matrix notation

$$\begin{bmatrix} \underline{F_1} \\ \underline{\upsilon_1} \end{bmatrix} = \underbrace{\begin{bmatrix} 1 & \text{j}\omega\, m \\ 0 & 1 \end{bmatrix}}_{=\boldsymbol{A}_\text{m}(\text{j}\omega)} \begin{bmatrix} \underline{F_2} \\ \underline{\upsilon_2} \end{bmatrix} \tag{2.41}$$

where $\boldsymbol{A}_\text{m}$ is the *transmission matrix of the mass.*
Basic 2-port determined by its *impedance-analogous parameter*:
A_{12} is the *mechanical impedance of the mass,* Z_m

$$A_{12} = \text{j}\omega\, m = \frac{\underline{F_\text{m}}}{\underline{\upsilon}} = Z_\text{m} \tag{2.42}$$

with the *mass (parameter) m.*

Damper. Fluid friction dissipator: *linear time-invariant damper*, Fig. 2.11c.
Force requirement (dynamic equilibrium)

$$\underline{F_1} = \underline{F_2} = \underline{F_\text{d}}$$

Dissipator element law (*linear viscous damping*)
force-motion performance equation

$$\underline{F_\text{d}} = c\left(\underline{\upsilon_1} - \underline{\upsilon_2}\right) = c\underline{\upsilon}$$

4-pole equations in matrix notation

$$\begin{bmatrix} \underline{F_1} \\ \underline{\upsilon_1} \end{bmatrix} = \underbrace{\begin{bmatrix} 1 & 0 \\ \frac{1}{c} & 1 \end{bmatrix}}_{=\boldsymbol{A}_\text{c}(\text{j}\omega)} \begin{bmatrix} \underline{F_2} \\ \underline{\upsilon_2} \end{bmatrix} \tag{2.43}$$

where $\boldsymbol{A}_\text{c}$ is the *transmission matrix of the damper.*
Basic 2-port determined by its *impedance-analogous parameter*:
A_{21} is the inverse of the mechanical impedance of the damper Z_c, i.e., the *mobility of the damper,* Y_c

$$A_{21} = \frac{1}{c} = \frac{\underline{\upsilon}}{\underline{F_\text{d}}} = \underline{Y_{12}} = 1/Z_\text{c} = Y_\text{c} \tag{2.44}$$

with the *(viscous) damping coefficient c.*

Mechanical Composite 2-ports

The use of matrix technique for determining the 2-port parameters of composite mechanical two-port networks (structures) is shown by applying the before mentioned connection rules on transmission matrices of translational mechanical components.

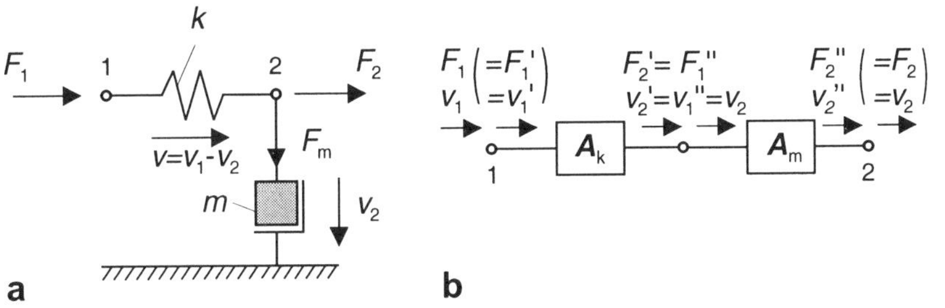

Fig. 2.12. Cascade-connected basic 2-ports *spring and mass*

Spring and Mass in Cascade. The connection in cascade or tandem, Fig. 2.12, is given by multiplying the component transmission matrices (non-commutative)

$$\boldsymbol{A}_\mathrm{k}\cdot\boldsymbol{A}_\mathrm{m}=\begin{bmatrix}1 & 0\\ \dfrac{\mathrm{j}\omega}{k} & 1\end{bmatrix}\begin{bmatrix}1 & \mathrm{j}\omega\, m\\ 0 & 1\end{bmatrix}=\underbrace{\begin{bmatrix}1 & \mathrm{j}\omega\, m\\ \dfrac{\mathrm{j}\omega}{k} & 1-\dfrac{m\omega^2}{k}\end{bmatrix}}_{=\boldsymbol{A}_{\mathrm{k,m}}(\mathrm{j}\omega)} \tag{2.44}$$

where $\boldsymbol{A}_{\mathrm{k.m}}$ is the *transmission* or *chain matrix of the cascade-connected spring-mass structure*.

Composite 2-port determined by its *impedance-analogous parameters*:

A_{12} is the *mechanical impedance of the mass,* Z_m

$$A_{12}=\mathrm{j}\omega\, m=Z_\mathrm{m} \tag{2.42}$$

A_{21} is the *mobility of the spring,* Y_k

$$A_{21}=\mathrm{j}\omega\,/\,k=1\,/\,Z_\mathrm{k}=Y_\mathrm{k} \tag{2.40}$$

A_{22} is a non-dimensional coefficient, called the *velocity transmissibility,* T^υ

$$A_{22}=1-m\omega^2\,/\,k=1+Z_\mathrm{m}\,/\,Z_\mathrm{k}=T^\upsilon \tag{2.45}$$

Spring and Damper in Parallel. Referring to the chosen force-to-voltage or impedance analogy it must be taken into account that mechanical 2-port networks connected in parallel, Fig. 2.13, are equivalent to electrical 2-ports in series being in line with the general two-port networks presented in 2.2.2. Hence, parallel con-

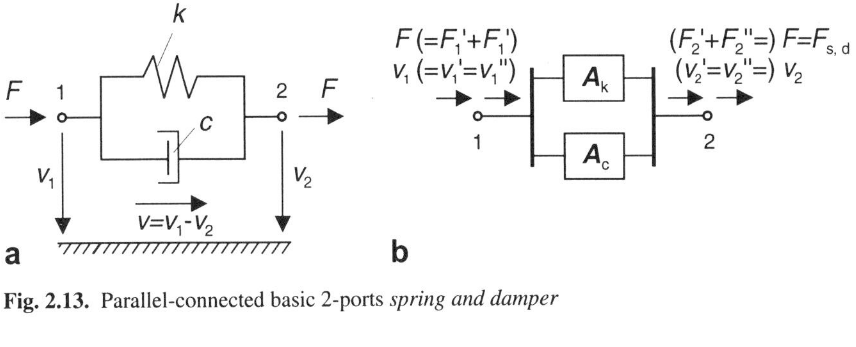

Fig. 2.13. Parallel-connected basic 2-ports *spring and damper*

nection results from adding the component impedance matrices, Eq. (2.31)

$$\boldsymbol{Z}_{\mathrm{k}} + \boldsymbol{Z}_{\mathrm{c}} = \begin{bmatrix} \frac{k}{\mathrm{j}\omega} & -\frac{k}{\mathrm{j}\omega} \\ \frac{k}{\mathrm{j}\omega} & -\frac{k}{\mathrm{j}\omega} \end{bmatrix} + \begin{bmatrix} c & -c \\ c & -c \end{bmatrix} = \underbrace{\begin{bmatrix} \frac{k}{\mathrm{j}\omega} + c & -\frac{k}{\mathrm{j}\omega} - c \\ \frac{k}{\mathrm{j}\omega} + c & -\frac{k}{\mathrm{j}\omega} - c \end{bmatrix}}_{=\boldsymbol{Z}_{\mathrm{k,c}}(\mathrm{j}\omega)} \tag{2.46}$$

where $\boldsymbol{Z}_{\mathrm{k,c}}$ is the *impedance* or *Z-matrix* of the structure with the antimetric attributes $Z_{21} = -Z_{12}$, $Z_{11} = -Z_{22}$ and the only two independent parameters

$$Z_{11} = Z_{21} = \frac{k}{\mathrm{j}\omega} + c = Z_{\mathrm{k}} + Z_{\mathrm{c}} = \underline{Z_{\mathrm{k,c}}}$$

$$Z_{12} = Z_{22} = -\frac{k}{\mathrm{j}\omega} - c = -(Z_{\mathrm{k}} + Z_{\mathrm{c}}) = -\underline{Z_{\mathrm{k,c}}}.$$

Computation of the transmission or chain matrix by conversion of the 2-port parameters $Z_{\ell\mathrm{m}}$ to A_{ik}

$$[A_{\mathrm{ik}}] = \frac{1}{Z_{21}} \begin{bmatrix} Z_{11} & -\det \boldsymbol{Z} \\ 1 & -Z_{22} \end{bmatrix} \quad \text{with } \det \boldsymbol{Z} = Z_{11}Z_{22} - Z_{21}Z_{12},$$

so that

$$A_{11} = Z_{11}/Z_{21} = 1$$

$$A_{12} = -Z_{11}\,Z_{22}/Z_{21} + Z_{12} = 0$$

$$A_{21} = 1/Z_{21} = 1/(k/\mathrm{j}\omega + c)$$

$$A_{22} = -Z_{22}/Z_{21} = 1$$

$$[A_{\mathrm{ik}}] = \underbrace{\begin{bmatrix} 1 & 0 \\ 1/(k/\mathrm{j}\omega + \mathrm{c}) & 1 \end{bmatrix}}_{=\boldsymbol{A}_{\mathrm{k,c}}} \tag{2.47}$$

where $\boldsymbol{A}_{\mathrm{k,c}}$ is the *transmission* or *chain matrix of the parallel-connected spring-damper structure* (basic model of visco-elastic structures).
Composite 2-port determined by its impedance-analogous parameter:
A_{21} is the equivalent *mobility of the spring-damper combination*, $\underline{Y}_{\mathrm{k,c}}$

$$A_{21} = \frac{1}{\frac{k}{\mathrm{j}\omega} + c} = \frac{1}{Z_{\mathrm{k}} + Z_{\mathrm{c}}} = \frac{1}{\underline{Z_{\mathrm{k,c}}}} = \underline{Y_{\mathrm{k,c}}} \tag{2.48}$$

2.2.5 Supplementary Mechanical Elements. *Couplers and Sources*

Complex mechanical objects (structures) are not represented by model systems resulting from only connecting fundamental elements among one another. Real system components taking part in power transmission are combined with different transforming and converting elements which will be defined as couplers and sources.

Couplers

Transformer. Physical objects which transmit power with neither storage nor absorption of energy are represented in model systems by an abstract object termed a coupler. As to physical interpretation coupling elements are called *transformers* because of retaining the initial form of energy up to the output. Examples of ideal mechanical transformers are rigid and lossless levers just as gears transforming the related motion at the input to a related motion at the output. Analogous devices denoting ideal fluid and electrical transformers are shown in Fig. 2.14.

Converter. Related changes in the state of physical objects also may be of a different type, hence, energy is converted from one form to another. The relationship between input and output variables of different physical type defines an abstract object termed a *converter*. An example of irreversible conversion of energy is given by the dissipator (or resistor). Reversibly converting active elements (generators) belong to the sources.

The performed transmission through measurement devices refers to signals not to power. Therefore, the relationship between signals of different type concerning a measuring element involves a power conversion being kept on a low level. Except, the measuring element is coupled with a separate power supply (auxiliary source) being implemented for amplifying the input signal. The first case defines a *transducer*, the latter one an *amplifier*.

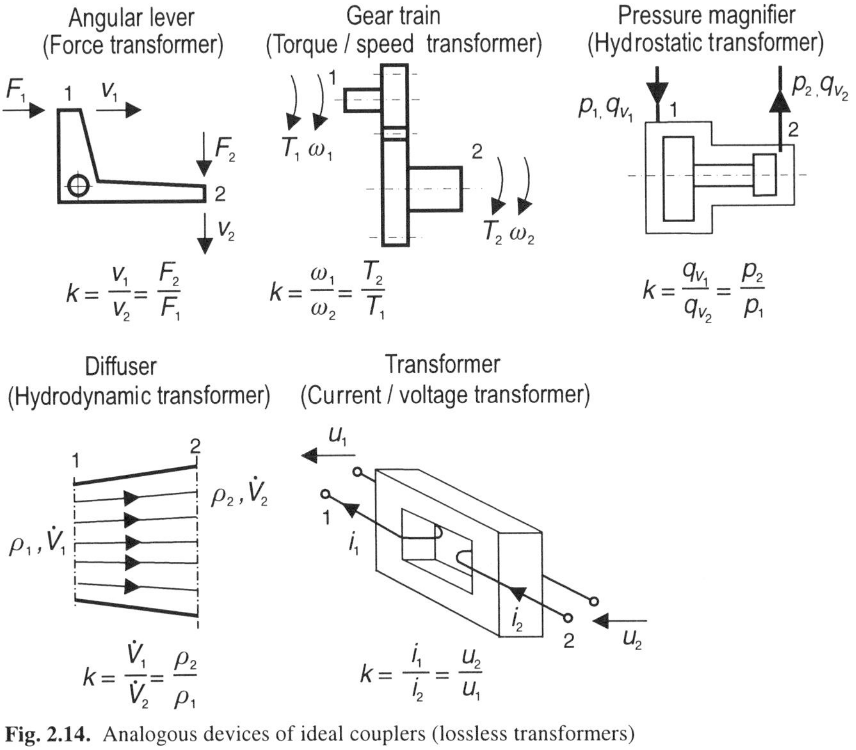

Fig. 2.14. Analogous devices of ideal couplers (lossless transformers)

For the sake of brevity, the input to output relationship will be referred to as the *transmission relationship* of an abstract object, here of the constituent object of the dynamical model idealized as a coupler.

The ideal coupler is a network component relating respectively a pair of T- and P-variable rates at opposite terminals, where the input and output powers are identically equal, i.e.

$$\alpha_1\tau_1 - \alpha_2\tau_2 \equiv 0$$

Thus, the transmission relationship of an ideal transformer reduces to a constant, defined as the *transformation ratio of the coupler k*

$$k = \frac{\alpha_1}{\alpha_2} = \frac{\tau_2}{\tau_1} \qquad (2.49).$$

In the case of a lossless mechanical transformer, e.g., the angular lever or the gear train, k is the *lever ratio* respectively the *gear ratio* defining the transmission relationship by the ratios either of velocities or of speeds, and of its inverses by those of forces or torques. In general, the ratios of appropriate flow and effort variables are inverse relationships of power variables.
Coupler: *ideal transformer*, Fig. 2.14.
4-pole equations in matrix notation

$$\begin{bmatrix} \alpha_1 \\ \tau_1 \end{bmatrix} = \underbrace{\begin{bmatrix} k & 0 \\ 0 & 1/k \end{bmatrix}}_{=\boldsymbol{A}_\mathrm{C}} \begin{bmatrix} \alpha_2 \\ \tau_2 \end{bmatrix} \qquad (2.50),$$

where $\boldsymbol{A}_\mathrm{C}$ is the *transmission matrix of the ideal transformer* with the *transformation ratio k*.

Sources

Energy conversion processes which activate physical systems are represented in terms of idealized rate sources. In model systems the sources are termed *active elements* of the model; by contrast, the stores, dissipators, and couplers are termed passive elements.

Corresponding to spatial relationship the class of the generated rate or power variable defines the type of idealized rate sources.

P-variable Source (flow variable source). This abstract object generates a flow variable τ_s which is a specified function of time but independent of the effort variable α across the two-terminal element. Examples are force, torque, and current sources, Fig. 2.15.

T-variable Source (effort variable source). This abstract object generates an effort variable α_s which is a specified function of time but independent of the flow variable τ through the two-terminal element. Examples are velocity, angular velocity, and voltage sources, Fig. 2.16.

The generally defined rate source may be regarded in particular as a source of electrical energy being included in the equivalent circuit of an active (circuit) element [16].

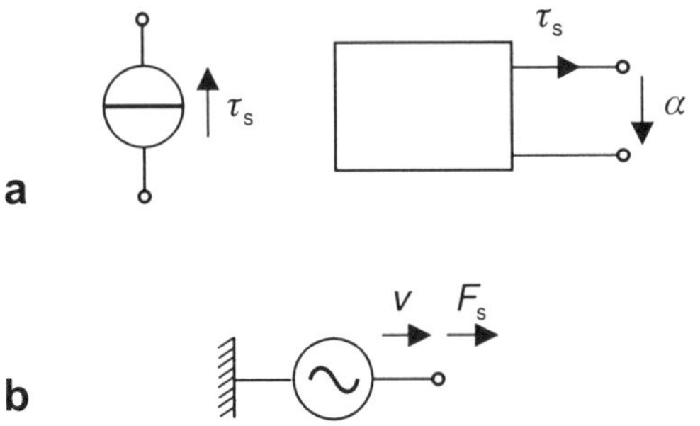

Fig. 2.15. Graphical symbols of ideal flow generators. **a** P-variable 2-terminal source (ideal current source and active one-port network); **b** ideal constant-force generator

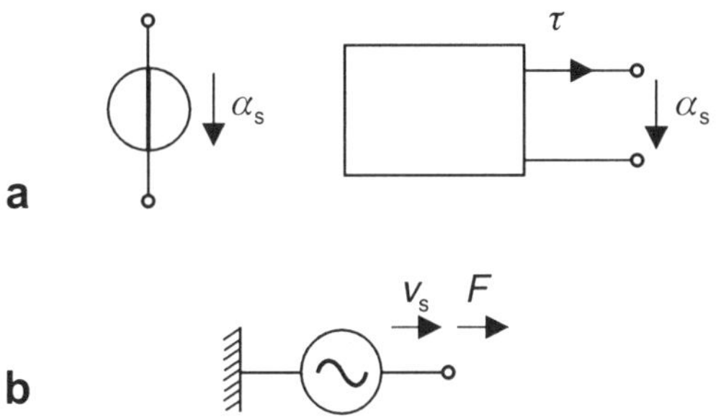

Fig. 2.16. Graphical symbols of ideal effort generators. **a** T-variable 2-terminal source (ideal voltage source and active one-port network); **b** ideal constant-velocity generator

Electrical Sources. Since the terminal current, called source current i_s, is independent of the terminal voltage across the element, the active element is characterized as an *ideal current source*

$$i(t) = i_s(t) \quad \textit{by definition} .$$

For the value $i_s(t) = 0$ the current source is identical with an open circuit; it is idle when short-circuited. Like an open circuit the current source is a constraint; but it is more general in that it constrains the current at its terminals to any desired value. In this sense a current source can be considered as a generalized open circuit.

In precisely an analogous manner the terminal voltage, called source voltage or electromotive force u_s, that is independent of the terminal current through the element, the active element turns out to be an *ideal voltage source*

$$u(t) = u_s(t) \quad \textit{by definition}$$

Specifying the value $e_s(t) = 0$, then the voltage source is identical with a short circuit; it is idle when open-circuited. Like a short circuit, the voltage source is a constraint maintaining the voltage at its terminals on any desired value. In this sense a voltage source renders as a generalized short circuit.

Active elements may be constructed by circuits of different arrangement superposing current and voltage sources. Replacing active elements by a 2-terminal network the internal structure is no more of interest but only the performance at the two terminals, considered as a port.

By measuring the performance at the 1-port in restrained-state description the behaviour of a 2-terminal source is completely described. Using complex representation of sinusoids (complex r.m.s. values of currents and voltages) the relationship of alternating terminal quantities results in the (transformed) equations

$$\underline{U} = \underline{U}_s - \underline{Z}_i \underline{I} \tag{2.51a}$$

$$\underline{I} = \underline{I}_s - \underline{Y}_i \underline{U} \tag{2.51b}$$

being identical because of

$$\underline{Y}_i \underline{Z}_i = 1 \quad ; \qquad \underline{I}_s = \underline{Y}_i \underline{U}_s \tag{2.52}$$

where $\underline{Z}_i$ is the *internal impedance*, $\underline{Y}_i$ the *internal admittance of the electrical source*.

The equivalence relations, Eq. (2.51a, b), correspond to the pair of source arrangements being thus externally equivalent, Fig. 2.17a, b.

The ideal current source in parallel with the internal admittance $\underline{Y}_i$ defines the equivalent circuit of an *independent current source*, providing the source current $\underline{I}_s$ when short-circuited, Fig. 2.17a

$$\underline{U} = 0 \quad ; \qquad \underline{I} = \underline{I}_s = \underline{I}_{sc} \quad .$$

The ideal voltage source in series with the internal impedance $\underline{Z}_i$ defines the equivalent circuit of an *independent voltage source*, producing the source voltage $\underline{U}_s$ when open-circuited, Fig. 2.17b

$$\underline{I} = 0 \quad ; \qquad \underline{U} = \underline{U}_s = \underline{U}_{oc} \quad .$$

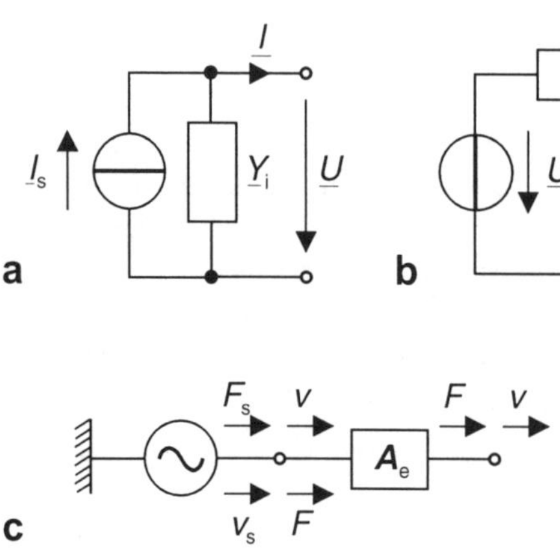

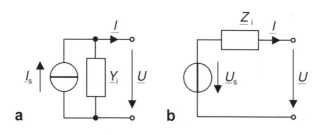

Fig. 2.17. Analogous arrangements of 2-terminal sources (practical generators). **a** Equivalent circuit of independent current source; **b** equivalent circuit of independent voltage source; **c** symbol for complete mechanical source

Mechanical Generators. The rate source now may be regarded as a source of mechanical energy supplying translational mechanical objects. Though being sometimes physically complicated mechanical sources in general behave in a rather simple manner. For this it is proper to represent physical sources by connecting an ideal mechanical generator and an elastic system of source. The resulting model, called a *complete mechanical source*, Fig. 2.17c, generates the prescribed sinusoidal force or velocity at prescribed frequencies on the input point of the elastic system which transmits its drive through for energy supplying at the output point.

Supposing that a constant sinusoidal force F_s is exerted at the input of the elastic system, regardless of the load which is coupled to the output the mechanical generator is characterized as an *ideal constant-force generator*, Fig. 2.15b

$$F(t) = F_s(t) = \mathrm{Re}\left[\underline{\hat{F}_s}\, e^{j\omega_f t}\right] = \sqrt{2}\,\mathrm{Re}\left[\underline{F_s}\, e^{j\omega_f t}\right] .$$

Provided that a constant sinusoidal velocity υ_s is maintained at the input point of the elastic system, independently of the load attached to the output the mechanical generator turns out to be an *ideal constant-velocity generator*, Fig. 2.16b

$$\upsilon(t) = \upsilon_s(t) = \mathrm{Re}\left[\underline{\hat{\upsilon}_s}\, e^{j\omega_f t}\right] = \sqrt{2}\,\mathrm{Re}\left[\underline{\upsilon_s}\, e^{j\omega_f t}\right] .$$

Frequently it is not feasible to physically separate the mechanical source into its components and for these cases it is desirable to have a means of describing the source in terms of quantities which can be measured at the only accessible junction, namely, the output point.

Output-terminal constraints, introduced in 2.2.3, will be applied to the mechanical source relating the system variables of the output $\underline{F}$, $\underline{\upsilon}$ to those of the input $\underline{F}_s$, $\underline{\upsilon}_s$. It is possible to determine the 2-port parameters of the elastic system by measuring the mechanical impedance looking into the source and also measuring either the "blocked force" or the "free velocity" of the source.

Ideal constant-force generator coupled to an elastic system

output point restrained from moving (zero velocity $\underline{\upsilon} = 0$)

$$\underline{F} = \underline{F_{oc}} - \left(A_{12}/A_{11}\right)\underline{\upsilon} \tag{2.53a}$$

output point free to move (no force exerted $\underline{F} = 0$)

$$\underline{\upsilon} = \underline{\upsilon_{sc}} - \left(A_{11}/A_{12}\right)\underline{F} \tag{2.53b}$$

Ideal constant-velocity generator coupled to an elastic system

output point restrained

$$\underline{F} = \underline{F_{oc}} - \left(A_{22}/A_{21}\right)\underline{\upsilon} \tag{2.54a}$$

output point free to move

$$\underline{\upsilon} = \underline{\upsilon_{sc}} - \left(A_{21}/A_{22}\right)\underline{F} \tag{2.54b}$$

Equations (2.53a, b) and (2.54a, b) forming two sets of performance equations describe alternatively the behaviour (performance) at two pairs of terminals of complete mechanical sources in terms either of the blocked force $\underline{F}_{oc}$ which it can generate or of the free velocity $\underline{\upsilon}_{sc}$ which it delivers and of related 2-port parameters ot its elastic system. The ratios of two distinct parameters relating to the present type of source remain to be determined.

Defining the parameter ratios A_{12} / A_{11} respectively A_{22} / A_{21} in terms of mechanical impedances of the elastic structure measured at point (2) either when point (1) is unrestrained (zero impedance) or when restrained (infinite impedance) the 2-port parameters may be eliminated by the following complexors:

free impedance $\underline{Z}_{sc} = A_{12} / A_{11}$ (2.55a)

blocked impedance $\underline{Z}_{oc} = A_{22} / A_{21}$ (2.55b).

The ideal constant-force generator would automatically terminate the elastic 4-pole in a zero impedance and also the ideal constant-velocity generator would necessarily terminate its elastic 4-pole in an infinite impedance. Thus, replacing $\underline{Z}_{sc}$ and $\underline{Z}_{oc}$ by the measured impedance $\underline{Z}_i$ the sets of performance equations (2.53), (2.54) will be reduced to the single pair of 4-pole equations

$$\underline{F} = \underline{F}_{oc} - \underline{Z}_i \underline{\upsilon} \quad (2.56a)$$

$$\underline{\upsilon} = \underline{\upsilon}_{sc} - \underline{Y}_i \underline{F} \quad (2.56b)$$

being identical because of

$$\underline{Z}_i \underline{Y}_i = 1; \qquad \underline{\upsilon}_{sc} = \underline{Y}_i \underline{F}_{oc} \quad (2.57)$$

where $\underline{Z}_i$ is the *internal impedance*, $\underline{Y}_i$ the *internal mobility of the mechanical source*.

The equivalence relations, Eq. (2.56a, b), correspond to both types of generators and describe the mechanical source in terms of the measured quantities $\underline{F}_{oc}$ and $\underline{Z}_i$ or of the measured quantities $\underline{\upsilon}_{sc}$ and $\underline{Y}_i$.

Two convenient electrical equivalent circuits exist for mechanical generators and either may be employed for the same mechanical source so long as the appropriate electrical quantities are inserted. Referring to the chosen force-to-voltage or impedance analogy the characteristic quantities of 2-terminal sources (lossy generators) are corresponding as follows

$$\underline{U}_s = \underline{U}_{oc} \mathrel{\hat{=}} \underline{F}_{oc} = \underline{F}_s; \qquad \underline{I}_s = \underline{I}_{sc} \mathrel{\hat{=}} \underline{\upsilon}_{sc} = \underline{\upsilon}_s \quad (2.58)$$

The representation associated with Eqs. (2.56a), (2.58), denoting the equivalent circuit of independent voltage source, Fig. 2.17b, is known as *Thevenin's theorem* in electrical circuit theory, and that associated with Eqs. (2.56b), (2.58), denoting the independent current-source, Fig. 2.17a, is known as *Norton's theorem.*

Mechanical sources exist whenever matter vibrates and by virtue of its contact with other matter causes a second portion of material to vibrate. Thus, mechanical sources are universal in their occurrence and are of prime importance to the vibration engineer. As pointed out by Eq. (2.57) two quantities are necessary to describe a mechanical source, the third one could be computed.

The source impedance $\underline{Z}_i$ of the structure is measured at the attachment points of the equipment. The free velocity $\underline{\upsilon}_{sc}$ is measured with vibration pick ups at the same points. During this measurement the structure is unloaded. These free impedance data are completely sufficient to determine the dynamic characteristics of the structure by experimental investigations. These data identify the structure as an unidirectional mechanical source and enable one to calculate the motion it will produce when driving any load [11], [17]. If 6 degrees of freedom motions are to be considered, the problem is more complicated but mechanical sources still play a basic role in the analysis [18].

2.2.6 Connections with Block Diagrams. *Transfer Function Block Diagrams*

Whereas in electrical circuit theory the realization of two-terminal pairs bases on their close relationship to the network representation, by contrast, in control system engineering connections between the two-port-diagram and the block-diagram representation obviously have priority.

Multivariable Systems

Applying transfer function block diagrams with multiple inputs and multiple outputs the advantages of the transfer function representation, 2.1.1, can be used for two-port diagrams being interpreted as *multivariable systems*. Thus, the general two port with conventional reference directions (chain arrow directions), Fig. 2.18a, may be considered as the equivalent system derived from the original *block diagram of a two-input, two-output system*, Fig. 2.18b, by means of block-diagram-reduction.

Signal Four Pole. The transforms of the input variables are related to the transforms of both output variables. Their interconnections are denoted by forward paths of the input signals U_1, U_2 branching out to each of the acting lines for the output signals V_1, V_2. The action relations are marked by 4 functional blocks wherein the corresponding transfer functions $B_{\ell m}(s)$ are depicted representing the

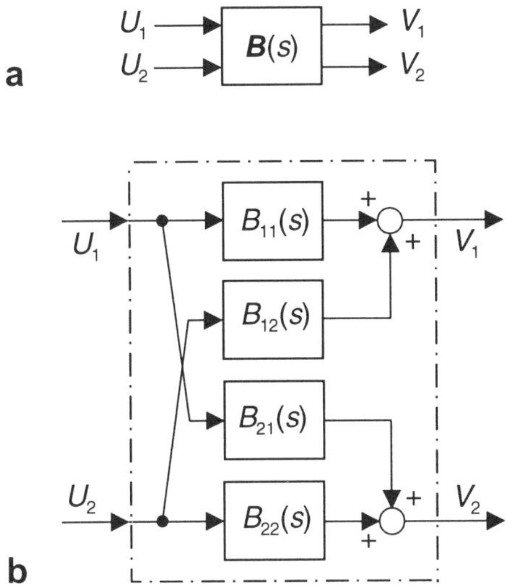

Fig. 2.18. Block diagram of a two-input, two-output system (basic structures of a signal 4-pole). **a** Reduced configuration; **b** original diagram in the canonical configuration

2-port parameters of the equivalent two-terminal-pair network. Corresponding to the signal flow along the connecting lines the single entity "2-port" is called *signal four pole*.

The resulting algebraic relationships between the (transforms of the) variables of this 4-pole are described by a pair of transfer function relations, where the output quantities are represented as functions of the input quantities

$$\begin{aligned} V_1(s) &= B_{11}(s)U_1(s) + B_{12}(s)U_2(s) \\ V_2(s) &= B_{21}(s)U_1(s) + B_{22}(s)U_2(s) \end{aligned} \qquad (2.59a)$$

or, in matrix notation

$$\begin{bmatrix} V_1(s) \\ V_2(s) \end{bmatrix} = \underbrace{\begin{bmatrix} B_{11}(s) & B_{12}(s) \\ B_{21}(s) & B_{22}(s) \end{bmatrix}}_{\boldsymbol{B}(s)} \begin{bmatrix} U_1(s) \\ U_2(s) \end{bmatrix} \qquad (2.59b)$$

where $\boldsymbol{B}(s)$ is the *transmission matrix of the signal 4-pole* with the transfer functions $B_{\ell m}(s)$ (ℓ,m = 1,2) as its 4 elements.

The transfer function relations, Eq. (2.59), referred to as the *canonical form* for the realization of a signal four pole, are corresponding with the general 4-pole equations in the inverse form (inverse transmission matrix $\boldsymbol{B}$), Eqs. (2.26), (2.27), introduced in 2.2.1.

Since for interconnections of basic multivariable systems other configurations are easier to work with, conversions of signal 4-poles are realized by equivalent block diagrams. Replacing in Fig. 2.18 the crossed pair of branches by a parallel pair and reversing the series branches as to reference direction, Fig. 2.19, the 4-pole is described by a pair of transfer function relations in matrix notation

$$\begin{bmatrix} U_2(s) \\ V_1(s) \end{bmatrix} = \underbrace{\begin{bmatrix} G_{11}(s) & G_{12}(s) \\ G_{21}(s) & G_{22}(s) \end{bmatrix}}_{\boldsymbol{G}(s)} \begin{bmatrix} U_1(s) \\ V_2(s) \end{bmatrix} \qquad (2.60)$$

where $\boldsymbol{G}(s)$ is the *transmission matrix of the converted signal 4-pole* with the transfer functions $G_{qr}(s)$ (q,r = 1,2) as its 4 elements.

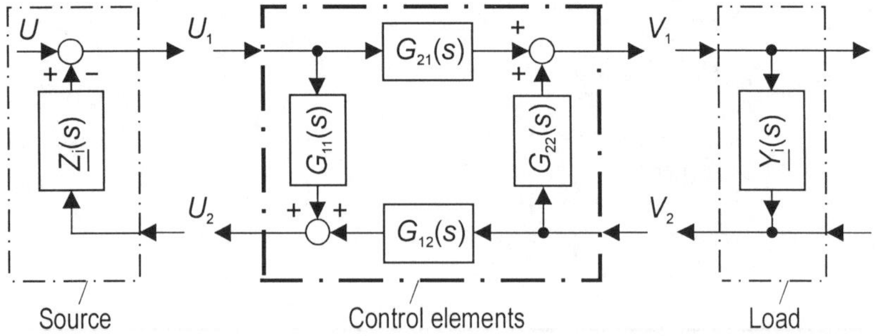

Fig. 2.19. Block diagram of a terminated two-input, two-output system (basic structure of a signal 4-pole in the field configuration connected to a source and a load)

The transfer function relations, Eq. (2.60), referred to as the *field form* of a signal four pole, are related to the general 4-pole equations for parallel-series connections of component 4-poles, Eq. (2.36).

Computation (of the inverse hybrid or $\boldsymbol{G}$-Matrix) by conversion of $\boldsymbol{B}(s)$

$$\boldsymbol{G}(s) = \left[G_{qr}(s)\right] = \frac{1}{B_{22}(s)}\begin{bmatrix} -B_{21}(s) & 1 \\ \det \boldsymbol{B}(s) & B_{12}(s) \end{bmatrix} \tag{2.61}$$

Control system design makes practical use of linear multivariable systems by the signal-four-pole representation. Unacceptable performance of a control system often results when simplified models are used to design controllers for complex systems in which several variables are to be controlled. If these variables affect each other the control loops are said to be interacting.

Systems with Interacting Loops (back effects)

The design of controllers for each loop as if it where a separate system must be regarded as an approach which may be sufficient in cases of slight interaction. Thus, the basic structure of a signal 4-pole may be considered as a specific block diagram for a *system with interacting loops*. For example, the field configuration, Fig. 2.19, gives the complete representation of a control element being connected with a two-variable source and a two-variable load. Hence, this configuration is qualified for indicating how subsystems are coupled, consequently how far the related variables are to be treated as coupled variables.

Total Noninteraction. The absence of interaction, especially the prevention of active *back effects* caused by the output variable acting on the input variable may be indicated by the transfer function block in the backward path, $G_{12}(s)$.

(Total) backward noninteraction: $G_{12}(s) = 0$ (2.62a)

Input U_1 affects output V_1, output V_2 has no effect on input U_2 (possibly V_1 independent of V_2, if in addition $G_{22}(s) = 0$).

Backward noninteraction is realized by *active 4-poles* (active elements) through *disconnection of energy flows*, Fig. 2.20.

It should be noticed that power and *signal amplifiers* as well as *modulators* are to be considered with more rigor as realizations of the basic device of a *three-port element* resulting from significant energy interaction at a minimum of three ports, e.g., low power input, high power output, energy source, [22], outlined in 2.5.1.

In electrical terminology, the disconnection mentioned above requires the input impedance of the second element to be infinite when connecting it with the first element, which means that no power is being withdrawn from the first element. Of course, a suchlike connection of components in cascade (or tandem) is an idealization characterized by the assumption that interaction of control elements is represented only by *nonloading elements*. In electrical devices, nonloading may be achieved by inserting an *isolation amplifier* with high impedance between the two circuits being combined and thus uncoupled from one another.

In cases where a significant loading exists, the overall transfer function for components connected in cascade must be obtained from their 4-pole equations rather than from the product law for single-variable systems, Eq. (2.1).

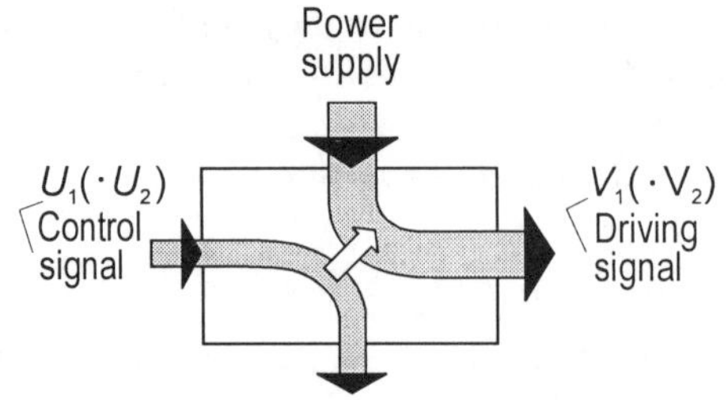

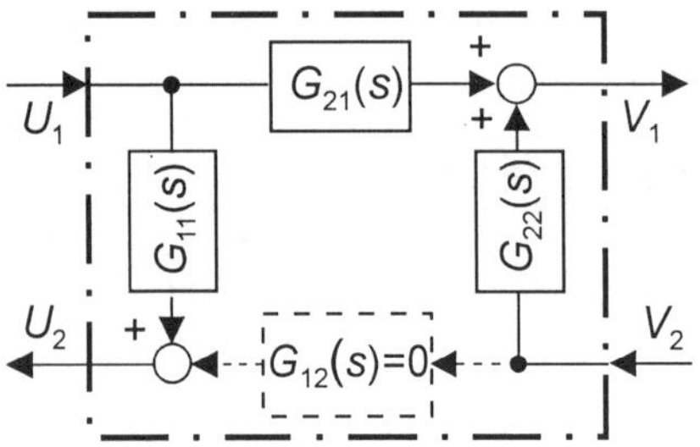

Fig. 2.20. Disconnected energy flows in active 4-poles. **a** Symbol for an actuator ("transmission of signal through" denoted by arrow); **b** basic structure of a signal 4-pole indicating backward non-interaction

Examples of active 4-poles are related to the definite type of an *actuator*.

Whereas sensors provide the measurements necessary for feedback elements actuators are required to control the energy flow of power supply. Acting on the plant the final control element operates on the low-level control signal to produce a signal of high intenseness covering the demanded driving power. This signal is called manipulating variable. For supplying the driving power requirements preferably electromechanical or hydraulic devices are applied.

Potentiometer system (regulating resistor). The potentiometer system generally is a position-to-voltage transducer very proper for the purpose of a feedback element (feedback potentiometer). As to input element (command potentiometer) there is a difference of use between the present class of control system being either a regulator or a follow-up system. In the latter case the controlled variable is kept near the command value, which is changing with time. In the first case of a regulator the controlled variable will be kept constant in spite of disturbances. For this purpose a command potentiometer is appropriated which primarily converts by an additional spring element command force into displacement representing the desired value of the constant controlled variable. The completion to a commonly known mass-damper-spring system points out that damping and inertia force may appear as disturbances affecting the force-position set point, Fig. 2.21.

The power supplies required for transducers and amplifiers usually are not shown in block diagrams of control systems, because they do not contribute to the control logic. However, their existence cannot be ignored. Concerning the posi-

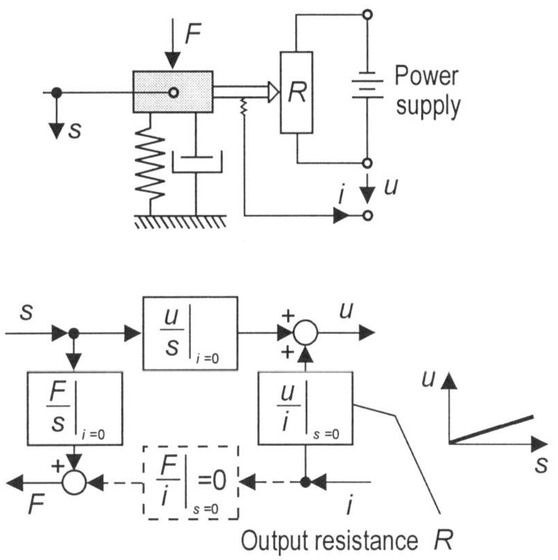

Fig. 2.21. Command potentiometer (force controlled input element for a regulator system); specific noninteracting signal 4-pole; characteristic curve

tion-to-voltage transducer it will be assumed that the output resistance of the potentiometer is low compared with the resistances of the combining network. Therefore, wiper voltage may be taken independent of wiper current. Besides, the current drawn by the wiper circuit does not affect the force required to move the sliding electrical contact, or wiper. Hence, the variables command force and drawn current are uncoupled.

Hydraulic servomechanism. The hydraulic servomotor provides fast response, high force and short stroke characteristics which meet many requirements of control system design. Fields of application exist in the aircraft industry using this type of actuator for power operated controls, antostabilizers and autopilots. In general industry it finds favour in speed control systems for prime movers, and in the operation of process control. Precision control systems in hydraulics have brought vast improvements in disseminating automatic control techniques. Hydraulic servomotors produce motion from a pressure source realized by a hydraulic power supply. Performing the opposite function of pumps, motors and actuators convert fluid energy back to mechanical energy. A *motor* usually produces a continuous rotary output, whilst an *actuator* delivers a limited linear or rotary output, Fig. 2.22.

The servovalve, denoted only by its basic stage being termed a five-port spool valve, controls both the direction of the flow and the flow rate of the working medium. The fluid is metered through the control ports when the spool uncovers a segment of the cylinder-sided orifice. When the initial movement caused by an interacting controller displaces the spool to the right the fluid enters the right-hand piston chamber of the receiving unit, a double acting cylinder (linear actuator). Its

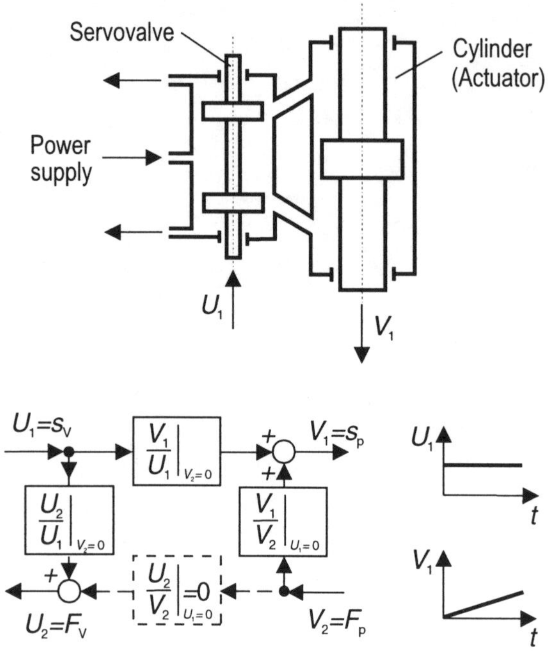

Fig. 2.22. Hydraulic servomotor (servoactuator); specific noninteracting signal 4-pole in restrained-state description; step input and step response functions

piston will be pushed to the left. This action is reversed for a valve displacement to the left.

By disconnection of energy flows the piston force F_P may be taken independent of piston displacement s_P. Besides, the piston force F_P applied to the appreciable load does not affect the force F_V required to displace the pilot valve. Hence, the variables displacing force F_V and driving force F_P are uncoupled.

The hydraulic servomechanism acts as an integrator, since the flow rate, controlled by the spool valve position, produces a rate of movement of the output piston rod. Therefore, the response of the linear actuator to an unit-step valve displacement $s_V(t)$ is a linearly increasing function of time (unit-ramp piston displacement) $s_P(t)$.

Backward noninteraction further can be realized by *passive 4-poles* (passive elements) under *restraints on energy transmission*. Applied constraints are given either by irreversibly converting energy from one form into another or by intensively diminishing the performance with the direction of energy flow from input to output terminals. Since the conversion of energy by passing physical objects is irreversible the output signal does not affect the input signal.

Examples of passive 4-poles are related to the definite type of a *converter*.

Photo-diode. Optoelectrical systems are objects where the inductor current does not affect the light source.

Irreversible conversion from some other form into thermal energy indicates a definite physical object being termed *dissipator*.

Many energy conversion devices are examples of thermal systems applied for *process control systems* where the controlled variable describes a thermodynamic process. Typically, such variables are temperature, pressure, flow rate, liquid level, chemical concentration, and so forth. On the contrary the before mentioned servomechanism is a control system whose controlled variable is mechanical position, velocity, or acceleration.

Other examples of passive 4-poles concerning conversion of energy are to class as transducers. Being used in electrical devices to make measurements those elements are termed *sensor*. To prevent errors in measurement passive sensors must have a low received performance to such a degree that energy transmission through the object to be measured is not affected. This requires a load to be infinite when attached to the output terminal-pair. In electrical terminology, a high-impedance input of the following element makes quite sure that no power is being withdrawn, Fig. 2.23.

In accordance with the before introduced idealization of nonloading elements the passive type of a sensor may be replaced by an active sensor. Measuring devices of that kind imply the demanded high impedance between the two circuits by inserting an isolation amplifier.

Reciprocity theorem is to be applied to passive 4-poles representing interacting components of electrical or mechanical systems. For input and output variables of the same type that theorem states the presence of

total backward interaction: $G_{21}(s) = G_{12}(s)$ (2.62b).

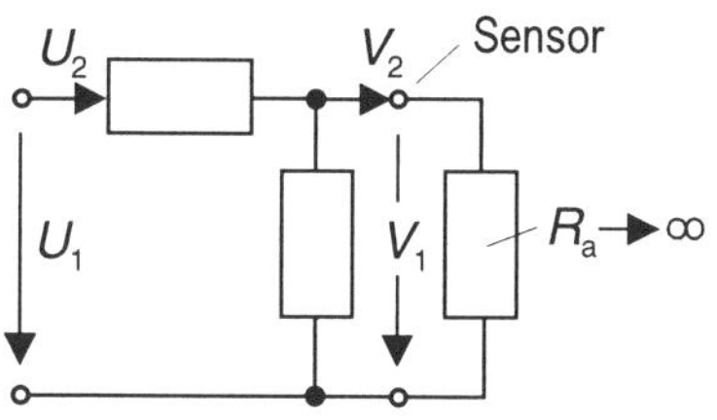

Fig. 2.23. Restraint on energy transmission by a passive 4-pole with infinite load; high-impedance input of the combining electrical circuit (passive sensor)

Partial Noninteraction. All kinds of usual passive elements accordingly converters such as electric motors or electrodynamic transducers have active effects. Because of lower supplied power than received at the input of passive elements the design of control loops requires at least one actuator to raise transmitted performance to the original level.

Depart from the idealization expressed by Eq. (2.61) many control elements are characterized by a

partial noninteraction: $G_{12}(s) \approx 0$ (2.62c).

The measuring transducer whose input can't be adjusted to infinite but only to high impedance, Fig. 2.23, is to be considered as an arrangement retaining partial active effects, for instance.

2.2.7 Analysis of Complex System Structures. *Fluid System*

When being satisfied with the validity of the chosen component models they can be used to predict the performance of the system in question. Predicting the performance from a model is a basic demand on systems *analysis*. A lot of types of analytical techniques has been developed whose applicability to complex system structures depends on the purpose of the analysis. For studying the dynamic characteristics of large-scale power transmissions it is convenient to use multivariable model systems. System components of different physical type are treated as multivariable subsystems with interacting loops. Thus, power-loading effects of connected elements are taken into account. Especially the two-input, two-output system represents interrelationships between the power variables at opposite terminals by 4 individual transfer function blocks for each component.

For example, *hydraulic power transmissions* are constituted by a pump, the main component of a hydraulic supply, which drives a hydraulic output device by transmitting fluid energy. Producing either a translational or a rotational motion the power output device in hydraulics is known as a linear or a rotary actuator. Latter one, called a motor, can achieve higher torque levels than electric motors. The simplest form of a hydraulic power drive for machinery is shown in Fig. 2.24a.

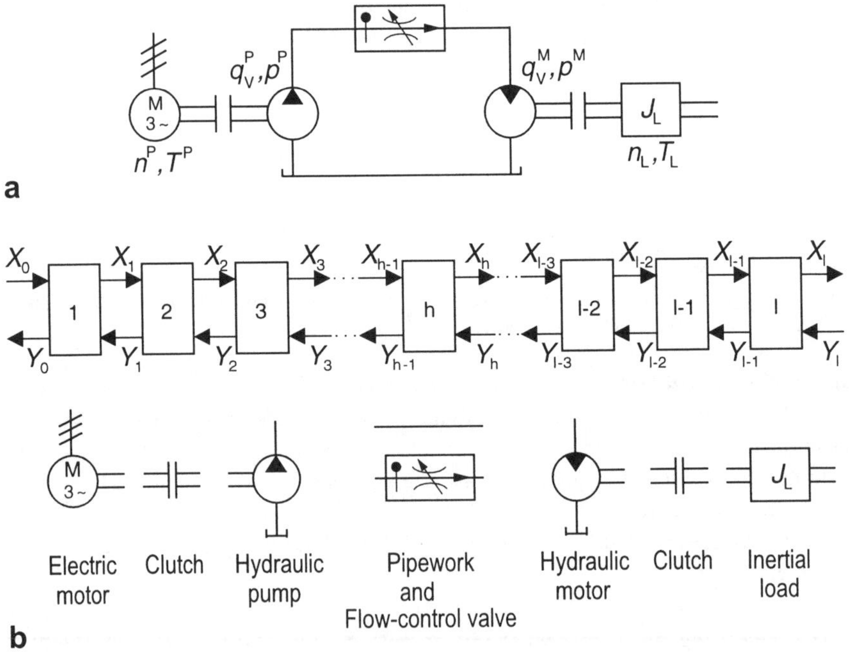

Fig. 2.24. Hydrostatic drive. **a** Basic hydrostatic system (rotary circuit, open-center flow-controlled by valve); **b** overall 2-port by signal 4-pole cascade connection

Detailed Models. *Mechanical systems with distributed parameters* (continuous systems) are not required for control system design. The analysis techniques must realize a model, that describes the dominant dynamic properties of the system to be controlled. Therefore, instead of covering the details of the fluid motion patterns it will be a sufficient approach to predict only the system behaviour in the gross by an *adjusted model system.*

Hydrostatic Power Transmission. Subdividing the transmission device into components each of them may be described by 4-pole equations relating to an individual signal 4-pole. To determine the actual 2-port parameters for the useful basic structure, Fig. 2.19, the hydrostatic component will be regarded as an analogous arrangement of supplementary elements, for example a 2-terminal source or a 4-terminal coupler. Contrary to the idealization of previously treated two-port devices the *transmission loss* cannot be ignored for the analysis procedure in systems engineering, for example in fluid power analysis.

Thus, a valid modelling of *fluid-mechanical converters*, i.e. of the pump and the motor device, requires characteristics of power loss quantified by measuring the influences of leakage and friction. For a sufficient estimation of parameters the linearization of steady-state operating curves is performed representing volumetric and torque efficiency of hydrostatic components. Hence, an approximately characterizing set of flows and pressures as well as of torques and speeds is associated with the input port respectively with the output port [19], [20].

Hydrostatic power transmissions also have energy storage characteristics in the form of fluid and mechanical mass, together with structural and fluid elasticity. These storage elements are very sensitive to excitation frequencies of a transmitted power. Therefore, fluid power design needs the prediction of disturbance influences like

- Excitation of model resonance
- Standing wave phenomena resulting in pipework damage
- Radiation of noise.

The pipework is a *fluid conductor* (continuous coupler) with interconnected storage elements being represented by a distributed-parameter model. For predicting the mentioned influences it is convenient to execute a lumped parameter approach. Depart from the known resistive characteristic involving a pressure drop, termed the *fluid resistance* R_h, there are analogous characteristics depending on excitation frequency. As equivalents to electric circuit elements they are defined as follows. The *fluid inductance* L_h characterizes inertia forces developed when a *fluid capacitance* C_h arises from compression of the liquid added by elastic deformation of the pipe wall. In general, therefore R_h, L_h, and C_h are composite terms of the *fluid impedance* Z_h of a fluid system. Thus, the interaction between pressure and flow rate results in the general transfer function relation

$$p(s) = Z_h(s) q_\upsilon(s) \tag{2.63a}$$

with the input impedance of the fluid conductor

$$Z_h(s) = R_h + L_h s + 1/(C_h s) \tag{2.63b}$$

and its composite terms

$$R_{\mathrm{h}} = \Delta p / q_{\upsilon}; \quad L_{\mathrm{h}} = \rho L / A; \quad C_{\mathrm{h}} = AL / \beta_{\mathrm{e}} \tag{2.63c}$$

determined by measuring the physical quantities: Δp pressure loss; q_{υ} flow rate, ρ density of the fluid; L length of pipe, A cross sectional area, β_{e} effective bulk modulus (reciprocal of compressibility).

Under dynamic conditions a *lumped-parameter equivalent system* consisting in a fluid resistor, an inductor, and a capacitor concentrated or "lumped" at one point is valid since the time required for a pressure wave travel the length of the hydraulic line is short with respect to the period of the highest frequency wave that is to be transmitted. This approach is certainly not suitable for step or impulse responses for example, because that types of aperiodic test functions are composed by a wide frequency range.

To evaluate dynamic characteristics of power transmissions preferably the *frequency response* is caused by applying a harmonic excitation sweeping over a representative frequency range. The *response characteristics* for steady forced oscillations along the fluid portion can be estimated by a numerical calculation of frequency responses related to input and output variables [21].

Defining the terminal pair at the output port h as functions of the terminal pair at the input port $h - 1$ the individual signal 4-pole is described by the transfer function relations in matrix notation

$$\begin{bmatrix} X_{\mathrm{h}}(s) \\ Y_{\mathrm{h}}(s) \end{bmatrix} = \overset{(\mathrm{h})}{\boldsymbol{B}(s)} \begin{bmatrix} X_{\mathrm{h-1}}(s) \\ Y_{\mathrm{h-1}}(s) \end{bmatrix} \tag{2.64}$$

where $\overset{(\mathrm{h})}{\boldsymbol{B}(s)}$ is the *transmission matrix of a hydrostatic component* marked by the running index h. Cascade connection by successively multiplying all individual transfer matrices from input component to termination component results in the algebraic relationship over the transmission line

$$\begin{bmatrix} X_{\ell}(s) \\ Y_{\ell}(s) \end{bmatrix} = \overset{(\ell)}{\boldsymbol{B}(s)} \overset{(\ell-1)}{\boldsymbol{B}(s)} \cdots \overset{(\mathrm{h})}{\boldsymbol{B}(s)} \cdots \overset{(2)}{\boldsymbol{B}(s)} \overset{(1)}{\boldsymbol{B}(s)} \begin{bmatrix} X_0(s) \\ Y_0(s) \end{bmatrix} = \underbrace{\begin{bmatrix} B_{11}(s) & B_{12}(s) \\ B_{21}(s) & B_{22}(s) \end{bmatrix}}_{=\boldsymbol{B}(s)} \begin{bmatrix} X_0(s) \\ Y_0(s) \end{bmatrix} \tag{2.65}$$

where $\boldsymbol{B}(s)$ is the *transmission matrix of the overall signal 4-pole* with the transfer functions $B_{\mathrm{ik}}(s)$ (i,k = 1,2) as its 4 composite elements.

2.3 Network-diagram Representation (Circuit Diagrams)

Lumped dynamic model systems portray graphically the interrelation of various components and sources which constitute the system. This type of model system being a well proved approach to describe electrical and electromechanical devices is termed the network or circuit representation [9], [15], [16], [27], [51].

Network diagrams represent the spatial flow of energy in the physical system in terms of an interconnected set of components or subsystems. Network elements restricted to defined types those components represent spatially localized energy storage and conversion processes in the physical system.

The element models relate the corresponding rates or power variables for each component. The equations relating the across and through power variables, preferably termed effort and flow variables, defined in 1.2.2, are postulated on the basis of measurements performed on physical systems. In mechanical model systems corresponding power variables, such as velocity υ and force F, are associated with an elastic, a damping, or an inertia element. When the elements are interconnected to form a model system, the effort and flow variables must satisfy interconnection requirements (statements of dynamic equilibrium and of restriction).

2.3.1 Direct Representation of Simple Systems by Networks

To describe complex systems the overall behaviour is made up of basic behaviour patterns that are contributed by each element. For simple physical systems a direct systems approach is possible by converting a previously depicted schematic diagram into the desired network diagram.

For example the translational mechanical system, Fig. 2.25, may be regarded as a set of distinct, interacting mechanical objects whose behaviour may be defined in terms of pairs of measurements.

The mass m_1 is connected to the suspension by a linear spring which has the elastic constant k_1 while the two masses are connected through the linear spring with k_2. Since the masses m_1 and m_2 are constrained only to translate in vertical direction the mechanical structure is characterized as a two-degree-of-freedom system. Each *component* of the physical system has to be *analysed in isolation* from the remainder of the physical system. The four physical objects to be considered are two springs, and two masses on which act the weight forces F_{g1}, F_{g2}. If the

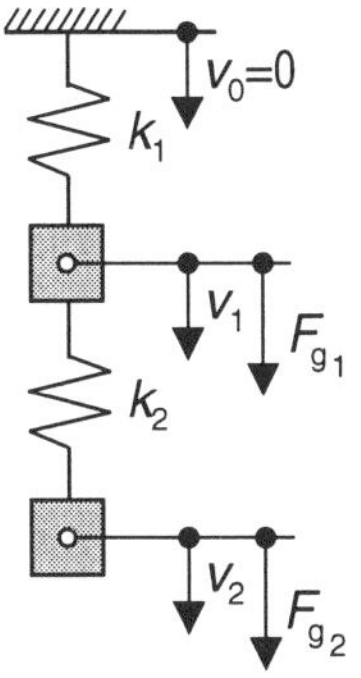

Fig. 2.25. Schematic diagram of a two degree of freedom mass-spring system (two-mass system) under the influence of gravity

rest of the physical system is removed the dynamic behaviour of spring 1 may be defined in terms of a measured relationship between the spring force and the total deflection of the spring. If the fixed point of suspension serves as the inertial reference point, the dynamic behaviour of the isolated mass 1 may be defined by a measured relationship between the inertial force and the acceleration.

Similar measurements with respect to definite connection points or between connection and inertial reference point define the isolated dynamic behaviour of spring 2 and of mass 2. By the effect of gravitational field the local forces of gravity (weight forces) act on the masses as a static load. Due to elongations in the springs the masses take up their equilibrium positions. Suspending the masses from an ideal force-measuring instrument the weight forces also may be defined.

2.3.2 Fundamental and Supplementary Mechanical Elements

The sets of ideal measurements required to specify the individual behaviour of the constituent objects of the physical system may be represented by the *network symbols* for the *mechanical elements* introduced in 2.2.4 and 2.2.5.

Elementary Mechanical Elements

The physical springs and masses are two different types to describe systems behaviour in terms of the two forms of stored energy in elements. While the *T*-stored energy in the spring coincides with the potential energy the *P*-stored energy in the mass corresponds to the kinetic energy. Both storage elements are represented by dynamic model symbols of *passive 2-terminal elements*, Fig. 2.26a.

The physical spring is concentrated on its elasticity, its principal property, and represented by the following idealized model.

Spring. The potential energy storage element is the *linear time-invariant spring*, Fig. 2.26a. The spring yields the defined or measured relationship between a compressive *elastic* (or spring) *force* F_s (*P*-variable rate or through power variable) and the relative deflection s (*T*-variable state or across energy variable) due to the displacements s_1, s_2 at the two connection points or terminals:

Hooke's law

$$F_s(t) = k(s_1 - s_2) = ks(t) \tag{2.66a}$$

with the *elastic (spring) constant* or *stiffness k*.

By the time rate of change of displacement s as the "state (or energy) variable" it follows the velocity υ as the "rate (or power) variable". Thus, according to continuity of space relationship, 1.2.1, Fig. 1.2, the *T-storage element law* of Eq. (2.66a) can be expressed by a relation between the applied force F_s and the resulting motion in terms of the velocity υ

$$F_s(t) = k\int_0^t(\upsilon_1 - \upsilon_2)\,\mathrm{d}t = k\int_0^t \upsilon(t)\,\mathrm{d}t \tag{2.66b}$$

presuming zero state (no energy stored in the initial instant $t = 0$).

The physical mass assumed to be a particle of mass perfectly rigid is concentrated on its inertia, its principal property, and represents the following idealized model.

Mass. The kinetic energy storage element is the *time-invariant mass*, Fig. 2.26a. The two terminals of the mass represent two points in space, where the *object terminal* - identical to the upper connection point being free and associated with the motion variable – is to be considered as a specific point on a massive physical object being free of friction of any kind, and the *reference terminal* –corresponding with the lower connection point at the *L*-shaped guide – represents a fixed point on the inertial reference system.

The mass defines a measured relationship between a momentum p (P-variable state or through energy variable) and a velocity υ (T-variable rate or through power variable)

$$p(t) = m\frac{\mathrm{d}s}{\mathrm{d}t} = m\upsilon(t) \tag{2.67a}$$

By the time rate of change of momentum p as the "state (or energy) variable" it follows the force F_{m} as the "rate (or power) variable". Thus, according to a time-invariant mass the *P-storage element law* of Eq. (2.67a) can be expressed by a relation between the impressed *inertial* (or mass) *force* F_{m} and the resulting motion in terms of the acceleration a:

Newton's second law (law of motion)

$$F_{\mathrm{m}}(t) = \frac{\mathrm{d}p}{\mathrm{d}t} = \frac{\mathrm{d}}{\mathrm{d}t}(m\upsilon) = m\frac{\mathrm{d}\upsilon}{\mathrm{d}t} = ma(t) \tag{2.67b}$$

with the *mass (parameter) m*.

The forgoing relations between force and a pertinent motion variable refer to reversible processes by storing energy in either of two forms in elements that serve as reservoirs. This behaviour identifying a *conservative system* is not valid for physical objects taking part in nonreversible conversion, e.g., caused by phenomena of friction.

Assuming energy dissipation – or, more precisely, energy conversion from the translational mechanical state into the thermal state, a *non-conservative system* is indicated. The frictional force in liquid friction contacts implying a lubrication film, or in solid friction contacts, is different and depends on the velocity in a complicated way. In certain cases, and over restricted ranges of velocity, the frictional force may approximate a linear proportionality with velocity. In that case the physical dissipator is concentrated on a type of fluid friction, its principal property, involving the energy conversion into heat, and will be represented by the following idealized model.

Damper. The fluid friction dissipator is the *linear time-invariant damper*, 2.2.4, Fig. 2.11c. The symbol for the dissipator element, shown in 2.2.4, Fig. 2.11c, is a *viscous damper* or *dashpot*, chosen because physical dashpots are commonly used for the intentional insertion of damping. As in the case of the spring element, both terminals of the damper are free to move independently.

This type of dissipator represents a defined or measured relationship between a resistive *damping* (or damper) *force* F_d (P-variable rate or through power variable) and the relative velocity υ (T-variable rate or across power variable) due to the velocities υ_1, υ_2 at the two terminals:

$$F_d(t) = c(\upsilon_1 - \upsilon_2) = c\upsilon(t) \tag{2.68}$$

with the *viscous damping coefficient c*.

The passive 2-terminal element symbols represent the storage or the irreversible conversion of energy in an object constrained to have only one degree of freedom in translational movement or the energy stored in one degree of freedom of a less constrained object.

Equations (2.66b), (2.67b), (2.68) define the types of elements by a pair of rate (or power) variables, i.e., by the instantaneous values of force and velocity. Their product is equal to the time rate of change in translational mechanical state (or energy) and represents

the *instantaneous power P*(t)

$$P(t) = F(t)\,\upsilon(t) \tag{2.69}$$

The relations between force and a pertinent motion variable quantity define idealized element laws by a descriptive quantity such as k, m, c, called *element parameters*.

The *local force of gravity* (weight force) F_g is rendered by an *active 2-terminal element* being concerned in the reversible conversion of energy. Thus, the effect of gravity may be represented by an:

Ideal mechanical source: *ideal force generator*, Fig. 2.26a.

This type of generator is appropriated to exert a *static load* on a mass m

$$F(t) = F_g = mg \tag{2.70}$$

due to the *acceleration of gravity g*.

Mechanical Generators

Ideal source elements serve as the conceptual means by which the introduction of energy or a signal into the system is represented. Because of the instantaneous power, Eq. (2.69), referring to the change of translational mechanical state, the active element delivering energy to the system will be defined as a *force* or a *velocity source* (or *generator*).

The source symbols, introduced in 2.2.5, take pattern from active elements of electric circuits. Like the symbols for the three passive elements, the active element symbols are shown as having two terminals, so that the sources can be interconnected with the passive elements to form closed circuits

The *complete description of the dynamic behaviour* of the physical system requires the specification of

- the relationships between the measured variable pairs for each of the sets of measurements, and
- the way in which the physical objects comprising the system are connected together.

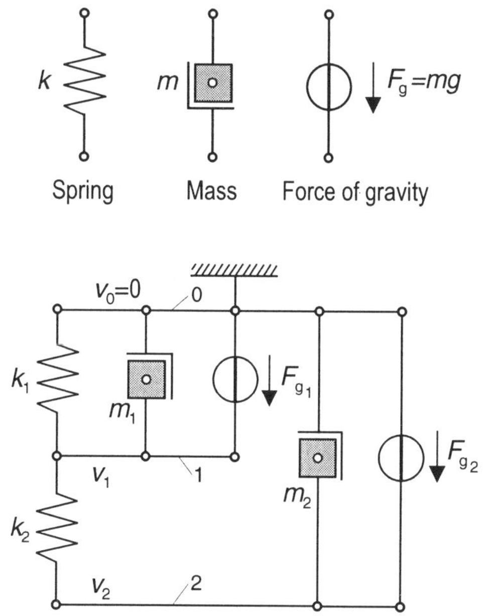

a

b

Fig. 2.26. Two-mass system. **a** Network symbols for basic linear time-invariant mechanical elements and ideal force generator; **b** system network diagram *mechanical circuit*

The overall system behaviour would be investigated by connecting the appropriate set of instruments to the composite system. This set of pairs of measurements and the stated relationships between the corresponding individual variable pairs may be represented by the *system network diagram (mechanical circuit)*, Fig. 2.26b, illustrating the overall behaviour of a two-mass system example.

2.3.3 Construction of Mechanical Network Diagram. *Mechanical Circuit*

This book is primarily concerned with relatively simple model systems in which each element of the network diagram corresponds to an obvious component in the actual mechanical system, and expressed in a different way, with systems modelling by drawing the mechanical network diagram directly from an understanding of the way in which the actual system operates.

The mechanical circuit is defined as a clothed path in space which includes one point in the inertial reference framework.

Illustrated by the two-mass-system example in 2.3.2, the *construction of mechanical circuits* will be reduced for simple systems being directly represented to the following *steps*:

- Each distinct velocity must be identified – including the reference, which is usually the *inertial frame* (or *inertial reference system*). Each velocity is

marked by a node, or junction point, in the network diagram (υ_1, υ_2 and the suspension reference);
- the passive elements (k, m, c) are inserted in the network diagram between the appropriate pairs of velocity nodes. In almost every case, each mass appears with reference as one of the two nodes (m_2 is drawn from υ_2 to the reference, since a force applied to m_2 depends on the acceleration (the derivative of υ_2), k_2 is inserted between υ_1 and υ_2 because a force applied to k_2 depends on the total deflection $s_1 - s_2$ (the integral of the relative velocity $\upsilon_1 - \upsilon_2$);
- the sources are inserted between the appropriate nodes (F_{g2} applied to m_2 is connected to the node moving at the velocity υ_2, F_{g1} applied to m_1 is connected to the node at the velocity υ_1, both remaining terminals of force generators must be connected to the frame).

2.3.4
Derivation of Mechanical Network Equations. *Equations of Motion*

If the assumptions made about the spatial flow of energy in the physical system are accurate, the network diagram will provide a complete representation of the system, from which the dynamic behaviour may be derived *systematically* in terms of mathematical relations. They are called *network (or circuit) equations* and define the mathematical model of a physical system. In many cases of mechanical systems modelling the equations of motion can be written directly by inspection from the network diagram.

By means of the three types of passive elements and the two types of ideal sources (or generators), introduced in the preceding Sects. 2.2.4, 2.2.5, 2.3.1, the relationships between the corresponding individual variable pairs are stated for mechanical translational systems. To describe completely the dynamic behaviour compatible interconnection requirements must be applied as well to the inspection of the mechanical network diagram as to its construction performed before.

Dynamic Equilibrium Statement. The simplest approach among several synthetic methods for the derivation of equations of motion bases on *D'Alembert's principle*. That statement of dynamic equilibrium, termed *force interconnection requirement*, is defined as follows:

- *The algebraic sum of the forces leaving (entering) a common junction (node) of a mechanical network model equals zero.*

Taking this rule as basis of inspection a set of motion equations will be obtained for each node in turn. Any one node (usually the reference node) can be omitted, since the equation at this node is simply the sum of all the other equations.
The simplicity of that method can be illustrated by inspecting the network diagram of the referred two-mass-system example, Fig. 2.26b.

Since two velocities, υ_1 and υ_2, are unknown two equations are obtained – one written for the sum of forces leaving the υ_1 node, the other for the υ_2 node.

There are four branches attached to the υ_1 *node*, the source F_{g1} and the three passive elements k_1, k_2, m_1:

$$F_{m_1} + F_{s_1} - F_{s_2} - F_{g_1} = 0 \qquad (2.71a).$$

The corresponding relation for the υ_2 *node* can be written by considering the attached three branches:

$$F_{m_2} + F_{s_2} - F_{g_2} = 0 \qquad (2.72a).$$

Performance Equations. Using the component relationships between power variables F; υ for each of the fundamental elements, Eqs. (2.66b), (2.67b), the interconnective relation among the system power variables υ_1 and F_{g_1} or υ_2 and F_{g_2} yields the

force-motion performance equations

$$m_1\dot{\upsilon}_1 + k_1\int_0^t \upsilon_1 \mathrm{d}\upsilon + k_2\int_0^t (\upsilon_1 - \upsilon_2)\mathrm{d}\upsilon - F_{g_1} = 0 \qquad (2.71b)$$

$$m_2\dot{\upsilon}_2 + k_2\int_0^t (\upsilon_2 - \upsilon_1)\mathrm{d}\upsilon - F_{g_2} = 0 \qquad (2.72b)$$

In vibration theory the displacement s is considered as the appropriate motional variable. Relating force and displacement as the system variables F; s the interconnection results in the

differential equations of motion (nonhomogeneous)

$$m_1\ddot{s}_1 + (k_1 + k_2)s_1 - k_2 s_2 = F_{g_1} \qquad (2.71c)$$

$$m_2\ddot{s}_2 - k_2 s_2 + k_2 s_2 = F_{g_2} \qquad (2.72c)$$

Exerting static loads on the masses, Eq. (2.70), the spring forces balance the weights at the static equilibrium position. Since the weight forces equal the spring forces due to the static deflections the *static equilibrium condition* is valid

$$\begin{aligned} m_1 g_1 - k s_{\mathrm{stat}_1} + k_2\left(s_{\mathrm{stat}_2} - s_{\mathrm{stat}_1}\right) &= 0 \\ m_2 g_2 - k_2\left(s_{\mathrm{stat}_2} - s_{\mathrm{stat}_1}\right) &= 0 \end{aligned} \qquad (2.73a,b)$$

where s_{stat_1}, and s_{stat_2} are the static deflections of the springs. Arbitrary positions s_1; s_2 from the static equilibrium positions are given by the equations of the vibratory motion

$$\begin{aligned} m_1\ddot{s}_1 &= m_1 g_1 - k_1\left(s_1 + s_{\mathrm{stat}_1}\right) + k_2\left(s_2 + s_{\mathrm{stat}_2} - s_1 - s_{\mathrm{stat}_1}\right) \\ m_2\ddot{s}_2 &= m_2 g_2 - k_2\left(s_2 + s_{\mathrm{stat}_2} - s_1 - s_{\mathrm{stat}_1}\right) \end{aligned} \qquad (2.74a,b)$$

Using the static equilibrium conditions, Eqs. (2.73a,b), the given relations (2.74a,b) reduce to the governing equations of the *free vibration* of an undamped two-degree-of freedom system:

Differential equations of motion (homogeneous)

$$\begin{aligned} m_1\ddot{s}_1 + (k_1 + k_2)s_1 - k_2 s_2 &= 0 \\ m_2\ddot{s}_2 - k_2 s_1 + k_2 s_2 &= 0 \end{aligned} \qquad (2.75a,b)$$

in matrix notation

$$\underbrace{\begin{bmatrix} m_1 & 0 \\ 0 & m_2 \end{bmatrix}}_{=M}\begin{bmatrix} \ddot{s}_1 \\ \ddot{s}_2 \end{bmatrix}+\underbrace{\begin{bmatrix} k_1+k_2 & -k_2 \\ -k_2 & +k_2 \end{bmatrix}}_{=K}\begin{bmatrix} s_1 \\ s_2 \end{bmatrix}=\begin{bmatrix} 0 \\ 0 \end{bmatrix} \tag{2.75c}$$

in compact matrix form

$$\boldsymbol{M}\ddot{\boldsymbol{s}}+\boldsymbol{K}\boldsymbol{s}=0$$

where $\boldsymbol{s}$ and $\ddot{\boldsymbol{s}}$ are column matrices, the displacement vector and the acceleration vector, respectively, and $\boldsymbol{M}$ and $\boldsymbol{K}$ are square matrices, the *mass matrix* being a diagonal matrix, and the *stiffness matrix*, respectively

$$\boldsymbol{M}=\left[m_{\text{ij}}\right]=\begin{bmatrix} m_1 & 0 \\ 0 & m_2 \end{bmatrix} \; ; \quad \boldsymbol{K}=\left[k_{\text{ij}}\right]=\begin{bmatrix} k_1+k_2 & -k_2 \\ -k_2 & k_2 \end{bmatrix} \tag{2.76}$$

with the *mass* or *inertia coefficients* m_{ii} (i,i = 1,2)
as its 2 elements:

$$m_{11}=m_1 \;, \quad m_{22}=m_2 \;, \quad \text{whereas} \quad m_{12}=m_{21}=0 \tag{2.77a}$$

and the *stiffness* or *elastic coefficients* k_{ij} (i,j = 1,2)
as its 4 elements

$$k_{11}=k_1+k_2 \;, \quad k_{22}=k_2 \;, \quad k_{12}=k_{21}=-k_2 \tag{2.77b}$$

Because the coefficients m_{12} and m_{21} are equal to zero, the two coordinates s_1 and s_2 are said to be *dynamically decoupled.* On the other hand, the coefficients k_{12} and k_{21} are not equal to zero, so that the coordinates s_1 and s_2 are said to be *elastically coupled.*

2.3.5 Connections with Signal-flow Diagrams. *Oriented Linear Graphs*

A representation of systems behaviour alternatively to networks (or circuits) is given by linear graphs. Replacing the introduced network symbols (2-terminal elements) by coded line segments called branches, and connecting the nodes to a *coded linear graph*, the topological relationship between the system variables will be equal to the presented network diagram (mechanical circuit). However, to represent the measurement pairs of the network an arbitrary set of orientations may be allocated corresponding to an arbitrarily oriented set of measuring instrument pairs. To agree upon positive directions of force and velocity (and hence displacement and acceleration) an orientation convention will be adopted consistently for force- and velocity-transducers measuring power absorbed in a network element. Thus, each measurement pair may be represented by an *oriented line segment*; the complete set of oriented line segments for the connected network is said to be an *oriented linear graph* for the network or for the physical system being represented by the network.

Signal-flow Graph Definitions. This type of systematic diagram representation applies oriented linear graphs on which nodes (vertices) represent variables and branches represent specific relationships between the vertex variables, 2.1.4.

The representation of power-variables measurement pairs bases on the *general orientation convention* that if both of the element power (or rate) variables are positive the element is absorbing power. Hence, connecting a complete set of power-measuring transducers with a consistent orientation to every network element a complete power balance will be obtained.

Tracing back to the two-mass-system example the network elements pertained are represented by single oriented line segments, Fig. 2.27a, whereas the interrelation of network variables is shown by the system oriented linear graph, Fig. 2.27b.

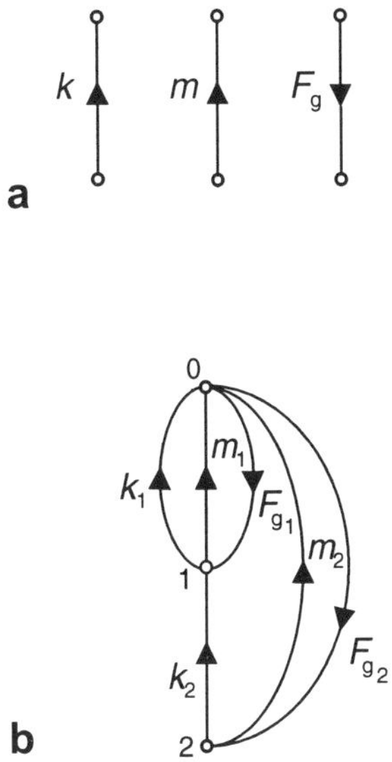

Fig. 2.27. Two-mass system. **a** Oriented line segments for basic linear time-invariant mechanical elements and ideal force generator; **b** system oriented linear graph *mechanical signal flow graph*

Summary of Graph Properties. Signal flow graphs (SFG) are proper for illustrating power transactions in networks by defining the spatial or interconnective relationships among the system power variables. The analysis of translational mechanical systems is based on the following *pair of postulates*:

Incidence (or vertex) relationship for flow variables (through power variables):

- The algebraic sum of all forces exerted on a point of connection (mechanical vertex) is identically zero (d'Alembert's principle).

Boundary (or circuit) relationship for effort variables (across power variables):

- The algebraic sum of all component velocities taken around any closed boundary (mechanical circuit) of a network is identically zero.

These two *fundamental laws* of mechanical network analysis are both necessary and sufficient for the *conservation of energy* in a network model. They may be regarded as a natural generalization of Kirchhoff's laws in electrical circuits.

Relationships which depend on the connection of the elements in the network diagram and not on the exact geometrical properties will be termed *topological relationships* being mainly used in network analysis by the following definitions:

- *Branch* (or path) is an element of a network (circuit element) or of an oriented linear graph.
- *Node* (or vertex) is the junction (point) at which two or more branches meet.
- *Loop* (or circuit) is a set of branches forming a closed path (subnetwork or subgraph) passing only once through any node.
- *Tree* is a set of branches joining all the nodes of a network (subnetwork or subgraph) without forming a loop.
- *Cut-set* of a network (or a graph) is a set of branches of a graph such that cutting all its branches increases the number of unconnected parts of the graph, but the retention of any one of these branches does not increase that number.

Some of the above terms may be illustrated by reference to the signal flow graph of Fig. 2.27b. The nodes are marked as 1, 2 and 3 and the branches as k_1, k_2, m_1, m_2, F_{g1}, and F_{g2}. Loops are formed by branches k_1 and m_1, by branches m_1 and F_{g1}, by branches k_1, m_2, and k_2, by branches F_{g1}, F_{g2} , and k_2, and so on. The branches k_1 and k_2 form a tree of the graph since they connect all the nodes of the graph without forming any loops. For this tree the remaining elements m_1, m_2, F_{g1}, and F_{g2} are the corresponding chords. If the branches k_2, m_2, and F_{g2} were removed from the graph, then node 3 would be separated from the rest of the graph and the connectivity of the remaining graph would be grater than that of the original graph. Further cut-sets of the graph are formed by k_1, k_2, m_1, and F_{g1}; by k_1, m_1, F_{g1}, m_2, and F_{g2}, and so on.

2.4 Combined-flow-diagram Representation

There are two concepts which are basic in systems approach by diagram representation. The first, and more fundamental, is the network (or circuit) concept, the visualization of a combination of physical components as a topological configuration of basic elements which obey simple natural laws. The second is the block-diagram (or signal-flow-diagram) approach to the analysis of complex systems.

To represent the behaviour of engineering systems in which two or more physical domains are coupled, particularly of feedback control systems containing an energy conversion process, a model system may be useful which is a combination of basic diagram concepts. The *combined -flow diagram* being a combination of network diagram and block diagram represents the behaviour of a physical system in terms of a combined *flow of energy and signals* in the model system preferably of *mixed domain dynamic systems*.

2.4.1 Symbolic Measurement of Dynamic System Variables

In order to represent systems by a combined-flow diagram, two symbolic conventions are required in addition to the conventions for the representation of dynamic behaviour in network terms and the representation of functional relationships in block diagram terms:

- Conventions for the symbolic extraction of signals from a network
- Conventions for the symbolic control of rate source outputs by signal variables.

Signal extraction conventions. In many cases, some dynamic system variable is a function of a *network rate variable* (intensity or power variable). To show this relationship symbolically the stated diagram representation requires a convention for the symbolic measurement of a network power variable.

The convention adopted for the symbolic extraction of a flow variable (through power variable) is marked by a circle at the end of a line segment being crossed by a terminal line of the appropriate network element. This symbol emphasizes the "through-propagating" nature of the power variable being measured and the 1-point nature of the measurement.

The convention adopted for the symbolic extraction of an effort variable (across power variable) is marked by a single crossline at either end of a branched out line segment being attached to the appropriate pair of nodes. That symbol emphasizes the "across-acting" nature of the power variable being measured and the 2-point nature of the measurement, shown in the treated example of 2.4.2, Fig. 2.29b.

As shown in 1.2.2, storage element state (or energy) variables are related by their time rate of change to the storage element rate (or power) variables. A *network state variable* (quantity or energy variable) for a storage element is thus obtained from the appropriate rate signal by use of an ideal integrating operator.

The conventions adopted for the symbolic extraction of storage element *P*-variable state and *T*-variable state signals (respectively through energy and across energy variables) are according to those for the appropriate rate signals supplemented by inserting the transfer function block of an integrator with the block symbol 1/p in the appropriate line segment, Fig. 2.29b.

2.4.2 Symbolic Regulation of Supplementary Elements. *Couplers and Controlled Sources*

Controlled Sources. Certain energy conversion and control processes may be very conveniently represented on a combined-flow diagram by the use of ideal controlled rate sources, also called dependent sources. The value of the rate (intensity or power) output is taken to be proportionally equal to the value of the controlling signal input. The symbolic conventions adopted for controlled *T*-variable or *P*-

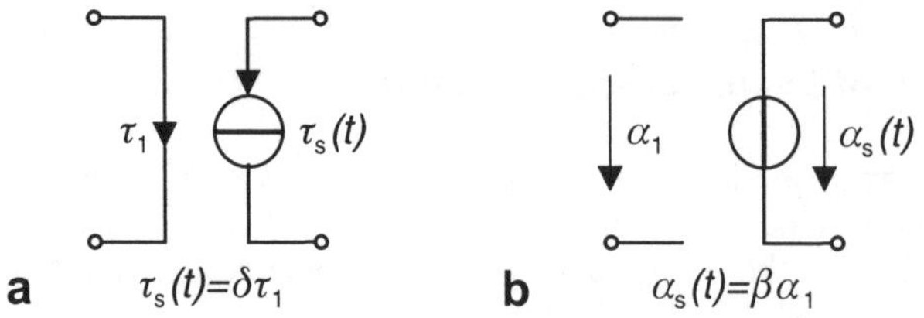

Fig. 2.28. Graphical symbols of ideal controlled generators. **a** Controlled *P*-variable 2-terminal source (current-controlled current source); **b** controlled *T*-variable 2-terminal source (voltage-controlled voltage source)

variable sources (respectively effort of flow variable sources) are corresponding to the graphical symbols for sources of electrical energy, Fig. 2.28.

Couplers. The way in which physical systems in which power is transmitted without storage or absorption of energy may be represented in terms of a network model termed a coupler. The symbolic conventions adopted for couplers are corresponding to the graphical symbols for electrical transformers or converters.

2.4.3 Connections between Network and Block Diagrams. *Mixed Domain System Structures*

Having laid down special symbolic conventions to combine systematic diagrams of different type the *combined-flow-diagram representation* can be applied to engineering systems incorporating both, energy conversion and automatic control. As an example of mixed-domain dynamic systems the electromechanical servomechanism acting as a position control system may be taken up as introduced in 2.1.3, Fig. 2.29.

The schematic diagram, Fig. 2.29a, illustrates a remote-positioning system suitable for large inertial loads, e.g., to control the rotational positioning of a telescope from a remote location.

The DC motor (actuator) is driven by an error voltage depending on the difference between desired Θ_r and actual angular position Θ. Applying this electrical signal to the stationary field structure to provide a motor control field (field resistance control) a torque T_S is developed, tending to rotate the motor output shaft with the variable speed ω. Thus, the DC motor and the separate power supply form together with the amplifier unit a controlled source configuration acting as an ideal voltage-controlled torque source. The potentiometers are converters with power conversion on a low level which act as ideal mechanical-electrical transducers yielding directly electrical voltages according to angular positions.

The servomechanism provides the possibilities of remote positioning and power amplification in a system with characteristics dependent primarily on the low-power elements (the potentiometers).

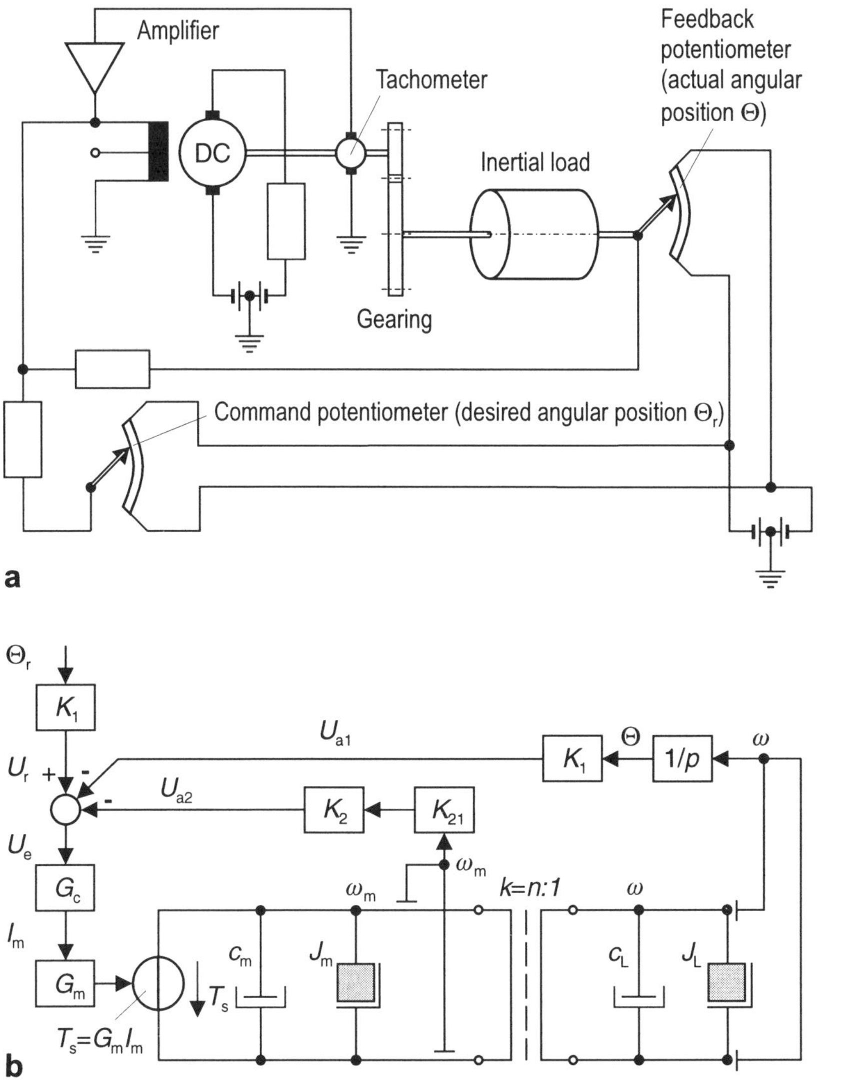

Fig. 2.29. Electromechanical servomechanism position controlled, following McFarlane, [9]. **a** Schematic diagram; **b** *combined flow structure* network block diagram

Combined-flow Diagram. The model representation of the position control system, Fig. 2.29b, will be constructed on the two main suppositions that the system is linear and that the only significant energy storage occurs in the rotational inertia of the motor armature, gearing, and load. This part of the control system will consequently be represented on the combined-flow diagram by a *network*; the remainder of the control system will be represented by a *block diagram* of the appropriate functional relationships.

Differing from the configuration of a basic control system illustrated in 2.1.3, Fig. 2.3, the alternate control system outlined in Fig. 2.29b, makes use of an *internal feedback compensation.* By this configuration system performance is improved. Being adapted to an intermediate plant variable the motor shaft speed ω_m is extracted in addition to the actual speed ω, and fed back for use by the controller.

System Equation Set. The transmission constants for the block diagram must be defined, in particular the controlled source must be specified by the motor's torque-current relation

$$G_m = T_s / I_m \tag{2.78}$$

Furthermore the system parameters of the network have to be determined. Especially the transformation ratio of the coupler is ascertained as an indefinite gear ratio

$$k = n : 1 \tag{2.79}$$

The relevant dynamic and signal variables of the model system are denoted by

Θ	actual angular position	(controlled variable)
Θ_r	desired angular position	(command input)
ω	actual speed	(extracted signal variable)
ω_m	motor shaft speed	(error variable)
U_e	error voltage	
U_r	generated voltage	

By inspection of the combined-flow-diagram the following interconnective relations can be derived:

Error voltage (amplifier input voltage)

$$U_e = U_r - U_{a1} - U_{a2} = K_1\Theta_r - K_1\Theta - K_2K_{21}\omega_m \tag{2.80}$$

Servomotor's torque

$$\begin{aligned} T_s &= G_c G_m U_e = G_c G_m \left(K_1\Theta_r - K_1\Theta - K_2K_{21}\omega_m\right) \\ &= c_m\omega_m + J_m \frac{d\omega_m}{dt} + \frac{1}{k}\left(c_L\omega + J_L\frac{d\omega}{dt}\right) \end{aligned} \tag{2.81}$$

Since $\omega_m = k\omega$, and putting $\omega = d\Theta/dt$ the differential equation is satisfied by the system response Θ

$$\left(J_L + k^2 J_m\right)\frac{d^2\Theta}{dt^2} + \left(c_L + k^2 c_m + k^2 G_m K_{21} K_2 G_c\right)\frac{d\Theta}{dt} + kG_c K_1 G_m \Theta = kG_c K_1 G_m \Theta_r \tag{2.88a}$$

Summarizing transmission constants in parentheses to corresponding constant coefficients the system differential equation will be reduced to

$$J\ddot{\Theta} + c\dot{\Theta} + k_t\Theta = k_t\Theta_r(t) \tag{2.88b}$$

thus representing a second-order model. The reduced form, Eq. (288b), can be rewritten

$$\frac{d\omega}{dt} = -\frac{k_t}{J}\Theta - \frac{c}{J}\omega + \frac{k_t}{J}\Theta_r \tag{2.89a}$$

and combined with

$$\frac{\mathrm{d}\Theta}{\mathrm{d}t} = \omega \tag{2.89b}$$

to give an equivalent pair of first-order differential equations governing the system behaviour

$$\begin{bmatrix} \frac{\mathrm{d}\Theta}{\mathrm{d}t} \\ \frac{\mathrm{d}\omega}{\mathrm{d}t} \end{bmatrix} = \begin{bmatrix} 0 & 1 \\ -\frac{k_t}{J} & -\frac{c}{J} \end{bmatrix} \begin{bmatrix} \Theta \\ \omega \end{bmatrix} + \begin{bmatrix} 0 \\ \frac{k_t}{J}\Theta_r \end{bmatrix} \tag{2.90}$$

2.5 Bond-graph Representation (Multiports)

Derived from the theory of linear graphs the bond graph method uses *modified graphs.*

Attempts to set up a structure for system analysis based on the electric circuit or graph concept will prove severely limited in usefulness for significant problems in many other fields. Basing on the implications of *energy flow* in physical systems the concepts and notation for bond graphs aim at providing a uniform mechanism for the description of a wide variety of physical systems. Whereas a model system such as the network diagram is well adapted to electromagnetic devices the main purpose of bond graphs is to represent systems involving *mixed physical domains.*

Focusing attention on energy exchange with the environment and also on the particular internal power transmission, problems will be formulated in terms of power variables first, and then proceeded to general purpose techniques involving nonelectric systems and their representation in differential equation form. In many cases, the means for direct physical modelling are provided (i.e., system synthesis) by directly accounting for energy storage, supply, and dissipation effects throughout the system based on fundamental physical considerations.

Although *pure signal flows* (i.e., "zero" power information transmission) are permitted in the scope of bond graph techniques, these signal flows will appear as a special case of general power interactions and will be used only as a result of a conscious act in the modelling process. By this, some fairly common errors and inconsistencies that result from an inadvertent neglect of back effects may be avoided.

Viewing system behaviour from the standpoint of energy continuity and power balance bond graph techniques are compatible with standard analysis methods of proven usefulness. Nevertheless, bond graph representation is quite apart from theoretical implications of the study of energy flow using linear graphs. The bond graph approach may be rather regarded as a practical method for generating a mathematical model. Applications will certainly be in the simulation of complex dynamic systems in which the physical logic of the subsystem models is not lost, and a certain flexibility in the choice of subsystem models is permissible, so that a balanced model system may be formed and the effects of system design changes may be readily exhibited.

Review of Bond Graph Approach. The term *multiport* – now standard in circuitry – originated with *H. A. Wheeler* in a conscious effort to extend energy-based network techniques to microwave bands and beyond. Generalizing on these ideas to analyse mechanical and other physical systems several researchers began to employ the terms *energy ports*, *power bonds*, and *multiport elements*. A former option favoured a form of duplex block diagram or signal flow graph being needlessly confusing for complex systems. Tracing back to earlier structure graphs in chemistry *W. K. Clifford* established the abstraction of a linear graph of *nodes* and *branches* as a mathematical system in its own right. Basing on graph theory it was possible to represent power flow as a single line – the power bond – to obtain the generalization of the power engineers *one-line diagram*. Taking the nodes to represent the *relations*, and the branches, the *terms*, *C. S. Peirce* expressed systems of relations and of proportional functions as simple structural formulas. *H. M. Paynter*, [22], adopting the convention of graphing each multiport as a *nodal* element settled the two ideal 3-port energy junctions to render the system of bond graphs a complete and formal discipline.

Attempts to present bond graph techniques to professional engineers have been made by *D. C. Karnopp* and *R. C. Rosenberg*, [23]. Besides an introduction to computer simulation of engineering systems the bond graph approach has been applied especially to fluid power systems and hydrostatic drives by *J. U. Thoma*, [24].

Multiport Components

The external variables selected for physical component description are directly related to power flow (power variables). The identification of variables may be regarded as the identification of a port at which power interactions between the component and its environment can occur. A way of describing a component is then to call it a *multiport*, i.e., a subsystem that may interact with other systems through one or more ports.

A bond graph will show all the ports explicitly whether or not the port is associated with a spatial location on a component. Though power is commonly expressed in terms of a product of two variables, diagrams will be simplified by restriction to product representations of power. Accordingly, components are represented with words or letters and ports or power interactions with single lines or bonds. The latter are of primary interest since the expression of power flow could, in principle take many forms with respect to the concerned pair of physical variables. Besides an internal description of each multiport though being quite unrestricted the term *bond* becomes significant for configurations of interconnected or bonded multiports (system bond graphs).

Effort and Flow Variables

Though scalar power flow need not be expressed as a product of two variables it is convenient to split into particular factors that are easily measured and that can be given physical interpretations.

For modelling purposes it is convenient to think of the intensive variable as an *effort*, e, and the extensive variable as a *flow*, f, so that their product will yield the *instantaneous power* exchanged

$$P(t) = e(t) \cdot f(t) \tag{2.91}$$

Many attempts have been made in associating analogously pairs of mechanical with electrical variable and in claiming in several domains to be of a certain natural similarity, as treated in 2.2.3. It also has been argued that the power variables split naturally into intensive-extensive or across-through pairs and that this provides a rationale for considering one member of the pair as the equivalent of an effort and the other as a flow. In the bond graph approach, however, this causes no difficulty because the treatment of effort and flow is symmetrical and thus the selection of effort-flow pairs is quite arbitrary. None of the concepts and techniques would change in any essential way if the roles of effort and flow were interchanged in any domain.

2.5.1 Classification of Multiports

Any physical system may be conceived as a multiported device with multiported elements. Using highly idealized versions of physical elements a basic set of multiports has been defined, other multiports may be formed by combining the basic multiports, Fig. 2.30a.

Basic Multiports

One-port. A 1-port element being defined in terms of a single pair of power variables at a port may be thought of as a *generalized impedance*. Some specific examples of 1-ports in which energy is stored are the

- capacity element (*C*-element or capacitor);
- inertance element (*I*-element or inductor);

or in which energy is dissipated being called the

- resistance element (*R*-element or resistor).

Active elements also can be defined by a 1-port device. An *ideal source* is assumed to provide one variable as a function of time while the other one is arbitrary without regard to the power delivered or absorbed, thus being defined as the

- effort source (*E*- or S_e-source);
- flow source (*F*- or S_f-source).

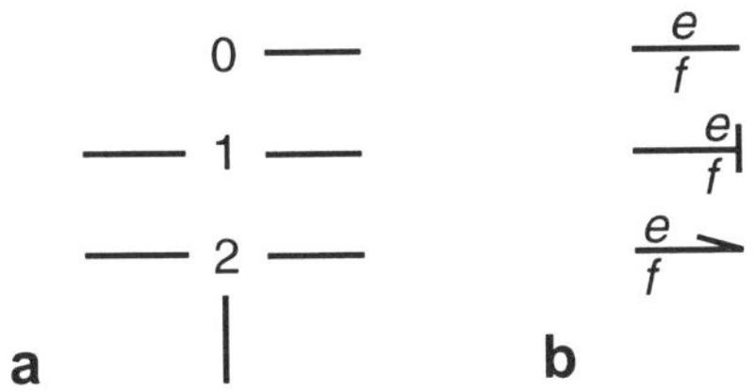

Fig. 2.30. Bond graph symbols and assignments. **a** Set of basic multiports: 1-port, 2-port, and 3-port; **b** power bond with indicated effort-flow signal pair: acausal bond, causal bond, and positively directed power flow

Two-port. A 2-port element may be conceived as a *generalized transport process*, i.e., a process by which energy is transformed, transmitted, or transduced. An electrical or mechanical oscillator formed by interconnecting a 1-port inertance and a 1-port capacitance may be interpreted by a 2-port device.

Also couplers are in general described by power-conserving 2-port elements acting as a *transformer*, a *gyrator*, or a *transducer*.

Three-port. A 3-port element may be thought of as a *generalized modulator*, including (triportal) ideal energy junctions, see 2.5.3.

This implies power and signal *modulators*, power and signal *amplifiers*, and *power exchangers* as proper examples for engineering devices.

2.5.2 Conventions for Interconnected Multiports. *Augmented Bond Graphs*

When components share common power variables the components are bonded together with a single line. Power interactions can occur and hence energy may flow from one multiport to the other. A vexing problem in all system investigations is the question of sign convention that also arises in bond graph modelling.

Sign Convention for Power

Half Arrow. Multiports being bonded together in such a way that common efforts and flows exist at the two ports the power flow out of one multiport is the power flow into the other. Furthermore, the *direction of power flow* along the bond must be considered. Conventional notations for positive values of the effort and flow quantities are so arranged that energy flows from left to right. On the bond graph this is indicated by a half arrow on the right end of the bond.

Causality Convention for Effort and Flow Variables

Causal Stroke. Causality implies the existence of two variables, one *independent* and the other *dependent*. Once a power bond has been described in terms of an effort-flow couple then it may be uniformly assumed that the effort and flow signals on a signal bond are always oppositely directed. Thus a transverse stroke at the one end of the bond, the so-called causal stroke, can serve to indicate simultaneously the direction of both, the effort and the flow signal. A useful mnemonic is the association of the flow variable, f, with a direction parallel to or along the bond, and the effort variable, e, with the transverse stroke, [22].

2.5.3 Fundamental Interconnective Relationships. *Generalized Kirchhoff's Laws*

The 3-port is a singular and most essential element. Classical mechanics recognizes but a single 3-port, namely the triportal energy junction; in this realm all systems are conceived as interconnected sets of 1-ports (generalized impedances) and ideal energy junctions.

For a generic multiported element restricted to be ideal, that means a flux of energy into this element is neither stored nor dissipated,
the *equation of energy continuity* states

$$\sum_{i=1}^{n} P_i = \sum_{i=1}^{n} e_i f_i = 0 \qquad (2.92).$$

A large class of energetic elements approximately satisfy this *fundamental principle* on which the overall theoretical analysis of model systems is based.

Several of the most useful ideal elements as couplers are already treated in 2.2.5. In particular, two basic 3-ports are presented being ideal in the power-conserving sense and that allow bond graph representations for a large number of physical systems analysed by conventional methods.

Ideal Energy Junctions

From this point onward a duality of interconnective relations must be emphasized, for there exist *two conjugate energy junctions*, the effort junction and the flow junction. These two ideal 3-ports form much of the basis for bond graph methods and largely set the bond graph apart from conventional representations. The fundamental idea is to represent as ideal multiports the two special types of connection structures (fundamental configurations) known as "parallel" and "series".

Both junctions are characterized by the condition that one of the two conjugate variables is common to all bonds, whilst the other ones sum up to zero.

Flow Junction (0-junction, or common-effort junction). The symbol for this element is a zero with three bonds emanating from it. Using the power sign convention for the junction with 3 bonds the *conjugate relationships* are defined as

$$e_i = e \quad (2.93a) \quad ; \quad \sum_{i=1}^{3} f_i = 0 \quad (i = 1,2,3) \quad (2.93b).$$

The 0-junction has a single effort on all its bonds and the sum of the flows to the element vanishes.

Effort Junction (1-junction, or common-flow junction). This dual of the 0-junction is represented by the symbol "1" with three attached bonds. With the indicated power sign convention for this element
the *conjugate relationships* result in

$$f_i = f \quad (2.94a) \quad ; \quad \sum_{i=1}^{3} e_i = 0 \quad (i = 1,2,3) \quad (2.94b).$$

A single flow exists for a 1-junction and the efforts on the bonds sum to zero. These equations bear a dual relationship to Eqs. (2.93a,b), but both the 0-junction and the 1-junction satisfy the power conservation relation, Eq. (2.92).

The conjugate relationships (junction laws) are simple generalizations to *Kirchhoff's* loop and node laws in the electrical case, and, borrowing *C. Ferrari's* terminology, the laws of velocity and equilibrium in the mechanical case.

2.5.4
Construction of Mechanical Bond Graph. *Mechanical Multiport*

The method for finding valid bond graphs may be applied to systems that are composed of circuitlike elements in a single energy domain. An example to study vibration problems is given by the elementary mechanical system (simple oscillator), Fig. 2.31.

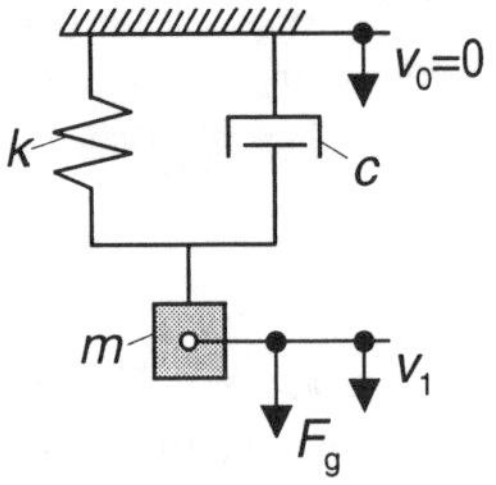

Fig. 2.31. Schematic diagram of a one degree of freedom mass-spring system (one-mass system) under the influence of gravity

Specifying the multiport structure for this common type of mechanical system amounts to translating standard system representation, i.e., the mechanical schematic diagram, into bond graphs using the basic set of multiports already defined in 2.5.1.

Partly due to tradition, one typically attempts to describe the dynamics in terms of velocities or displacements. Thus the development of a bond graph is started by establishing 1-junctions for each velocity of interest and particularly for every mechanical node in the system. The forces on the nodes may then be found by using 0-junctions to find relative velocities across the 1-port elements.

It must be taken care of inertia elements as the force does not pass through the mass and the velocity of the mass is measured with respect to an inertial frame. Despite these restrictions for mechanical circuits the bond graph representation of the example system is easily verified by relating the forces (efforts) of the *C*-element *spring k* and the *R*-element *damper c* to the proper relative velocity (flow) υ, further by applying the element forces to the node (1-junction), and thus to the *I*-element *mass m*. The gravity force F_g is an effort source being also connected to the corresponding node, Fig. 2.32.

Features of System Bond Graphs

For lending insight to a variety of mechanical systems certain features of bond graphs appear frequently leading to proved concepts.

Concept of a Junction Structure. By bonding together sets of 1-ports and ideal energy junctions any number of ideal multiports may be formed which are designated *junction structures*. In the case of elementary mechanical systems junction structures composed of 0- and 1-junctions are common, Fig. 2.32.

For lumped parameter systems with moving reference points (or points of attack of 1-port elements) one shall be advised to follow a *systematic procedure*:

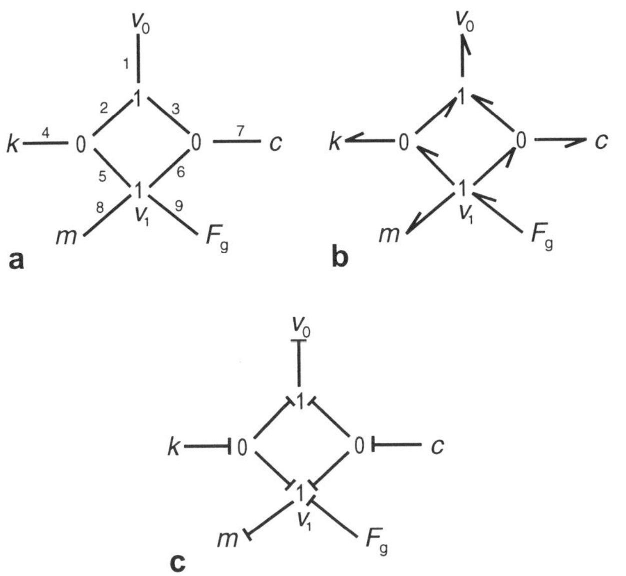

Fig. 2.32. One-mass system. System bond graph *mechanical junction structure*. **a** Acausal graph with explicit variables and parameters shown; **b** graph with sign conventions for power flow; **c** causally augmented graph

- Assign a 1-junction to all points with a definite *velocity* (flow) v;
- put all 1-port elements, except I-elements *masses* m, between the appropriate 1-junctions behind a 0-junction;
- I-elements *masses* m are directly attached to the 1-junction representing the speed of the appropriate point or moving element.

For many problems, the finding of a junction structure provides a satisfying check on the consistency of the formulation.

A slight generalization of the 3-port 0- and 1-junctions is possible by noting that two similar 3-port junctions joined by a single bond form a 4-port junction.

Concept of a Field. Generally, transactions of energy are related to *multiport fields*. The system configuration is composed of C-, I, and R-elements, respectively. This partitioning of a system, which is mainly of conceptual interest, has great practical significance for automatic simulation, [22], [23], [24].

2.6 Comparison of Diagram Representations (References to Applications)

For visualizing the input-output behaviour of a physical system being verified by a valid model system a variety of diagram representations is in common use. For the

Model system representation (levels of abstraction)	Verification criteria: Static behaviour	Dynamic behaviour	Signal flow	Inter-action (loading)	Energy flow	Non-linearity	Design paramet. estimat.
Block diagram Signal flow graph	x*	x*	x	(x)	(x)	–	–
Two-port diagram (bilateral signal flow)	x*	x*	x	x	x	–	–
Bond-graph representation (multiports)	x	x	x	x	x	–	–
Network diagram (circuit)	x*	x*	x	x	x	–	x
Schematic diagram (pictorial representation)	x	–	–	–	–	x	x**

Degree of abstraction (increasing upward)

* valid only for linear systems, otherwise for systems linearized at equilibrium point
**valid only for static or steady-state behaviour of systems

Fig. 2.33. Specifications of diagram representations related to model verification

the individual case a choice between the diagram representations is made in consideration of the special features in the type of transactions and the complexity of interconnected structures. The types of diagrams representing dynamic systems as outlined in 2.1 to 2.5 by a comprehensible survey rather differ in the *degree of abstraction* being embodied and in *verification criteria* being satisfied by the model system Fig. 2.33.

2.6.1
Schematic Diagrams. *Visually Descriptive Diagrams*

Pictorial-schematic Diagram. A systematic approach to the development of technical products (machines and devices) includes design engineering methods, in particular the calculation for designing. Thus, the dynamic analysis being performed on an engineering system generally implies the *task of design*.

Dynamic system design involves virtually all phases of dynamic analysis. The only difference is that in design the analysis can be repeated several times. Indeed, if the designed system does not perform as desired, then improving changes must be made until it does.

A *pictorial representation* looking somewhat simular to the engineering system yields an image of physical objects thought to be realized. Though the symbols on this visually descriptive type of diagram are not well standardized essential design parameters for a systematic embodiment are frequently depicted. Parametric informations are only concerned with the *static behaviour* including stiffness effects, at best with an approximate steady-state behaviour since somewhat inertial and damping effects are assumed. The purpose of this type of representation is to display information needed to construct a *schematic diagram*. Mechanical schematic diagrams composed of basic elements are typical for the overall theoretical analysis in vibrations.

2.6.2 Systematic Diagrams. *Interconnection Diagrams*

To visualize the *dynamic behaviour* of interconnected structures specific types of *systematic diagram* will be put into action. The modelling process for lumped systems includes a mathematical description of each of the components plus a complete description of the manner in which the components are interconnected to form the system. A systematic diagram consists of a set of component mathematical models which

- indicate the pairs of points in the system at which measurements would be made to correspond with the system variables;
- indicate the polarity of those measurements.

An interconnection diagram joins the model of any component – its measurement diagram and terminal characteristic – and the model of the system interconnection pattern together to derive *corresponding sets of equations* required for an unique solution.

Block Diagram and Signal Flow Graph. By transforming related input-output variables the *block diagram* allows an algebraic and graphical representation of the cause-and-effect relationships in a given system. It is convenient to maintain the functional uniqueness of each of the physical parameters. This permits the direct manipulation of the block diagram to effect corresponding changes in physical components or values.

The principal advantage of a block diagram is that the system's physical components are themselves represented by blocks indicating their operations rather than by line segments. Thus, the block diagram more closely resembles the physical structure of the system.

Though the block-diagram approach is a useful tool in the analysis of the stability and dynamic performance of feedback systems there are several fundamental limitations and disadvantages. An alternate representation of the system which pictures the system in more detail than a block diagram, but which retains the visual representation of the flow signals through the system is given by the *signal-flow diagram.* It provides the representation of system performance and proves an efficient aid in the analysis of complex feed back systems. It further simplifies the derivation of specific rules and techniques for the analysis of feedback systems. Once the signal-flow diagram is drawn and the preliminary analysis of the detailed system behaviour is accomplished reduction formulas may be applied.

The very nature of the block diagram (or signal flow graph) assumes that the transfer function of the tandem combination of two blocks is the product of the individual transfer functions. Thus, the primary advantage of the block-diagram approach is, however, the ease with which the contributions of various components to overall system performance can be evaluated. The full realization of this advantage usually demands that the component blocks are assumed to be non-loading. that means without any interaction.

Thus, the isolated single-coupled representation is an idealization satisfactorily suited to low-energy transactions basing on the indications of signal flow.

Two-port Diagram. To cover systems behaviour with interacting loops (back effects) a *bilaterial signal-flow diagram* may be applied representing an interaction conceived in two variables, thus attributing a direction of causality in the interaction. Two-port networks are standardized two-line diagrams interconnecting two-terminal pairs (2-port components) which are separate entities of many common engineering components being modelled on a higher level of abstraction.

Various internal reticulations of 2-port models have been schematically depicted in electrical engineering originating in the theory of long power transmission lines.

Linear 2-ports may be mathematically represented by way of 2 x 2 transformation matrices in standard form. In addition to the four causal matrices the transmission matrix (or chain matrix) has been developed which establishes a direct spatial correspondence to the ports themselves. By this the transmission of power through is described for the resultant two-terminal-pair network.

Bond-graph Representation. For predominantly passive systems composed of energetically coupled physical components rather than of isolated signal-coupled functional boxes the bond graph is to be preferred. This type of diagram inherently maintains the proper pairing of signals to give actual powers, whereas the same signal pairs are apt to be separated and somewhat dispersed in manipulating a block diagram or a signal flow graph.

A *power bond* may be conceived as an interaction; associated with each bond are two variables, the first pertaining to an effort and the second to a flow, their product yielding the power or energy flow rate.

The bond graph approach is compatible with common analysis methods. In particular, the bond graph techniques are well suited to represent the power flow through various energy domains (mixed-domain dynamic systems), thus being an alternative for replacing combined-flow diagrams. Using configurations of interconnected (bonded) multiports and graphing them as nodal elements the preceding concept of 2-ports has been carried on. Together with a modified graph the power flow is represented by the *power engineers one-line diagram*, thus being a generalization on a higher degree of abstraction.

Network Diagram. Complementary to the block diagram concept by network diagrams a more fundamental concept is established, also called *circuit concept* as being well approved in electrical circuitry. This type of diagram bases on visualizing a combination of lumped-parameter components as a topological configuration of network elements which obey elementary physical laws. Lumped systems or networks are interconnection diagrams being configurations of two-terminal elements (1-port components). Since sources and couplers are inserted and sign conventions introduced mechanical circuit diagrams can be constructed usually by converting the more convenient schematic diagrams. Thus, for a wide variety of practical problems network equations (equations of motion) can be derived systematically by inspection without founding on advanced principles of analytical dynamics.

By transforming the pertinent force-motion relations into the frequency domain the *concepts of mechanical mobility* as well as *of dynamic compliance* permit the steady-state approach in vibration analysis under bypassing differential equations.

In more complicated mechanical systems the circuit equations can be derived by the application of the power-balance concept (in somewhat more generalized form termed Lagrange's formulation in complex situations).

3 System Representation by Equations (Mathematical Model)

The model representation through diagrams corresponds to the mathematical relations between the quantities (system variables) of the model system.

Mechanical systems with discrete parameters (lumped parameter models) are described by ordinary differential equations delivering in most applications a sufficient approach to predict the dynamic system behaviour.

3.1 Representation of Mechanical Systems by Differential Equations of Motion

The forced *mass-damper-spring system* commonly referred to as a *single degree-of-freedom system* (simple oscillator) will be used as a *generic system* for which only one coordinate is required to define completely the configuration of the system at any instant.

Variables correspond to mechanical quantities describing excitation and response, and they are functions of time. For translational mechanical systems the *system variables* can be identified with the *force F* (input; excitation) and the *displacement s* (output; response) as the pertinent motion variable quantity.

Element Parameters (lumped parameters). Mechanical components, or elements, refer to parts (subsystems) of the mechanical system. The passive elements are identified with the *element parameters*:

elastic (spring) constant (stiffness) : k
linear (viscous) damping coefficient : c
mass : m

Those element parameters characterize the model system as linear, time-invariant, and they are measured or defined by the three types of idealized element laws treated in the preceding chapter (linear characteristic curves depending on material properties and geometrical fundamentals), Fig. 3.1.

The model system behaviour is governed by the *equation of motion* (vibration equation)

$$L[s] \equiv \sum_{\nu=0}^{n=2} a_\nu s^{(\nu)}(t) = m\ddot{s} + c\dot{s} + ks = F(t) \tag{3.1}$$

being classified in mathematics as a linear, nonhomogeneous, second-order ordinary *differential equation* with constant coefficients. The differential equation can

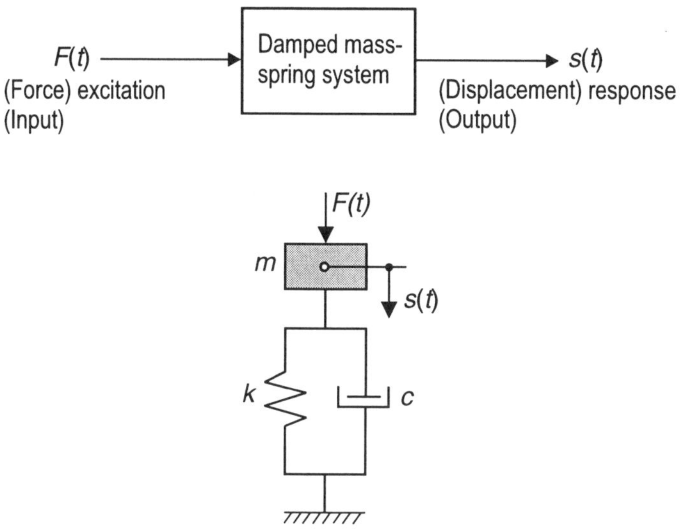

Fig. 3.1. Damped single degree-of-freedom system. **a** System functional block diagram, **b** schematic diagram

be subdivided into the

left-hand side:
linear differential expression of the *output* of the system or the *response*:
Displacement $s(t)$
(and its time derivatives)

right-hand side:
source function acting directly on the system by the *input* or the *excitation*:
Forcing function $F(t)$
(external force)

In this case, the translational coordinate x is equal to s as the convenient symbol for the physical variable *displacement* (translational mechanical system).

3.1.1 System Specifications by Normalization of the Differential Equation. *Time-response Analysis*

Normalized differential equation used in *vibration theory*

$$\ddot{s} + \frac{c}{m}\dot{s} + \frac{k}{m}s = \frac{1}{m}F(t) \tag{3.2a}$$

makes evident the model system behaviour by the following specifications.

System Parameters (vibratory specifications). Quantities which characterize a system by interrelation effects of passive elements. The parameters formed by ratios of the element parameters k, c, m are constants. Definite values can be assigned to each constant ratio.

Damping coefficient (attenuation coefficient):

$$\delta = \frac{c}{2m} = \omega_0\zeta \tag{3.3}$$

time constant (relaxation time):

$$T_{\mathrm{r}} = 1/\delta = 2m/c \tag{3.4}$$

damping ratio (fraction of critical damping):

$$\zeta(\text{or } \vartheta) = \frac{\delta}{\omega_0} = \frac{c}{2m\omega_0} = \frac{c}{2\sqrt{mk}} = \frac{c}{c_{\mathrm{c}}} \tag{3.5}$$

critical damping coefficient (critical viscous damping):

$$c_{\mathrm{c}} = 2m\omega_0 = 2\sqrt{mk} \tag{3.5a}$$

(undamped) *natural frequency* as an angular (circular) frequency:

$$\omega_0 = \sqrt{k/m} \tag{3.6a}$$

as a (cyclic) frequency

$$f_0 = \omega_0/(2\pi) = \sqrt{k/m}/(2\pi) \tag{3.6b}$$

being the reciprocal of the *natural period* of oscillation:

$$T_0 = 2\pi/\omega_0 = 1/f_0 \tag{3.7}$$,

and describing the frequency of free vibration resulting from only elastic and inertial forces of the system.

By substitution Eq. (3.2a) is converted into

$$\ddot{s} + 2\delta\dot{s} + \omega_0^2 s = \frac{1}{m}F(t) \tag{3.2b}$$

$$\ddot{s} + 2\zeta\omega_0\dot{s} + \omega_0^2 s = \frac{\omega_0^2}{k}F(t) \tag{3.2c}$$.

A change in the independent variable t by applying the non-dimensional time (natural time)

$$\tau = \omega_0 t \tag{3.2d}$$,

introduced by *K. Magnus*, [25], and developing the first two derivatives of s with respect to τ

$$\dot{s} = \frac{\mathrm{d}s}{\mathrm{d}t} = \frac{\mathrm{d}s}{\mathrm{d}\tau}\frac{\mathrm{d}\tau}{\mathrm{d}t} = \omega_0\frac{\mathrm{d}s}{\mathrm{d}\tau} = \omega_0 s'$$

$$\ddot{s} = \frac{\mathrm{d}\dot{s}}{\mathrm{d}t} = \frac{\mathrm{d}\dot{s}}{\mathrm{d}\tau}\frac{\mathrm{d}\tau}{\mathrm{d}t} = \omega_0^2 s''$$

permits together with the vibratory specifications a conversion of Eq. (3.2a) into the "*non-dimensional form*" *of the equation of motion*

$$s'' + 2\zeta s' + s = \frac{1}{k}F(\tau) \tag{3.2e}$$.

The constant coefficients are one or quantities of the dimension one in the left-hand-side expression. The non-dimensional system representation by use of "natural time" τ will be applied to the *time-response analysis* in the following sections. Especially the graphical representation of actual responses takes advantage of normalized dependency on time because *sets* (*or families*) *of time-history plots* easily can be traced by varying the only one system parameter introduced by *E.Lehr*, [26], that is the *damping ratio* ζ. This parameter defined by the ratio of two quantities of the same kind, and that by relating the actual damping coefficient c to the critical damping coefficient c_{c}, solely specifies the system behaviour of a single degree-of-freedom oscillator, see Appendix A.

Standardized differential equation used in *control theory*

$$\frac{m}{k}\ddot{s} + \frac{c}{k}\dot{s} + s = \frac{1}{k}F(t) \tag{3.8a}$$

makes evident the transmission behaviour, e.g., of a controlled mechanical system, also termed a *mechanical plant*, by the following specifications.

***System Parameters* (*control specifications*).** Quantities which characterize a system by the speed of attaining equilibrium state (delay time) after removal of excitation or restraint. The parameters are constants and have the dimension of time.
Time constant due to damping:

$$T_1 = \frac{c}{k} = \frac{2\delta}{\omega_0^2} = \frac{2\zeta}{\omega_0} = 2\zeta T \tag{3.9}$$

time constant due to inertia:

$$T_2 = \sqrt{\frac{m}{k}} = \frac{1}{\omega_0} = T \tag{3.10}$$

time constant related to natural period of oscillation

$$T = \frac{1}{2\pi}T_0 \tag{3.11}$$

proportional action coefficient (proportional gain):

$$K_p = \frac{1}{k} \tag{3.12}$$

With the control specifications follows the
time-constant representation of the equation of motion

$$T_2^2\ddot{s} + T_1\dot{s} + s = K_p F(t) \tag{3.8b}$$

$$T^2\ddot{s} + 2\zeta T\dot{s} + s = \frac{1}{k}F(t) \tag{3.8c}$$

which describes the proportional behaviour of a *2nd-order delay (or lag) element* (*P*-T_2-element), acting as a phase-shifting section, [27] to [31].

3.1.2 Free and Forced Response of Damped Second-order Systems

Time response of the vibrating system variable, the displacement, is obtained as the general (complete) solution of the differential equation of motion. Some general properties of solutions of linear differential equations may be recited.

Superposition principle (linearity principle). The general or complete solution of a nonhomogeneous linear differential equation is a combination of the solution of the corresponding homogeneous differential equation and an arbitrary *particular solution* of the nonhomogeneous linear differential equation, [32] to [35].

Free Response of a Single Degree-of-freedom System

Corresponding homogeneous differential equation describes the linear oscillator in the absence of an external force as excitation (forcing function) $F(t) \equiv 0$

$$L[s] \equiv \sum_{\nu=0}^{n=2} a_\nu s^{(\nu)}(t) = 0 \tag{3.13a}$$

By assuming an exponential function in the form

$$s_{\mathrm{h}}(t) = \hat{s}\mathrm{e}^{pt} \tag{3.14a}$$

$$\dot{s}_{\mathrm{h}}(t) = \hat{s}p\mathrm{e}^{pt} \quad , \quad \ddot{s}_{\mathrm{h}}(t) = \hat{s}p^2\mathrm{e}^{pt} \tag{3.14b}$$

with the constants $\hat{s}$ and p as trial solution and substituting Eqs. (3.14a,b) into Eq. (3.13a) it is obvious that nontrivial solutions only will be obtained by satisfying the *characteristic equation*, in this case a quadratic equation

$$\mathrm{P}(p) \equiv mp^2 + cp + k = 0 \qquad (3.13\mathrm{b});$$

two roots (solutions): eigenvalues

$$\begin{aligned} p_{1,2} &= -\frac{c}{2m} \pm \sqrt{\left(\frac{c}{2m}\right)^2 - \frac{k}{m}} \\ &= -\delta \pm \sqrt{\delta^2 - \omega_0^2} = -\zeta\omega_0 \pm \omega_0\sqrt{\zeta^2 - 1}\;. \end{aligned} \tag{3.15}$$

According to the theory of differential equations for two distinct solutions, that form a fundamental system, i.e., $s_{\mathrm{h}_{1,2}}$ are linearly independent, the Wronskian determinant is non-zero.

General solution of the homogeneous differential equation is the complementary function $s_{\mathrm{h}}(t)$ being the sum of the two exponential quantities and describing the *natural (eigen-) motion of the damped oscillator*

$$s_{\mathrm{h}}(t) = C_1\mathrm{e}^{p_1 t} + C_2\mathrm{e}^{p_2 t} \qquad (3.16).$$

The roots of the characteristic equation are specifications to distinguish natural vibrations with regard to stability and periodicity of motion

root criterion for stability: sign of the real part: $c/(2m) = \delta$

If the real part is negative the solution function decreases with time and the represented physical system is said to be a *stable system.*
Vice versa the solution function increases and the oscillator is an *unstable system*

root criterion for periodicity: sign and value of the radical being called *discriminant* $\Delta = [(c/(2m))^2 - (k/m)] = (\delta^2 - \omega_0^2)$.

Three cases concerning the discriminant differ from oneanother. If the discriminant is negative and not zero the solution function describes an oscillatory vibration being said a *periodic motion.* Vice versa the solution function describes a non-oscillatory vibration being said an *aperiodic motion.*

Case 1: $\Delta < 1$; $p_{1,2}$ ***two complex conjugate roots*** (always in a pair due to real element parameters).
Less-than-critical damping (underdamped system):

$$c < c_{\mathrm{c}} = 2\sqrt{mk} \quad ; \quad \delta < \omega_0 \quad ; \quad \zeta < 1 \tag{3.17}$$

$$\begin{aligned} p_{1,2} &= -\delta \pm \mathrm{j}\omega_{\mathrm{d}} = -\delta \pm \mathrm{j}\sqrt{\omega_0^2 - \delta^2} \\ &= -\zeta\omega_0 \pm \mathrm{j}\omega_0\sqrt{1-\zeta^2}\;. \end{aligned} \tag{3.18}$$

System Parameters. Quantities which describe the free vibration of a damped linear system.

Damped natural frequency as an angular (circular) frequency

$$\omega_d = \sqrt{\omega_0^2 - \delta^2} = \omega_0 \sqrt{1-\zeta^2} = \omega_0 \sqrt{1-\left(\frac{c}{2mk}\right)^2} \tag{3.19a}$$

as a (cyclic) frequency, the reciprocal to *damped natural period*

$$f_d = \frac{\omega_d}{2\pi}\;; \qquad T_d = \frac{2\pi}{\omega_d} \tag{3.19b}$$

Both solutions are complex

$$s_h(t) = C_1 e^{(-\delta + j\omega_d)t} + C_2 e^{(-\delta - j\omega_d)t} = e^{-\delta t}\left(C_1 e^{j\omega_d t} + C_2 e^{-j\omega_d t}\right) \tag{3.20a}$$

by applying Euler's formula (identity) for both positive and negative exponents

$$e^{j\alpha} \equiv \cos\alpha + j\sin\alpha$$

$$e^{-j\alpha} \equiv \cos\alpha - j\sin\alpha$$

the two real solutions or trigonometric functions are obtained

$$s_h(t) = \underbrace{(C_1 + C_2)}_{=C_1^*}\cos\omega_d t + \underbrace{(jC_1 - jC_2)}_{=C_2^*}\sin\omega_d t \tag{3.20b}$$

trigonometric theorem

$$a\cos\alpha - b\sin\alpha = c\cdot\{\cos(\alpha - \chi)\ \mathit{bzw.}\ \sin(\alpha + \theta)\}$$

$$c = \sqrt{a^2 + b^2}\;; \quad \tan\chi = b/a\;; \quad \tan\theta = a/b$$

fundamental (natural) vibration of the underdamped system

$$s_h(t) = Ce^{-\delta t}\cos(\omega_d t - \chi) \tag{3.20c}$$

Constants of integration

$$C = \sqrt{C_1^{*2} + C_2^{*2}}\;; \quad \chi = \arctan\left(C_2^*/C_1^*\right).$$

The natural vibration is of oscillatory nature but not periodic as its amplitude decays exponentially with time. The free damped oscillator is permitted to vibrate after being displaced from its equilibrium position (zero-driving response). By this the energy storage elements are loaded, thus, the *initial state* is defined by the non-zero initial condition at time $t = 0$

$$|s_0| + |\dot{s}_0| > 0 \tag{3.21}$$

The constants of integration C, χ can be determined by using initial displacement; initial velocity

$$s(0^-) = s_0\;; \quad \dot{s}(0^-) = \dot{s}_0 \tag{3.22}$$

so that $C = \sqrt{s_0^2 + \left(\frac{\dot{s}_0 + \delta s_0}{\omega_d}\right)^2}\;; \qquad -\chi = \arctan\frac{\dot{s}_0 + \delta s_0}{s_0\omega_d}$.

Initial-condition free vibration (transient response) of the underdamped system:

$$s(t) = \sqrt{s_0^2 + \left(\frac{\dot{s}_0 + \delta s_0}{\omega_d}\right)^2}\, e^{-\delta t}\cos(\omega_d t - \chi) \tag{3.23a}$$

normalized form by relating free displacement s to initial displacement s_0

$$\frac{s(\tau)}{s_0} = \sqrt{1 + \frac{(s_0'/s_0 + \zeta)^2}{1-\zeta^2}}\; \mathrm{e}^{-\zeta\tau} \cos\left(\sqrt{1-\zeta^2}\,\tau - \chi\right) \tag{3.23b}$$

with normalized initial velocity; (trigonometric) initial phase

$$s_0' = \dot{s}_0 / \omega_0 \; ; \quad -\chi = \arctan\frac{s_0'/s_0 + \zeta}{\sqrt{1-\zeta^2}} \; .$$

Natural Data Plotting. The displacement time history of the oscillatory free vibration (transient motion) started from specific initial values can be seen in Fig. A.1 of the Appendix A.

The system parameters are determined in general not by direct measuring of response time history (displacement-time curve) but by deduction using measurable quantities which base on time or length (transient-response specifications).

So, the determination of *natural* (cyclic) *frequency* f_0, (Eq. 3.6b), applies the static equilibrium condition introduced in 2.3.4, to turn over to a measurement of length. Substituting Eq. (2.73) in Eq. (3.6b) yields:

$$f_0 = \sqrt{g/s_{\mathrm{stat}}}\,/(2\pi) \tag{3.6c}$$

which implies that the natural (cyclic) frequency can be obtained, consequently the natural period of oscillation, Eq. (3.7), once the static deflection of the spring s_{stat} caused by the local force of gravity F_g, Eq. (2.70), is known.

For a determination of the (viscous) *damping ratio* ζ, Eq. (3.5), the measuring of the time constant (relaxation time) T_r, Eq. (3.4), or of damped natural period T_d, Eq. (3.19b), would be possible. The boundary within which the decaying response curve oscillates is a decaying exponential termed envelope of "amplitude". The time constant is specified by the time range which would be required for the envelope to reach zero if the initial slope (or rate of decay) did not change. The damped natural period is specified by the time range for one cycle of the oscillatory decay curve. Taking up the subtangent adjacent to the initial tangent of the envelope respectively the zero crossings in the same direction of the response curve the following relations result from measurement of time intervals (durations):

$$T_r = 1/\delta = 1/(\omega_0\zeta) \; ; \qquad \text{normalized:} \quad \tau_r = 1/\zeta \tag{3.4a}$$

$$T_d = 2\pi/\omega_d = 2\pi\Big/\left(\omega_0\sqrt{1-\zeta^2}\right) \qquad \text{normalized:} \quad \tau_d = 2\pi\Big/\left(\sqrt{1-\zeta^2}\right) \tag{3.19c}$$

One of the simplest and most frequently used technique of vibration-measuring systems is the experimental determination of damping by picking up distinct peaks in vibratory motion history. The ratio of two successive displacement amplitudes (peaks) of like sign in the decay s_i, s_{i+1} is a constant

$$\frac{s_i}{s_{i+1}} = \frac{\sqrt{1-\zeta^2}}{\sqrt{1-\zeta^2}}\,\frac{\hat{s}}{\hat{s}}\,\frac{\mathrm{e}^{-\zeta\omega_0 t_i}}{\mathrm{e}^{-\zeta\omega_0 (t_i + T_d)}} = \mathrm{e}^{\zeta\omega_0 T_d} = \mathrm{e}^{T_d/T_r} \, ,$$

and the *logarithmic decrement* is defined

$$\Lambda\;(\text{or } \delta) = \ln(s_i/s_{i+1}) = \zeta\omega_0 T_d = T_d/T_r = \tau_d/\tau_r = \frac{2\pi\zeta}{\sqrt{1-\zeta^2}} \tag{3.5c}$$

It should be noted that this analysis assumes that the point of maximum displacement in a cycle and the point where the envelope of the oscillatory decay curve touches the decay curve itself, are coincident. This is commonly very nearly so, and the error in making this assumption is usually negligible. This proves true for a damping being very small, so that

$$\tau_d \approx 2\pi \;\; ; \qquad \Lambda \approx 2\pi/\tau_r = 2\pi\zeta \qquad \text{for } \zeta \ll 1 \tag{3.5d}$$

Measuring two nonsuccessive displacement peaks n cycles apart s_i, s_{i+n}, where n is a integer, the logarithmic decrement is defined

$$\Lambda\ (\text{or}\ \delta) = \ln\left(s_i / s_{i+n}\right) = n\zeta\omega_0 T_d = nT_d / T_r = n\tau_d / \tau_r = n\zeta\tau_d \qquad (3.5e).$$

Thus, the damping ratio can be determined according to Eqs. (3.5c), (3.5e) as

$$\zeta = \frac{\Lambda}{\sqrt{(2\pi)^2 + \Lambda^2}} = \frac{T_d / T_r}{\sqrt{(2\pi)^2 + \left(T_d / T_r\right)^2}} \qquad (3.5f).$$

The *equivalent viscous damping coefficient* c_{eq} can be determined as

$$c_{eq} = \zeta\, c_c = \zeta\left(2m\omega_0\right) = 2\zeta m\omega_0 \qquad (3.5b)$$

where c_c is the critical damping coefficient, m the mass, and ω_0 the natural frequency.

Case 2*: $\Delta > 1$; $p_{1,2}$ *two distinct real roots.

Greater-than-critical damping (overdamped system)

$$c > c_c = 2\sqrt{mk}\ ; \quad \delta > \omega_0\ ; \quad \zeta > 1 \qquad (3.24)$$

$$p_{1,2} = -\delta \pm \lambda = -\delta \pm \sqrt{\delta^2 - \omega_0^2} \qquad (3.25).$$

System Parameter. In addition to Eqs. (3.19a) to (3.19c) the overdamped case is characterized by the term *radical of the real eigenvalues*

$$\lambda = \sqrt{\delta^2 - \omega_0^2} = \omega_0\sqrt{\zeta^2 - 1} = \omega_0\sqrt{\left(\frac{c}{2mk}\right)^2 - 1} \qquad (3.19d).$$

Both solutions are real

$$s_h(t) = C_1 e^{(-\delta+\lambda)t} + C_2 e^{(-\delta-\lambda)t} = e^{-\delta t}\left(C_1 e^{\lambda t} + C_2 e^{-\lambda t}\right) \qquad (3.26a);$$

by applying the hyperbolic identity

$$\cosh\alpha = \frac{e^{\alpha} + e^{-\alpha}}{2} = \cos j\alpha$$

$$\sinh\alpha = \frac{e^{\alpha} - e^{-\alpha}}{2} = -j\sin j\alpha$$

the two real solutions of hyperbolic functions are obtained

$$s_h(t) = \underbrace{\left(C_1 + C_2\right)}_{=C_1^*}\cosh\lambda t + \underbrace{\left(C_1 - C_2\right)}_{=C_2^*}\sinh\lambda t \qquad (3.26b);$$

hyperbolic addition theorem

fundamental (natural) vibration of the overdamped system

$$s_h(t) = Ce^{-\delta t}\sinh\left(\lambda t + \theta\right) \qquad (3.26c).$$

Constants of integration

$$C = \sqrt{C_1^{*2} + C_2^{*2}}\ ; \quad \theta = \operatorname{artanh}\left(C_1^* / C_2^*\right).$$

The natural vibration is non-oscillatory and is often referred to as a "subsidence" (it subsides). If the system is displaced and released corresponding to the non-zero initial condition (zero-driving response), Eq. (3.21), the constants of integration C, Θ will be determined.

$$C=\sqrt{s_0^2+\left(\frac{\dot{s}_0+\delta s_0}{\lambda}\right)^2}\;;\quad \theta=\operatorname{artanh}\frac{s_0\lambda}{\dot{s}_0+\delta s}$$

Initial-condition free vibration (transient response) of the overdamped system:

$$s(t)=\sqrt{s_0^2+\left(\frac{\dot{s}_0+\delta s_0}{\lambda}\right)^2}\,\mathrm{e}^{-\delta t}\sinh(\lambda t+\theta) \qquad (3.27a);$$

normalized form

$$\frac{s(\tau)}{s_0}=\sqrt{1-\frac{(s_0'/s_0-\zeta)^2}{\zeta^2-1}}\,\mathrm{e}^{-\zeta\tau}\sinh\left(\sqrt{\zeta^2-1}\;\tau+\theta\right) \qquad (3.27b)$$

with normalized initial velocity; hyperbolic initial phase

$$s_0'=\dot{s}_0/\omega_0;\quad \theta=\operatorname{artanh}\frac{\sqrt{\zeta^2-1}}{s_0'/s_0+\zeta}\;.$$

Another form to represent the aperiodic approach to equilibrium position of a free damped oscillator is preferably applied in controls.

System Parameters (control specifications). Parameters characterizing an overdamped oscillator by the speed of attaining equilibrium state are given by the *time constants*, also termed *delay times*:

$$T_{s_1}=-\frac{1}{p_1}=\frac{1}{\delta-\lambda}=\frac{1}{\omega_0^2}(\delta+\lambda)\;=\frac{T_1}{2}+\sqrt{\left(\frac{T_1}{2}\right)^2-T_2^2}$$
$$T_{s_2}=-\frac{1}{p_2}=\frac{1}{\delta+\lambda}=\frac{1}{\omega_0^2}(\delta-\lambda)\;=\frac{T_1}{2}-\sqrt{\left(\frac{T_1}{2}\right)^2-T_2^2}\;. \qquad (3.28a)$$

Time-constant representation is common for two 1st-order storage elements in series (P-T_1-elements):

$$s(t)=\frac{T_{s_1}\left(s_0+T_{s_2}\dot{s}_0\right)}{T_{s_1}-T_{s_2}}\,\mathrm{e}^{-t/T_{s_1}}-\frac{T_{s_2}\left(s_0+T_{s_1}\dot{s}_0\right)}{T_{s_1}-T_{s_2}}\,\mathrm{e}^{-t/T_{s_2}} \qquad (3.29a);$$

normalized form by relating free displacement s to initial displacement s_0

$$\frac{s(\tau)}{s_0}=\frac{\tau_{s_1}\left(1+\tau_{s_2}s_0'/s_0\right)}{\tau_{s_1}-\tau_{s_2}}\,\mathrm{e}^{-\tau/\tau_{s_1}}-\frac{\tau_{s_2}\left(1+\tau_{s_1}s_0'/s_0\right)}{\tau_{s_1}-\tau_{s_2}}\,\mathrm{e}^{-\tau/\tau_{s_2}} \qquad (3.29b)$$

with normalized time constants (delay times)

$$\tau_{s_1}=T_{s_1}\omega_0=\zeta+\sqrt{\zeta^2-1}$$
$$\tau_{s_2}=T_{s_2}\omega_0=\zeta-\sqrt{\zeta^2-1} \qquad (3.28b).$$

Natural Data Plotting. The displacement time history of the non-oscillatory free vibration (transient motion) started from specific initial values can be seen in Fig. A.2 of the Appendix A.

The subsiding character of the decaying response curve results from a superposition of two decaying exponential components. The normalized time constants τ_{s1}, τ_{s2} are specified by the time ranges bringing the components to zero if their decaying would be continued at initial slopes. Taking up the corresponding subtangents the relations to the damping ratio ζ are given by Eq. (3.28b). Each of the decaying exponentials is regarded effectively zero after a time equal to four time constants. This approximate criterion is applied to the dominant component with the largest of both time constants by which the duration of transient response can be valued.

Figure A.3 of the Appendix A shows the effect on the non-oscillatory free displacement response caused by varying the given set of initial conditions. The decay curves approach equilibrium position without passing zero except that the signs of initial velocity and initial displacement are opposite.

Case 3*: $\Delta = 0$; $p_1 = p_2$ *a real double root.

Critical damping (critically damped system)

$$c = c_c = 2\sqrt{mk}\,; \quad \delta = \omega_0\,; \quad \zeta = 1 \tag{3.30}$$

$$p_1 = p_2 = -\delta = -\omega_0 \tag{3.31}$$

In this case Eq. (3.16) yields only one solution

$$s_{h_1}(t) = C_1^* \mathrm{e}^{-\omega_0 t} \tag{3.32a}$$

From the theory of differential equations, it is shown that another solution $s_{h_2}(t)$ can be assumed in the following form

$$s_{h_2}(t) = s_{h_1}(t) \cdot u(t) \qquad \text{where} \quad u(t) = \mathrm{e}^{-\omega_0 t} \tag{3.32b}$$

Equations (3.13b), (3.32) are solutions of (3.13a). They constitute a fundamental system. The corresponding general solution is

$$s_h(t) = \left(C_1 + C_2 t\right)\mathrm{e}^{-\omega_0 t} \tag{3.32c}$$

The solution function (3.22c) is similar to (3.26c) belonging to case 2. The two arbitrary constants C_1, C_2 can be determined from the initial conditions

$$C_1 = s_0\,; \quad C_2 = \dot{s}_0 + \omega_0 s_0\,.$$

This yields the corresponding form (zero-driving response) to (3.27a,b).

Initial-condition free vibration (transient response) of the critically damped system

$$s(t) = \left[s_0 + \left(\dot{s}_0 + \omega_0 s_0\right)t\right]\mathrm{e}^{-\omega_0 t} \tag{3.33a}$$

normalized form by relating free displacement s to initial displacement s_0

$$\frac{s(\tau)}{s_0} = \left[1 + \left(\frac{s_0'}{s_0} + \tau\right)\right]\mathrm{e}^{-\tau} \tag{3.33b}$$

Figure A.3 of the Appendix A shows in addition to overdamped system responses the free response of a critically damped system. The decay curve pertaining to $\zeta = 1$ represents the limit of oscillatory motion and yields an equilibrium position approach being the fastest without any oscillations. That is, a critically damped system has the smallest amount of damping required for non-oscillatory motion. Many devices, particularly electrical instruments in control systems are critically damped to take advantage of this property.

Forced Response of a Single Degree-of-freedom System to Harmonic Excitation

Particular solution of the nonhomogeneous linear differential equation. To find such a function three approaches are possible:

- *Method of reduction*: differential equation reducible to linear first-order differential equations;

- *Variation of parameters (Lagrange)*:
$$s_p(t) = C_1(t) s_{h_1}(t) + C_2(t) s_{h_2}(t) \tag{3.34}$$
trial functions $C_1(t)$, $C_2(t)$ (parameters replaced by functions that are to be determined);
- *Method of undetermined coefficients*
suitable when forcing function is such that the form of a particular solution may be guessed by assuming for it a type of function that is congruous to the excitation.

The *simple harmonic excitation* as the most common external force (forcing function) acting on a system and characterizing a forced vibration is given by
$$F(t) = \hat{F} \cos \omega_f t \tag{3.35a}$$
with the (real) force-excitation amplitude: $\hat{F}$, (single) forcing angular frequency: ω_f (or Ω). The harmonic excitation $F(t)$ has a periodic time denoted as the *forcing period* T_f and defined as
$$T_f = \frac{2\pi}{\omega_f} \tag{3.35b}$$
The trigonometric trial function
$$s_p(t) = s_1 \cos \omega_f t + s_2 \sin \omega_f t = \hat{s} \cos\left(\omega_f t - \varphi_{0_s}\right) \tag{3.36}$$
$$\dot{s}_p(t) = -\hat{s}\omega_f \sin\left(\omega_f t - \varphi_{0_s}\right)$$
$$\ddot{s}_p(t) = -\hat{s}\omega_f^2 \cos\left(\omega_f t - \varphi_{0_s}\right)$$
satisfies by substituting the trigonometrical theorem
$$\cos \omega_f t \left\{ \hat{s}\left(k - m\omega_f^2\right) \cos \varphi_{0_s} + c\omega_f \hat{s} \sin \varphi_{0_s} - \hat{F} \right\}$$
$$+ \sin \omega_f t \left\{ \hat{s}\left(k - m\omega_f^2\right) \sin \varphi_{0_s} - c\omega_f \hat{s} \cos \varphi_{0_s} \right\} = 0$$
the equation of motion, Eq. (3.1), if the solution function representing a forced response takes the following characteristic parameters of a sinusoidal quantity:
Displacement amplitude $\hat{s}$ as the forced peak value
$$\begin{aligned}\hat{s} &= \frac{\hat{F}}{\left(k - m\omega_f^2\right)\cos\varphi_{0_s} + c\omega_f \sin\varphi_{0_s}} = \hat{F} \underbrace{\frac{1}{\sqrt{(k - m\omega_f^2)^2 + (c\omega_f)^2}}}_{=A(\omega_f)} \\ &= \underbrace{\frac{\hat{F}}{k}}_{=\hat{s}_{stat}} \underbrace{\frac{1}{\sqrt{(1-\eta_1^2)^2 + (2\zeta\eta_1)^2}}}_{=A(\eta_1)/A(0)} .\end{aligned} \tag{3.37a}$$
Displacement phase angle (or initial phase) φ_{0_s} as the forced argument
$$\begin{aligned}\varphi_{0_s} &= \underbrace{\arctan \frac{c\omega_f}{k - m\omega_f^2}}_{=\psi_1} + \varphi_{0_F} \\ &= \psi_1 \quad ; \quad \text{provided that } \varphi_{0_F} = 0 ,\end{aligned} \tag{3.37b}$$

where φ_{0s} equals ψ_1 in the case of usually chosen zero-crossing excitation, thus omitting the initial phase of excitation force φ_{0F}.

System Parameters. Real-valued quantities which describe a system characteristic by the relationship between excitation and forced response.
Amplitude (-frequency) response $A(\omega)$ at $\omega = \omega_\mathrm{f}$, called
the *displacement response factor*

$$A(\omega_\mathrm{f}) = 1 \Big/ \left| \sqrt{(k - m\omega_\mathrm{f}^2)^2 + (c\omega_\mathrm{f})^2} \right| = \omega_0^2 \Big/ \left| \sqrt{\left(\omega_0^2 - \omega_\mathrm{f}^2\right)^2 + \left(2\delta\omega_\mathrm{f}\right)^2} \right| \\ = \hat{s} / \hat{F} \qquad (3.38\mathrm{a})$$

signifying the amplitude ratio of displacement response $\hat{s}$ to force excitation $\hat{F}$, and in normalized form being called
the *magnification factor* (or non-dimensional displacement response factor)

$$\frac{A(\eta_1)}{A(0)} = 1 \Big/ \left| \sqrt{\left(1 - \eta_1^2\right)^2 + \left(2\zeta\eta_1\right)^2} \right| = R_\mathrm{d} = \hat{s} / s_\mathrm{stat} \qquad (3.38\mathrm{b})$$

at the frequency ratio (ratio of forcing frequency to undamped natural frequency):

$$\eta_1 = \frac{\omega_\mathrm{f}}{\omega_0} \qquad (3.39),$$

being defined by the amplitude ratio of displacement response $\hat{s}$ to spring displacement that would occur if force $\hat{F}$ were applied statically (static deflection) $s_\mathrm{stat} = \left(\hat{F}/k\right)$.
Phase (-frequency) response $\varphi(\omega)$ at $\omega = \omega_\mathrm{f}$, called
the *phase difference*

$$\psi(\omega_\mathrm{f}) = \psi_1 = \arctan\frac{c\omega_\mathrm{f}}{k - m\omega_\mathrm{f}^2} = \arctan\frac{2\delta\omega_\mathrm{f}}{\omega_0^2 - \omega_\mathrm{f}^2} \\ = \arctan\frac{2\zeta\eta_1}{1 - \eta_1^2} = \varphi_{0_\mathrm{s}} - \varphi_{0_\mathrm{F}} \qquad (3.40\mathrm{a})$$

signifying the phase angle difference between displacement response φ_{0_s} and force excitation φ_{0_F} measured from the same (zero-time) origin; in particular identical with the displacement phase angle φ_{0_s}.

Being identified as the phase difference of a second quantity (the response) with respect to the first (the excitation) the argument parameter will be termed the *phase lag* (phase shift)

$$-\psi_1 = \varphi_{0_\mathrm{F}} - \varphi_{0_\mathrm{s}} = -\varphi_{0_\mathrm{s}} \; ; \qquad \varphi_{0_\mathrm{F}} = 0 \qquad (3.40\mathrm{b}).$$

Forced vibration (steady-state response) of the underdamped system

$$s_\mathrm{p}(t) = \frac{\hat{F}}{\sqrt{(k - m\omega_\mathrm{f}^2)^2 + (c\omega_\mathrm{f})^2}} \cos\left(\omega_\mathrm{f} t - \arctan\frac{c\omega_\mathrm{f}}{k - m\omega_\mathrm{f}^2}\right) \qquad (3.41\mathrm{a});$$

normalized form by relating forced displacement s_p to static deflection s_{stat}

$$\frac{s_p(\tau)}{s_{stat}} = \frac{1}{\sqrt{(1-\eta_1^2)^2 + (2\zeta\eta_1)^2}} \cos\left(\eta_1\tau - \arctan\frac{2\zeta\eta_1}{1-\eta_1^2}\right) \quad (3.41b).$$

The steady-state response is thus a continuing harmonic vibration with the forcing frequency ω_f, Eq. (3.35a), respectively the forcing period T_f, Eq. (3.35b), or in normalized form with the frequency ratio η_1, Eq. (3.39), respectively the *normalized forcing period* τ_1 defined as

$$\tau_1 = \frac{2\pi}{\eta_1} = T_f\omega_0 \quad (3.35c).$$

Total or Complete Response of a Single Degree-of-freedom System

General solution of the nonhomogeneous differential equation describes the response (output) of the system to both an excitation (external force) and the initial state. Applying the superposition principle to the linear system the homogeneous solution $s_h(t)$, called complementary function, is added to the particular solution $s_p(t)$ to obtain the total or complete response.

Complete displacement response of the underdamped system

$$s_{ges}(t) = s_h(t) + s_p(t) = Ce^{-\delta t}\cos(\omega_d t - \chi) + \hat{s}\cos(\omega_f t - \psi_1) \quad (3.42).$$

The constants of integration, C, χ, are determined from the initial conditions. For simplicity, all initial values vanish defining thereby the *zero state*:

$$s(0^-) = \dot{s}(0^-) = 0 \quad (3.43);$$

i.e., the oscillator with energy storages being empty is assumed to start at $t = 0$ from rest (zero-state response).

Complete displacement response for the oscillator initially at rest

$$s_{ges}(t) = \hat{F}\frac{1}{\sqrt{(k - m\omega_f^2)^2 + (c\omega_f)^2}}\left\{\cos(\omega_f t - \psi_1) - \frac{\omega_0}{\omega_d}e^{-\delta t}\cos(\omega_d t - \chi)\right\} \quad (3.44a)$$

where ψ_1 equals Eq. (3.40a) ; $\chi = \arctan\left(\frac{\delta}{\omega_d}\frac{\delta^2 + \omega_d^2 + \omega_f^2}{\delta^2 + \omega_d^2 - \omega_f^2}\right)$.

The above classification is somewhat artificial because initial conditions are generally produced by the removal of excitation or restraint. Thus, the response depends of the characteristic parameters of the excitation quantity in addition to the system parameters itself. In this regard, it is convenient to distinguish between the *steady-state response* $s_s(t)$ representing a forced vibration and the *transient response* $s_t(t)$ characterizing the change from one steady state to another by a free vibration, so that:

$$s_p(t) = s_s(t) \quad ; \quad s_h(t) = s_t(t)$$

normalized form by relating displacement s to static deflection s_{stat}

$$\frac{s(\tau)}{s_{\mathrm{stat}}} = \underbrace{\frac{1}{\sqrt{(1-\eta_1^2)^2 + (2\zeta\eta_1)^2}}\left\{\cos(\eta_1\tau - \psi_1)\right.}_{=s_{\mathrm{s}}(\tau)/s_{\mathrm{stat}}} \cdots \underbrace{\left. - \frac{1}{\sqrt{1-\zeta^2}} e^{-\zeta\tau} \cos(\sqrt{1-\zeta^2}\,\tau - \chi)\right\}}_{= s_{\mathrm{t}}(\tau)/s_{\mathrm{stat}}}$$

(3.44b)

with the non-dimensional steady-state vibration $s_{\mathrm{s}}(\tau)/s_{\mathrm{stat}}$, respectively the non-dimensional transient vibration $s_{\mathrm{t}}(\tau)/s_{\mathrm{stat}}$,

and where ψ_1 equals Eq. (3.40a) ; $\chi = \arctan\left(\frac{\zeta}{\sqrt{1-\zeta^2}} \frac{1+\eta_1^2}{1-\eta_1^2}\right)$.

The complete response, also called *combined motion*, behaves differently in approaching the steady state since the dissipation of energy per cycle increases. A large amount of damping results in a change from oscillatory into non-oscillatory transient state of free vibration.

Complete displacement response of the overdamped system initially at rest

$$s_{\mathrm{ges}}(t) = s_{\mathrm{h}}(t) + s_{\mathrm{p}}(t) = C e^{-\delta t} \sinh(\lambda t - \theta) + \hat{s}\cos(\omega_{\mathrm{f}} t - \psi_1)$$

$$= \hat{F}\frac{1}{\sqrt{(k - m\omega_{\mathrm{f}}^2)^2 + (c\omega_{\mathrm{f}})^2}}\left\{\cos(\omega_{\mathrm{f}} t - \psi_1) - \frac{\omega_0}{\lambda} e^{-\delta t} \sinh(\lambda t + \theta)\right\}$$

(3.45a)

where ψ_1 equals Eq. (3.40a) ; $\theta = -\mathrm{artanh}\left(\frac{\lambda}{d}\frac{\delta^2 - \lambda^2 - \omega_{\mathrm{f}}^2}{\delta^2 - \lambda^2 + \omega_{\mathrm{f}}^2}\right)$;

normalized form

$$\frac{s(\tau)}{s_{\mathrm{stat}}} = \underbrace{\frac{1}{\sqrt{(1-\eta_1^2)^2 + (2\zeta\eta_1)^2}}\left\{\cos(\eta_1\tau - \psi_1)\right.}_{=s_{\mathrm{s}}(\tau)/s_{\mathrm{stat}}} \cdots \underbrace{\left. - \frac{1}{\sqrt{\zeta^2 - 1}} e^{-\zeta\tau} \sinh(\sqrt{\zeta - 1}\,\tau + \theta)\right\}}_{=s_{\mathrm{t}}(\tau)/s_{\mathrm{stat}}}$$

(3.45b)

where ψ_1 equals Eq. (3.40a) ; $\theta = -\mathrm{artanh}\left(\frac{\sqrt{\zeta^2 - 1}}{\zeta}\frac{1-\eta_1^2}{1+\eta_1^2}\right)$.

Response Data Plotting. The time history of the oscillatory likewise of the non-oscillatory displacement response (combined motion) is given in Fig. A.4 and A.5 of the Appendix A, apart from fitting the curve to zero phase angle $\varphi_{0\mathrm{F}} = 0$.

In general, *steady-state response* is one in which the system achieves a certain type of equilibrium, such as a constant response or a response that repeats itself ad infinitum, without approaching zero or without growing indefinitely with time. In describing the steady-state response, time becomes an incidental factor. In fact, quite often the steady-state response can be obtained from the total response by letting t approach infinity. On the other hand, the *transient response* depends strongly on time.

Considering the combined motion of complete displacement responses to a harmonic excitation, Fig. A.4 and A.5, Appendix A, the forced vibrations do not differ on time history, although the magnitude of damping has changed over from less, Fig. A.4, to greater than critical damping, Fig. A.5. Certainly, a difference in response amplitude and phase angle can be observed. Those characteristic parameters related to the previously specified *system parameters* are used to characterize the steady-state behaviour in *vibration theory*. The system parameters gathered from forced vibration characteristics are furthermore correlated with *frequency-domain or steady-state specifications* required for system analysis in vibrations and controls.

On the contrary the free vibrations inherent in the combined motions are showing quite different shapes of time history due to a change in the damping magnitude. To characterize the combined motion by a change in equilibrium, for example from zero state of a system being initially at rest to the steady state of forced vibration, in addition to the above characteristics *time-domain or transient-response specifications* are stated, see Fig. A.1 and A.2, Appendix A, belonging to the performance requirements for system analysis in *control theory*.

Broadly speaking, steady-state response occurs in the case of constant, harmonic, or periodic excitation, and transient response occurs in the case of initial excitation and in the case of external excitation other than the ones just mentioned. This external excitation is often called transient excitation [2].

Resonance and Beating

Within a small frequency range where forcing frequency meets natural frequency the forced vibration response tends to singularity phenomena since actual damping is considerably less than critical damping. However, the complete responses appropriate to the following two singular frequency ratios will demonstrate, that even for an undamped vibration those phenomena can not occur immediately.

Resonance. A singularity phenomenon occurs by coinciding of forcing frequency with the undamped natural frequency, accordingly since the requency ratio is equal to one. Thus,
the *condition of resonance* is defined by

$$\omega_f = \omega_0 \qquad ; \qquad \eta_1 = \omega_0 / \omega_0 = \eta_0 = 1 \qquad (3.46a).$$

Resonance of a system in forced oscillation exists when any change, however small, in forcing frequency causes a decrease in a response of the system. Even though Eq. (3.37a) shows that the response amplitude respecting the absence of damping goes to infinity for any value of time a valid physical behaviour requires that the amplitude takes time to grow. Analogous with the solution method concerning free critically damped vibration, Eqs. (3.32b,c), the solution at resonance can be found as the product of a sinusoidal function with a function that depends linearly on time.
For the oscillator starting at $t = 0$ from rest it follows:
Complete displacement response of the undamped system at resonance

$$s_{ges}(t) = \frac{\hat{F}}{2k}\omega_0 t \sin \omega_0 t \qquad \text{for } t \geqq 0 \qquad (3.46b)$$

normalized form by relating displacement s to static deflection s_{stat}

$$\frac{s(\tau)}{s_{stat}} = \frac{1}{2}\tau \sin \tau \qquad \text{for } \tau \geqq 0 \qquad (3.46c).$$

Beating. Beating of a system in forced oscillation exists when periodic variations in the response amplitude arise resulting from the combination of two oscillations of slightly different frequencies. Consequently the appearance of a further phenomenon is caused by a forcing frequency slightly different from undamped natural frequency, accordingly since the frequency ratio is close to one. Thus, the *condition of beating* is defined by

$$\omega_f \approx \omega_0 \quad ; \quad \eta_1 = \omega_f / \omega_0 \approx 1 \tag{3.47a}$$

The beats occur at the difference frequency, called the *beat frequency*

$$\Delta\omega_f = \left|\omega_0 - \omega_f\right| \tag{3.47b}$$

the *beat frequency ratio*:

$$\Delta\eta_1 = \frac{\Delta\omega_f}{\omega_0} = \frac{\left|\omega_0 - \omega_f\right|}{2\omega_f} = \frac{\left|1-\eta_1\right|}{2} \quad ; \quad 0 < \Delta\eta_1 \ll 1 \tag{3.47c}$$

Complete displacement response of the undamped system beating normalized form by relating displacement s to static deflection s_{stat}

$$\begin{aligned} \frac{s(\tau)}{s_{stat}} &= \frac{1}{1-\eta_1^2}(\sin \eta_1 \tau - \sin \tau) = \mp \frac{2}{1-\eta_1^2} \cos \frac{\eta_1 + 1}{2} \tau \sin\left(\Delta\eta_1 \tau\right) \\ &= -\frac{1}{\Delta\eta_1 (1+\eta_1)} \cos \frac{\eta_1 + 1}{2} \tau \sin\left(\Delta\eta_1 \tau\right) , \end{aligned} \tag{3.48a}$$

so that a sinusoidal beating will be stated

$$\frac{s(\tau)}{s_{stat}} \approx \underbrace{-\frac{1}{2\eta_1 \Delta\eta_1} \sin\left(\Delta\eta_1 \tau\right)}_{=A(\tau)} \cos\left(\eta_1 \tau\right) \tag{3.48b}$$

The harmonic function $\cos(\eta_1\tau)$ has a period $\tau_m = 2\pi/\eta_1$, while the harmonic function $\sin(\Delta\eta_1\tau)$ has a period $\tau_s = 2\pi/\Delta\eta_1$. Since $\Delta\eta_1$ is a very small number, the harmonic function in the parentheses varies more rapidly than the harmonic function forming their boundaries. The beats occur at any half period $\tau_s/2$ equal to $\pi/\Delta\eta_1$.

Response Data Plotting. The time history of the displacement response at frequencies distinguished by resonance or beating are shown in the Appendix A.

Subjected to resonance the system attains infinite displacement, but not instantaneously, as shown in Fig. A.6.

The occurance of beating characterized by a periodically time-varying amlitude is illustrated in Fig. A.7.

3.1.3 Forced Response of a Single Degree-of-freedom System to Complex Excitation. *Phasor-response Analysis*

For simplification of finding solutions of differential equations with periodic inputs the *algebra of complex numbers* is used. In case of a sinusoidal input the trigonometric function is replaced by an exponential function with the imaginary number $j\omega_f t$. Thus, the sinusoidal forced or steady-state response may be obtained through consideration of a *complex constituent* alone, called the *complex excitation*.

The concept of complex excitations and responses involves the advantages in making evident the coherency between mathematically modelling and system behaviour by

- the geometric interpretation of forced responses representing vectors in the complex plane (*phasors*);
- the derivation of frequency-response functions defining phasor ratios (*complexors*).

The *actual excitation* is given by the simple harmonic (or sinusoidal) excitation, Eq. (3.35).

The *complex excitation* is an input having the real and imaginary parts:

$$\underline{F}(t) = \hat{F}e^{j\omega_f t} \equiv \underbrace{\hat{F}\cos\omega_f t}_{=\mathrm{Re}[\hat{F}e^{j\omega_f t}]} + j\,\underbrace{\hat{F}\sin\omega_f t}_{=\mathrm{Im}[\hat{F}e^{j\omega_f t}]} . \tag{3.49}$$

Thus, the simple harmonic excitation can be written either in real notation or, using Euler's formula, Eq. (3.49), as the real part of a complex notation indicated by the symbol Re.

A solution of the differential equation, Eq. (3.1), under complex excitation (forcing function), Eq. (3.49):

$$L[s] \equiv \sum_{\nu=0}^{n=2} a_\nu s^{(\nu)}(t) = m\ddot{s} + c\dot{s} + ks = \hat{F}e^{j\omega_f t} \tag{3.50}$$

will be found using the method of undetermined coefficients. A type of function is assumed being congruent to that of the given complex excitation. That means, the trial function is also a complex constituent replacing sinusoidal functions, Eq. (3.36), applied in 3.1.2:

$$\begin{aligned} \underline{s}_p(t) &= \hat{\underline{s}}e^{j\omega_f t} = \hat{s}e^{-j\psi}e^{j\omega_f t}\,; \quad \text{displacement phasor } \hat{\underline{s}} = \hat{s}\,e^{-j\psi} \\ &= \hat{s}e^{j(\omega_f t-\psi)} \end{aligned} \tag{3.51}$$

$$\dot{\underline{s}}_p(t) = \underbrace{j\omega_f\hat{\underline{s}}}_{=\hat{\dot{\underline{s}}}}\,e^{j\omega_f t}\,; \quad \text{velocity phasor } \hat{\dot{\underline{s}}} = j\omega_f\hat{s}\,e^{-j\psi} = \omega_f\hat{s}\,e^{-j(\psi-\pi/2)}$$

$$\ddot{\underline{s}}_p(t) = \underbrace{-\omega_f^2\hat{\underline{s}}}_{=\hat{\ddot{\underline{s}}}}\,e^{j\omega_f t}\,; \quad \text{acceleration phasor } \hat{\ddot{\underline{s}}} = -\omega_f^2\hat{s}\,e^{-j\psi} = \omega_f^2\hat{s}\,e^{-j(\psi-\pi)}$$

Phasor. A phasor is defined as a complex quantity the modulus (magnitude) of which is the amplitude (or alternatively the r.m.s. value) and the argument (angle) of which is the initial phase of a sinusoidal quantity.

Assuming a sinusoidal forced vibration the complex trial function $\underline{s}_p(t)$, Eq. (3.51), is characterized by the displacement phasor $\hat{\underline{s}}$, whereas the time derivatives $\dot{\underline{s}}_p(t)$, $\ddot{\underline{s}}_p(t)$ are respectively marked by the velocity or the acceleration phasor $\hat{\dot{\underline{s}}}$, $\hat{\ddot{\underline{s}}}$.

The supposed displacement phasor is distinguished by a displacement amplitude $\hat{s}$ and an initial phase φ_{0_s} which is equal to the phase difference ψ_1 using zero-crossing excitation. The phase difference is indicated as a phase lag (phase shift) $-\psi_1$.

The different motion phasors are related to oneanother by use of repeated multiplying by the differential factor $j\omega_f$. The multiple of $j\omega_f$ constitutes the order of the derivatives defining the pertinent motion variable quantity. The vector representation in the complex plane illustrates the phasor relations by a counter-clockwise rotation through the angle $+\pi/2$ rad combined with a stretch in length about the multiplier ω. So, having a phase difference of $+\pi/2$ rad respectively of $+\pi$ rad the velocity phasor is in quadrature, or else the acceleration phasor in opposition to the displacement phasor, Fig. 3.2.

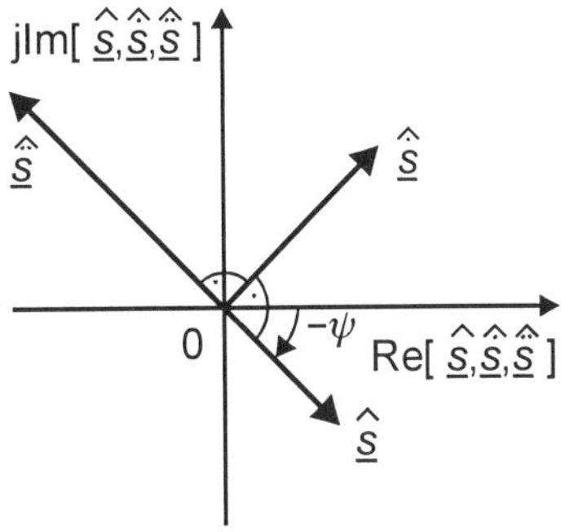

Fig. 3.2. Complex representation of sinusoidal trial functions of motion by phasors: displacement, velocity, and acceleration phasors

Phasor Method

The phasor representation of sinusoids is the most convenient method for obtaining a particular solution of the nonhomogeneous differential equation to harmonic excitation.

Instead of a trigonometric trial function the complex trial function, Eq. (3.51), is substituted leading to a factored equation which expresses its dependency on time only in terms of a complex exponential function as a multiplier

$$\left\{-m\omega_f^2\,\hat{\underline{s}} + jc\omega_f\,\hat{\underline{s}} + k\hat{\underline{s}}\right\} e^{j\omega_f t} = \hat{F}e^{j\omega_f t} \qquad (3.52a).$$

Being of like angular frequency $\omega = \omega_f$ and non-zero for all finite times $|t| < \infty$, the exponential factor can be cancelled on both sides of the equation. Thus, the original differential equation will be converted into an algebraic equation being called

the *phasor equation* (vector equation) of the differential equation

$$-m\omega_f^2\,\hat{\underline{s}} + jc\omega_f\,\hat{\underline{s}} + k\hat{\underline{s}} = \hat{F} \qquad (3.52b).$$

The only real problem is to calculate the amplitude and phase of the sinusoid as a particular solution by solving an algebraic equation. This will be favourably done by using phasors, and the method is called the *phasor method.*

The complex trial function, Eq. (3.51), satisfies the equation of motion, Eq. (3.1), if the complex solution function representing a forced response takes the following complex sinusoidal quantity

the *displacement response phasor* $\underline{\hat{s}}$

$$\underline{\hat{s}} = \hat{F} \underbrace{\frac{1}{k + \mathrm{j}c\omega_{\mathrm{f}} - m\omega_{\mathrm{f}}^2}}_{=G(\mathrm{j}\omega_{\mathrm{f}})} = \underbrace{\frac{\hat{F}}{c}}_{=s_{\mathrm{stat}}} \underbrace{\frac{1}{1 + \mathrm{j}2\zeta\eta_1 - \eta_1^2}}_{=G(\mathrm{j}\eta_1)/G(0)} \quad . \tag{3.53}$$

Complex System Parameter. Complex quantity which describes a system characteristic by the relationship between excitation and forced vibration in complex notation.

Frequency-response function $G\,(\mathrm{j}\omega)$ at $\omega = \omega_{\mathrm{f}}$ termed

the *displacement response complexor* $G(\mathrm{j}\omega_{\mathrm{f}})$

$$G(\mathrm{j}\omega_{\mathrm{f}}) = |G(\mathrm{j}\omega_{\mathrm{f}})|\mathrm{e}^{\mathrm{j}\,\mathrm{arc}\,G(\mathrm{j}\omega_{\mathrm{f}})} = A(\omega_{\mathrm{f}})\mathrm{e}^{-\mathrm{j}\psi_1} = \underline{\hat{s}} / \underline{\hat{F}} = \underline{\hat{s}} / \hat{F} \tag{3.54}$$

Complexor. A complexor, introduced in electric circuit theory, is defined as a complex quantity equal to the *quotient of two phasors* representing two sinusoidal quantities of like angular frequency. In the actual case of a harmonic vibration the complex ratio of the displacement response phasor $\underline{\hat{s}}$ to the phasor of the excitation force $\underline{\hat{F}} = \hat{F}$ may be identified with the displacement response complexor $G(\mathrm{j}\omega_{\mathrm{f}})$, whose modulus is equal to the amplitude (frequency-) response at $\omega = \omega_{\mathrm{f}}$, called

the *displacement response factor*, often referred to as *amplitude ratio* (gain)

$$|G(\mathrm{j}\omega_{\mathrm{f}})| = A(\omega_{\mathrm{f}}) = \frac{1}{\left|\sqrt{(k - m\omega_{\mathrm{f}}^2)^2 + (c\omega_{\mathrm{f}})^2}\right|} = |\underline{\hat{s}}| / |\underline{\hat{F}}| = \hat{s} / \hat{F} \tag{3.55a}$$

and whose argument equals the phase (frequency-) response at $\omega = \omega_{\mathrm{f}}$, called the *phase difference (phase shift)*

$$\begin{aligned} \mathrm{arc}\,G(\mathrm{j}\omega_{\mathrm{f}}) &= -\psi_1 = -\arctan\frac{c\omega_{\mathrm{f}}}{k - m\omega_{\mathrm{f}}^2} = \mathrm{arc}\,\underline{\hat{s}} - \mathrm{arc}\,\underline{\hat{F}} \\ &= -\varphi_{0_{\mathrm{s}}} - \left(-\varphi_{0_{\mathrm{F}}}\right) = -\varphi_{0_{\mathrm{s}}}\ ; \qquad \varphi_{0_{\mathrm{F}}} = 0 \end{aligned} \tag{3.56}$$

Single value of the frequency-response function in non-dimensional form

$$G\left(\mathrm{j}\frac{\omega_{\mathrm{f}}}{\omega_0}\right) \Big/ G(0) \quad \text{at} \quad \frac{\omega_{\mathrm{f}}}{\omega_0} = \eta_1$$

being called the *complex magnification factor*

$$G(\mathrm{j}\eta_1) / G(0) (= \underline{Y_{\mathrm{c}}}) = k\underline{\hat{s}} / \underline{\hat{F}} = k\underline{\hat{s}} / \hat{F} = \underline{\hat{s}} / s_{\mathrm{stat}} \tag{3.55b}$$

gives the characteristic response parameters by combining the normalized displacement amplitude, Eq. (3.38b), with the phase lag (phase shift), Eq. (3.40b), to a sole complex quantity at frequency ratio $\eta = \eta_1$.

The particular solution of the nonhomogeneous differential equation describes the forced vibration given by the *complex response*:

$$\underline{s}_{\mathrm{p}}(t) = \frac{\hat{F}}{k + \mathrm{j}c\omega_{\mathrm{f}} - m\omega_{\mathrm{f}}^2}\mathrm{e}^{\mathrm{j}\omega_{\mathrm{f}}t} = \hat{F}G(\mathrm{j}\omega_{\mathrm{f}})\mathrm{e}^{\mathrm{j}\omega_{\mathrm{f}}t} \tag{3.57}$$

caused by the complex excitation, Eq. (3.49).

The *actual response*, easily to calculate by the phasor method, can be written either in real notation or as the real part of the complex notation:
Forced vibration (steady-state response)

$$\begin{aligned} s_{\mathrm{p}}(t) &= \mathrm{Re}\left[\underline{s}_{\mathrm{p}}(t)\right] = \mathrm{Re}\left[\hat{F}G(\mathrm{j}\omega_{\mathrm{f}})\mathrm{e}^{\mathrm{j}\omega_{\mathrm{f}}t}\right] \\ &= \hat{F}A(\omega_{\mathrm{f}})\mathrm{Re}\left[\mathrm{e}^{\mathrm{j}(\omega_{\mathrm{f}}t-\psi_1)}\right] = \hat{F}A(\omega_{\mathrm{f}})\cos(\omega_{\mathrm{f}}t - \psi_1) \end{aligned} \tag{3.58}$$

Vector Representation of Sinusoids. The geometric interpretation of calculating forced responses by phasor method can be gathered from the vector representation in the complex plane, Fig. 3.3.

The complex sinusoids are represented by vectors rotating on circles of the radius $\hat{F}$ or $\hat{s}$ at the same forcing angular velocity ω_{f} in the counterclockwise direction. Thus, $\underline{F}(t)$ and $\underline{s}_{\mathrm{p}}(t)$ may be called time-varying (rotating) phasors. The actual excitation and response can be interpreted geometrically by taking the projection on the real axis.

The displacement phasor lags the force phasor by ψ_1, that is the motion response occurs after the force excitation has been applied (causal principle in real systems affected by time delay).

For system analysis neither complex nor actual responses caused by sinusoidal excitation are of decisive interest. For characterizing the sinusoidal steady-state behaviour it is dispensable to describe system variable quantities at a given instant, i.e., by their *instantaneous value*. The steady-state quantities being significant of system variables are called the *characteristic parameters*. Those parameters are given only by the magnitudes (amplitudes or r.m.s. values) and the relative phase angles (phase differences), and they are assigned to a single frequency, the forcing (angular) frequency $\omega = \omega_{\mathrm{f}}$. Complex notation yet involves the advantage of combining both sinusoidal response parameters to one *complex system parameter*, i.e., to the *response phasor*. Thus, in diagram representation not the time-varying (rotating) phasors, in Fig. 3.3 marked by dotted lines, but the *constant (resting) phasors*, in Fig. 3.3 marked by solid lines, are of use to be figured for illustrating the forced or *sinusoidal steady-state interrelations* between the system variables.

The phasor representation can be applied to mechanical components or elements by reiterating the relations defining idealized element laws, Eqs. (2.66) to (2.68). Using the phasor relations for the pertinent motion variable quantities, Eq.

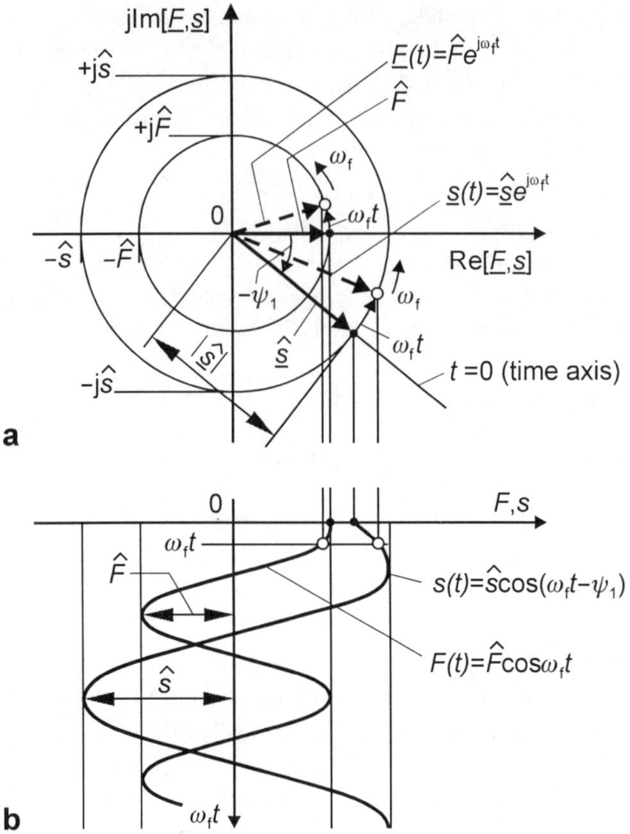

Fig. 3.3. Complex representation of sinusoidal response and excitation by time-varying (rotating) phasors and their projection on the real axis. **a** Phasor diagram; **b** time history

(3.51), the steady-state interrelation of parts (subsystems) can be illustrated by the *force phasor polygon* balancing element forces and excitation force, Fig. 3.4.

Corresponding to the vector representation of the motion variable the derivative of which is of zero order continued to second order the elastic force phasor $\hat{\underline{F}}_s$ is in phase, whereas the damping force phasor $\hat{\underline{F}}_d$ is in quadrature, respectively the inertial force phasor $\hat{\underline{F}}_m$ in opposition to the displacement response phasor $\hat{\underline{s}}$.

The *frequency-dependent change in phasor configuration* can easily be studied by varying the forcing angular frequency ω_f. At very low frequencies compared to the undamped natural frequency (resonance frequency): $\omega_f \ll \omega_0$ equivalent to $\eta_1 \ll 1$, the elastic force approximately equals the exciting force in amplitude $\hat{F}_s \approx \hat{F}$; whereas at very high frequencies: $\omega_f \gg \omega_0$, or $\eta_1 \gg 1$, the inertial force clearly approaches the exciting force in amplitude $\hat{F}_m \approx \hat{F}$.

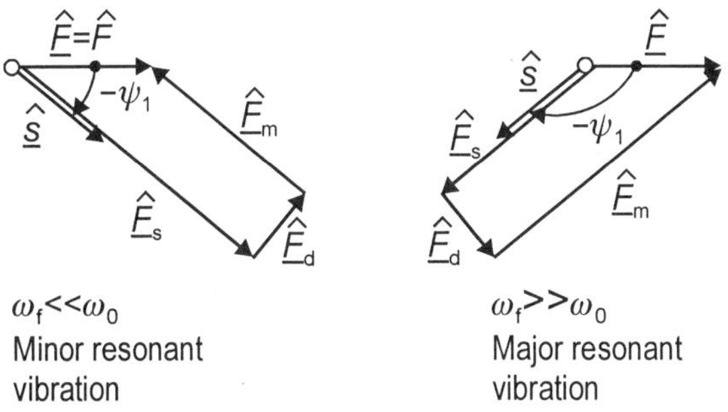

Fig. 3.4. Phasor diagram of element forces and excitation force related to low respectively to high forcing angular frequency

3.2 Representation of Mechanical Systems by Integral-transformed Models (Transform Methods)

General Input-Output Relation by a linear Operator

The *transmission system*, commonly represented by a single-variable system, e.g., a mechanical single degree-of-freedom system, relates one dependent output variable $\upsilon(t)$ to one independent input variable $u(t)$ by the *operator* T, that defines the functional dependence in the symbolic form:

$$\upsilon(t) = \mathrm{T}\left[u(t)\right] \tag{3.59}$$

signifying the transmission behaviour of the system. The operator must satisfy the properties on linearity (superposition principle), time-invariance (independence of time-shifting), causality (response lags excitation), memory (the output variable $\upsilon(t)$ at an arbitrary moment, t_1, depends on all previous values of $u(t)$, from $-\infty$ to t_1), Fig. 3.5.

The operator T contains in a compact form all the dynamic characteristics of the system and is usually given implicitly by functional dependences, e.g., differential equations, that can be evaluated either theoretically (by physical laws) or determined experimentally (by measurements).

The calculus of *linear functional transformation* or *integral operators* yields a method of computation that proves to be a specification of Eq. (3.59) concerning linear systems.

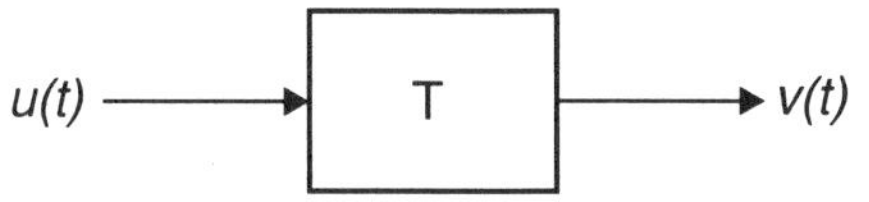

Fig. 3.5. Graphical symbol of the operator of a transmission system

3.2.1
Periodic Vibration. *Fourier Series Analysis*

In the case of vibration drives or rotating drives being affected by disturbances of different origin, the mechanical structure is frequently excited by more than one sinusoid each having a frequency that may be an integral multiple of the forcing fundamental frequency (multi-sinusoidal vibration). Hence, it exists a *periodic excitation* $F(t)$, that can be decomposed into sinusoids that are simple harmonics with the periods $T_n = T/n$, where T is the fundamental period and n is the order, accordingly with the (cyclic) frequencies $f_n = nf_1$ (harmonic analysis).

Superposition of the sinusoidal terms leads to the periodic excitation function (harmonic synthesis), that approximates a periodic forcing function of arbitrary shape by the appropriate Fourier series (Fourier expansion) with sufficient performance (with respect to Gibb's phenomenon, presupposing arbitrary periodic function in compliance with Dirichlet's condition).

Fourier Series

A Fourier expansion of $F(t)$ into a *Fourier series* is given as

$$F(t) = \frac{a_0}{2} + \sum_{n=1}^{\infty}(a_n \cos\omega_n t + b_n \sin\omega_n t) \quad \text{for } n = 1,2,3... \qquad (3.60a)$$

where a_n and b_n are the *Fourier coefficients*, or by combining them through goniometric addition theorems to form single terms as

$$F(t) = \frac{c_0}{2} + \sum_{n=1}^{\infty}\hat{c}_n \cos(\omega_n t - \varphi_n) \quad \text{for } n = 1,2,3... \qquad (3.60b)$$

with the *Fourier amplitude* of the nth harmonic (component)

$$\hat{c}_n = \sqrt{a_n^2 + b_n^2} \qquad (3.61a),$$

and its *Fourier phase angle*,

$$\varphi_n = \arctan b_n / a_n \qquad (3.61b).$$

Exponential Fourier Series. Following the theorem of Moivre from complex calculus (or from complex- function theory considering the Fourier series as the special form of a Laurent series by change of variable: $z - z_0 = e^{j\phi}$) a Fourier expansion of $F(t)$ into a *complex Fourier series* can be described as

$$F(t) = \sum_{n=-\infty}^{+\infty} \underline{c_n} e^{j\omega_n t} = \sum_{n=-\infty}^{+\infty} \underline{c_n} e^{-j2\pi nt/T} \quad \text{for } n = 0, \pm 1, \pm 2, ... \qquad (3.60c)$$

where $\underline{c_n}$ is the *complex Fourier coefficient* of the value

$$\underline{c_n} = \frac{1}{2\pi}\int_{-\pi}^{+\pi} F(t) e^{-j\omega_n t} dt = \frac{1}{T}\int_{-\frac{T}{2}}^{+\frac{T}{2}} F(t) e^{-j2\pi nt/T} dt \qquad (3.62a)$$

with the(angular) frequency of the nth harmonic, called *harmonic frequency*

$$\omega_n = n\omega_1 = 2\pi n/T \qquad (3.62b)$$

being an integral multiple of the *fundamental (angular) frequency*

$$\omega_1 = 2\pi / T$$

defined by the product of the reciprocal of the fundamental period $1/T$ and the factor 2π, whereat the multiple n constitutes the order of the harmonic.

For real-valued periodic quantities (actual excitations) $F(t)$ the complex Fourier coefficients of each order with opposite sign are complex conjugates: $\underline{c}_{-n} = \underline{c}_n^*$.

Thus, the complex Fourier series, Eq. (3.60c), is completely equivalent to the Fourier series in real notation, Eq. (3.60b), since the exponential factors associated with each (angular) frequency occur in complex conjugate pairs: $e^{+j\omega_n t}$, $e^{-j\omega_n t}$.

$$F(t) = \underline{c}_0 + \begin{Bmatrix} \underline{c}_1 e^{j\omega_1 t} + \underline{c}_2 e^{j\omega_2 t} + \underline{c}_3 e^{j\omega_3 t} + \ldots \\ + \underline{c}_{-1} e^{-j\omega_1 t} + \underline{c}_{-2} e^{-j\omega_2 t} + \underline{c}_{-3} e^{-j\omega_3 t} + \ldots \end{Bmatrix} \tag{3.62c}$$

$$= \underline{c}_0 + \sum_{n=1}^{\infty} \left(\underline{c}_n e^{j\omega_n t} + \underline{c}_n^* e^{-j\omega_n t} \right)$$

so that, using Euler's formula, the Fourier series can be written as the real part of the complex notation :

$$F(t) = \underline{c}_0 + 2\sum_{n=1}^{\infty} \mathrm{Re}\left[\underline{c}_n e^{j\omega_n t} \right] \qquad \text{for } n = 1,2,3,\ldots \tag{3.62d}$$

Writing the Fourier coefficients in the polar form of complex conjugate numbers the phasor representation is useful

$$\begin{aligned} \underline{c}_n &= |\underline{c}_n| e^{-j\varphi_n} = \frac{\hat{c}_n}{2} e^{-j\varphi_n} \; ; \quad \underline{c}_0 = |\underline{c}_0| = \frac{c_0}{2} \\ \underline{c}_n^* &= |\underline{c}_n| e^{j\varphi_n} = \frac{\hat{c}_n}{2} e^{j\varphi_n} \; . \end{aligned} \tag{3.63a}$$

Finally it follows according to Eq. (3.60b)

$$F(t) = |\underline{c}_0| + 2\sum_{n=1}^{\infty} |\underline{c}_n| \, \mathrm{Re}\left[e^{j(\omega_n t + \mathrm{arc}\, c_n)} \right] = \frac{c_0}{2} + \sum_{n=1}^{\infty} \hat{c}_n \mathrm{Re}\left[e^{j(\omega_n t - \varphi_n)} \right] \tag{3.63b}$$

Vector Representation of Harmonics. The geometric interpretation of complex Fourier series can be deduced from the vector representation in the complex plane, Fig. 3.6.

Associated to the complex conjugate numbers $\underline{c}_n$, $\underline{c}_n^*$, exists a phasor pair of the modulus $|\underline{c}_n|$ and the argument $\pm\varphi_n$, in Fig. 3.6 noted by the solid lines. The double of the modulus is equal to the amplitude to the nth harmonic:

$$2|\underline{c}_n| = \hat{c}_n \tag{3.63c}$$

The complex conjugate pair of exponential harmonics

$$\underline{c}_n e^{j\omega_n t} \; ; \qquad \underline{c}_n^* e^{-j\omega_n t}$$

constituting the nth term of the complex Fourier series will be represented by a phasor pair co- and counter-rotating on a circle of the radius $|\underline{c}_n|$ at the same angular velocity opposite in sign $\pm\omega_n$, in Fig. 3.6 noted by dotted lines.

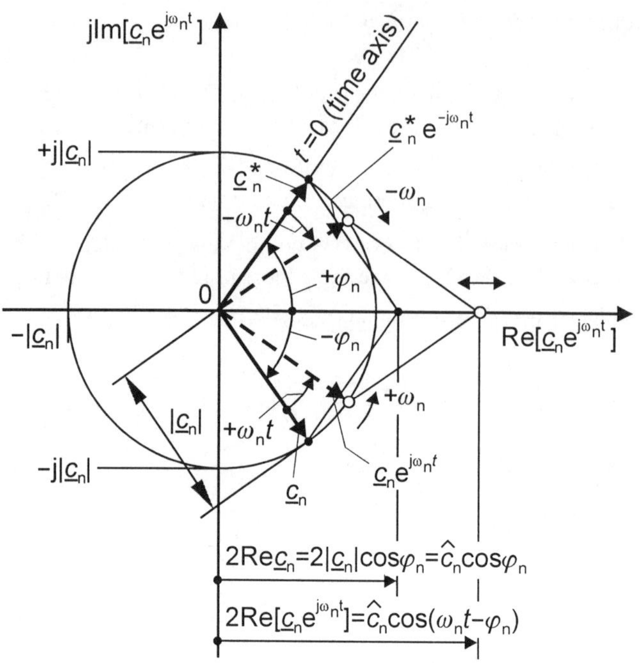

Fig. 3.6. Complex representation of the nth harmonic of a periodic quantity by time-varying (rotating) phasors and their vectorial addition on the real axis

The nth harmonic or frequency component can be interpreted geometrically by the resultant of the time-varying (rotating) phasor pair on the real axis following the parallelogram law of vector addition:

$$\underline{c}_n e^{j\omega_n t} + \underline{c}_n^* e^{-j\omega_n t} = \hat{c}_n \cos(\omega_n t - \varphi_n) \tag{3.63d}$$

The complex (or exponential) Fourier series has an elegant mathematical form and is useful in many theoretical developments.

The Fourier expansion of a periodic function described by repeated simple time functions (ideal periodic excitations) is easily carried out by a complex Fourier coefficient evaluation. In cases of periodic functions for which the repeated time-history curve has an arbitrary shape the harmonic analysis whether in tabular form or in graphic representation will be carried out with aid of numerical techniques. The computer algorithm known as the Fast Fourier Transform (FFT) reduces considerably the computational time.

Rectangular Shock Pulsating (pulse-train excitation). The periodic forcing function of *repeated rectangular pulses*, each of duration τ_0, Fig. 3.7, is defined as

$$F_{R_1}(t) = \begin{cases} 0 \ \text{for} \ \dfrac{2\nu T + \tau_0}{2} < t < \dfrac{2(\nu+1)T - \tau_0}{2} \\ F_0 \ \text{for} \ \dfrac{2\nu T - \tau_0}{2} < t < \dfrac{2\nu T + \tau_0}{2} \end{cases} \qquad \nu = 0, \pm 1, \pm 2, \ldots \tag{3.64a}$$

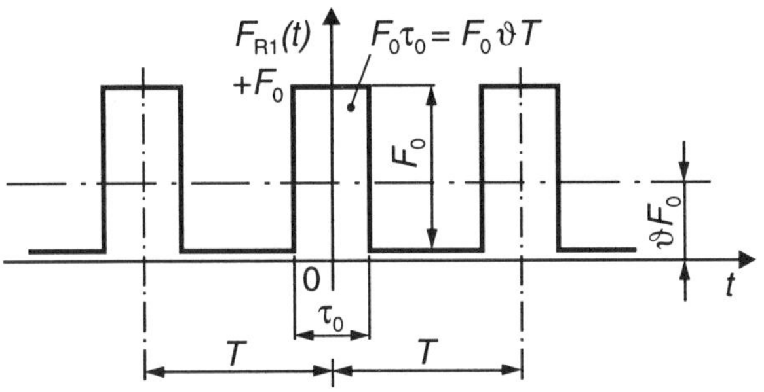

Fig. 3.7. Periodic rectangular pulse train with even symmetry and non-zero average

with the parameters characterizing a pulsed quantity:

maximum height F_0; single pulse duration τ_0

fundamental period T

duty cycle (or pulse control factor) $\vartheta = \tau_0 / T$

impulse (single pulse area) $I_{R_1T} = F_0\tau_0 = F_0\vartheta T$;

and being of non-zero mean value:

$$\overline{F_{R_1}(t)} = \frac{1}{T}\int_0^T F_{R_1}(t)\,dt = \vartheta F_0 \qquad (3.65a).$$

Evaluation of the nth complex Fourier coefficient applying Eq. (3.62a) to the ideal periodic excitation defined by Eq. (3.64a):

$$\underline{c_{R_1n}} = \frac{1}{T}\int_{-\frac{T}{2}}^{+\frac{T}{2}} F_{R_1}(t)\,e^{-j\omega_n t}dt = F_0\frac{1}{T}\int_{-\frac{\tau_0}{2}}^{+\frac{\tau_0}{2}} e^{-j\omega_n t}dt =$$

$$= F_0\frac{1}{T\omega_n}\frac{e^{j\tau_0\omega_n/2} - e^{-j\tau_0\omega_n/2}}{j} = 2F_0\frac{\vartheta}{\tau_0\omega_n}\sin\left(\frac{\tau_0}{2}\omega_n\right) \qquad (3.65b)$$

$$= \underbrace{F_0\vartheta}_{=\overline{F_{R_1}(t)}}\frac{\sin\left(\frac{\tau_0}{2}\omega_n\right)}{\frac{\tau_0}{2}\omega_n} = \overline{F_{R_1}(t)}\cdot \mathrm{si}\left(\frac{\tau_0}{2}\omega_n\right) \quad \text{for } \omega_n = n\omega_1, \; n = 0, \pm 1, \pm 2, \ldots$$

the distinct values of which are expressed by the mean value $\overline{F_{R_1}(t)}$ multiplied by a frequency function of the type (sin x) /x that is commonly denoted by the symbol si (*si*-function).

Replacing the specifying constant parameters ω_n, τ_0 by the equivalent parameters multiple n and duty cycle ϑ the coefficient will be modified into a term of the frequency ratio ω_n/ω_1 constituting the order by its integral number, called *harmonic number n*

$$\underline{c_{R_1n}} = \underbrace{F_0\vartheta}_{=\overline{F_{R_1}(t)}}\frac{\sin\left(\vartheta\pi\frac{\omega_n}{\omega_1}\right)}{\vartheta\pi\frac{\omega_n}{\omega_1}} = \overline{F_{R_1}(t)}\cdot \mathrm{si}(\vartheta\pi n) \quad \text{for } n = 0, \pm 1, \pm 2, \ldots \qquad (3.65c).$$

Taking a duty cycle of the particular value $\vartheta = 1/2$ (pulse duration equals half the period) the evaluated Fourier coefficient represents the *square wave* with even symmetry and non-zero average:

$$\underline{c}_{\mathrm{R_1 n}} = \frac{F_0}{2}\,\mathrm{si}\left(\frac{\pi}{2}n\right) \tag{3.66a}$$

The *Fourier amplitude* of the nth harmonic (amplitude spectrum) is

$$\left|\underline{c}_{\mathrm{R_1 n}}\right| = \frac{\hat{c}_{\mathrm{R_1 n}}}{2} = \frac{F_0}{2}\begin{cases} 1 & \text{for } n & = 0 \\ 0 & \text{for } n = \pm 2\nu & = \pm 2, \pm 4, \pm 6, \ldots \\ \frac{2}{\pi}\left|\frac{1}{n}\right| & \text{for } n = \pm(2\nu - 1) & = \pm 1, \pm 3, \pm 5, \ldots \end{cases} \tag{3.66b}$$

The *Fourier phase angle* (phase spectrum) is

$$\mathrm{arc}\,\underline{c}_{\mathrm{R_1 n}} = \varphi_{\mathrm{R_1 n}} = \begin{cases} 0 & \text{for } n = \begin{cases} \pm(2\nu - 2) = 0, \pm 2, \pm 4, \ldots \\ \pm(4\nu - 3) = \pm 1, \pm 5, \pm 9, \ldots \end{cases} \\ \pm\pi & \text{for } n = \pm(4\nu - 1) = \pm 3, \pm 7, \pm 11, \ldots \end{cases} \quad \text{if} \quad \nu = 1, 2, 3, \ldots \tag{3.66c}$$

Fourier Spectrum (line spectrum)

The description of the harmonic components (Fourier amplitudes) as a function of frequency, given by the complex Fourier coefficient $\underline{c}_{\mathrm{n}}$, Eqs. (3.62a), (3.63a), defines a Fourier spectrum the components of which only occur at discrete frequencies ω_{n} given by the integral multiples of the fundamental frequency ω_1. Thereby, a *line spectrum* is depicted. A Fourier spectrum representation requires a pair of complementary spectra described either by spectra of the "amplitudes" of the real and the imaginary part, or by a spectrum of the moduli, called *amplitude spectrum*, completed of a spectrum of the arguments, respectively called *phase spectrum*.

Response Data Plotting. Those diagram pairs of Fourier spectra being exemplified for the square wave are portrayed in Fig. A.8 of the Appendix A.

The Fourier expansion of periodic functions takes advantage of simplifying the evaluation due to symmetries. In the special case of a square wave possessing even symmetry the complex spectrum of amplitudes is composed only of a real part having alternating sign, so that the phase angle takes the alternating value 0 or $\pm\pi$.

The envelope of the line spectrum is described by a decaying sinusoidal function, the so-called *si-function*, signifying a periodic cut-off frequency function which coincides with the spectral density (continuous spectrum) appointed to the specifying rectangular pulse $\underline{F}_{\mathrm{R10}}(\omega)$, Eq. (3.71b), multiplied by the reciprocal of the fundamental period, i.e., the fundamental frequency $f_1 = 1/T$

$$\overline{F_{\mathrm{R_1}}(t)}\,\mathrm{si}\left(\frac{\tau_0}{2}\omega\right) = \frac{F_0}{2}\,\mathrm{si}\left(\frac{\pi}{2\omega_1}\omega\right) = \frac{1}{T}\underline{F}_{\mathrm{R_1 0}}(\omega) \tag{3.67a}$$

The spectrum of the absolute amplitude values, plotted in Fig. A.9 of the Appendix A, is given by the *amplitude function* defined as

$$\left|\underline{c}_{\mathrm{R_1 n}}\right| = \overline{F_{\mathrm{R_1}}(t)}\left|\mathrm{si}\left(\frac{\pi}{2}n\right)\right| = \frac{F_0}{2}\frac{2}{\pi}\left|\frac{1}{n}\right| = c_{\mathrm{R_1 0}}\frac{2}{\pi}\left|\frac{1}{n}\right| \tag{3.67b}$$

that indicates the decrease of the harmonic amplitudes with increasing harmonic number (1st order hyperbola).
The effect of the varying parameter *duty cycle* (or pulse control factor) ϑ on the amplitude spectrum is shown in Fig. A.10 of the Appendix A.

Whereas the maximum height F_0 changes only the absolute amplitude values linearly, a change in the pulse duration $\tau = 0$ as well varies the absolute values as shifts the spectrum lines. Both variations do not affect the shape of the Fourier spectrum. The smaller the duty cycle ϑ, i.e., the shorter the pulse duration, the smaller becomes the relative distance between the spectrum lines, the distance coincides with the duty cycle if the discrete variable $(\vartheta\, n)$ is chosen as abscissa.

The Fourier expansion of the square wave into the complex Fourier series

$$F_{R_1}(t) = \underbrace{\frac{F_0}{2}}_{=\overline{F_{R_1}(t)}} \sum_{n=-\infty}^{+\infty} \left[\mathrm{si}\left(\frac{\pi}{2} n\right) \right] e^{j2\pi nt/T} \qquad \text{for } n = 0, \pm 1, \pm 2, \ldots \quad (3.64b)$$

can be converted into its real notation by combining conjugate complex pairs of the same order n:

$$F_{R_1}(t) = \overline{F_{R_1}(t)} \left\{ 1 + \left[\left(\mathrm{si}\frac{\pi}{2}\right)\left(e^{j\frac{2\pi}{T}t} + e^{-j\frac{2\pi}{T}t}\right) + (\mathrm{si}\,\pi)\left(e^{j\frac{4\pi}{T}t} + e^{-j\frac{4\pi}{T}t}\right) + \cdots \right] \right\}$$

$$= \overline{F_{R_1}(t)} \left\{ 1 + 2 \sum_{n=1}^{\infty} \left[\mathrm{si}\left(\frac{\pi}{2} n\right) \right] \cos\left(\frac{2\pi n}{T} t\right) \right\} \qquad \text{for } n = 1, 2, 3, \ldots \; .$$

The first four terms of the Fourier series for the forcing function are then

$$F_{R_1}(t) = \frac{F_0}{2} \left\{ 1 + \frac{4}{\pi} \left[1 \cos\left(\frac{2\pi}{T} t\right) - \frac{1}{3} \cos\left(\frac{6\pi}{T} t\right) + \frac{1}{5} \cos\left(\frac{10\pi}{T} t\right) - \cdots \right] \right\}$$

$$= \underbrace{\frac{c_{R_1 0}}{2}}_{=\left|\underline{c_{R_1 0}}\right| = \overline{F_{R_1}(t)}} + \hat{c}_{R_1 0} \cos\omega_1 t - \hat{c}_{R_1 3} \cos\omega_3 t + \hat{c}_{R_1 5} \cos\omega_5 t - \cdots \qquad (3.64c).$$

The Fourier series which represents a square wave, defined by Eq. (3.64a) for $\vartheta = 1/2$, only contains cosine terms and possesses terms of odd multiples, the function $F(t)$ thus is said to have even and half-wave symmetry.

3.2.2 Non-periodic Vibration. *The Fourier Integral*

Shock excitations are in effect short-time phenomena with a loading and a restitution phase and with changes in position and velocity. Shock motions of finite duration cannot be decomposed into harmonic components, i.e., they cannot be described by a discrete amplitude spectrum. Non-periodic functions of time are related to continuous functions of frequency.

Simularities in Fourier Series and Fourier Integral

One can consider the continuous spectrum as the limiting case of the discrete spectrum with the distance between the discrete components, ω_1, approaching zero, and subsequently, the "period" of the aperiodic function tending towards infinity.

Rewriting the complex Fourier series, Eq. (3.60c), and the nth Fourier coefficient, Eq. (3.62a), as a discrete spectral function

$$F(t)=\frac{1}{2\pi}\sum_{\omega_n=-\infty}^{+\infty}\left(\frac{2\pi c_n}{\omega_1}\right)e^{jt\omega_n}\Delta\omega_n$$

$$\left(\frac{2\pi c_n}{\omega_1}\right)=\int_{t=-\frac{T}{2}}^{+\frac{T}{2}}F(t)e^{-j\omega_n t}dt \tag{3.68a}$$

and finally forming the limits when ω_1 becomes small as T becomes large

$$\omega_1\to 0\;;\qquad T\to\infty$$

$$\left(n\to\infty;\ \omega_n(=n\omega_1)\to\omega;\ \Delta\omega_n(=\omega_1)\to d\omega;\ c_n\to 0;\left(2\pi\frac{c_n}{\omega_1}\right)=(Tc_n)\to\underline{F}(\omega)\right) \tag{3.68b}$$

the Fourier series tends to the Fourier integral which normally exists for non-periodic functions such as pulses of nonrepetitive nature. This limiting process does not claim to be a rigorous derivation but serves as a heuristic deduction for intuitively introducing the Fourier integral.

The Fourier Integral

The representation of the function $F(t)$ by an integral of the following form defines the

inverse Fourier transform; Fourier integral

$$\mathrm{F}^{-1}[\underline{F}(\omega)]=F(t)=\frac{1}{2\pi}\int_{\omega=-\infty}^{+\infty}\underline{F}(\omega)e^{jt\omega}d\omega \tag{3.69a}$$

which is indicated by the symbol F^{-1}.

The transformation of the function $F(t)$ into a function of the real variable ω defines the

Fourier transform

$$\mathrm{F}[F(t)]=\underline{F}(\omega)=\int_{t=-\infty}^{+\infty}F(t)e^{-j\omega t}dt \tag{3.69b}$$

being indicated by the symbol F.

The spectral amplitudes of a non-periodic function $F(t)$ may be represented by $\underline{F}(\omega)d\omega$ as the contribution of the "harmonics" for an adequately small bandwidth ω to $\omega + d\omega$. Marked by the differential $d\omega$ the spectral amplitudes are infinitesimally small quantities in analogy to the boundary value of the discrete frequency component c_n for $T\to\infty$.

The Fourier transform $\underline{F}(\omega)$ having in general complex function values is a function continuously dependent on the *angular frequency* ω. This real variable is also termed *pulsatance*. The F-transform can be interpreted as the excitation *spectral density*. That is the first derivation of the associate *spectral contribution* $\int\underline{F}(\omega)d\omega$ with respect to ω.

Definition of the Fourier Integral Equation. The integral operator, in this case the Fourier transform, Eq. (3.69b), is defined by:

1. The *kernel of the transformation*, here given by $\mathrm{K}(\mathrm{j}\omega, t) = \mathrm{e}^{-\mathrm{j}\omega t}$ being a function of the time *t* as the original variable, and of (j-times) the angular frequency ω as the subsidiary variable which is a real quantity;
2. The *limits of integration*, here $(-\infty, +\infty)$ ranging from minus infinity to plus infinity (two-sided transformation), which coincide with the definition range of time functions (original functions) permitted to exist prior to $t = 0$ and to be unrestrained in duration (non-causal functions $F(t) \neq 0$ for $t < 0$);
3. The *class of time functions*, here of forcing functions $F(t)$, to which the integral transformation given by Eqs. (3.69a,b) is true (existence theorem) [36], [37].

As the integral transformation can be considered as a functional mapping, one says that the function *F*(*t*) is being mapped (or imaged) from the original (time or t-) domain onto the corresponding subsidiary (frequency or ω-) domain.

The transform $\underline{F}(\omega)$ and its inverse $F(t)$ constitute a *Fourier transform pair.* The corresponding functional relation is called a *correspondence*

$$F(t) \circ\!\!\stackrel{\mathrm{F}}{-\!\!-\!\!-}\!\!\bullet\ \underline{F}(\omega) \qquad (3.69c)$$

being indicated by the correspondence sign and the operator symbol F above it in case of confusion with other forms of integral transformations.

Restriction of convergence (existence theorem). The Fourier transform exists under restrictions similar to those of the nth Fourier coefficient (Dirichlet's conditions), though the basic interval is extended on both sides to infinity, and the *absolute convergence criterion* must be satisfied

$$\int_{-\infty}^{+\infty} |F(t)| \mathrm{d}t < \infty \qquad (3.70),$$

i.e., the time function $F(t)$ must be absolutely integrable over the interval $(-\infty, +\infty)$ (or the integral of the absolute time function must be of bounded variation). This describes the class of Fourier-transformable functions, which covers only a part of time-varying functions occuring in practice, so not the frequently used harmonic type of excitation function (sinusoidal excitation). For nearly any forcing function approaching zero for large values of time in both directions, now as before being of interest for analysing vibrating systems, the Fourier transform exists. This proves true in particular for excitations involving shock and transient vibrations (time-limited functions), consequently bringing shock pulses into focus.

It is relatively simple to determine $\underline{F}(\omega)$ from *F*(*t*) and vice versa if the integrals can be transformed into series of rapid convergence. This is the case for time-limited functions or functions with frequency components belonging to a finite bandwidth. On condition that the ranges of the variables are bounded the integral transformation will be reduced to finite limits of the integral, that points at the *finite Fourier transform* preferably applied in vibration analysis [40], [41].

An extension of the class of Fourier-transformable time functions to periodic excitations (time-unlimited sinusoidal and multi-sinusoidal functions) being important to steady-state analysis was achieved by the *theory of generalized functions* or *distributions* according to Schwartz [38], [39]. By this the transformation theorems of the Fourier transformation can be extended to distributions of slow growth, called *tempered distributions*, which satisfies the rather restrictive convergence criterion of the Fourier transform. Tempered distributions are distributions that define a functional of bounded support or with testing functions of rapid descent. Such a testing function always can be found for (time-unlimited) periodic quantities.

Rectangular Shock Pulse (single-pulse excitation). The non-periodic forcing function of a single *rectangular pulse* of duration τ_0, Fig. 3.8, is defined as

$$F_{\mathrm{R}_1 0}(t) = \begin{cases} 0 & \text{for } |t| > \dfrac{\tau_0}{2} \\ F_0 & \text{for } |t| < \dfrac{\tau_0}{2} \end{cases} \tag{3.71a}$$

with the parameters characterizing a pulse:

maximum height F_0 ; pulse duration τ_0

impulse (pulse area) $I_{\mathrm{R}_1 0} = F_0 \tau_0$.

Evaluation of the Fourier transform applying Eq. (3.69b) to the ideal shock pulse defined by Eq. (3.71a):

$$\begin{aligned} \underline{F_{\mathrm{R}_1 0}(\omega)} &= \int_{-\infty}^{+\infty} F_{\mathrm{R}_1 0}(t) \mathrm{e}^{-\mathrm{j}\omega t} \mathrm{d}t = F_0 \int_{-\frac{\tau_0}{2}}^{+\frac{\tau_0}{2}} \mathrm{e}^{-\mathrm{j}\omega t} \mathrm{d}t \\ &= F_0 \frac{1}{\omega} \frac{\mathrm{e}^{\mathrm{j}\frac{\tau_0}{2}\omega} - \mathrm{e}^{-\mathrm{j}\frac{\tau_0}{2}\omega}}{\mathrm{j}} = 2F_0 \frac{1}{\omega} \sin\left(\frac{\tau_0}{2}\omega\right) \\ &= \underbrace{F_0 \tau_0}_{= I_{\mathrm{R}_1 0}} \frac{\sin\left(\frac{\tau_0}{2}\omega\right)}{\frac{\tau_0}{2}\omega} = I_{\mathrm{R}_1 0} \,\mathrm{si}\left(\frac{\tau_0}{2}\omega\right) \end{aligned} \tag{3.71b}$$

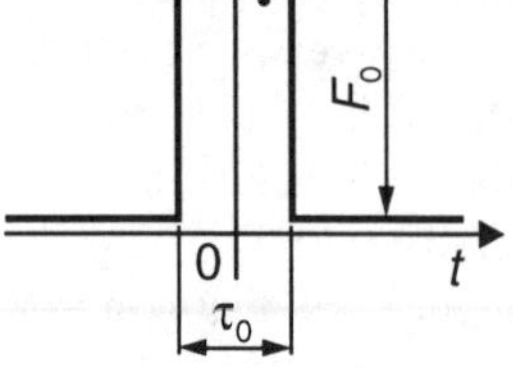

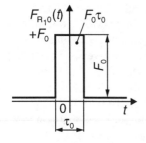

Fig. 3.8. Rectangular pulse with even symmetry

the continuous values of which are expressed by the impulse I_{R10} multiplied by the frequency *si*-function denoting the previously mentioned type (sin x)/x.

The absolute value of the Fourier spectrum (amplitude density spectrum) is

$$\left|\underline{F}_{R_10}(\omega)\right| = A_{R_10}(\omega) = F_0\tau_0\left|\mathrm{si}\left(\frac{\tau_0}{2}\omega\right)\right| \qquad (3.72a).$$

The *phase angle of the Fourier spectrum* (phase spectrum) is

$$\mathrm{arc}\,\underline{F}_{R_10}(\omega) = \Phi_{R_10} = \begin{cases} 0 \;\; \text{for} \begin{cases} 4\nu \cdot \frac{\pi}{\tau_0} < \omega < 2(2\nu-1)\frac{\pi}{\tau_0};\ \nu = 0,+1,+2,\ldots \\ 2(2\nu-1)\frac{\pi}{\tau_0} < \omega < 2(2\nu-2)\frac{\pi}{\tau_0};\nu = -1,-2,\ldots \end{cases} \\ \pm\pi \;\; \text{for} \begin{cases} 2(2\nu-1)\frac{\pi}{\tau_0} < \omega < 2(2\nu-2)\frac{\pi}{\tau_0};\nu = 0,+1,+2,\ldots \\ 4\nu \cdot \frac{\pi}{\tau_0} < \omega < 2(2\nu-1)\frac{\pi}{\tau_0};\nu = -1,-2,\ldots \end{cases} \end{cases}$$

(3.72b).

Fourier Spectrum (continuous spectrum)

The description of the spectral density as a function of frequency, given by the (complex) Fourier transform $\underline{F}\,(\omega)$, Eq. (3.69b), defines a Fourier spectrum the components of which are continuously distributed over a frequency range. Thereby a *continuous spectrum* is depicted. A Fourier spectral density representation requires a pair of complementary spectra described by spectra of the real and the imaginary part, alternatively by a spectrum of the absolute value, called *amplitude density spectrum*, completed with a spectrum of the Fourier phase angle, respectively called *phase spectrum*.

Response Data Plotting. Those diagram pairs of Fourier spectra being exemplified for the rectangular pulse are portrayed in Fig. A.11 of the Appendix A.

In the special case of a rectangular pulse possessing even symmetry the amplitude density spectrum is composed only of a real part having alternating sign, so that the phase angle takes the alternating value 0 or $\pm\pi$.

The Fourier spectrum of the absolute values plotted in Fig. A.12 of the Appendix A, is given by the *amplitude density function* defined as

$$A_{R_10}(\omega) = I_{R_10}\left|\mathrm{si}\left(\frac{1}{2}\tau_0\omega\right)\right| = F_0\tau_0\left|\mathrm{si}\left(\pi\frac{\omega}{\omega_1}\right)\right| \qquad (3.73).$$

This function to be considered as the generic spectrum typical for single force pulses is composed of a main lobe at low frequency followed by higher-frequency side-lobes whose magnitudes decrease rapidly with frequency.

The effect of the varying parameters *maximum height* F_0 and *pulse duration* τ_0 on the amplitude density is shown in Fig. A.13 of the Appendix A.

While the maximum height F_0 changes linearly the absolute values only, a change in pulse duration τ_0 as well varies the absolute values as shifts the bandwidths of the frequency components. Regarding the latter one of the outlined force spectrum characteristics it is obvious that the finite usable bandwidth is inversely proportional to the pulse duration. For instance, in experimental techniques of *impact testing* this provides a useful means of concentrating the excitation energy below the maximum frequency of interest in order to make optimum use of the dynamic range of the measurement system, [46].

Conclusions of the Fourier Spectrum Representation. In vibration system analysis the following statements are of practical use:

The *harmonic analysis*		The *spectral analysis* (decomposition)
	permits through	
the complex Fourier coefficients $\underline{c}_n$		the Fourier transform $\mathcal{F}\,[F(t)]$
	of	
periodic (multi-sinusoidal) vibrations e.g., pulse-train excitations		non-periodic vibrations mechanical shock, e.g., single-pulse excitations
	the representation by	
a (discrete) complex *amplitude spectrum* (line spectrum)		a (continuous) complex *spectral density* (continuous spectrum)
$\underline{c}(\omega_n)$ or $\underline{c}(n)$		$\underline{F}(\omega)$ or $\underline{F}\left(\frac{\omega}{\omega_1}\right)$.

In vibration analysis the Fourier spectrum representation has the particular signification of a *data processing* causing a change in the original information by transforming time functions. The determination of corresponding spectra is to be considered as a *data reduction to the frequency domain* by means of which time-varying quantities (signals or data records) can be analysed by extracting from the great many of instantaneous values (original time histories) an indicative set of specific parameters related to the frequency components (characteristic frequency parameters). Thus, the requirements on the dynamic system behaviour are more efficiently specified (*frequency-response specifications*).

3.2.3 Fourier Transform Method. *Frequency-response Analysis*

The Fourier integral proves its usefulness not only for representing non-periodic functions (shock excitations) by their corresponding Fourier spectrum but also for determining analytic solutions of equations of motion. Frequency concepts used in response calculation aim above all at specifying efficiently the frequency-response characteristics required for mechanical systems. Looking at frequency components transient responses (shock motions) are determined by calculation or measurement (data processing).

In general the importance of applying integral transformation to linear system analysis is based on modelling lumped parameter systems in the frequency domain by making use of algebraic equation calculus. Starting from time-domain representation by the governing differential equation of motion one must be able to express the transform of the derivatives in terms of the transform of the system variable, e.g., of the displacement $s(t)$ as the pertinent motion variable quantity.

Fourier transform of derivatives (differentiation theorem). Let the Fourier transform of a function $s(t)$ exist and let $s(t) \to 0$ as $|\,t\,| \to \infty$, then if

$(\mathrm{d}/\mathrm{d}t)s(t)$ exists

$$\mathrm{F}\left[\frac{\mathrm{d}}{\mathrm{d}t}s(t)\right] = \mathrm{j}\omega\,\mathrm{F}\left[s(t)\right]$$

denoted as a transform pair $\dot{s}(t)$, $\mathrm{j}\omega\underline{s}(\omega)$ symbolically by the *correspondence*

$$\dot{s}(t) \;\circ\!\!\overset{\mathrm{F}}{-\!\!-\!\!-}\!\!\bullet\; \mathrm{j}\omega\,\underline{s}(\omega) \tag{3.74a}$$

in general, for higher derivatives

$$s^{(\mathrm{n})}(t) \;\circ\!\!\overset{\mathrm{F}}{-\!\!-\!\!-}\!\!\bullet\; (\mathrm{j}\omega)^{\mathrm{n}}\,\underline{s}(\omega) \tag{3.74b}$$

(as long as $s(t)$ is $n$times differentiable, and the absolute convergence criterion, Eq. (3.70), for $s^{(\mathrm{n})}(t)$ is satisfied).

Fourier Transform Method

The solution of a differential equation by integral transformation, here by the Fourier transformation, will be obtained by the following general steps, visualized as the scheme of Fig. 3.9.

The *first step* is to transform the differential equation with s and F as functions of the one original variable t on both sides into an algebraic equation with s and F being functions of a subsidiary variable, here of the real frequency variable ω. By

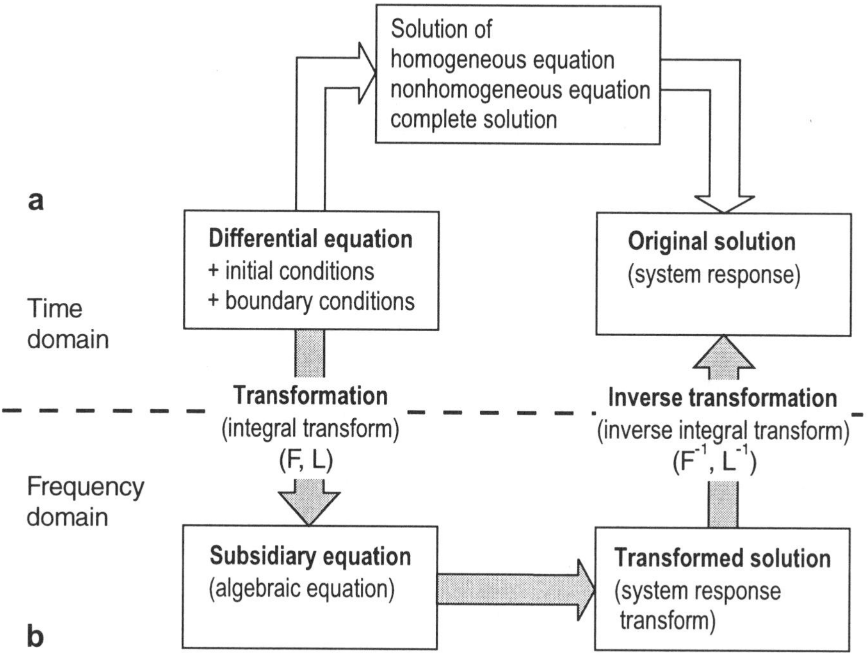

Fig. 3.9. Solution scheme for ordinary linear differential equations. **a** Classical method; **b** by means of transform method

applying integral transform equations the initial conditions or boundary conditions must be implied. *Then*, the algebraic equation, called subsidiary equation, can be solved without much difficulty compared to the classical (or conventional) method of solving differential equations. The *last step* consists in performing the inverse transformation of the transformed solution (ω-domain solution), here implying an evaluation of the Fourier integral, Eq. (3.69a), yet for the function $\underline{s}(\omega)$, since an original solution (t-domain solution) is required. The case may be whenever the time history of the pertinent motion response, e.g., the *actual displacement response* $s(t)$, is of essential interest (*real-time* or *time-response analysis*).

In vibration data analysis that applies especially to shock motions exemplified later on by the time-response calculation referred to a rectangular shock pulse (single-pulse excitation).

Nevertheless, for a lot of problems concerning system analysis and design the transform of the unknown response, e.g., the *displacement-response transform* $\underline{s}(\omega)$, already provides a sufficient description of system behaviour by the corresponding frequency components (*spectral density* or *frequency-response analysis*).

If still being of essential interest the inverse transformation possibly involves a rather complicated step of calculation that may be facilitated by the use of tables of transform pairs. They are readily available for a variety of basic functions. In case, the evaluation may be carried out by numerical integration with respect to ω, or experimentally by use of signal-processing methods basing on the *discrete Fourier transform* (DFT). Confining oneself to time-response calculations for idealized types of transient excitation acting on lumped parameter systems the evaluation of the inverse transform, here of the Fourier integral, reduces to a few fundamental rules of mathematics being of use for practical applications, see 3.2.4, [36], [37], [40], [41].

Response to Rectangular Shock Pulse. The transformation of the differential equation governing a forced mass-damper-spring system, Eq. (3.1), under a nonperiodic excitation (single-pulse excitation), Eq. (3.71a),

$$m\ddot{s} + c\dot{s} + ks = F(t) = F_{\mathrm{R}_1 0}(t) \tag{3.75}$$

by applying the Fourier transformation to both sides of the differential equation

$$m\mathrm{F}[\ddot{s}] + c\,\mathrm{F}[\dot{s}] + k\,\mathrm{F}[s] = \mathrm{F}\left[F_{\mathrm{R}_1 0}(t)\right] \tag{3.76a}$$

and using the differentiation theorem, Eq. (3.74b), successively for the first two derivatives yields an algebraic equation of the transforms being called
the *subsidiary equation* of the differential equation

$$-m\omega^2 \underline{s}(\omega) + \mathrm{j}c\omega\underline{s}(\omega) + k\underline{s}(\omega) = \underline{F_{\mathrm{R}_1 0}}(\omega) \tag{3.76b}$$

Solving the algebraic equation with ease for $\underline{s}(\omega)$ the transformed solution is obtained representing
the *displacement-response transform* $\underline{s}(\omega)$

$$\underline{s}(\omega) = \underline{s_{R_1 0}}(\omega) = \underbrace{\frac{1}{k + \mathrm{j}c\omega - m\omega^2}}_{=G(\omega)} \underline{F_{R_1 0}}(\omega) = G(\omega)\underline{F}(\omega) \tag{3.77a}$$

Fourier- (or ω-)domain-response Characteristic. The input-output relation of the compact type of an algebraic product, Eq. (3.77a), is one with considerable significance in practical application that relates transform method with block diagram representation, see 2.1. The response transform $\underline{s}(\omega)$ results from multiplying the excitation transform $\underline{F}_{R_1 0}(\omega)$ by a complex quantity termed

the *displacement frequency response* $G(\omega)$

$$G(\omega)=\frac{1}{(k-m\omega^2)+\mathrm{j}c\omega}=\frac{1}{m}\frac{1}{(\omega_0^2-\omega^2)+\mathrm{j}2\delta\omega}=\frac{1}{k}\frac{1}{(1-\eta^2)+\mathrm{j}2\zeta\eta} \quad (3.78a).$$

Frequency-response Function. This response characteristic is defined as a frequency-dependent property equal to the *quotient of two Fourier transforms* representing the actual response and the excitation function by their Fourier spectral densities (continuous spectra), Eq. (3.79). Contrary to the complexor, defined in 3.1.3 as a quotient of phasors representing harmonic quantities which are related to one single (forcing) frequency $\omega = \omega_{\mathrm{f}}$, the present quotient is a *complex system parameter* varying in ω over the definition range of the related Fourier transforms. Being a property of linear dynamic systems the frequency-response function does not depend on the type of excitation function. Excitation can be a harmonic, random, or a transient function of time. In the case of *motion response* by a displacement the ratio of the displacement-response transform $\underline{s}(\omega)$ to the transform of the excitation force $\underline{F}(\omega)$ is identified with

the *displacement frequency-response function* $G(\omega)$

$$G(\omega)=|G(\omega)|\mathrm{e}^{\mathrm{j}\operatorname{arc}G(\omega)}=A(\omega)\mathrm{e}^{-\mathrm{j}\psi(\omega)}=\frac{\mathrm{F}[s(t)]}{\mathrm{F}[F(t)]}=\frac{\underline{s}(\omega)}{\underline{F}(\omega)} \quad (3.79),$$

whose modulus is equal with

the *amplitude (frequency-) response* (gain response)

$$|G(\omega)|=A(\omega)=\frac{|\underline{s}(\omega)|}{|\underline{F}(\omega)|} \quad (3.80a),$$

and whose argument equals

the *phase (frequency-) response*

$$\operatorname{arc}G(\omega)=-\psi(\omega)=\operatorname{arc}\underline{s}(\omega)-\operatorname{arc}\underline{F}(\omega) \quad (3.80b).$$

Time-response Characteristic. An important concept in linear system analysis is to apply a pulse-type excitation idealized by approximating a rectangular pulse shape of unit pulse area (impulse value of unity) and zero pulse duration, called *unit pulse, Dirac* or *delta "function"*, denoted $\delta(t)$. Indeed, forcing functions suddenly applied for a short time and then removed (short-duration force pulses) are often used as reference input (nominal shock pulse) in vibrations to check the response of mechanical structures. For instance, translational impulsive forces can be generated by use of an exciter (impactor) which is not attached to the structure under test. Approximating the unit pulse by an equivalent impact excitation certain system parameters may be experimentally determined by measuring the *actual response to the unit pulse*, i.e., by the response characteristic $g(t)$.

In practice, impact measuments are associated with signal-processing methods basing on the discrete Fourier transform (DFT). Used as an approximation of the continuous Fourier transform the transformation of the response characteristic $g(t)$ thus may be performed by a digital Fourier transform system or analyser. However, the pulse of infinitesimal duration $\delta(t)$ containing equal energy at all frequencies only possesses theoretical signification. The actual force pulse in contrast has a spectrum of finite usable bandwidth. With respect to the accuracy of measurements the extension of usable frequency range must be fitted to the response characteristics of the structure, that means, the force spectrum characteristics of shock pulse excitations must be taken into consideration, [46].

Unit Pulse Response (weighting function). The response characteristic in the time domain, in control theory called *weighting function* and denoted $g(t)$, is related to the characteristic in the frequency domain previously dealt with, the frequency-response function $G(\omega)$. Both response characteristics constitute a Fourier transform pair denoted by the *correspondence*

$$G(\omega) \bullet\!\!\stackrel{\mathrm{F}^{-1}}{\longrightarrow}\!\!\circ\ g(t) \tag{3.81a}$$

The evaluation performed by the Fourier integral for $G(\omega)$

$$g(t) = \mathrm{F}^{-1}[G(\omega)] = \frac{1}{2\pi} \int_{-\infty}^{+\infty} G(\omega)\,\mathrm{e}^{\mathrm{j}t\omega}\,\mathrm{d}\omega \tag{3.81b}$$

may be facilitated due to the causality principle:

$$g(t) \equiv 0 \quad \text{for } t < 0 \tag{3.82}$$

by applying the inverse Fourier cosine transformation

$$g(t) = \frac{1}{\pi} \int_{0}^{\infty} G(\omega) \cos t\omega\,\mathrm{d}\omega \tag{3.81c}$$

The *unit pulse response*, thus defined as the inverse Fourier transform of the frequency-response function $G(\omega)$, and termed in mathematical notation Green's function

- accounts for the "weight" by which a value of the excitation preexistent at the time τ before the observation time t still takes effect on the response (related to convolution or Duhamel's integral, Eq. (3.87a,b));
- represents the general solution (complete response) of a system subjected to the specific pulse type "unit pulse excitation" at $t = 0$ with zero initial conditions, Eq. (3.21);
- or else represents the free vibration (transient response) of a system starting from the specific "non-zero initial conditions" defined by $s_0 = 0$; $s_0' = 1$, Eq. (3.22), (3.23b);
- contains the system parameters in the time domain (*transient-response specifications*) to be received from a specific transient vibration at natural frequency, i.e., from an "idealized force pulse-motion history".

Unit pulse response of the underdamped system

$$g(t) = \frac{1}{m\omega_d} e^{-\delta t} \sin \omega_d t = \frac{T_d}{2\pi m} e^{-t/T_r} \sin\left(2\pi \frac{t}{T_d}\right); \quad t \geqq 0 \qquad (3.83a)$$

with the system parameters δ, T_r, ω_d, T_d see 3.1.1, Eqs. (3.3), (3.4), and 3.1.2, Eqs. (3.19a), (3.19b).

Substituting the non-dimensional time τ, Eq. (3.2d), then it follows the *normalized form* by relating unit pulse response $g(t)$ to the reciprocal of the indicial (or characteristic) mechanical impedance $1/\sqrt{mk}$

$$\sqrt{mk}\, g(\tau) = \frac{1}{\sqrt{1-\zeta^2}} e^{-\zeta\tau} \sin\left(\sqrt{1-\zeta^2}\,\tau\right) = \frac{\tau_d}{2\pi} e^{-\tau/\tau_r} \sin\left(2\pi \frac{\tau}{\tau_d}\right); \; \tau \geqq 0 \quad (3.83b)$$

Unit pulse response of the overdamped system

$$g(t) = \frac{1}{m\lambda} e^{-\delta t} \sin h\lambda t = \frac{1}{m\lambda} e^{-t/T_r} \sin h\lambda t; \quad t \geqq 0 \qquad (3.84a)$$

system parameters δ, T_r, λ see 3.1.1, Eqs. (3.3), (3.4), and 3.1.2, Eqs. (3.19d); the *normalized form*

$$\sqrt{mk}\, g(\tau) = \frac{1}{\sqrt{\zeta^2-1}} e^{-\zeta\tau} \sinh\left(\sqrt{\zeta^2-1}\,\tau\right) = \frac{1}{\sqrt{\zeta^2-1}} e^{-\tau/\tau_r} \sinh\left(\sqrt{\zeta^2-1}\,\tau\right); \; \tau \geqq 0$$

(3.84b).

Response Data Plotting. The time history *unit pulse response* is shown in Fig. A.17 of the Appendix A.

The *effect* of the system parameter *damping ratio* ζ on the *time-response characteristic* (unit pulse response or weighting function $g(t)$) is represented graphically in the normalized form $\sqrt{mk}\, s_\delta(\tau) = \sqrt{mk}\, g(\tau)$ by a set (or family) of time-history curves for various amounts of damping. The damping effect on the vibratory nature of a structure is illustrated by the *oscillatory transient state of motion* for $\zeta < 1$ (indicating an underdamped mechanical system) differing apparently from the *non-oscillatory transient state of motion* for $\zeta > 1$ (of an overdamped system).

Response Calculation with the Fourier Transform

The remaining problem is to determine the system response in the time domain (actual displacement response) $s(t)$ corresponding to the transformed solution (displacement-response transform) $\underline{s}(\omega)$ already being calculated, Eq. (3.77a). There are *two approaches* to perform the inverse transformation:

- to *evaluate the Fourier integral* by substituting the response transform $\underline{s}(\omega)$ for $\underline{F}(\omega)$ into the inversion formula of the (complex) Fourier transform, Eq. (3.69a),

$$\mathrm{F}^{-1}\left[\underline{s}(\omega)\right] = s(t) = \frac{1}{2\pi} \int_{\omega=-\infty}^{+\infty} \underline{s}(\omega) e^{jt\omega}\, d\omega \qquad (3.85a)$$

(analytic continuation by extending the real ω-domain to the complex z-plane, and applying integral theorems to the corresponding complex function);

- to *apply the convolution integral* (superposition integral) to unit pulse response $g(t)$, Eqs. (3.83a), (3.84a), and to actual excitation $F(t)$, Eq. (3.71a) (integral transformation in the t-domain).

Convolution Approach. The latter approach is performed by a time response calculation using a corresponding time-domain relation of equivalent significance to the general algebraic product $G(\omega)\underline{F}(\omega)$ of Eq. (3.77a) representing the displacement-response transform $\underline{s}(\omega)$.

Inversion formula of algebraic product (time convolution theorem). Taking into account the Fourier transform of derivatives (differentiation theorem) the algebraic multiplying of the complex system parameter varying in $\omega, G(\omega)$, by the excitation transform $\underline{F}(\omega)$ being differentiated w.r.t. frequency involves a transform pair (differentiation theorem of the convolution integral) denoted by the *correspondence*

$$\underline{s}(\omega) = \mathrm{j}\omega\left[\frac{G(\omega)}{\mathrm{j}\omega}\cdot \underline{F}(\omega)\right] \bullet\!\!\xrightarrow{\;\mathrm{F}^{-1}\;}\!\!\circ \frac{\mathrm{d}}{\mathrm{d}t}\left[h(t) \underset{-\infty}{\overset{+\infty}{*}} F(t)\right] = s(t) \qquad (3.86a).$$

The derivative of the related time functions in brackets w.r.t. time implies an integral expression known as *Duhamel's integral*

$$s(t) = \frac{\mathrm{d}}{\mathrm{d}t}\left[h(t) \underset{-\infty}{\overset{+\infty}{*}} F(t)\right] = \dot{h}(t) \underset{-\infty}{\overset{+\infty}{*}} F(t) + h(0^+)F(t) \qquad (3.86b).$$

Evaluation by parts results in a sum of two terms the prime of which is defined by the convolution of the functions $\dot{h}(t)$ and $F(t)$ called *convolution* (or *Faltung) integral* (convolution theorem)

$$s(t) = \int_{-\infty}^{+\infty} \dot{h}(\tau)F(t-\tau)\,\mathrm{d}\tau + h(0^+)F(t) \qquad (3.87a)$$

or, equivalently, due to the given symmetry in $\dot{h}(t)$ and $F(t)$ (commutativity of convolution)

$$s(t) = \int_{-\infty}^{+\infty} F(\tau)\dot{h}(t-\tau)\,\mathrm{d}\tau + h(0^+)F(t) \qquad (3.87b).$$

To perform the integration given by the time convolution theorem, Eqs. (3.87a), (3.87b) a few distinct system parameters related to the time-domain concept are to be determined.

Time-response Characteristic. An important concept being an alternative to unit pulse response in linear system analysis is to apply a step-type excitation idealized by approximating a vertically fronted step of unit height (constant maximum height of unity) and zero rise time, called *unit step* or *Heaviside function* $u_0(t)$. Forcing functions applied as a constant force excitation (simple step force) experimentally easier to realize than short-duration force pulses are a reference input appropriate in case for vibrations to check the response of mechanical structures. Thus, distinct system parameters may be determined by the *actual response to the unit step*, i.e., by the response characteristic $h(t)$.

Unit Step Response. The response characteristic in the time domain, in control theory denoted $h(t)$, is related to the time-response characteristic previously dealt with, i.e., $g(t)$. With respect to the differentiation theorem the Fourier transform pair of system response characteristics, Eq. (3.81a), may be completed with the transform of unit step response $\{[G(\omega)]/(\mathrm{j}\omega)\} = H_{\bullet}(\omega)$, used as a system function being supplementary to the known complex system parameter $G(\omega)$, denoted by the *approximate correspondence*

$$G(\omega) = \mathrm{j}\omega H_{\bullet}(\omega) \;\bullet\!\!\xrightarrow{\;F^{-1}\;}\!\!\circ\; \frac{\mathrm{d}}{\mathrm{d}t} h(t) = g(t) \qquad (3.88).$$

Hence, the first derivative w.r.t. time of the unit step response, $\dot{h}(t)$, is equal to the unit pulse response (weighting function) $g(t)$

$$\dot{h}(t) = g(t) \qquad (3.89a).$$

The *unit step response*, thus defined as the inverse transform of the system function given by $H_{\bullet}(\omega)$ (being the conventional approximation to the exact transform with the additional term $\pi\, G(0)\delta(\omega)$, i.e., which involves the delta functional at $\omega = 0$, [36], yet to be considered as a negligible spectral component)

- represents the general solution (complete response) of a system subjected to the specific step type "unit step excitation" at $t = 0$ with zero initial conditions;
- contains the system parameters in the time domain (*transient-response specifications*) to be determined from a specific transient vibration at natural frequency, i.e., from an "idealized constant force-motion history".

Unit step response of the underdamped system in explicit form may be taken from the primary response to a rectangular shock pulse, Eq. (3.104a), if deviding by the maximum height F_0 and shifting by half the pulse duration $-\tau_0/2$ prior to 0.

Response Data Plotting. The time history *unit step response* is shown in Fig. A.18 of the Appendix A.

The *effect* of the system parameter *damping ratio* ζ on the *time-response characteristic* (unit step response or Heaviside response function $h(t)$) is represented graphically in the normalized form $ks_0(\tau) = kh(\tau)$ by a set (or family) of time-history curves for various amounts of damping analogous to Fig. A.17. Contrary to the unit pulse response the present time-response characteristic approaching unity final value is particularly suited to state *performance criteria* (transient-response specifications). In the *oscillatory transient state of motion* for $\zeta < 1$ the maximum overshoot decreases, whereas the rise time increases with increasing damping ratio. In the *non-oscillatory transient state of motion* for $\zeta > 1$ no more an overshoot does occur, while the rise time continues growing on with increasing damping ratio.

Schedule of terms and symbols

initial value of unit step response: $h(0^+)$

dummy variable in time (variable of integration): τ

time of observation (running parameter): t

convolution operator symbol: $*$

(with indicated limits of integration).

In view of mathematics the prime term, e.g. Eq. (3.87b), represents an integral transformation which has as its kernel $\dot{h}(t - \tau)$ and which transforms $F(t)$ into $s(t)$ (linear operational calculus) to be conceived as a calculus for specifying the operator T, see 3.2, Eq. (3.59), [39].

On account of physical interpretation the prime term being the convolution integral represents the delayed system response, on the other hand the incidental term implies the immediate response proportional to the excitation $F(t)$. Both terms together form the total system response.

The convolution integrals in Eqs. (3.87a,b) have infinite limits, which can cause difficulties in the evaluation of the integrals. Such difficulties can be obviated if the infinite limits are replaced by finite ones.

Due to *causality principle* (real physical system, response lags excitation), Eq. (3.82), the lower and higher limit of integration may be changed to 0 respectively t (as the integral is equal to zero for all earlier $\tau < t$).

Hence, it follows alternately

$$s(t) = \left\{ \begin{array}{c} \int\limits_{\tau=0}^{\infty} \dot{h}(\tau) F(t-\tau) \mathrm{d}\tau \\ \text{or} \\ \int\limits_{\tau=-\infty}^{t} F(\tau) \dot{h}(t-\tau) \mathrm{d}\tau \end{array} \right\} + h(0^+) F(t) \qquad (3.90\text{a,b}),$$

for short symbolically

$$s(t) = \left\{ \begin{array}{c} \dot{h}(t) \overset{\infty}{\underset{0}{*}} F(t) \\ \text{or} \\ F(t) \overset{t}{\underset{-\infty}{*}} \dot{h}(t) \end{array} \right\} + h(0^+) F(t) \qquad (3.91\text{a,b}).$$

Furthermore, following Eq. (3.89a), $\dot{h}(t)$ may be replaced by $g(t)$.

For mechanical systems being passive and *predominated by inertia* the frequency-response function $G(\omega)$ is a rational function, the numerator polynomial of which $B(\omega)$ is of lower order m than the denominator polynomial $H(\omega)$ being of order n, whereby the *common case* is indicated:

$$G(\omega) = B(\omega)/H(\omega) \quad ; \qquad m < n \qquad (3.89\text{b})$$

thus, the initial value of unit step response vanishes (Abel's or initial-value theorem, loosely phrased, state $G(\omega)$ being asymptotic for large ω, then $h(t)$ is zero at $t = 0$):

$$h(0^+) = G(\infty) = 0 \qquad (3.89\text{c}).$$

When additionally considering only *causal excitation functions* (starting at $t = 0$):

$$F(t) \equiv 0 \quad \text{for} \quad t < 0 \qquad (3.93)$$

the convolution theorem, Eq. (3.86a), reduces to the algebraic product of frequency-response characteristic and excitation transform corresponding with the convolution product of time-response characteristic and actual excitation. This relation constitutes the *standard transform pair,* denoted by the *correspondence*

$$\underline{s}(\omega) = G(\omega) \cdot \underline{F}(\omega) \quad \bullet\!\!\xrightarrow{\;F^{-1}\;}\!\!\circ \quad g(t) \overset{t}{\underset{0}{*}} F(t) = s(t) \tag{3.92a}$$

(The limits of integration fitted to boundedness of observation time can be dropped as the convolution symbol is unique).
The convolution integral in detail

$$s(t) = \int_{\tau=0}^{t} g(\tau)F(t-\tau)\,d\tau = \int_{\tau=0}^{t} F(\tau)g(t-\tau)\,d\tau \tag{3.92b}$$

corresponds formally to the convolution theorem of the Laplace-transformation, 3.2.8, Eq. (3.153), for the passive system with dominant inertia.

According to the alternate form of convolution integral the excitation is a function of τ and the unit pulse response is a function of $t-\tau$; that is, the weighting function is shifted. It is pointed out that the same result follows from the excitation being shifted instead of the unit pulse response.

For a system function with non-dominant inertia so that $B(\omega)$ and $H(\omega)$ are of equal order $m = n$, Eq. (3.90a,b) has to be treated not as a conventional but as a generalized convolution product with which the first time derivative of unit step response is defined by the regular generalized function $h'(t)$, Eq. (3.94), involving a (Dirac) delta functional $\delta(t)$:

$$h'(t) = \dot{h}(t) + h(0^+)\delta(t)\ ; \qquad h'(t) = g(t) \tag{3.94}$$

The convolution integral is thus not defined in the analytical (Riemann) sense, therefore the convolution theorem has been extended to generalized functions (distributions) [36], [38], [39].

Response Data Plotting. To point out the meaning of the convolution therorem it may be illustrated by both the graphical interpretation and the approximate evaluation.

Graphical interpretation of the convolution integral. The convolution consists of reflecting ("folding") $g(\tau)$ in (respective about) the ordinate to $g(-\tau)$, shifting $g(-\tau)$ by the time interval of observation t to $g(t-\tau)$, and an integration of the product $g(t-\tau)F(\tau)$ between the limits 0 and t, as shown in Fig. A.14 of Appendix A.

Exemplified for an unit ramp excitation $u_1(\tau)$, Eq. (3.174), and an underdamped weighting function $g(\tau)$, Eq. (3.83a), the shaded area is equal to the value of unit ramp response $s(t)$ at a particular instant t. Any instantaneous value of system response $s(t)$ depends on the excitation value $F(\tau)$ at the same instant $\tau = t$, and at all preceding instants $\tau \leq t$, multiplied ("weighted") by the weighting function that is shifted by the observation time t, i.e., by $g(t-\tau)$. The equivalence of both shaded areas whether the weighting function or the excitation function has been shifted according to the integral, Eq. (3.92b), demonstrates the symmetry of integral transformation in $F(t)$ and $g(t)$ (commutativity of convolution).

Approximate evaluation of the convolution integral. In order to visualize the meaning of the convolution transformation it is possible to derive the continuous-time responses of linear systems by sequences of sample values. At first an actual force history is approximated by a series of pulses of equally short duration respectively by a series of steps of small height, as shown in Fig. A.15 and A.16 of the Appendix A.

The former approximation bases on Eq. (3.86b), the latter one on the commutative form using the first derivative $\dot{F}(t)$ of the excitation function $F(t)$

$$s(t) = \frac{\mathrm{d}}{\mathrm{d}t}\left[F(t) \overset{+\infty}{\underset{-\infty}{*}} h(t)\right] = \dot{F}(t) \overset{+\infty}{\underset{-\infty}{*}} h(t) + F(-\infty)h(\infty) \qquad (3.86c).$$

Considering a single rectangular pulse of the impulse $f(\tau_\upsilon)\Delta\tau$ applied at $t = \tau_\upsilon$ an arbitrary input function may be regarded as

a *series of n rectangular pulses* (pulse approximation):

$$f(t) = \lim_{\Delta\tau \to 0} \sum_{\tau_\upsilon = 0}^{\tau_n} f(\tau_\upsilon) r_{\Delta\tau}(t - \tau_\upsilon)\Delta\tau = \int_{\tau=0^-}^{t} f(\tau)\delta(t-\tau)\mathrm{d}\tau \qquad (3.95a)$$

becoming in the limit, as $\Delta\tau \to 0$, $\tau_n = t$, an integral or continuous-time representation of $f(t)$ by delta pulses. Looking at a single rectangular step function of the maximum height $\Delta f(\tau_\upsilon)$ applied at $t = \tau_\upsilon$ an arbitrary input function may be regarded as

a *series of n rectangular steps* (step approximation):

$$f(t) = \lim_{\Delta\tau \to 0} \sum_{\tau_\upsilon = 0}^{\tau_n} \frac{\Delta f(\tau_\upsilon)}{\Delta\tau} u_0(t - \tau_\upsilon)\Delta\tau = \int_{\tau=0^-}^{t} \frac{\mathrm{d}}{\mathrm{d}t} f(\tau) u_0(t-\tau)\mathrm{d}\tau \qquad (3.95b)$$

becoming in the limit, as $\Delta\tau \to 0$, $\tau_n = t$, an alternative continuous-time representation of $f(t)$ by step functions. The response to unit impulse at $t = \tau_\upsilon$ is the weighting function delayed by the time interval t, that is $g(t - \tau_\upsilon)$. Hence, the single pulse contributes to the output an amount given by

$$y_\upsilon(t) = f(\tau_\upsilon) y_{r\Delta\tau}(t - \tau_\upsilon).$$

The sum of n delta pulse responses is approximately equal to the output caused by the arbitrary input. Letting $\Delta\tau \to 0$, $\tau_n = t$, and replacing the summation by integration, the approximate output results in

the *continuous-time system response* $y(t)$ *by delta pulse responses*:

$$y(t) = \lim_{\Delta\tau \to 0} \sum_{\tau_\upsilon = 0}^{\tau_n} f(\tau_\upsilon) y_{r_{\Delta\tau}}(t - \tau_\upsilon)\Delta\tau = \int_{\tau=0^-}^{t} f(t) \underbrace{y_\delta(t-\tau)}_{=g(t-\tau)} \mathrm{d}\tau = f(t) * g(t) \qquad (3.96a),$$

respectively in that *by rectangular step responses*:

$$y(t) = \lim_{\Delta\tau \to 0} \sum_{\tau_\upsilon = 0}^{\tau_n} \frac{\Delta f(\tau_\upsilon)}{\Delta\tau} y_{u_0}(t - \tau_\upsilon)\Delta\tau = \int_{\tau=0^-}^{t} \frac{\mathrm{d}}{\mathrm{d}\tau} f(\tau) \underbrace{y_{u_0}(t-\tau)}_{=h(t-\tau)} \mathrm{d}\tau = \dot{f}(t) * h(t) \qquad (3.96b).$$

Together with the continuous-time approximation by sequences of singularity functions a derivation of the convolution integral is performed according to Eqs. (3.86c) and (3.86b).

3.2.4
System Response to Transient Excitation. *Pulse-type Functions*

The Fourier transform method will be utilized for determining the system response in the time domain as follows. Taking up the *non-periodic forcing function of pulse-type* as a transient excitation applied to the mass-damper-spring system the inverse Fourier transformation will be performed by both of the approaches pointed out before, that is the inversion formula and the convolution integral. In any case the calculation of the unknown time response will be carried out by virtue of the *frequency-*, or else the *time-response characteristic*, both being known with reference to the already calculated frequency-response function $G(\omega)$, respectively the unit pulse response $g(t)$, Eqs. (3.78a), (3.83a), (3.84a).

Inversion Formula Approach

Writing the displacement-response transform $\underline{s}(\omega)$ in 3.2.3, Eq. (3.77a), in an explicit form by use of the known Fourier transform of the rectangular shock pulse, Eq. (3.71b), the transformed solution of the differential equation, Eq. (3.75), results in

$$\underline{s}(\omega)=\underline{s_{\mathrm{R_I}0}}(\omega)=\underbrace{\frac{\omega_0^2}{k}\,\frac{1}{(\omega_0^2-\omega^2)+\mathrm{j}2\delta\omega}}_{=G(\omega)}\;\underbrace{F_0\frac{1}{\mathrm{j}\omega}\left(\mathrm{e}^{\mathrm{j}\tau_0\omega/2}-\mathrm{e}^{-\mathrm{j}\tau_0\omega/2}\right)}_{=\underline{F_{\mathrm{R_I}0}}(\omega)} \qquad (3.77b).$$

$$=\frac{F_0\omega_0^2}{k}\;\frac{\mathrm{e}^{\mathrm{j}\tau_0\omega/2}-\mathrm{e}^{-\mathrm{j}\tau_0\omega/2}}{\mathrm{j}\omega\left[(\omega_0^2-\omega^2)+\mathrm{j}2\delta\omega\right]}$$

Applying the inverse Fourier transform, Eq. (3.69a), to the response transform $\underline{s}(\omega)$ by substituting Eq. (3.77b) for $\underline{F}(\omega)$ the inversion formula of the complex Fourier transform, Eq. (3.85a), yields the explicit form

$$s(t)=s_{\mathrm{R_I}0}(t)=\frac{F_0\omega_0^2}{k}\;\frac{1}{2\pi}\int_{-\infty}^{+\infty}\frac{\mathrm{e}^{\mathrm{j}\tau_0\omega/2}-\mathrm{e}^{-\mathrm{j}\tau_0\omega/2}}{\mathrm{j}\omega\left[(\omega_0^2-\omega^2)+\mathrm{j}2\delta\omega\right]}\,\mathrm{e}^{\mathrm{j}t\omega}\,\mathrm{d}\omega \qquad (3.85b).$$

To evaluate the Fourier integral being a real integral, for which the interval of integration is not finite, thus denoting an *improper integral*, the technique of *analytic continuation* is commonly used. By introducing the complex variable z

$$z=\omega+\mathrm{j}\gamma \qquad (3.97)$$

the response transform $\underline{s}(\omega)$ will be continued over all parts of the complex z-plane in which $\underline{s}(z)$ is analytic. Thus, the *theory of analytic functions* (complex variable functions) can be utilized by means of theorems for complex integration. Complex line integrals being integrated along a closed path in the z-plane are known as contour integrals. *Cauchy's integral theorem* enables to evaluate that type of complex integral. Analytic functions are, to a certain extent, essentially characterized by their singularities. Thus, if the transform $\underline{s}(z)$ is analytic on and inside a contour C, except at a finite number n of interiour isolated singularities z_k, then the value of the contour integral is given by $2\pi\mathrm{j}$ times the sum of the residues of the integrand at those points. This result is known as *Cauchy's residue theorem*:

$$2\pi\, s(t)=\int_{-\infty}^{+\infty}\underline{s}(\omega)\mathrm{e}^{\mathrm{j}t\omega}\,\mathrm{d}\omega=2\pi\mathrm{j}\sum_{k=1}^{n}\operatorname*{Res}_{z_k}\left[\underline{s}(z)\mathrm{e}^{\mathrm{j}tz}\right]\quad \text{for } t>0 \qquad (3.98).$$

The calculus of residues is useful in evaluating certain classes of complicated real integrals. Indeed, first the real improper integral defined for integrating along the real axis has to be completed in a *corresponding contour integral.* The path of integration thus will be closed around a contour by tracing a real axis segment from $-R$ to $+R$ and a semicircle in the counterclockwise direction in the upper half z-plane, and let $R \to \infty$.

The complete contour integral can be written as the sum of two parts of line integrals

$$\int_{-\infty}^{+\infty} = \int\limits_{\rightarrow} + \underbrace{\int\limits_{\curvearrowleft}}_{\equiv 0} = \oint$$

The first part is the integral to start with when $R \to \infty$. The second one tends to zero as $R \to \infty$, being a condition (Jordan's lemma) to be proved in the individual case for validating the complex integration.

The proof follows separately for different ranges of the variable time parameter t being of positive or negative value. Therefore it is thought proper to divide the actual displacement response (shock-motion history) into the response which occurs during the time in which the shock acts, called the *initial* or *primary response* $s_{\mathrm{Init}}(t)$ and the response which occurs during the free vibration existing after the shock has terminated, and that is called the *residual response* $s_{\mathrm{Res}}(t)$.

Residual Response. For the *residual shock period* $t > \tau_0/2$ the contour is closed owing to only positive values of time $t > 0$ in the upper half-plane. Thus, the integrand with particular regard to its kernel function $\mathrm{e}^{\mathrm{j}tz}$ converges to zero let $z \to \infty$:

$$\lim_{z\to\infty} \mathrm{e}^{\mathrm{j}tz} = \lim_{\gamma\to\infty} \left[\mathrm{e}^{\mathrm{j}t(\omega+\mathrm{j}\gamma)}\right] = \lim_{\gamma\to\infty} \left[\mathrm{e}^{-\gamma t}\mathrm{e}^{\mathrm{j}t\omega}\right] = 0 \qquad \text{for } t > 0$$

The frequency-response function $G(z)$ remains finite (is analytic) everywhere except at isolated singular points, called *singularities* or *poles*, such poles occurring where the denominator polynomial vanishes. The "singularities of the system" are identical with the known roots (solutions) or eigenvalues, compared to Eq. (3.15), consisting in case of less-than-critical damping in two first-order or *simple poles* (a pair of complex conjugate roots), denoting the *system-poles* $z_{1,2}$:

$$\begin{aligned} z_{1,2} &= \mathrm{j}\frac{c}{2m} \pm \sqrt{\frac{k}{m} - \left(\frac{c}{2m}\right)^2} \\ &= \mathrm{j}\delta \pm \sqrt{\omega_0^2 - \delta^2} = \mathrm{j}\zeta\omega_0 \pm \omega_0\sqrt{1-\zeta^2} \end{aligned} \tag{3.18a}$$

lying for real (causal) systems in the upper half-plane, Fig. 3.10.

The excitation transform $\underline{F}_{\mathrm{R_1}0}(z)$ being undefined at the "singularity of the excitation" $z_3 = 0$ can suitably be defined by approaching the constant τ_0 as $z \to 0$:

$$\underline{F}_{\mathrm{R_1}0}(0) = \lim_{\omega\to 0} \frac{\mathrm{e}^{\mathrm{j}\frac{\tau_0}{2}\omega} - \mathrm{e}^{-\mathrm{j}\frac{\tau_0}{2}\omega}}{\mathrm{j}\omega} = \tau_0 \,\mathrm{si}\!\left(\frac{\tau_0}{2}\omega\right)\Bigg|_{\omega=0} = \tau_0 .$$

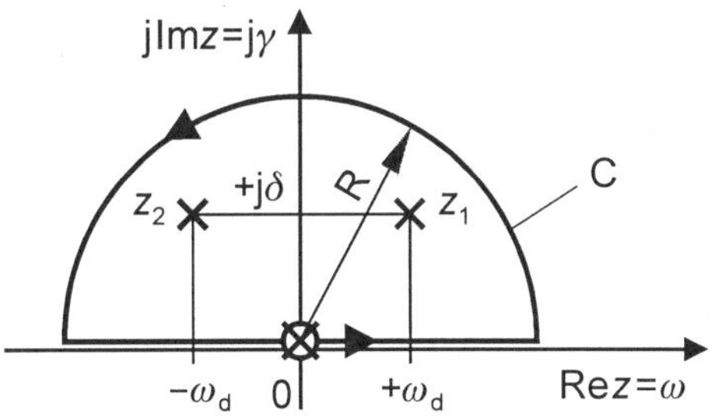

Fig. 3.10. Complex z-plane with a closed curve C surrounding the system-poles $z_{1,2}$ in the upper half-plane for a contour integration evaluating the inversion formula over the residual shock period

As a result the function $\underline{F}_{R_1 0}(z)$ becomes analytic at the origin coinciding with an *excitation-pole* $z_3 = 0$ that can be removed, denoting a *removable singularity*.

For the integrand $f(z)$ thus being analytic on the contour C with the two simple poles $z_k = z_{1,2}$ inside the contour the value of the corresponding contour integral follows from the sum taken over the residues of $f(z)$, that is the response transform $\underline{s}_{R_1 0}(z)$ multiplied by the kernel e^{jtz}, at $z = z_{1,2}$, denoted by $\text{Res}(z_{1,2}) = R_{1,2}$.

In view of mathematics the residue $\text{Res}(z_k)$ is the coefficient $A_{k,-1}$ of $(z - z_k)^{-1}$ in the corresponding Laurent series expansion of $f(z)$, in powers of $z - z_k$, which is valid near the center of convergence $z = z_k$. In the case of a simple pole z_k the *residue by evaluating the limit as* $z \to z_k$ is obtained as

$$\operatorname*{Res}_{z=z_k} f(z) = R_k = \lim_{z \to z_k} (z - z_k) f(z) = \left[(z - z_k) f(z)\right]_{z=z_k} \qquad (3.99).$$

Before carrying out the residue calculus the denominator polynomial of the frequency-response function $G(z)$ should be factored uniquely into the linear factors $(z - z_k)$, where z_k are the known roots (eigenvalues) of the characteristic equation for $k = 1, \ldots, n$; $n = 2$

$$G(z) = \frac{1}{m} \frac{1}{(\omega_0^2 - z^2) + j2\delta z} = -\frac{\omega_0^2}{z} \frac{1}{(z - z_1)(z - z_2)} \qquad (3.78b).$$

Thus, by use of factoring process, Eq. (3.78b), the residues R_1 and R_2 are easily obtained from Eqs. (3.18a), (3.85b), (3.99):

$$R_1 = \lim_{z \to z_1} \left[(z - z_1) \underline{s_{R_1 0}(z)} e^{jtz}\right] = \frac{F_0 \omega_0^2}{kj} \left[-(z - z_1) \frac{e^{j\frac{\tau_0}{2} z} - e^{-j\frac{\tau_0}{2} z}}{z(z - z_1)(z - z_2)} e^{jtz} \right]_{z=z_1}$$

$$= -\frac{F_0 \omega_0^2}{kj} \frac{e^{j\frac{\tau_0}{2} z_1} - e^{-j\frac{\tau_0}{2} z_1}}{z_1(z_1 - z_2)} e^{jtz_1} = -\frac{F_0 \omega_0^2}{kj} \frac{e^{j\frac{\tau_0}{2}(j\delta + \omega_d)} - e^{-j\frac{\tau_0}{2}(j\delta + \omega_d)}}{2(j\delta + \omega_d)\omega_d} e^{jt(j\delta + \omega_d)} \qquad (3.100a)$$

$$R_2 = \lim_{z\to z_2}\left[(z-z_2)\underline{s_{R_10}}(z)e^{jtz}\right] = \frac{F_0\omega_0^2}{kj}\,\frac{e^{j\frac{\tau_0}{2}(j\delta-\omega_d)} - e^{-j\frac{\tau_0}{2}(j\delta-\omega_d)}}{2(j\delta-\omega_d)\omega_d}\,e^{jt(j\delta-\omega_d)}$$

(3.100b)

so that, according to Eq. (3.98):

$$s_{Res}(t)=j[R_1+R_2]=j\frac{F_0\omega_0^2}{2jk\omega_d\underbrace{(\delta^2+\omega_d^2)}_{=\omega_0^2}}\left\{(j\delta-\omega_d)\left[e^{j\frac{\tau_0}{2}(j\delta+\omega_d)}-e^{-j\frac{\tau_0}{2}(j\delta+\omega_d)}\right]e^{jt(j\delta+\omega_d)}\right.$$

$$\left.-(j\delta+\omega_d)\left[e^{j\frac{\tau_0}{2}\omega(j\delta-\omega_d)}-e^{-j\frac{\tau_0}{2}(j\delta-\omega_d)}\right]e^{jt(j\delta-\omega_d)}\right\}$$

(3.101a)

and by summarizing the exponential functions of a complex and a conjugate number to circular functions of a real number the sum of residues finally results in:

Residual response of the underdamped system

$$s_{Res}(t)=\underbrace{F_0\underbrace{\frac{1}{k}}_{=G(0)}}_{=s_{stat}}\left\langle e^{-\delta(t-\frac{\tau_0}{2})}\left\{\cos\left[\omega_d(t-\frac{\tau_0}{2})\right]+\frac{\delta}{\omega_d}\left[\sin\omega_d(t-\frac{\tau_0}{2})\right]\right\}\right.$$

(3.101b)

$$\left.-e^{-\delta(t+\frac{\tau_0}{2})}\left\{\cos\left[\omega_d(t+\frac{\tau_0}{2})\right]+\frac{\delta}{\omega_d}\sin\left[\omega_d(t+\frac{\tau_0}{2})\right]\right\}\right\rangle;\quad t>\frac{\tau_0}{2}.$$

Introducing the non-dimensional frequency parameter:

$$\tau'=\omega_0\tau_0=2\pi\tau_0/T_0 \qquad (3.2f),$$

and the non-dimensional time (excitation time) by modifying Eq. (3.2d):

$$\tau/\tau'=\omega_0 t/(\omega_0\tau_0)=t/\tau_0 \qquad (3.2g)$$

where τ_0 is the pulse duration, and ω_0, T_0 are system parameters, see Eqs. (3.6a), (3.7), then it follows

the *normalized form* by relating displacement *s* to static deflection s_{stat}

$$\frac{s_{Res}(\frac{t}{\tau_0})}{s_{stat}}=e^{-\zeta 2\pi(\frac{\tau_0}{T_0})(\frac{t}{\tau_0}-\frac{1}{2})}\left\{\frac{\zeta}{\sqrt{1-\zeta^2}}\sin\left[\sqrt{1-\zeta^2}\,2\pi\left(\frac{\tau_0}{T_0}\right)\left(\frac{t}{\tau_0}-\frac{1}{2}\right)\right]\right.$$

$$\left.+\cos\left[\sqrt{1-\zeta^2}\,2\pi\left(\frac{\tau_0}{T_0}\right)\left(\frac{t}{\tau_0}-\frac{1}{2}\right)\right]\right\}$$

(3.101c)

$$-e^{-\zeta 2\pi(\frac{\tau_0}{T_0})(\frac{t}{\tau_0}+\frac{1}{2})}\left\{\frac{\zeta}{\sqrt{1-\zeta^2}}\sin\left[\sqrt{1-\zeta^2}\,2\pi\left(\frac{\tau_0}{T_0}\right)\left(\frac{t}{\tau_0}+\frac{1}{2}\right)\right]\right.$$

$$\left.+\cos\left[\sqrt{1-\zeta^2}\,2\pi\left(\frac{\tau_0}{T_0}\right)\left(\frac{t}{\tau_0}+\frac{1}{2}\right)\right]\right\};\quad \frac{t}{\tau_0}>\frac{1}{2}.$$

and when $\zeta = 0$, the *residual response of the undamped system*

$$\frac{s_{\mathrm{Res}}\left(\frac{t}{\tau_0}\right)}{s_{\mathrm{stat}}} = \cos\left[2\pi\left(\frac{\tau_0}{T_0}\right)\left(\frac{t}{\tau_0}-\frac{1}{2}\right)\right] - \cos\left[2\pi\left(\frac{\tau_0}{T_0}\right)\left(\frac{t}{\tau_0}+\frac{1}{2}\right)\right]$$
$$= \underbrace{\left[2\sin\left(\pi\frac{\tau_0}{T_0}\right)\right]}_{=\hat{s}_{\mathrm{Res}}/s_{\mathrm{stat}}} \cdot \sin\left[2\pi\left(\frac{\tau_0}{T_0}\right)\frac{t}{\tau_0}\right]; \quad \frac{t}{\tau_0} > \frac{1}{2} \tag{3.101d}$$

where the maximum value of the specified response is given by the *residual response amplitude factor*

$$\left(\hat{s}_{\mathrm{Res}}/s_{\mathrm{stat}}\right) = 2\sin\left(\pi\tau_0/T_0\right) \tag{3.102a}$$

This response parameter and thus the residual response of the undamped system vanishes for integer multiples of the *period ratio* τ_0/T_0 (interference phenomenon).

Primary Response. For the *initial shock period* $-\tau_0/2 < t < +\tau_0/2$ the integrand does not behave suitably at infinity because of time taking not only positive but also negative values $t < 0$. Accordingly, the evaluation of the Fourier integral, Eq. (3.85b), will be performed in two parts enclosing separately the upper half-plane for the first part and the lower half-plane for the second part, Fig. 3.11.

Now, however, the integrand of each part having a complex exponential in the numerator includes a pole at the origin, denoting the *excitation-pole* $z_3 = 0$. To avoid the integration through a singularity an *intended contour* should be traced by introducing a small semicircle of radius r with center at the singular point. It may be reserved to a rather sophisticated proof to show that the integrand is bounded in absolute value as $r \to 0$ [36], [37], [41]:

$$s_{\mathrm{Init}}(t) = \frac{F_0\omega_0^2}{k}\frac{1}{2\pi}\left\{\int_{-\infty}^{+\infty}\frac{\mathrm{e}^{\mathrm{j}(t+\frac{\tau_0}{2})\omega}}{\mathrm{j}\omega\left[(\omega_0^2-\omega^2)+\mathrm{j}2\delta\omega\right]}\mathrm{d}\omega - \int_{-\infty}^{+\infty}\frac{\mathrm{e}^{\mathrm{j}(t-\frac{\tau_0}{2})\omega}}{\mathrm{j}\omega\left[(\omega_0^2-\omega^2)+\mathrm{j}2\delta\omega\right]}\mathrm{d}\omega\right\} \tag{3.85c}$$

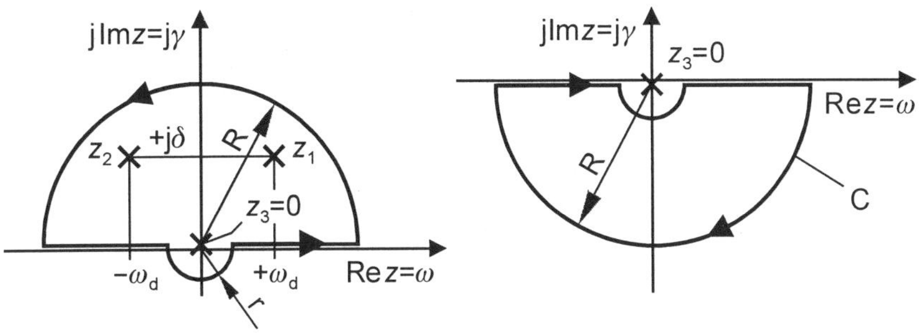

Fig. 3.11. Complex z-plane with indented closed curves surrounding either the system-poles $z_{1,2}$ and the excitation-pole $z_3 = 0$ in the upper half-plane or enclosing the lower half-plane for a contour integration evaluating over the initial shock period

Thus there will be no contribution from the second integral as its contour does not enclose any pole.

According to Eq. (3.98) the contribution from the first integral is given by the sum of the residues at the excitation-pole z_3 and at the two system-poles $z_{1,2}$.

By use of factoring process, Eq. (3.78b), the residues R_1, R_2, and R_3 are easily obtained from Eqs. (3.18a), (3.85c), (3.99):

$$R_1 = \lim_{z\to z_1}\left[(z-z_1)I_1(z)\right] = \frac{F_0\omega_0^2}{kj}\left[-(z-z_1)\frac{e^{j(t+\frac{\tau_0}{2})z}}{z(z-z_1)(z-z_2)}\right]_{z=z_1} \tag{3.103a}$$

$$= -\frac{F_0\omega_0^2}{kj}\frac{e^{j(t+\frac{\tau_0}{2})z_1}}{z_1(z_1-z_2)}$$

$$R_2 = \lim_{z\to z_2}\left[(z-z_2)I_1(z)\right] = \frac{F_0\omega_0^2}{kj}\left[-(z-z_2)\frac{e^{j(t+\frac{\tau_0}{2})z}}{z(z-z_1)(z-z_2)}\right]_{z=z_2} \tag{3.103b}$$

$$= \frac{F_0\omega_0^2}{kj}\frac{e^{j(t+\frac{\tau_0}{2})z_2}}{z_2(z_1-z_2)}$$

$$R_3 = \lim_{z\to z_3}\left[z I_1(z)\right] = \frac{F_0\omega_0^2}{kj}\left[-z\frac{e^{j(t+\frac{\tau_0}{2})z}}{z(z-z_1)(z-z_2)}\right]_{z=z_3} \tag{3.103c}$$

$$= -\frac{F_0\omega_0^2}{kj}\frac{1}{z_1 z_2}$$

so that, according to Eq. (3.98), and by summarizing the exponential functions to circular functions the sum of the residues finally results in:

Primary response of the underdamped system

$$s_{\text{Init}}(t) = \frac{F_0}{k}\left\langle 1 - e^{-\delta(t+\frac{\tau_0}{2})}\left\{\cos\left[\omega_d\left(t+\frac{\tau_0}{2}\right)\right] + \frac{\delta}{\omega_d}\sin\left[\omega_d\left(t+\frac{\tau_0}{2}\right)\right]\right\}\right\rangle; \tag{3.104a}$$

$$-\frac{\tau_0}{2} < t < \frac{\tau_0}{2}$$

normalized form

$$\frac{s_{\text{Init}}(\frac{t}{\tau_0})}{s_{\text{stat}}} = 1 - e^{-\zeta 2\pi(\frac{\tau_0}{T_0})(\frac{t}{\tau_0}+\frac{1}{2})}\left\{\cos\left[\sqrt{1-\zeta^2}\,2\pi\left(\frac{\tau_0}{T_0}\right)\left(\frac{t}{\tau_0}+\frac{1}{2}\right)\right]\right.$$

$$\left. + \frac{\zeta}{\sqrt{1-\zeta^2}}\sin\left[\sqrt{1-\zeta^2}\,2\pi\left(\frac{\tau_0}{T_0}\right)\left(\frac{t}{\tau_0}-\frac{1}{2}\right)\right]\right\}; \tag{3.104b}$$

$$-\frac{1}{2} < \frac{t}{\tau_0} < \frac{1}{2}$$

and when $\zeta = 0$, the *primary response of the undamped system*

$$\frac{s_{\text{Init}}(\frac{t}{\tau_0})}{s_{\text{stat}}} = 1 - \cos\left[2\pi\left(\frac{\tau_0}{T_0}\right)\left(\frac{t}{\tau_0}+\frac{1}{2}\right)\right] = \underbrace{2}_{=\hat{s}_{\text{Init}}/s_{\text{stat}}} \cdot \sin^2\left[\pi\left(\frac{\tau_0}{T_0}\right)\left(\frac{t}{\tau_0}+\frac{1}{2}\right)\right]; \qquad -\frac{1}{2} < \frac{t}{\tau_0} < \frac{1}{2} \tag{3.104c}$$

where the maximum value of the specified response is given by the *primary response amplitude factor*

$$\left(\hat{s}_{\text{Init}}/s_{\text{stat}}\right) = 2 \tag{3.102b}$$

being a constant for all period ratios τ_0/T_0.

Convolution Approach

Due to a excitation function starting prior to $t = 0$ (non-causal function $F(t) \neq 0$ for $t < 0$) the convolution integral in the limits of Eq. (3.90a,b) holds true, whereat the incidental term implying the immediate response will be dropped because of vanishing initial value of the unit step response, Eq. (3.89c), for the proper case of $G(\omega)$ with $m < n$ (dominant inertia), Eq. (3.89b), so that:

$$s_{\text{R}_1 0}(t) = \int_{\tau=-\infty}^{t} F_{\text{R}_1 0}(\tau) g(t-\tau)\mathrm{d}\tau = F_{\text{R}_1 0}(t) \underset{-\infty}{\overset{t}{*}} g(t) \tag{3.105}$$

Writing the actual displacement response $s_{\text{R}_1 0}(t)$, Eq. (3.105), in an explicit form by use of the actual rectangular shock pulse defined over specified time intervals, Eq. (3.71a), the evaluation of the convolution integral results in the *primary response of the system*

$$s_{\text{Init}}(t) = F_0 \int_{\tau=-\frac{\tau_0}{2}}^{t} 1 \cdot g(t-\tau)\mathrm{d}\tau = -F_0 \int_{u=t+\frac{\tau_0}{2}}^{0} g(u)\mathrm{d}u = F_0 \int_{0}^{t+\frac{\tau_0}{2}} g(u)\mathrm{d}u = F_0\left[h\left(t+\frac{\tau_0}{2}\right) - h(0^+)\right]; \quad \text{where } h(0^+) = 0 \tag{3.106}$$

which equals the response to a constant force excitation (simple step force), coincident with F_0 times the unit step response shifted by half the pulse duration $-\tau_0/2$ prior to $t = 0$, identical with Eq. (3.104a).

By use of integration term by term, the first one over the initial shock period from $-\tau_0/2$ to $+\tau_0/2$, and the second one over the residual shock period from $+\tau_0/2$ to *t*, last substituting the unit pulse response, Eq. (3.83a), the convolution integral becomes the *residual response of the underdamped system*

$$s_{\text{Res}}(t) = F_0 \left\{ \int_{-\frac{\tau_0}{2}}^{+\frac{\tau_0}{2}} 1 \cdot \frac{1}{m\omega_{\text{d}}} \mathrm{e}^{-\delta(t-\tau)} \sin[\omega_{\text{d}}(t-\tau)]\mathrm{d}\tau + \underbrace{\int_{\tau=+\frac{\tau_0}{2}}^{t} 0 \cdot \frac{1}{m\omega_{\text{d}}} \mathrm{e}^{-\delta(t-\tau)} \sin[\omega_{\text{d}}(t-\tau)]\mathrm{d}\tau}_{\equiv 0} \right\} \tag{3.107}$$

where the second integral does not contribute because of a forcing function being zero over the residual period. Only the integration over the initial period yields a contribution by the first definite integral being evaluated by *integration by parts*

$$s_{\mathrm{Res}}(t)=F_0\frac{\omega_0^2}{k\omega_\mathrm{d}}\frac{\omega_\mathrm{d}^2}{\omega_\mathrm{d}^2+\delta^2}\left|\mathrm{e}^{-\delta(t-\tau)}\left\{\cos\left[\omega_\mathrm{d}(t-\tau)\right]+\frac{\delta}{\omega_\mathrm{d}}\sin\left[\omega_\mathrm{d}(t-\tau)\right]\right\}\right|_{-\frac{\tau_0}{2}}^{+\frac{\tau_0}{2}}\quad;\ t>\frac{\tau_0}{2}$$

with the result identical to Eq. (3.101b).

Response Data Plotting. The actual displacement response (shock-motion history) $s(t)$ as the result of time-response calculation is shown in Fig. A.19 and A.20 of the Appendix A.

The effect of the system parameter damping ratio ζ on vibratory motion is represented graphically for a mechanical system being subjected to a rectangular shock pulse, and is portrayed as for selected fractions of critical damping as for two different values of the period ratio τ_0/T_0. In the response of systems to a shock excitation several kinds of maximum values are of considerable physical significance. The *residual response amplitude* $\hat{s}_{\mathrm{Res}}$ is the amplitude of the free vibration after the removal of a constant force excitation. The *maximax response* s_{Max} is the maximum value of the greatest magnitude attained at any time during the shock motion. The maximax response may occur either within the duration of the pulse or during the residual vibration era. In the latter case the maximax response is equal to the residual response amplitude. This is generally true in the case of short-duration pulses, e.g., for the small period ratio of $\tau_0/T_0 = 1/2$, delivering the response curve in Fig. A.19.

To the contrary the maximax response occurs in the initial vibration era, e.g., for the large period ratio $\tau_0/T_0 = 2$. Being casually an integer multiple, the residual response thus vanishes for the undamped system, see Fig. A.20.

Shock Response Spectrum

Measurement of shock-motion history is not useful directly for engineering purposes. A method of reducing the time history is then necessary depending upon the purpose for which the data will be used. There are several *concepts of data reduction.* A description in terms of the effect on structures when the shock acts as a specified excitation is designated *reduction to the response domain.* The usual concept of the shock response spectrum is based upon the single degree-of-freedom system, commonly considered linear and undamped. With only two system parameters involved, the (undamped) natural frequency ω_0 and the damping ratio ζ, it is feasible to obtain from the shock measurement a systematic presentation of the *response maxima* of many simple structures. The shock response spectrum is a graphical presentation of a selected quantity in the response. As one of various characteristics the response peaks may be depicted and plotted as a function of a non-dimensional frequency parameter. For that commonly the ratio of pulse duration to natural period is employed, called the *period ratio* τ_0/T_0.

Response Data Plotting. The shock response spectrum calculated form the time history of the total displacement response $s_{\mathrm{Init}}(t)$, $s_{\mathrm{Res}}(t)$ to a rectangular pulse of force $F_{\mathrm{R_1}0}(t)$ is expressed in terms of the equivalent static displacement s_{stat} for the undamped responding structure, $\zeta = 0$, Eqs. (3.101d), (1.104c), shown in Fig. A.21 of the Appendix A.

The response spectrum representation is subdivided into a *primary shock spectrum* being a straight line of value 2 for $\tau_0/T_0 > 1/2$, and a *residual shock spectrum* forming a half-sine range vanishing at integer multiples of τ_0/T_0 (interference phenomena), both depicted as solid lines. *The maximax shock spectrum*, depicted as dotted line, either coincides with the primary spectrum or with the residual spectrum. The latter case, however, is reserved to short-duration pulses given by $\tau_0/T_0 < 1/2$.

To attenuate the transmission of shock the admissable range of period ratios is limited to $\tau_0/T_0 < 1/6$. Provided that a mechanical structure has to be protected from a series of shock forces of given pulse duration τ_0 a *shock isolation* aims at an appropriate elongation of the natural period T_0 by element parameter variations (tuning of energy storage elements).

Shock Response Spectrum and Fourier Spectrum in Contrast. A description of the shock in terms of its inherent properties, viz., of the amplitudes and phase relations of its frequency components, is given by a continuous function of frequency, and the composite function is evaluated by integration. Thus, the *Fourier spectrum* may be considered as a capable *data reduction to the frequency domain* to represent the time history of shock excitations by their corresponding spectral densities. Moreover, the Fourier spectrum is an important aid used in frequency-domain analysis. Measurements involving *signal-processing methods* base on the discrete Fourier transform (DFT). For instance, the *impact-excitation technique* has become a popular method to determine frequency-response functions of structures (mobility or dynamic compliance measurements), [46]. For this purpose the actual force pulse and the transient motion response are simultaneously measured. Computing for each of both signals the discrete Fourier transform and forming the response ratio of Fourier spectra the complex system parameter $G(\omega)$ thus will be obtained in conformity with Eq. (3.79), see 3.2.3..

In contrast the *shock response spectrum* describes the effect of the shock upon a structure in terms of peak responses of shock vibrations. Thus, the time history of a shock cannot be determined from the knowledge of the peak responses, i.e., the *data reduction to the response domain* by calculation of peak responses is an irreversible operation. Further relations see references, e.g., [42].

3.2.5 Random Vibration. *Data Processing*

Collecting records of vibrations (time histories) and extracting informations from those records for practical purposes of vibration data analysis is not only referred to non-periodic vibrations (shock motions). *Data processing* much more includes the handling and reduction of original informations extracted from stochastic time histories (random vibrations) the associate time functions of which (random process) are characterized through statistical properties.

Types of Vibration

The two principal categories of vibrations differ from oneanother whether a particular vibration is deterministic or random.

Deterministic Vibration. The class of vibrations for which the instantaneous value of the vibration at a specified time is determined precisely by its time history is called *deterministic*. The value of a deterministic function, e.g., a time-varying function of previously treated periodic or transient type (periodic respectively non-periodic vibration) can be predicted from knowledge of its behaviour at previous times.

Random Vibration. The class of vibrations for which the instantaneous value of the vibration at a specified time *cannot* be determined by its time history is designated as *random*. Though the magnitude of a random vibration cannot be predicted for any given instant of time, data analysis aims at specifying the probability that the magnitude of an expected motion variable is within a given range.

Examples of random functions are the noise intensity caused by jet and rocket engines (random noise) in acustics, or representative samples of load to which airplanes taxing on rough runways are subjected (random motion) as data ensembles in structure design basing on fatigue limit analysis or fracture mechanics analysis (cumulative damage).

A collection of signals picked up just as recorded is classed with a *random* or *stochastic process* if the set (ensemble) of time functions can be characterized through statistical properties, such as the probability distribution function of specified vibration magnitudes, the mean-square value, or the frequency-averaged value of vibration magnitudes.

Random Process Classification

A set of time functions is ordinarily termed a process rather than an ensemble when it should be emphasized that the informations (properties) represented by signals are associated with a group.

Stationary Process. A process characterized by an ensemble of time histories such that their statistical properties are invariant with respect to translations to time is defined as *stationary*.

One may regard a stationary random process as somewhat analogous to the steady-state vibration in the case of deterministic functions.

For a stationary random vibration, the probability that the magnitude will be within a given magnitude range is taken to be equal to the ratio of the time that the vibration is within that range to the total time of observation.

A random vibration being not stationary defines a non-stationary process (transient random vibration).

Ergodic Process. A stationary process containing an ensemble of time histories where the time averages are the same for every time history is stated *ergodic*.

It follows that these time averages from any time history then will be equal to corresponding statistical averages over the ensemble.

Normal or Gaussian Distribution. A random vibration the magnitudes of which are accumulated by a probability distribution of Gaussian type specifies a *normal distribution* (Gaussian random vibration).

Random vibration analysis basing on fundamental statistic properties is confined to a random process which is stationary, ergodic and of normal distribution. Statistical properties of such a kind of random process base on *time averages* from time history ranges of interest (representative vibration-time records). Due to ergodicity assumption only one sample function can be chosen being a representative of the corresponding process in calculating averages instead of using the entire ensemble.

Correlation Function (correlation analysis)

Autocorrelation Function. Various averages used in random vibration analysis and probability distributions. A random variable x is a real-valued variable being discrete or continuous which is defined on a sample space (event space). Taking account of an ergodic process and evaluating time averages over sample functions of long duration two representative functions $F_1(t)$ and $F_2(t)$ define a *joint* or *second-order probability distribution function* $P(F_1, F_2, t_1, t_2)$

$$P(F_1, F_2, t_1, t_2) = \text{Prob}\left[F_1(t) \le F(t_1), F_2(t) \le F(t_2) = F(t_1 + \tau)\right],$$

associated with the probability that the magnitude of a quantity F at any particular time of observation t_1 will be less than (or equal to) a given value of this function $F(t_2)$. The statistical properties of a stationary process representing the function $F(t)$ do not depend on the particular times t_1, t_2 but only on the delay τ. Thus, a *second-order probability density function* $p(F_1, F_2, \tau)$ can be defined

$$p(F_1, F_2, \tau) = \frac{\partial^2}{\partial F_1 \partial F_2} P(F_1, F_2, \tau)$$

the first moment of which is the *autocorrelation function* of the excitation $F(t)$

$$R_{FF}(\tau) = \overline{F(t)F(t+\tau)} = \lim_{T \to \infty} \frac{1}{2T} \int_{-T}^{+T} F(t)F(t+\tau)\,\mathrm{d}t \qquad (3.108a).$$

Physically $R_{FF}(\tau)$ may be considered as the temporal mean value of the product of the value of forcing function $F(t)$ at time t with its value at time $(t+\tau)$.

An autocorrelation function is independent of the choice of origin and thus an even function:

$$R_{FF}(\tau) = \overline{F(t)F(t+\tau)} = \overline{F(t-\tau)F(t)} = R_{FF}(-\tau) \qquad (3.108b).$$

Assuming zero mean value (average value) the autocorrelation at zero delay $(\tau = 0)$ is equal to

the excitation *mean-square value*

$$R_{FF}(0) = \overline{F^2(t)} \qquad (3.109a)$$

defined by the mean of the squared forcing function values over the time interval $2T$ approaching infinity. In practice, the averaging time T is finite and Eqs. (3.108a), (3.109a) only give an estimate with a certain statistical uncertainty which increases as T decreases. Since a quantity cannot be more coherent to another quantity than it is by itself, the maximum value of $R_{FF}(\tau)$ occurs when $\tau = 0$.

Cross-correlation Function. A combined second-order probability density function can be defined between two vibration magnitudes being physically distinct and specified by the representative functions of excitation $F(t)$ and its caused displacement $s(t)$. The associated first moment is given by
the *cross-correlation function* between $F(t)$ and $s(t)$

$$R_{\mathrm{Fs}}(\tau) = \overline{F(t)s(t+\tau)} = \lim_{T\to\infty} \frac{1}{2T}\int_{-T}^{+T} F(t)s(t+\tau)\,\mathrm{d}t \qquad (3.110a).$$

Thus, $R_{\mathrm{Fs}}(\tau)$ may be interpreted as the temporal mean value of the product of the value of forcing function $F(t)$ at time *t* and the value of the displacement $s(t)$ at time $(t+\tau)$.

A cross-correlation function in general is not even as to point out by its reverse form between $s(t)$ and $F(t)$

$$R_{\mathrm{sF}}(\tau) = \overline{s(t)F(t+\tau)} = \lim_{T\to\infty} \int_{-T}^{+T} s(t)F(t+\tau)\,\mathrm{d}t \qquad (3.110b),$$

and by use of the substitution $t+\tau = u$

$$R_{\mathrm{sF}}(\tau) = \lim_{T\to\infty} \int_{-T}^{+T} s(u-\tau)F(u)\,\mathrm{d}u\,,$$

so that

$$R_{\mathrm{sF}}(\tau) = R_{\mathrm{Fs}}(-\tau) \qquad (3.110c).$$

No configuration of symmetry turns up at zero delay $(\tau = 0)$ and
the *mean-product value*

$$R_{\mathrm{Fs}}(0) = \overline{F(t)s(t)} \qquad (3.111a)$$

does not necessarily coincide with the maximum value of $R_{\mathrm{Fs}}(\tau)$ when $\tau = 0$. Contrary to autocorrelation function the cross-correlation function includes certain phase informations being of use in correlation analysis.

Power Spectral Density (spectral density analysis)

Definition of the Fourier Transform. Correlation analysis of vibration data has many uses. The autocorrelation function permits to determine both the real-time nature of a quantity $F(t)$ and its frequency or spectral properties. This is performed by integral transformation defining the autocorrelation function $R_{\mathrm{FF}}(\tau)$ as the inverse Fourier transform of the *power spectral density* (auto-spectral density or auto-spectrum) of the excitation $S_{\mathrm{FF}}(\omega)$ $\left(\text{or } G_{\mathrm{FF}}(\omega)\right)$. Thus, both statistical functions being in correspondence constitute
the *Fourier transform pair* (Wiener-Khinchin equations)

$$S_{\mathrm{FF}}(\omega) = \int_{-\infty}^{+\infty} R_{\mathrm{FF}}(\tau)\mathrm{e}^{-\mathrm{j}\omega\tau}\,\mathrm{d}\tau \qquad (3.112a)$$

$$R_{\mathrm{FF}}(\tau) = \frac{1}{2\pi}\int_{-\infty}^{+\infty} S_{\mathrm{FF}}(\omega)\mathrm{e}^{\mathrm{j}\tau\omega}\,\mathrm{d}\omega \qquad (3.112b).$$

Physically $S_{FF}(\omega)$ may be considered as the mean-square value of that part of the forcing function passed by a narrow-band filter of centre frequency ω devided by the bandwidth $\Delta\omega$ of the filter, as the bandwidth approaches zero.

In case of zero delay $\tau = 0$ it follows the excitation *mean-square value*

$$\overline{F^2(t)} = \sigma_F^2 = R_{FF}(0) = \frac{1}{2\pi}\int_{-\infty}^{+\infty} S_{FF}(\omega)\mathrm{d}\omega = \frac{1}{\pi}\int_0^{\infty} S_{FF}(\omega)\mathrm{d}\omega \quad (3.109b).$$

This points out a way of data processing for obtaining the mean-square value by the excitation auto-spectrum. This spectral representation of a random forcing function either will be known or may be determined by the corresponding correlation function $R_{FF}(\tau)$ performing its Fourier transform. Hereat the existence of $S_{FF}(\omega)$ must be noticed in view of the *absolute convergence criterion*

$$\int_{-\infty}^{+\infty} |R_{FF}(\tau)| \mathrm{d}\tau < \infty \quad (3.112c).$$

Under that condition $S_{FF}(\omega)$ is bounded similar to the spectral density of Fourier-transformable deterministic functions, Eq. (3.70). $S_{FF}(\omega)$ being an even, real function of ω the inverse Fourier cosine transformation, Eq. (3.81c),

$$R_{FF}(\tau) = \frac{1}{\pi}\int_0^{+\infty} S_{FF}(\omega)\cos\tau\omega\mathrm{d}\omega \quad (3.112d)$$

may be applied with integration reduced to one-sided infinity. With regard to the resulting term of Eq. (3.109b), the calculation of the mean-square value thus will be facilitated by data reduction.

Response to Random Excitation

Correlation analysis of vibration data can be extended to two different points of a mechanical structure for studying the transmission of vibration through structures. Taking pattern from the frequency concept used for response calculations in deterministic vibration, 3.2.3, the Fourier transform method also proves its usefulness in random vibration. Contrary to periodic or non-periodic vibrations it is impossible to determine the time history of the pertinent motion response (actual displacement response). Nevertheless, *random vibration analysis* takes advantage of the transform method in predicting the probability of response events or in estimating mean responses (response time averages). The determination of analytic solutions of the equation of motion thus is confined to statistical properties from which significant parameters can be gathered (*statistical response specifications*).

Response Calculation with the Fourier Transform. Applying the convolution integral, Eq. (3.92b), to time-varying functions of an ergodic process the autocorrelation function of the excitation and the caused effect upon a structure are related by convolution with the time-response characteristic. The given effect in a subsidiary time domain (time-delay- or τ-domain) is a correlation function between the applied force and the displacement response, known as the *cross-correlation function* at $F(t)$ and $s(t)$ by $g(t)$

$$R_{Fs}(\tau)=\overline{F(t)s(t+\tau)}=\int_{0^-}^{\infty} g(t)R_{FF}(\tau-t)\mathrm{d}t=g(\tau)\underset{0^-}{\overset{\infty}{*}}R_{FF}(\tau) \qquad (3.113a)$$

or in its reverse form between the displacement response and the applied force as the *cross-correlation function* at $s(t)$ and $F(t)$ by $g(t)$

$$R_{sF}(\tau)=\overline{s(t)F(t+\tau)}=\int_{0^-}^{\infty} g(t)R_{FF}(\tau+t)\mathrm{d}t=g(-\tau)\underset{0^-}{\overset{\infty}{*}}R_{FF}(\tau) \qquad (3.113b),$$

so that with respect to Eq. (3.110c) $R_{sF}(\tau)=R_{Fs}(-\tau)$.

Corresponding with Eq. (3.113a) the correlation-function transforms of the excitation and the caused effect are related by multiplying the excitation auto-spectrum with the frequency-response characteristic. The given effect in the frequency domain (ω-domain) is represented by the frequency or spectral components of the correlation between applied force and the displacement response $S_{Fs}(\omega)$ $\left(\text{or } G_{Fs}(\omega)\right)$, respectively by its reverse $S_{sF}(\omega)$ $\left(\text{or } G_{sF}(\omega)\right)$, known as the *cross-spectral density* (cross-spectrum)

$$S_{Fs}(\omega)=G(\omega)\cdot S_{FF}(\omega) \qquad (3.114a)$$

$$S_{sF}(\omega)=G^{*}(\omega)\cdot S_{FF}(\omega) \qquad (3.114b)$$

where the companion form is the complex conjugate of the original one

$$S_{sF}(\omega)=S_{Fs}^{*}(\omega) \qquad (3.114c).$$

The response characteristic defined in 3.2.3 with respect to deterministic vibration, Eq. (3.79), is equal to the quotient of two Fourier transforms representing the actual response and the excitation function. Being a property of linear dynamic systems that is independent of the type of excitation by definition the signification of this frequency-dependent property may be extended to random data processing. In the actual case of a random vibration the ratio of the cross-correlation function to the autocorrelation function of the excitation, both represented by their power spectral densities $S_{Fs}(\omega)$ $\left(\text{or } G_{Fs}(\omega)\right)$, respectively $S_{FF}(\omega)$ $\left(\text{or } G_{FF}(\omega)\right)$, is identified with
the *displacement frequency-response function* $G(\omega)$

$$G(\omega)=|G(\omega)|\mathrm{e}^{\mathrm{j\,arc}G(\omega)}=\frac{\mathrm{F}[R_{Fs}(\tau)]}{\mathrm{F}[R_{FF}(\tau)]}=\frac{S_{Fs}(\omega)}{S_{FF}(\omega)} \qquad (3.115).$$

The caused effect is described by the cross-spectral density $S_{sF}(\omega)$ $\left(\text{or } G_{sF}(\omega)\right)$, a complex-valued function of frequency in which the phase information is retained.

In spectral analysis the argument of the frequency-response function $G(\omega)$ results from relating both of the cross-spectral densities to be determined by data processing:

$$\frac{S_{Fs}(\omega)}{S_{sF}(\omega)}=\frac{G(\omega)}{G^{*}(\omega)}=|G(\omega)|\mathrm{e}^{\mathrm{j2\,arc}G(\omega)}=A(\omega)\mathrm{e}^{-\mathrm{j}2\psi(\omega)} \qquad (3.116a),$$

and its real part

$$\mathrm{Re}\left[S_{\mathrm{Fs}}(\omega)/S_{\mathrm{sF}}(\omega)\right]=\cos\left[2\psi(\omega)\right]$$

where the *phase (frequency-)response* is given by

$$\operatorname{arc}G(\omega)=\psi(\omega)=(1/2)\arccos\left\{\mathrm{Re}\left[S_{\mathrm{Fs}}(\omega)/S_{\mathrm{sF}}(\omega)\right]\right\} \qquad (3.116\mathrm{b}),$$

usually represented in the logarithmic form

$$\psi(\omega)=(1/2)\left[\lg S_{\mathrm{Fs}}(\omega)-\lg S_{\mathrm{sF}}(\omega)\right] \qquad (3.116\mathrm{c}).$$

The inverse Fourier transformation will be performed by the two approaches introduced for the class of deterministic vibrations, the inversion formula and the convolution integral. In any case the calculation of the unknown response in the subsidiary time domain (τ-domain) will be carried out by virtue of the known response characteristics, the frequency-response function $G(\omega)$ and the unit pulse response $g(\tau)$, Eqs. (3.78a), (3.83a), (3.84a).

Convolution Approach. Basing on the already determined correlation function $R_{\mathrm{Fs}}(\tau)$ and its reverse $R_{\mathrm{sF}}(\tau)$, Eqs. (3.113a,b), the convolution integral, Eq. (3.92a), or its commutative form, Eq. (3.92b) may be applied again. Changing the time of observation t in the time response characteristic $g(t)$ by the particular delay parameters τ_1 or τ_2 and considering the temporal mean values at time $\tau-t$ respectively $\tau+t$:

$$R_{\mathrm{sF}}(\tau-t)=\int_{0^-}^{\infty}g(\tau_1)R_{\mathrm{FF}}\left[(\tau-t)+\tau_1\right]\mathrm{d}\tau_1 \qquad (3.117\mathrm{a})$$

$$R_{\mathrm{Fs}}(\tau+t)=\int_{0^-}^{\infty}g(\tau_2)R_{\mathrm{FF}}\left[(\tau+t)-\tau_2\right]\mathrm{d}\tau_2 \qquad (3.117\mathrm{b})$$

the caused effect can be derived from the above cross-correlation functions by convoluting them with the unit pulse response $g(\tau)$ or its reverse $g(-\tau)$ giving the autocorrelation function at the displacement response $s(t)$

$$R_{\mathrm{ss}}(\tau)=\int_{0^-}^{\infty}g(t)R_{\mathrm{sF}}(\tau-t)\,\mathrm{d}t=g(\tau)\underset{0^-}{\overset{\infty}{*}}R_{\mathrm{sF}}(\tau) \qquad (3.118\mathrm{a}).$$

$$R_{\mathrm{ss}}(\tau)=\int_{0^-}^{\infty}g(t)R_{\mathrm{Fs}}(\tau+t)\,\mathrm{d}t=g(-\tau)\underset{0^-}{\overset{\infty}{*}}R_{\mathrm{Fs}}(\tau) \qquad (3.118\mathrm{b})$$

By substitution of Eq. (3.117a) in Eq. (3.118a) respectively of Eq. (3.117b) in (3.118b) a repeated convolution of the excitation autocorrelation function with the unit pulse response yields the *response autocorrelation function* at $s(t)$

$$\begin{aligned}R_{\mathrm{ss}}(\tau)=\overline{s(t)s(t+\tau)}&=\int_{-\infty}^{+\infty}g(\tau_1)\int_{0^-}^{\infty}g(\tau_2)R_{\mathrm{FF}}(\tau+\tau_1-\tau_2)\mathrm{d}\tau_1\mathrm{d}\tau_2\\&=g(\tau)*R_{\mathrm{FF}}(\tau)*g(-\tau)\end{aligned} \qquad (3.119\mathrm{a}).$$

Corresponding to zero mean value of the forcing function the response mean value is also assumed to be zero, then Eq. (3.119a) yields at zero delay (τ=0) the *response mean-square value*

$$R_{ss}(0) = \overline{s^2(t)} \quad (3.120a)$$

defined by the mean of the squared displacement response values over a sufficiently large averaging time T.

The response autocorrelation function, Eq. (3.119a), can be converted to

$$R_{ss}(\tau) = \int_{-\infty}^{+\infty} R_{FF}(\tau - t)\mathrm{d}t \int_{0^-}^{\infty} g(\tau_1)g(t+\tau_1)\mathrm{d}\tau_1 \quad (3.121a)$$

wherein the second term represents an autocorrelation function at the deterministic, aperiodic time characteristic (weighting function) $g(t)$. This property used in communication and control sciences is termed
the *filter correlation function*

$$R_{gg}(t) = \int_{0^-}^{\infty} g(\tau_1)g(t+\tau_1)\mathrm{d}\tau_1 = g(-t) \underset{0^-}{\overset{\infty}{*}} g(t) \quad (3.122),$$

and corresponds in the ω-domain with the square of the frequency-response function. In spectral density analysis that property is called
the *square of transmission characteristic* or of *frequency-response characteristic*

$$G(\omega)G^*(\omega) = |G(\omega)|^2 = \int_{-\infty}^{+\infty} \mathrm{e}^{-\mathrm{j}\omega t} R_{gg}(t)\mathrm{d}t \quad (3.123a)$$

to turn out the filtering property of a system.
Being a *real system parameter* and an even function of ω the Fourier cosine transformation is valid

$$|G(\omega)|^2 = 2\int_0^{\infty} [\cos\omega t] R_{gg}(t)\mathrm{d}t \quad (3.123b).$$

Writing the response autocorrelation function as a convolution of the filter correlation function with the excitation autocorrelation function

$$R_{ss}(\tau) = \int_{-\infty}^{+\infty} R_{gg}(t)R_{FF}(\tau - t)\mathrm{d}t = R_{gg}(\tau) \underset{-\infty}{\overset{+\infty}{*}} R_{FF}(\tau) \quad (3.121b)$$

a form of solution in the τ-domain can be found that is equivalent to Eq. (3.119a).

The repeated convolution product of excitation autocorrelation function $R_{FF}(\tau)$ and time-response characteristic $g(\tau)$ as of its reverse $g(-\tau)$ corresponds with the multiple algebraic product of excitation auto-spectrum $S_{FF}(\tau)$ and frequency-response characteristic $G(\omega)$ as of its complex conjugate $G^*(\omega)$. This transformed solution represented by
the *response power spectral density*

$$S_{ss}(\omega) = G(\omega)S_{FF}(\omega)G^*(\omega) = |G(\omega)|^2 S_{FF}(\omega) \quad (3.119b)$$

constitutes together with Eq. (3.119a) a transform pair.

In spectral analysis the modulus of the frequency-response function $G(\omega)$ results from relating both of the auto-spectral densities to be determined by data processing:

$$\frac{S_{ss}(\omega)}{S_{FF}(\omega)} = |G(\omega)|^2 = A^2(\omega) \tag{3.116d}$$

where the *amplitude (frequency-)response* (gain response) is given by

$$|G(\omega)| = A(\omega) = \left| \sqrt{\frac{S_{ss}(\omega)}{S_{FF}(\omega)}} \right| \tag{3.116e}$$

usually represented in the logarithmic form

$$\lg A(\omega) = (1/2)\left[\lg S_{ss}(\omega) - \lg S_{FF}(\omega)\right] \tag{3.116f}$$

Applying the inverse Fourier transform, Eq. (3.69a), to Eq. (3.119b), alternatively can be found the response autocorrelation function by evaluating the Fourier integral

$$R_{ss}(\tau) = \frac{1}{2\pi}\int_{-\infty}^{+\infty} e^{j\tau\omega} S_{FF}(\omega)d\omega = \frac{1}{2\pi}\int_{-\infty}^{+\infty} e^{-j\tau\omega} |G(\omega)|^2 S_{FF}(\omega)d\omega \tag{3.124a}$$

or due to the Fourier transform $S_{FF}(\omega)$ being a real and even function of ω by use of the inverse Fourier cosine transformation

$$R_{ss}(\tau) = \frac{1}{\pi}\int_{0}^{\infty}[\cos\tau\omega]S_{FF}(\omega)d\omega = \frac{1}{\pi}\int_{0}^{\infty}[\cos\tau\omega]|G(\omega)|^2 S_{FF}(\omega)d\omega \tag{3.124b}$$

In case of zero delay $\tau = 0$ it follows the *response mean-square value*

$$\overline{s^2(t)} = \sigma_s^2 = R_{ss}(0) = \frac{1}{2\pi}\int_{-\infty}^{+\infty}|G(\omega)|^2 S_{FF}(\omega)d\omega = \frac{1}{\pi}\int_{0}^{\infty}|G(\omega)|^2 S_{FF}(\omega)d\omega \tag{3.120b}$$

This points out a way of data processing for obtaining the response mean-square value by the known frequency or spectral properties of the excitation process.. If the excitation random process is ergodic, it follows that the response random process is also ergodic. The appertaining statistical average equal to the time average $\overline{s^2(t)}$ is associated with the time history of the random function $s(t)$ (actual displacement response).

For a normal or Gaussian process it is sufficient to know the statistical average $\overline{s^2(t)}$, to determine the response probability density function associated with the distribution of expected displacement values.

3.2.6 System Response to Random Excitation. *White Noise*

Broad-band, stationary random variables with Gaussian distribution are commonly termed *white noise*. Ideally, white noise has equal energy for any frequency band of constant width (or per unit bandwidth) over the spectrum to infinity. Though being used as nominal random excitation, in data processing white noise is band-limited to the frequency range of interest. To evaluate the statistical response properties of *white random vibration* an "idealized random excitation" may be applied as reference input to the forced mass-damper spring system introduced in 3.1.

The frequency-response characteristic $G(\omega)$ generally defined in random vibration by the ratio of power spectral densities, Eq. (3.115), gets a distinct signification in the particular case of white random vibration.
The *frequency-response function* is identified with

$$G(\omega) = \frac{1}{S_0} S_{\mathrm{Fs}}(\omega) \tag{3.125a}$$

being in correspondence with the *unit pulse response* (weighting function)

$$g(\tau) = \frac{1}{S_0} R_{\mathrm{Fs}}(\tau) \tag{3.125b}$$

Both response characteristics constitute a transform pair being of practical use in vibration data analysis. As an alternative to the idealized pulse-type excitation (unit pulse) a random excitation with flat spectrum serves as reference input (ideal white noise) to check the response of mechanical structures.

Inversion Formula Approach

By use of the known *auto-spectrum of excitation* specified by white noise and being independent of frequency

$$S_{\mathrm{FF}}(\omega) = S_{\mathrm{FF,w}}(\omega) = S_0 \tag{3.126}$$

the transformed solution in the ω-domain is given by Eq. (3.119b) as the *response auto-spectrum* to white noise excitation

$$S_{\mathrm{ss}}(\omega) = S_{\mathrm{ss,w}}(\omega) = \left|G(\omega)^2\right| S_0 \tag{3.127a}$$

in *normalized form*

$$\frac{S_{\mathrm{ss}}(\omega)}{S_0/k^2} = \sigma_{\mathrm{ss,w}}(\eta) = \left|\frac{G(\omega)}{G(0)}\right|^2 = A^2(\eta)/A^2(0) \tag{3.127b}$$

by relating S_{ss} to $\left(S_{\mathrm{FF,w}} \cdot |G(0)|^2\right) = \left[S_0 A^2(0)\right] = \left(S_0/k^2\right)$.

The square of transmission characteristic can be written in an explicit form by getting the square of the known complex system parameter, Eq. (3.77a):

$$|G(\omega)|^2 = \frac{1}{(k-m\omega^2)^2 + (c\omega)^2} = \frac{1}{m^2}\frac{1}{(\omega_0^2-\omega^2)^2+(2\delta\omega)^2} = \frac{1}{k^2}\frac{1}{(1-\eta^2)^2+(2\zeta\eta)^2} \tag{3.126}$$

Applying the inverse Fourier transform, Eq. (3.69a), to the response transform $S_{\mathrm{ss}}(\omega)$ $\left(\text{or } G_{\mathrm{ss}}(\omega)\right)$ by substituting Eqs. (3.125a), (3.127a) the inversion formula of the complex Fourier transform, Eq. (3.124a), assumes the explicit form

$$R_{\mathrm{ss}}(\tau) = R_{\mathrm{ss,w}}(\tau) = \frac{S_0}{k^2}\frac{1}{2\pi}\int_{-\infty}^{+\infty} \frac{1}{\left(\omega_0^2-\omega^2\right)^2+(2\delta\omega)^2} e^{\mathrm{j}\tau\omega}\, \mathrm{d}\omega \tag{3.128a}$$

To evaluate the Fourier integral the previously mentioned integral theorems of the theory of analytic functions will be utilized. By use of
Cauchy's residue theorem

$$2\pi R_{\mathrm{ss}}(\tau) = S_0 \int_{-\infty}^{+\infty} |G(\omega)|^2 e^{\mathrm{j}\tau\omega}\, \mathrm{d}\omega = 2\pi S_0 \mathrm{j} \sum_{k=1}^{n=2} \operatorname*{Res}_{z_k}\left[|G(z)|^2 e^{\mathrm{j}\tau z}\right] \tag{3.128b}$$

the corresponding contour integral enclosing two singularities (system-poles) in the upper half-plane will be evaluated as shown in 3.2.4, [37], [40].

Convolution Approach

Applying the double convolution integral, Eq. (3.119a), to a white noise excitation the autocorrelation of which is S_0 times the unit pulse (Dirac or delta „function") the convolution theorem must be extended to generalized functions (distributions) [38], [39].

Due to the fact that the convolution of a generalized function f with a delta functional δ yields f

$$f * \delta = f \tag{3.129a}$$

the generalized convolution product thus results in

$$\begin{aligned} R_{ss}(\tau) = R_{ss,w}(\tau) &= S_0 \int_{-\infty}^{+\infty} g(\tau_1) \int_{0^-}^{\infty} g(\tau_2)\delta(\tau+\tau_1-\tau_2)\mathrm{d}\tau_1 \mathrm{d}\tau_2 \\ &= S_0 \int_{0^-}^{\infty} g(t)g(\tau+t)\mathrm{d}t = S_0 g(-\tau) \underset{0^-}{\overset{\infty}{*}} g(\tau) \end{aligned} \tag{3.129b}$$

The same result may be obtained by use of Eq. (3.121b)

$$R_{ss}(\tau) = R_{ss,w}(\tau) = S_0 \int_{-\infty}^{+\infty} R_{gg}(t)\delta(\tau - t)\,\mathrm{d}t \tag{3.129c}$$

by the substitution: $\tau - t = u$, and Eq. (3.129a), it follows

$$\int_{-\infty}^{+\infty} \delta(u) R_{gg}(\tau - u)\,\mathrm{d}u = R_{gg}(\tau)\,.$$

According to Eq. (3.122) the effect to white noise excitation in the τ-domain is given by S_0 times the *filter correlation function*

$$R_{ss,w}(\tau) = S_0 R_{gg}(\tau) = S_0 g(-\tau) \underset{0^-}{\overset{\infty}{*}} g(\tau) \tag{3.129d}$$

Hence, both of the outlined approaches reduce to evaluate either the inverse transform of the square of transmission characteristic by following the residue theorem, Eq. (3.98), or to evaluate the filter correlation function by applying the integration by parts comparatively with Eq. (3.107), though to be done twice, so that finally results:

the *response autocorrelation function* at $s(t)$ *of the underdamped system*

$$R_{ss}(\tau) = \frac{1}{mk}\frac{S_0}{4\delta} \mathrm{e}^{-\delta|\tau|}\left[\frac{\delta}{\omega_d}\sin(\omega_d|\tau|) + \cos(\omega_d|\tau|)\right] \tag{3.130a}$$

in *normalized form*

$$\begin{aligned} \frac{R_{ss}(\tau)}{S_0/\left(4k\sqrt{mk}\right)} = \varphi_{ss,w}(\omega_0|\tau|) = \frac{1}{\zeta}\mathrm{e}^{-\zeta\omega_0|\tau|}\Bigg[&\frac{\zeta}{\sqrt{1-\zeta^2}}\sin\left(1-\zeta^2\right)\omega_0|\tau| \\ &- \cos\left(\sqrt{1-\zeta^2}\,\omega_0|\tau|\right)\Bigg] \end{aligned} \tag{3.130b}$$

by relating R_{ss} to $[\zeta R_{ss}(0)]=(\zeta\sigma_s^2)=[S_0/(4k\sqrt{mk})]$.
In case of zero delay $\tau = 0$ it follows
the *response mean-square value* to white noise excitation

$$R_{ss}(0)=\overline{s^2(t)}=\sigma_s^2=\frac{1}{mk}\frac{S_0}{4\delta} \qquad (3.131a).$$

Statistical System Parameters. The response calculation with the Fourier transform allows to describe the process of white random vibration by its statistical parameters which are associated with the probability distribution of random vibration magnitudes, $s = s_w$, of the displacement response $s(t)$. The response process being ergodic, the ensemble averages equal the corresponding time averages.Taking in random vibration zero-mean value (of excitation accordingly of response) for granted, the response mean-square value, Eq. (3.129), is thus interchangeable with
the *response variance*

$$\sigma_s^2 = \overline{s^2(t)} = \frac{1}{mk}\frac{S_0}{4\delta} = \frac{S_0}{4k\sqrt{mk}}\frac{1}{\zeta} \qquad (3.131b).$$

The square root of the response variance is
the *response standard deviation*

$$\sigma_s = \left|\sqrt{\overline{s^2(t)}}\right| = s_{\text{eff}} = \frac{1}{2}\left|\sqrt{\frac{S_0}{mk\delta}}\right| = \frac{1}{2}\left|\sqrt{\frac{S_0}{k\sqrt{mk}}}\right|\frac{1}{\left|\sqrt{\zeta}\right|} \qquad (3.132)$$

being identical with the square root of the time average of squared vibration magnitudes (displacements) s, that is termed the root-mean-square value (r.m.s. value).

The response process being normal, or Gaussian, the response variance, Eq. (3.130), equals the second moment of the probability distribution. The response variance thus comes up to determine
the *response probability density function*

$$p(s) = \frac{1}{\sqrt{2\pi}\,\sigma_s}\,e^{-\frac{1}{2\sigma_s^2}s^2} = \left|\sqrt{\frac{2}{\pi}\frac{mk\delta}{S_0}}\right|\,e^{-2\frac{mk\delta}{S_0}s^2} \qquad (3.133a),$$

in *normalized form*

$$p(\rho) = \frac{1}{\sqrt{2\pi}\lambda_\rho}\,e^{-\frac{1}{2\lambda_\rho^2}\rho^2} = \left|\frac{1}{\sqrt{2\pi}}\sqrt{\zeta}\right|\,e^{-\frac{\zeta}{2}\rho^2} \qquad (3.133b).$$

The normalization is carried out by introducing
the *non-dimensional (displacement) vibration magnitude* $\rho = \rho_w$

$$\rho_w = \frac{s_w}{\frac{1}{2}\left|\sqrt{\frac{S_0}{k\sqrt{mk}}}\right|} \qquad (3.134a)$$

where $s = s_w$ is the response random variable related to a constant parameter ratio, and by determining the following statistical system parameters (*statistical response specifications*):
the *non-dimensional response standard deviation* $\lambda_\rho = \lambda_{\rho w}$

$$\lambda_{\rho w} = \left|\sqrt{\overline{\rho_w^2(t)}}\right| = \rho_{w,eff} = \frac{\sigma_{sw}}{\frac{1}{2}\left|\sqrt{\frac{S_0}{k\sqrt{mk}}}\right|} = \frac{1}{\left|\sqrt{\zeta}\right|} = \frac{1}{\zeta}\left|\sqrt{\zeta}\right| \qquad (3.134b)$$

interchangeable with the non-dimensional response r.m.s. value $\left|\sqrt{\overline{\rho_w^2(t)}}\right|$, also called effective value $\rho_{w,eff}$, furthermore

the *non-dimensional response median* $m_\rho = m_{\rho w}$

$$m_{\rho w} = p(\rho_w = 0) = \frac{1}{\left|\sqrt{2\pi}\right|}\left|\sqrt{\zeta}\right| \qquad (3.134c).$$

Response Data Plotting. The *statistical response curves and specifications*, Eqs. (3.130b) to (3.133b), as the result of response calculation for an applied random process of white noise excitation – approximately of an equivalent broad-band random excitation – are shown in Fig. A.22 to A.24 of the Appendix A.

The influence of the system parameter damping ratio ζ on the response autocorrelation function is represented for a mechanical system being subjected to a white noise excitation, and is portrayed for selected fractions of critical damping. The statistical parameters of random vibration are given for the underdamped system by the set of decaying sinusoidal *autocorrelation curves*, Eq. (3.130b), plotted in Fig. A.22 versus the non-dimensional time-delay $\omega_0|\tau|$ over the positive half plane. This type of correlation curve points at a narrow-band random vibration. For a system lightly damped less than $\zeta \ll 1$ the autocorrelation curve decreases rapidly as τ increases, because there is little correlation between the instantaneous values of $s(t)$ at different times.

The *response mean-square values* $\overline{s^2(t)}$ being associated with representative function of the actual displacement responses $s(t)$ for different damping ratios ζ are given by $[\zeta R_{ss}(0)]$ times the maximum-values $(1/\zeta)$ at zero delay $\tau = 0$. From that distinct value of the response autocorrelation function $\varphi_{ss,w}(\omega_0\tau)$, Eq. (3.130b), results that the level of the responding time average $\overline{s^2(t)}$, Eq. (3.131b), is a reciprocal of the damping ratio ζ.

The frequency or spectral parameters of representative random vibration functions $s(t)$ are given by the set of *power spectral density curves*, Eq. (3.127b), plotted in Fig. A.23 versus the non-dimensional angular frequency ω/ω_0 (frequency ratio η), which are directly proportional to the square of the transmission characteristics $|G(\omega)|^2 = A^2(\omega)$ for different damping ratios ζ. For a system lightly damped less than $\zeta \ll 1$ this characteristic is sharply peaked by resonance at $\eta = 1$, so that a selective frequency band at resonance will be amplified and thus provide filter action. Evidently the oscillator acts as a *narrow-band filter*. Nevertheless, the transmission characteristic of mechanical filters employing resonances depends upon the magnitude of damping. If in contrast the actual damping amounts to larger magnitudes, by example to the particular fraction of critical damping $\zeta = (1/2)\sqrt{2}$, the oscillator approaches an *ideal bandpass filter* the response of which is a constant auto-spectrum within an effective bandwidth and zero elsewhere. This effect of damping on transmission characteristic is used in measurements to reduce distortion of transducers.

The *response mean-square values* $\overline{s^2(t)}$ are given by $\left(S_{FF,w}|F(0)|^2/\pi\right)$ times the area under the $A^2(\eta)$ -curves over the positive half plane, Eqs. (3.120b), (3.127b).

The statistical parameters of random vibration determine a set of *probability density distribution curves* of Gaussian type, also called *bell-shaped curves*, Eq. (3.133b), plotted in Fig. A.24 versus the continuous random variable ρ_w (non-dimensional vibration magnitude of displacement response *s*), Eq. (3.134a), for different damping ratios ζ. Concerning a system lightly damped less than $\zeta \ll 1$ the intensifying effect on magnitudes at resonance causes a large response standard deviation σ_s, so that the displacement responses are spread over a wide magnitude range. From the statistical parameter σ_s defining the response probability density $p(\rho)$, Eq. (3.133b), results that the level of the *response root-mean-square value* $\sqrt{s^2(t)}$ or effective value s_{eff} – associated with the concept of average power in network theory –, Eq. (3.132), corresponds with the reciprocal root of the damping ratio ζ.

By use of probability density curves the probability that a given magnitude of random variable will not be exceeded can be predicted.Thus, the probability that the expected instantaneous vibration magnitude $\rho = \rho_w$ will be within a certain range of magnitude values is equal to the integral of the response probability density function $p(\rho)$ integrated over that range $[\rho_{w1}, \rho_{w2}]$:

$$\Pr[\rho_{w1} \le \rho \le \rho_{w2}] = \int_{\rho_{w1}}^{\rho_{w2}} p(\rho)\mathrm{d}\rho = P(\rho_{w2}) - P(\rho_{w1}),$$

where $P(\rho)$ is the *cumulative probability distribution function*

$$P(\rho) = \int_{-\infty}^{\rho_w} p(\rho)\mathrm{d}\rho .$$

In random data processing the given magnitude range is commonly specified by the double r.m.s. value defining the total range of 68-percent defined value to the two- or threefold of that specified range termed the total range of 95-, or of 99-percent probability, T_{95}, respectively T_{99} (three-sigma limits).

Verifying a selected fraction of damping, e.g. $\zeta_1 = (1/2)\sqrt{2}$, a non-dimensional response r.m.s. value of

$$\lambda_{\rho w1} = \left|\sqrt{\rho_{w1}^2(t)}\right| = \left|\sqrt[4]{2}\right| = 1{,}1892$$

results, and the probability that the instantaneous vibration magnitudes fall under the specified range [– 1,1892, + 1,1892] definitely amounts to

$$\Pr[-1{,}1892 \le \rho \le +1{,}1892] = \int_{-1{,}1892}^{+1{,}1892} p(\rho)\mathrm{d}\rho = P(+1{,}1892) - P(-1{,}1892) = 0{,}683 .$$

According to the extended total ranges defined by the outlined multiples of the double r.m.s. value an appropriate probability of 0,954, respectively of 0,997 can be calculated.

In case of a lightly damped system, e.g. for $\zeta = 0{,}05$, the recording of displacement magnitudes presupposing the 68-percent probability requires a magnitude range given by the response r.m.s. value of

$$\lambda_{\rho w2} = \left|\sqrt{\rho_{w2}^2(t)}\right| = 10/\left|\sqrt{5}\right| = 4{,}47213$$

which is the 3,76 fold to $\lambda_{\rho w1}$ for $\zeta_1 = (1/2)\sqrt{2}$.

3.2.7
Transient Vibration. *The Laplace Integral*

Transient vibrations include motions of a system which are neither steady-state (periodic) nor random vibrations. This category is related to non-periodic vibrations, and the term transient is basically associated with mechanical shock. Shock motions of finite duration (pulse-type excitations) are presented in 3.2.2. Supplementary to this previously treated type of shock motions non-periodic vibrations are of interest which are caused by forcing functions suddenly applied (at zero time) and of unlimited duration (step-type excitations).

The Laplace Integral

The Laplace transformation corresponds with the Fourier transformation, Eqs. (3.69a, b), operating on the real variable ω associated with the angular frequency, to the effect that a complex variable p comes into operation.

Remark on Symbol p. In particular context of mechanical vibrations the symbol p may be given for this complex quantity though the symbol s is in use as Laplace variable. The main symbol s is recommended for displacement as the prime motion variable quantity in mechanics, therefore it will be justified to give the "reserve symbol" p for the complex frequency or Laplace domain variable.

The representation of the function $F(t)$ by an integral of the following form defines the

inverse Laplace transform; Laplace integral

$$\mathrm{L}^{-1}\left[\underline{F}(p)\right] = F(t) = \frac{1}{2\pi \mathrm{j}} \int_{\sigma_1 - \mathrm{j}\infty}^{\sigma_1 + \mathrm{j}\infty} \underline{F}(p) \mathrm{e}^{tp} \mathrm{d}p \qquad (3.135\mathrm{a})$$

where $\sigma_1 \geq c$, the abscissa of convergence of $F(p)$, and the inverse transform is indicated by the symbol L^{-1}. The transformation of the function $F(t)$ into a function of the complex variable p defines the

Laplace transform

$$\mathrm{L}\left[F(t)\right] = \underline{F}(p) = \int_0^{\infty} F(t) \mathrm{e}^{-pt} \mathrm{d}t \qquad (3.135\mathrm{b})$$

being indicated by the symbol L.

The Laplace transform $\underline{F}(p)$ is a function continuously dependent on the *complex angular frequency p*. This complex variable, also termed *complex pulsatance*, is sometimes denoted the "complex frequency of the drive" with reference to the corresponding special type of forcing function, 3.2.10. The L-transform can be interpreted as the *generalized spectral density* of the function $F(t)$.

Definition of the Laplace Integral Equation. The integral operator, in this case the Laplace transform, Eq. (3.135b), is defined by:

1. The *kernel of the transformation*, here given by $K(t, p) = \mathrm{e}^{-pt}$ being a function of the time t as the original variable, and of the "complex (angular) frequency" $p = \sigma + \mathrm{j}\omega$ as the subsidiary variable which in general is a complex quantity. The

presented kernel signifies an analytic continuation of the kernel of the Fourier integral $K(t, j\omega)$ through multiplying it by an exponential function that is defined by the real part of the complex frequency (growth coefficient) Re $p = \sigma$:

$$e^{-\sigma t} e^{-j\omega t} = e^{-(\sigma + j\omega)t} = e^{-pt} \tag{3.136}$$

This exponential term in the Laplace kernel acts as an *exponentially decaying (or "convergence") factor* inducing the convergence of time functions largely of those types which fail to satisfy the convergence criterion of the Fourier transform, Eq. (3.70). However, this only proves true for positive values of time as to be gathered from Fig. 3.12;

2. The *limits of integration*, here $(0, \infty)$ ranging from the origin to infinity (one-sided transformation), which coincide with the definition range of time functions (original functions) defined only for positive values of time: $t \geqq 0^-$, Eq. (3.135c), (causal functions);

3. The *class of time functions*, here of forcing functions $f(t)$, to which the integral transformation given by Eqs. (3.135a,b) is true (existence theorem) [37], [44], [47], [48].

As the integral transformation can be considered as a functional mapping, one says that the function $F(t)$ is being mapped (imaged) from the original (time or t-) domain onto the corresponding subsidiary (complex frequency or p-, elsewhere preferably s-)domain.

The transform $\underline{F}(p)$ (or $\underline{F}(s)$) and its inverse $F(t)$ constitute a *Laplace transform pair*. The corresponding functional relation is called a *correspondence*

$$F(t) \overset{L}{\circ\!\!-\!\!\!-\!\!\bullet} \underline{F}(p) \tag{3.135c}$$

being indicated by the correspondence sign and the operator symbol L above it if need be, see 3.2.2, Eq. (3.69c).

Restriction of convergence (existence theorem). The Laplace transform exists for time functions being cut off for times prior to $t = 0^-$

$$F(t) \equiv 0 \quad \text{for} \quad t < 0^- \tag{3.135d),}$$

but then under the more temperate restrictions as follows. Assuming the function $F(t)$ to be piecewise continuous the Laplace transform exists if

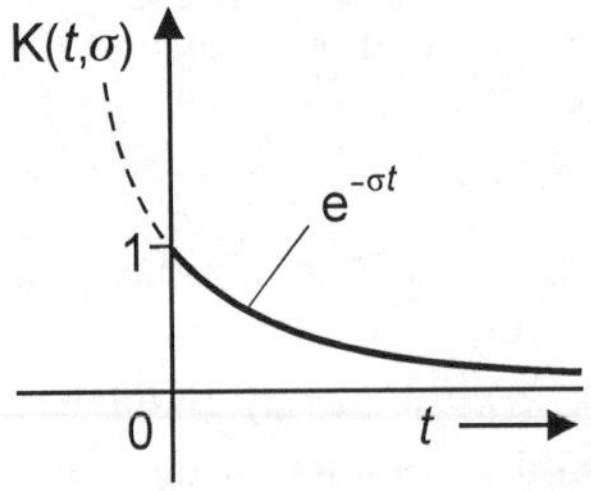

Fig. 3.12. Decaying exponential (convergence factor) in the kernel of the Laplace integral.

the *absolute convergence criterion* is satisfied

$$\int_0^\infty \left| e^{-p_0 t} F(t) \right| dt = \int_0^\infty e^{-\mathrm{Re}\, p_0 t} \left| F(t) \right| dt < \infty \qquad (3.136a),$$

i.e., whenever the Laplace integral, Eq. (3.135b), exists for some number $p = p_0$, then there exists a closed right-half plane of absolute convergence

$$\mathrm{Re}\, p = \sigma \geqq \mathrm{Re}\, p_0 = \sigma_0 = c \qquad (3.136b)$$

wherein $F(p)$ is analytic (or regular), Fig. 3.13.

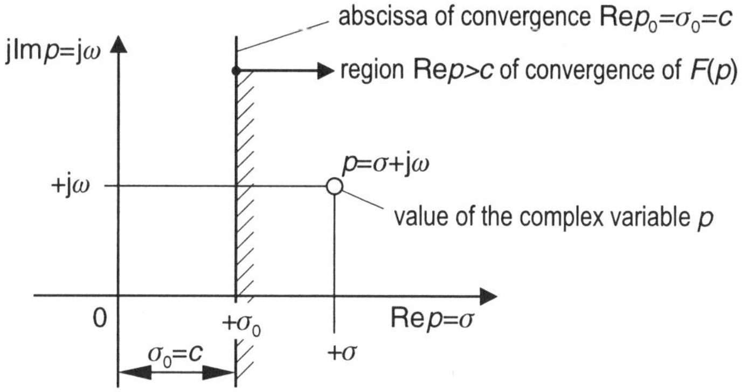

Fig. 3.13. Complex p-plane with region and abscissa of convergence

Condition (3.136a) implies that $F(t)$ though becoming infinitely large as $t \to \infty$, $|F(t)|$ does not "grow" more rapidly than a multiple M of some exponential function of t (or the product $e^{p_0 t}|F(t)|$ is bounded for large values of t):

$$e^{-p_0 t}|F(t)| < M \quad or \quad |F(t)| < M e^{p_0 t} \qquad (3.136c).$$

Such a function $F(t)$ is said to be of *exponential order* and is denoted $F(t) = O(e^{p_0 t})$. Compared to the Fourier transformation these restrictions are weak. Thus, the *class of Laplace-transformable functions* includes most of the time-varying functions occuring in practice, e.g., the steady sinusoidal excitation, as far as the varying values of time are only positive (causal functions). Subsequently for analysing vibrations caused by non-periodic, or rather transient excitations it becomes almost unnecessary to extend theorems on the Laplace transformation to the theory of generalized functions (distributions).

Transient Harmonic Excitation (sinusoidal step excitation). The forcing function of a harmonic excitation suddenly applied at $t = 0$, thus being a non-periodic sinusoid section abbreviately termed a *transient harmonic excitation*, Fig. 3.14, is defined as

$$F_\omega(t) = \begin{cases} 0 & \text{for } t < 0 \\ \hat{F}\cos(\omega_f t - \varphi_{0_F}) & \text{for } t > 0 \end{cases} \qquad (3.138a)$$

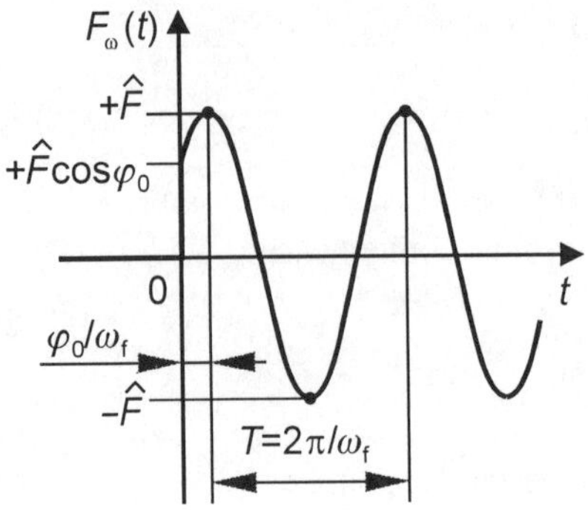

Fig. 3.14. Transient harmonic excitation (sinusoidal step)

with the parameters characterizing a sinusoidal step:
force-excitation amplitude $\hat{F}$, forcing (angular) frequency ω_f and force-excitation initial phase φ_{0F} not being omitted.

Evaluation of the Laplace transform applying Eq. (3.135b) to the transient harmonic excitation defined by Eq. (3.138a) is facilitated by use of Euler's formula, Eq. (3.49), thus substituting one-half the *sum of the complex constituents* $\hat{F}e^{-j\varphi_{0F}}e^{j\omega_f t}$ and $\hat{F}e^{j\varphi_{0F}}e^{-j\omega_f t}$ for the simple harmonic excitation:

$$\underline{F_\omega}(p) = \mathrm{L}[F_\omega(t)] = \frac{1}{2}\hat{F}\int_0^\infty \left[e^{-j\varphi_{0F}}\,e^{-(p-j\omega_f)t} + e^{+j\varphi_{0F}}\,e^{-(p+j\omega_f)t}\right]dt$$

$$= \frac{1}{2}\hat{F}\left\{e^{-j\varphi_{0F}}\frac{1}{-(p-j\omega_f)}\underbrace{e^{-(p-j\omega_f)t}\Big|_0^\infty}_{=-1} + e^{j\varphi_{0F}}\frac{1}{-(p+j\omega_f)}\underbrace{e^{-(p+j\omega_f)t}\Big|_0^\infty}_{=-1}\right\}$$

$$= \frac{1}{2}\hat{F}\left[\frac{(p+j\omega_f)e^{-j\varphi_{0F}} + (p-j\omega_f)e^{j\varphi_{0F}}}{p^2+\omega_f^2}\right] = \hat{F}\frac{p\cos\varphi_{0F} + \omega_f\sin\varphi_{0F}}{p^2+\omega_f^2}$$

(3.138b).

The direct calculation of Laplace transforms bases on the evaluation of the definition integral, Eq. (3.135b), which is an improper definite, and in particular a real integral. Definitively the variable of integration is the time t being a real variable, since the subsidiary variable p indicates only a constant being connected with the kernel of the definition integral.

In addition to evaluating procedure certain *properties of the L-transformation* are used, in consequence of which transforms of many functions can be obtained in a simple manner. The most important property, already used in above calculation, implies the *combination of functions by a linear operation* (linear property).

Moreover theorems on the transform of functions are associated with the direct calculation, perhaps that one on the *transform of delayed functions* (second or time shifting theorem). In determining the transform of the excitation that theorem provides for the shifting of functions, e.g., of the forcing function or of their component parts, on the t-axis in the positive direction. Using transform pairs of elementary functions and their known transforms, being listed in *transform tables*, and if so applying delayed-function transforms, the Laplace transformation may be carried out with moderate effort to definable time-varying functions as to be gathered from specialized references [37], [42], [44], [45], [111].

3.2.8 Laplace Transform Method. *Transfer-function Analysis*

The Laplace integral introduced to gain integral transforms of time-unlimited forcing functions (step-type excitations) proves its usefulness first of all for solving linear differential equations arising in engineering mathematics. Vibration analysis likewise takes advantage of this operational method to determine analytic solutions of equations of motion. The Laplace transform method appears well suited to *transient excitations*. The response calculating for this special type of excitation usually performed in the time (or t-domain), will be carried out in a generalized frequency domain, known as the Laplace (or p-) domain. Nevertheless, the corresponding operational method performing a response calculation continuously in the time (or t-domain) will be demonstrated by which a comparison with the classical method, exemplified in 3.1, becomes possible.

Starting algebraic equation calculus from time-domain representation by the governing differential equation of motion one must be able to express the transform of the derivatives in terms of the transform of the system variable, e.g., of the displacement $s(t)$ as the pertinent motion variable quantity.

Laplace transform of derivatives (differentiation theorem). Let the Laplace transform of a function $s(t)$ exist and let $\dot{s}(t)$ be continuous for all $t \geq 0$, then if $(\mathrm{d}/\mathrm{d}t)s(t)$ exists

$$\mathrm{L}\left[\frac{\mathrm{d}}{\mathrm{d}t}s(t)\right] = p\mathrm{L}[s(t)]$$

denoted as a transform pair $\dot{s}(t)$; $p\underline{s}(p)$, $s(0)$ symbolically by the correspondence

$$\dot{s}(t) \;\circ\!\!\xrightarrow{\;\mathrm{L}\;}\!\!\bullet\; p\underline{s}(p) - s(0) \qquad (3.139\mathrm{a}).$$

This theorem is proved through an integration by parts

$$\mathrm{L}[\dot{s}(t)] = \int_0^\infty \mathrm{e}^{-\mathrm{pt}}\dot{s}(t)\,\mathrm{d}t = \mathrm{e}^{-\mathrm{pt}}s(t)\Big|_0^\infty + p\int_0^\infty \mathrm{e}^{-\mathrm{pt}}s(t)\,\mathrm{d}t$$

$$= -s(0^+) + p\underline{s}(p),$$

wherein the initial value $s(0^+)$ is to be interpreted as a right-hand limit. This corresponds to the definition integral taking the lower limit as 0^+:

$$\mathrm{L}[s(t)] = \lim_{\varepsilon\to 0}\int_\varepsilon^\infty s(t)\,\mathrm{e}^{-\mathrm{pt}}\mathrm{d}t \quad , \varepsilon > 0 .$$

To exclude a discontinuity at $t = 0$, the origin will be approached on principle from the right, thus defining in the following

$$s(0) = s(0^+) .$$

Repeating the procedure $n$times the transform of the nth derivative will be obtained by induction

$$s^{(n)}(t) \circ\!\!\xrightarrow{\;L\;}\!\!\bullet\; p^{n}\underline{s}(p) - s(0)p^{n-1} - \dot{s}(0)p^{n-2} - \cdots - s^{(n-1)}(0) \qquad (3.139b)$$

with the initial values:

$$s^{(n-1)}(0),\cdots,\dot{s}(0),s(0)$$

(as long as $s^{(n)}(t)$ exists, in addition $s(t)$ and all its derivatives through the $(n-1)st$ derivative are continuous).

Laplace Transform Method

The solution of a differential equation by integral transformation, here by the Laplace transformation, will be obtained by general *steps* already pointed out in 3.2.3, just so visualized as the scheme of Fig. 3.9.

The Laplace transform method provides a most convenient means of solving linear differential equations of motion for lumped parameter systems. It can be used successfully also in the case of linear partial differential equations governing continuous systems as well as for equivalent systems in the state variable form (state space method). Both of the advanced methods are treated by more specialized references [15], [30], [35].

Although being similar to Heaviside's operational calculus the Laplace transform method is mathematically rigorous. In contrast to the classical method that requires the fitting of the general solution to the initial or boundary conditions, these conditions are automatically incorporated in the transformed solution for any arbitrary excitation. Even contrary to the Fourier transform method the first step oh which includes the implying of initial or boundary conditions, the Laplace transform method yields the *total or complete response at the first attempt.*

In general the complete response is a superposition of the system response to both the excitation and the initial conditions. In electrical circuit theory one solution part is distinguished from the other by the terms *zero-state response* and *zero-input response.* A system (or a circuit) is said to be in the zero state if all the initial conditions are zero. The response of a system starting from the zero state is due exclusively to the excitation (or input). In the other case, looking at a system without an applied excitation, the response is a function of the initial state defined by the non-zero initial values described at $t = 0$. Referred to mechanical applications the system response to initial conditions may be termed *zero-driving response* replacing the notion zero input.

The *last step* consists in performing the inverse transformation of the transformed solution (p-domain solution), here implying an evaluation of the Laplace integral, Eq. (3.135a), yet for the function $\underline{s}(p)$, since an original solution (t-domain solution) is required. The case may be whenever the time history of the pertinent motion response, e.g., the *actual displacement response s(t)*, is of essential interest (real-time or *time-response analysis*).

In vibration data analysis that applies especially to transient vibrations caused by shock loadings where a mechanical source is suddenly turned on and remains stationary thereafter. It will be exemplified later on by the time-response calculation referred to a harmonic shock excitation (sinusoidal step excitation) that the Laplace transform method is based on an integral transformation being well adapted to that type of a shock motion (step-type excitation).

Nevertheless, for a lot of problems concerning system analysis and design the system behaviour can be determined merely by examining the transform of the unknown response, e.g., the *displacement-response transform* $\underline{s}(p)$, without actually carrying out the inverse transformation (*transfer-function analysis*).

If still being of essential interest the inverse transformation possibly involves a rather complicated step of calculation that may be facilitated by the use of tables of transform pairs. Confining oneself to time-response calculations for idealized types of transient excitation acting on lumped parameter systems the evaluation of the inverse transform, here of the Laplace integral, reduces to some approved rules of mathematics being of use for practical applications, see 3.2.9, [37], [40], [41], [45], [47], [48].

Response to Harmonic Shock Excitation. The transformation of the differential equation governing a forced mass-damper-spring system, Eq. (3.1), under a transient harmonic excitation (sinusoidal step excitation), Eq. (3.138a),

$$m\ddot{s} + c\dot{s} + ks = F(t) = F_{\omega}(t) \tag{3.140}$$

by applying the Laplace transformation to both sides of the differential equation

$$m\mathrm{L}[\ddot{\mathrm{s}}] + c\mathrm{L}[\dot{\mathrm{s}}] + k\mathrm{L}[\mathrm{s}] = \mathrm{L}[\mathrm{F}_{\omega}(t)] \tag{3.141a}$$

and using the differentiation theorem, Eq. (3.139b), successively for the first two derivatives yields an algebraic equation of the transforms, being called the *subsidiary equation* of the differential equation

$$mp^2\underline{s}(p) + cp\underline{s}(p) + k\underline{s}(p) = \underline{F}_{\omega}(p) + ms(0^+)p + cs(0^+) + m\dot{s}(0^+) \tag{3.141b}$$

Solving the algebraic equation with ease for $\underline{s}(p)$ the transformed solution is obtained representing the *complete displacement-response transform* $\underline{s}(p)$

$$\underline{s}(p) = \underline{s}_{\omega}(p) = \underbrace{\frac{1}{k + cp + mp^2}}_{=G(p)}\underline{F}_{\omega}(p) + \frac{1}{k + cp + mp^2}\underbrace{[(mp + c)s(0) + m\dot{s}(0)]}_{=I(p)} \tag{3.142a}$$

To emphasize its physical meaning the transform of the unknown response $s(t)$, called the response transform $\underline{s}(p)$, is separated into two terms, one due to the transform of the known forcing function $F_{\omega}(t)$, sometimes called the *driving transform* $\underline{F}_{\omega}(p)$, and the other due to the known *function of the initial values* $I(p)$ with s and $\dot{s}$, i.e., the initial displacement $s(0)$ and the initial velocity $\dot{s}(0)$. Thus, as one of the fundamental properties of linear time-invariant systems can be stated, the complete displacement-response transform $\underline{s}(p)$ is the *sum* of the *transform of the zero-state response* $\underline{s}_{\mathrm{F}}(p)$ and the *transform of the zero-driving response* $\underline{s}_{\mathrm{I}}(p)$, i.e.:

$$\begin{aligned}\underline{s}(p) = \underline{s}_{\omega}(p) &= G(p)\underline{F}_{\omega}(p) + G(p)I(p)\\ &= \underline{s}_{\mathrm{F}}(p) + \underline{s}_{\mathrm{I}}(p)\end{aligned} \tag{3.142b}$$

Also, by the linearity of Laplace transforms the same holds for the corresponding actual response in the time domain.

For simplicity, all initial values are prescribed to vanish, Eq. (3.43), and the oscillator is assumed to start at $t = 0$ from rest. From the so defined zero state ensues $I(p) = 0$, thus, the second term drops out and the transform of the complete displacement response $\underline{s}(p)$, Eq. (3.142b), reduces to
the *displacement-response transform of zero-state response* $\underline{s}_F(p)$

$$\underline{s}(p) = \underline{s}_\omega(p) = G(p)\underline{F}_\omega(p) = \underline{s}_F(p) \qquad (3.142c).$$

Laplace- (or p-) domain-response Characteristic. The input-output relation of the compact type of an algebraic product, Eq. (3.142c), is one with considerable significance in practical application that relates transform method and block diagram representation, see 2.1. The response transform of zero-state response $\underline{s}(p) = \underline{s}_F(p)$ results from multiplying the excitation transform $\underline{F}_\omega(p)$ by a complex quantity termed
the *displacement transfer function G(p)*

$$G(p) = \frac{1}{mp^2 + cp + k} = \frac{1}{m}\frac{1}{p^2 + 2\delta p + \omega_0^2} = \frac{\omega_0^2}{k}\frac{1}{p^2 + 2\zeta\omega_0 p + \omega_0^2} \qquad (3.143).$$

Transfer Function. This response characteristic is defined as a property depending on the complex frequency p (or s), that is equal to the *quotient of two Laplace transforms* representing the actual zero-state response and the excitation function by their generalized spectral densities, Eq. (3.144). Contrary to the complexor appropriate to forced sinusoidal responses for steady-state analysis, 3.1.3, the present quotient is a *complex system parameter* varying in p over the definition range of the related Laplace transforms. To give a meaning of the complex variable sometimes the conception of a "complex frequency of the drive" accounts for p with respect to an exponentially growing sinusoidal excitation.

The transfer function G(p) being a property of linear dynamic systems is:

- a *rational function* of the complex frequency variable p (or s) with real coefficients;
- a *generalized frequency-response characteristic* taking account of the mechanical components or elements (element parameter values) and the interconnection of parts or subsystems (topology of the mechanical system);
- suited to more generalized types of *transient excitation functions.*

Particularly the Laplace transformation is connected with the *block diagram representation* being an effective tool to provide a great deal of insight into the physical system. By use of algebraic input-output relations between the component variables the overall behaviour of interconnected parts (subsystems) can be determined in the Laplace domain (transfer function block diagram), see Sec. 2.1.1.

In the case of *motion response* by a displacement the ratio of the displacement-response transform $\underline{s}(p)$ to the transform of the excitation force $\underline{F}(p)$ is identified

for the system initially at rest with
the *displacement transfer function G(p)*

$$G(p) = \frac{\mathrm{L}[s(t)]}{\mathrm{L}[F(t)]} = \frac{\underline{s}(p)}{\underline{F}(p)} \qquad (3.144).$$

The transfer function $G(p)$ is closely related to the frequency-response function $G(\omega)$, Eq. (3.79), being equal for $p = \mathrm{j}\omega$. However, certain analytic properties have to be observed concerning the relation of Laplace transforms to the Fourier integral [36]. The frequency-response function $G(\omega)$ can be considered as the boundary function $G(\mathrm{j}\omega)$ of the transfer function $G(p)$ on the imaginary axis, in short the equality

$$G(\omega) = G(p)\big|_{\mathrm{p=j\omega}} = G(\mathrm{j}\omega) \qquad (3.145a)$$

holds true since

- the region of convergence of $G(p)$ contains the $\mathrm{j}\omega$-axis in its interior, i.e., if $c < 0$;
- the $\mathrm{j}\omega$-axis being the boundary of the region of convergence of $G(p)$, i.e., if $c = 0$, no singular point (system-pole) p_{k} lies on the imaginary axis.

It can be shown that these conditions of the p-domain response characteristic $G(p)$ validating Eq. (3.145a) correspond with an absolutely integrable time response characteristic $g(t)$ the L-transform of which is identical with the F-transform of $g(t)$ for all $p = \mathrm{j}\omega$. This results in the equality

$$\mathrm{F}[\mathrm{g(t)}] = \mathrm{L}[\mathrm{g(t)}, \mathrm{j}\omega] \qquad (3.145b).$$

Apart from the property of causality that one of asymptotical stability is satisfied. This proves right in any case of passive vibrating systems with damping. By reason of the corresponding equalities, Eqs. (3.145a,b) the pertinent motion characteristic is indicated as well by the nature of the eigenvalues (system-poles), Eq. (3.15), as by the condition of integrability of $g(t)$, Eq. (3.149b).

Time-response Characteristic. An important concept in linear system analysis is to apply the *unit pulse*, *Dirac* or *delta "function"*, denoted $\delta(t)$ and already introduced in 3.2.3.

Unit Pulse Response (weighting function). The response characteristic in the time domain, in control theory called *weighting function* and denoted $g(t)$, is related to the characteristic in the frequency domain previously dealt with, the transfer function $G(p)$. Both response characteristics constitute a Laplace transform pair denoted by the *correspondence*

$$G(p) \;\bullet\!\!\xrightarrow{\;\mathrm{L}^{-1}\;}\!\!\circ\; g(t) \qquad (3.146a).$$

The evaluation performed by the Laplace integral for $G(p)$

$$g(t) = \mathrm{L}^{-1}[G(p)] = \frac{1}{2\pi \mathrm{j}} \int_{\sigma_1 - \mathrm{j}\infty}^{\sigma_1 + \mathrm{j}\infty} G(p)\,\mathrm{e}^{\mathrm{tp}} \mathrm{d}p \qquad (3.146b)$$

yields the unit pulse response being a causal function, Eq. (3.82).

The *unit pulse response*, thus defined as the inverse Laplace transform of the transfer function *G*(*p*), and termed in mathematical notation Green's function has already been interpreted as to different kinds of excitation as to the determination of system characteristics in the time domain (*transient-response specifications*) by the Fourier transform method, 3.2.3.

A set of time-history curves of the transient vibration "unit pulse response" is shown in Fig. A.17 of the Appendix A.

Stability of Motion. The transfer function *G*(*p*) being an algebraic response characteristic in the subsidiary (or *p*-) domain permits to make some assessment of the dynamic system's behaviour by analysing the denominator polynomial *H*(*p*). Being identical with the *characteristic equation* P(*p*), Eq. (3.13b),

$$H(p) \equiv \mathrm{P}(p) \tag{3.147}$$

the *roots* of the denominator polynomial represent the *characteristic values* (or eigenvalues), so the poles of the transfer function (system-poles) p_k are equal to the characteristic values.

The motion characteristics depend on the nature of the system-poles determining properties and system parameters such as system's stability, time constants, damping ratio and natural frequency. The root criterion (discriminant) already used by the classical solution method in 3.1.2 for demonstrating periodicity (oscillatory behaviour) and stability of the natural (eigen-) motion of the damped oscillator may be completed by relating the nature of the system-poles p_k to the *stability criteria*:

i. At least one of the poles is *real and positive* or they are *complex conjugates with positive real part*:

$$\mathrm{Re}\, p_k = \sigma_k > 0\,; \quad \mathrm{Im}\, p_k = \mathrm{j}\omega_k = 0 \quad \text{or} \quad \mathrm{j}\omega_k \neq 0 \tag{3.148a}$$

the fundamental (natural) vibration increases exponentially with time, hence, the motion is *unstable*.

ii. The poles are either *real and negative* or they are *complex conjugates with negative real part*:

$$\mathrm{Re}\, p_k = \sigma_k < 0\,; \quad \mathrm{Im}\, p_k = \mathrm{j}\omega_k = 0 \quad \text{or} \quad \mathrm{j}\omega_k \neq 0 \tag{3.148b}$$

the fundamental vibration approaches the equilibrium position as time increases, thus, the motion is *asymptotically stable*.

iii. The poles are *pure imaginary*:

$$\mathrm{Re}\, p_k = \sigma_k = 0\,; \quad \mathrm{Im}\, p_k = \mathrm{j}\omega_k \neq 0 \tag{3.148c}$$

then the fundamental vibration is a simple harmonic oscillation about the equilibrium position. This root location characterizes a limited stability, thus, the motion is termed *neutrally stable*.

Neutral stability also occurs since one pole is negative and the other one is zero. Excluding two multiple zero poles (zero double roots) as an actual root location for the considered mass-damper-spring system (second-order system) neutral stability generally can be stated since at least one pole has a zero real part and the other one is without a positive real part. Conversely, for all characteristic root locations, since at least one pole has a positive real part, the motion is unstable.

These observations hold true for systems of higher order. The conditions for stability of higher-order systems are summarized by the *Routh-Hurwitz stability criteria* used as a convenient stability test in control system design. Those criteria consist in several rules applied to the coefficients of the characteristic polynomial without solving for the roots [2], [3].

The response characteristic $g(t)$ in the original (or t-) domain, defined by Eq. (3.146b) as the inverse Laplace transform of the transfer function $G(p)$, has already been interpreted as a specific transient vibration at natural frequency, 3.2.3. Thus, the relationship existing between the pole configuration of the transfer function $G(p)$ and the corresponding unit pulse response $g(t)$ proves useful for determining properties and system parameters. The preceding relationships can be represented by plotting the location of the characteristic roots in the *complex frequency plane*, called the *p- (or s-) plane*, and the graphical interpretation is based on the *p- (or s-) plane geometry of the transfer function G(p)*.

For a preliminary study only the steady-state value of the time-response characteristic $g(t)$ is of interest. The behaviour of $g(t)$ in the neighbourhood of $t \to \infty$ can be derived from the behaviour of $pG(p)$ at the origin for $p = 0$ by use of the *equality* (final-value theorem)

$$\lim_{p \to 0} pG(p) = \lim_{t \to \infty} g(t) = 0 \qquad (3.149a).$$

The theorem's conditions are satisfied since $G(p)$ is analytic (or regular) in the right-half plane, that means, all the system-poles lie in the left-half plane. This restriction on the poles of $pG(p)$ corresponds to the criterion of asymptotic stability, and the system is thus said to possess *significant behaviour*.

In the case of system-poles lying on the imaginary axis (on the boundary of the left-half plane) $g(t)$ never dies out but remains bounded within the values $+\hat{g}$ and $-\hat{g}$ (peak-to-peak value or double amplitude of a harmonic vibration) as $t \to \infty$, and the system is thus said to possess *critical behaviour* corresponding with the criterion of neutral stability.

A more rigorous restriction on the behaviour of $g(t)$ as $t \to \infty$ is given by the *condition of absolute integrability* equalling the fundamental theorem of Eq. (3.70) that states the existence of a Fourier transform of the unit pulse response

$$\lim_{t \to \infty} \int_0^t |g(t)|\,dt = \int_0^\infty |g(t)|\,dt < \infty \qquad (3.149b).$$

The absolute convergence criterion is satisfied, thus the frequency function $G(\omega)$ exists according to Eq. (3.145), since the region of convergence of $G(p)$ contains the $j\omega$-axis in its interior, that means, no system-pole lies on the imaginary axis. This restriction on the convergence of $g(t)$ contains a strict conformity with the criterion of asymptotic stability.

Algebraic stability criteria, so the Routh-Hurwitz criterion for continuous-time models or the Jury criterion for discrete-time models, give informations about the *absolute stability* of higher-order systems represented by the transfer function of linear time-invariant models. For predicting a stable system in reality though the numerical values of the coefficients of the characteristic polynomial are not known exactly the stability test is reduced in practice to state the proximity of the characteristic roots to the imaginary axis, thus indicating the *relative stability of a system*.

Root Locus Study of Damping. The stability criteria based on the nature of the characteristic values (or eigenvalues) can be best visualized by the *p-* (or *s-*) *plane geometry of the transfer function G(p)*. Then, the left half-plane represents the region of asymptotic stability, the imaginary axis the region of neutral (or critical) stability, and the right half-plane the region of instability.

System design is not only interested in stability statements but also in response characteristics, and in particular how a response characteristic in the original (or

t-) domain changes as the system parameters change. Therefore, the locations of the roots (characteristic values) being equal to the system-poles are plotted in terms of some system parameter varying its value within the range of interest. This method is termed the *root-locus method* which turns out that the roots lie on smooth curves, known as loci, and the plots themselves are called *root-locus plots*. This method is widely used in control system design for determining the control logic elements, likewise for driving the characteristic values into the left half-plane from the right half-plane the imaginary axis included.

Generally some degree of damping is usually desired in vibration systems for safety, while in control systems a considerable margin of stability is essential, preferably with little or no oscillation. By example of a second-order system as the damped oscillator the following root-locus plot can be traced by plotting the imaginary parts versus the real parts of the characteristic values $p_{1,2}$, Eq. (3.15), where the system parameter damping coefficient (attenuation coefficient) δ, Eq. (3.3), varies, whilst the (undamped) natural frequency ω_0, Eq. (3.6a), is fixed, Fig. 3.15a.

In the case of *less-than-critical damping* $\delta < \omega_0$ the two loci (complex-roots case) move along a semicircle of radius ω_0 centered on the origin as δ increases, until the roots coalesce on the real axis as δ reaches ω_0.

In the case of *greater-than-critical damping* $\delta > \omega_0$ the two loci (real-roots case) split once again, moving along the negative real axis in opposite directions.

The distinct values of δ are appropriate to the two points on the imaginary axis and the splitting point on the negative real axis both representing limiting conditions, first the case of *absent damping* $\delta = 0$ (pure imaginary-roots case), second the case of *critical damping* $\delta = \omega_0$ (repeated-roots case).

By relating the characteristic values $p_{1,2}$ to natural frequency (angular) ω_0 the normalized form of root-locus plot will be obtained by varying the damping ratio ζ, Eq. (3.5), as the only system parameter, Fig. 3.15b.

Roots with the *same damping ratio* $\zeta = \delta / \omega_0$ lie on the same *line through the origin* making an angle of Θ with the imaginary axis.

Roots on a given *line parallel to the imaginary axis* have the *same damping coefficient* δ constituting the distance of the line from the border line of stability. As δ equals the reciprocal of the time constant (relaxation time) T_r, Eq. (3.4), the greater the distance, the smaller T_r. The relative stability of the system consequently depends on the root lying closest by the imaginary axis, that means the root with the largest time constant thus termed the *dominant root*. By use of the time constants T_{s1}, T_{s2}, Eq. (3.28a), the time-constant representation of non-oscillatory free vibration (real-roots case) demonstrates that in the case of greater-than-critical-damping the natural motion (transient motion) resulting from two decaying exponential components can be replaced by the approximate one with the largest time constant, that is T_{s1} associated with the dominant root $p_1 = \delta - \lambda$, Eq. (3.25), see Fig. A.2 of Appendix A.

The focusing on the root that plays the significant role in assessing dynamic system's behaviour simplifies the analysis of motion characteristics basing on the p- (or s-) plane geometry of the transfer function $G(p)$, and the method is called the *dominant-root concept* [3].

Time-response Characteristics and Root Location. To illustrate how the transient motion at natural frequency change as the system parameter (viscous) damping

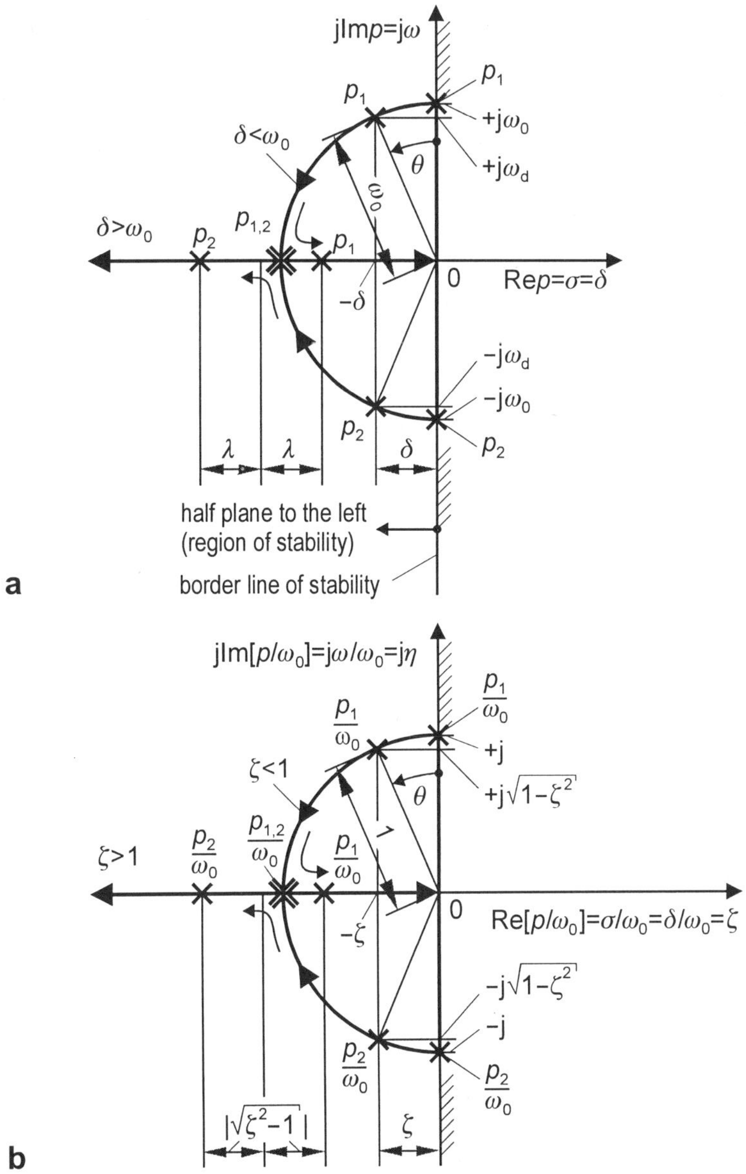

Fig. 3.15. Complex p-plane with the roots (system-poles) $p_{1,2}$ in terms of system parameters. **a** Root-locus plot; **b** normalized root-locus plot

ratio ζ changes the time-response characteristics are classified according to the locations of the characteristic roots (system-poles) in the p-plane on the left and the associate unit pulse response curves $\sqrt{km}\cdot g(\tau)$, Eqs. (3.83b), (3.84b), just as the unit step response curves $k\cdot h(\tau)$ on the right, Fig. 3.16.

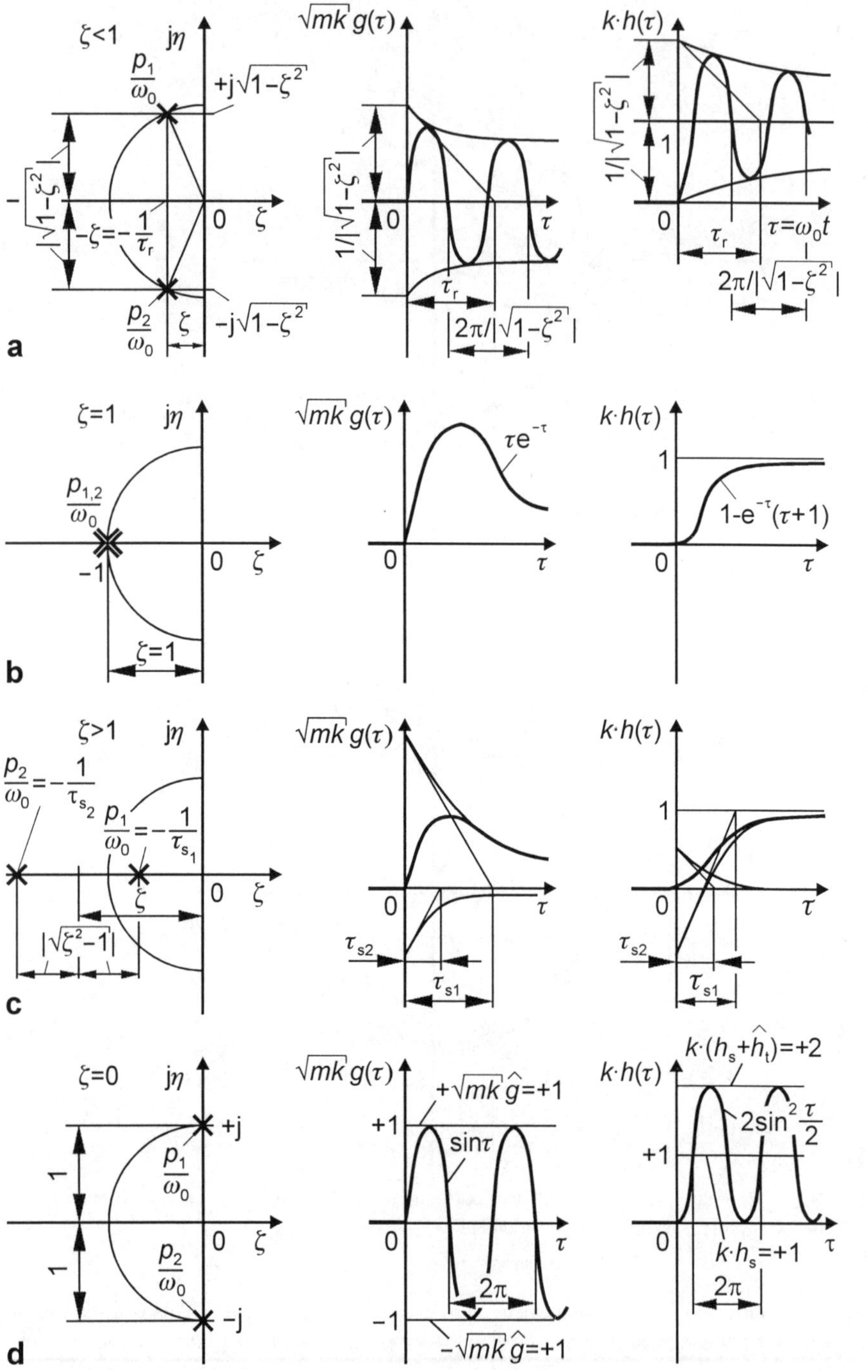

Fig. 3.16. Root locations and time-response characteristics in terms of the damping ratio ζ for a system being: **a** Underdamped; **b** critically damped; **c** overdamped; **d** undamped

The four cases of damping are related to both, the periodicity of motion (oscillatory behaviour) and the stability of motion.

The *damped oscillator* is an *(asymptotically) stable system* of significant behaviour portrayed by decaying response curves. The margin of stability given by the proximity of the roots to the imaginary axis differs with the actual damping, respectively with the pertinent time constant. The damping effect on nature of vibrating (periodicity) involves with increasing damping ratio a change in natural motion from the oscillatory to the non-oscillatory transient state, Fig. 3.16a to c, whereas at critical damping, Fig. 3.16b, the limiting transient state is passed.

The *undamped oscillator* possesses critical behaviour characterized by an oscillating response curve the amplitude of which does not increase nor decrease. Thus, the oscillator is a *neutrally stable system*, Fig. 3.16d.

Response Calculation with the Laplace Transform

The remaining problem is to determine the system response in the time domain (actual displacement response) $s(t)$ corresponding to the transformed solution (complete displacement-response transform) $\underline{s}(p)$ already being calculated, Eq. (3.142a).

There are two approaches to perform the inverse transformation:

- to *evaluate the Laplace integral* by substituting the response transform $\underline{s}(p)$ for $\underline{F}(p)$, into the inversion formula of the Laplace transform, Eq. (3.135a),

$$\mathrm{L}^{-1}\left[\underline{s}(p)\right] = s(t) = \frac{1}{2\pi \mathrm{j}} \int_{\sigma_1 - \mathrm{j}\infty}^{\sigma_1 + \mathrm{j}\infty} \underline{s}(p)\, \mathrm{e}^{tp} \mathrm{d}p \tag{3.150a}$$

  (inverse transformation in the p- (or s-) domain by using integral theorems of the theory of analytic functions);
- to *apply the convolution integral* (superposition integral) to unit pulse response $g(t)$, Eqs. (3.83a), (3.84a), and to actual excitation $F(t)$, Eq. (3.138a) (integral transformation in the t-domain).

Convolution Approach. The latter approach is performed by a time response calculation using a corresponding time-domain relation of equivalent significance to the general algebraic product $G(p)\underline{F}(p)$ of Eq. (3.142c) representing the displacement-response transform of zero-state response $\underline{s}(p) = \underline{s}_\mathrm{F}(p)$.

Inversion formula of algebraic product (time convolution theorem). Taking into account the Laplace transform of derivatives (differentiation theorem) the algebraic multiplying of the complex system parameter varying in p, $G(p)$, by the excitation transform $\underline{F}(p)$ being differentiated w.r.t. complex frequency involves a transform pair (differentiation theorem of the convolution integral) denoted by the *correspondence*

$$\underline{s}(p) = p\left[\frac{G(p)}{p} \cdot \underline{F}(p)\right] \bullet\!\!\stackrel{\mathrm{L}^{-1}}{\longrightarrow}\!\!\circ \frac{\mathrm{d}}{\mathrm{d}t}\left[h(t) \overset{t}{\underset{0}{*}} F(t)\right] = s(t) \tag{3.151a}$$

The derivative of the related time functions in brackets w.r.t. time implies an integral expression known as *Duhamel's integral*

$$s(t) = \frac{\mathrm{d}}{\mathrm{d}t}\left[h(t) \overset{\infty}{\underset{0}{*}} F(t)\right] = \dot{h}(t) \overset{\infty}{\underset{0}{*}} F(t) + h(0^+)F(t) \tag{3.151b}$$

Evaluation by parts results in a sum of two terms the prime of which is defined by the convolution of the functions $\dot{h}(t)$ and $F(t)$ called *convolution* (or *Faltung) integral* (convolution theorem)

$$s(t) = \int_{\tau=0}^{t} \dot{h}(\tau)\, F(t-\tau)\mathrm{d}\tau + h(0^+)F(t) \tag{3.152a}$$

or, equivalently, due to the given symmetry in $\dot{h}(t)$ and $F(t)$ (commutativity of convolution)

$$s(t) = \int_{\tau=0}^{t} F(\tau)\, \dot{h}(t-\tau)\mathrm{d}\tau + h(0^+)F(t) \tag{3.152b}$$,

wherein the unit step response $h(t)$ is the pertinent time-response characteristic the first derivative w.r.t. time, $\dot{h}(t)$, of which is equal to the unit pulse response (weighting function) $g(t)$, Eq. (3.89a).

Contrary to the previously treated transform method basing on Fourier transforms the Laplace transform method satisfies the restraints of causality principle, Eq. (3.82), and of causal excitation functions, Eq. (3.93), per definition (one-sided transformation). Furthermore, following Eq. (3.89a) $\dot{h}(t)$ may be replaced by g(t).

Presupposing for mechanical systems predominated by inertia a transfer function $G(p)$ being a rational algebraic fraction the order of which is indicated by the proper case $m < n$, Eq. (3.89b), the initial value of unit step response vanishes, Eq. (3.89c). Thus, the convolution theorem, Eq. (3.151a), reduces to the algebraic product of generalized frequency-response characteristic and excitation transform corresponding with the convolution product of time-response characteristic and actual excitation. This relation constitutes the *standard transform pair*, being associated with that of the Fourier transform method, Eq. (3.92a), and denotes the *correspondence*

$$\underline{s}(p) = G(p)\cdot \underline{F}(p) \;\bullet\!\!\xrightarrow{\;L^{-1}\;}\!\!\circ\; g(t) \overset{t}{\underset{0}{*}} F(t) = s(t) \tag{3.153}$$

(The limits of integration fitted to boundedness of observation time can be dropped as the convolution symbol is unique.)

The convolution integral in detail is identical with that of Eq. (3.92b), 3.2.3.

3.2.9 System Response to Transient Excitation. *Step-type Functions*

The Laplace transform method will be utilized for determining the system response in the time domain as follows. Taking up the *non-periodic forcing function of step-type* as a transient excitation applied to the mass-damper-spring system the inverse Laplace transformation will be performed by both of the approaches pointed out before, that is the inversion formula and the convolution integral. In any case the calculation of the unknown time response will be carried out by virtue of the *generalized frequency-*, or else the *time-response characteristic*, both being known with reference to the already calculated transfer function $G(p)$, and the unit pulse response $g(t)$, Eqs. (3.172), (3.38a), (3.84a).

Inversion Formula Approach

Writing the displacement-response transform of zero-state response $\underline{s}(p) = \underline{s}_{\mathrm{F}}(p)$ in 3.2.8, Eqs. (3.142a,c), in an explicit form by use of the known Laplace transform of the transient harmonic excitation, Eq. (3.138b), the transformed solution of the differential equation, Eq. (3.140), with special reference to the oscillator initially at rest results in

$$\underline{s}(p) = \underline{s_{\omega}}(p) = \underbrace{\frac{\omega_0^2}{k}\frac{1}{p^2+2\delta p+\omega_0^2}}_{=\,G(p)}\underbrace{\hat{F}\frac{p\cos\varphi_{0\mathrm{F}}+\omega_{\mathrm{f}}\sin\varphi_{0\mathrm{F}}}{p^2+\omega_{\mathrm{f}}^2}}_{=\,\underline{F_{\omega}}(p)} = \frac{1}{H(p)}\frac{D(p)}{N(p)}$$

$$= \frac{\hat{F}\omega_0^2\cos\varphi_{0\mathrm{F}}}{k}\frac{p+\omega_{\mathrm{f}}\tan\varphi_{0\mathrm{F}}}{\left(p^2+2\delta p+\omega_0^2\right)\left(p^2+\omega_{\mathrm{f}}^2\right)} = K_{\mathrm{s}}\frac{1}{h(p)}\frac{d(p)}{n(p)}\;. \tag{3.142d}$$

The response transform $\underline{s}(p)$ is a rational function of the variable p represented by the *ratio of two polynomials*

$$\underline{s_{\omega}}(p) = \frac{\sum_{i=0}^{r=1} b_i p^i}{\sum_{k=0}^{n=4} a_k p^k} = \frac{b_1 p + b_0}{a_4 p^4 + a_3 p^3 + a_2 p^2 + a_1 p + a_0} = \frac{P(p)}{Q(p)} \tag{3.142e}$$

or in form of two *reduced polynomials*

$$\underline{s_{\omega}}(p) = \frac{\sum_{i=0}^{r=1} \beta_i p^i}{\sum_{k=0}^{n=4} \alpha_k p^k} = K_{\mathrm{s}}\frac{\beta_1 p + \beta_0}{\alpha_4 p^4 + \alpha_3 p^3 + \alpha_2 p^2 + \alpha_1 p + \alpha_0} = K_{\mathrm{s}}\frac{p_{\bullet}(p)}{q(p)} \tag{3.142f}$$

where K_{s} is the (real) *response factor*

$$K_{\mathrm{s}} = b_1/a_4 = \frac{\hat{F}\omega_0^2\cos\varphi_{0\mathrm{F}}}{k}\;,$$

and the (real) coefficients b_i, a_k are reduced to the *response coefficients* β_i, α_k

$$\beta_1 = 1;\;\; \beta_0 = b_0/b_1 = \omega_{\mathrm{f}}\tan\varphi_{0\mathrm{F}}$$

$$\alpha_4 = 1;\;\; \alpha_3 = a_3/a_4 = 2\delta\;;\qquad \alpha_2 = a_2/a_4 = \omega_0^2 + \omega_{\mathrm{f}}^2$$

$$\alpha_1 = a_1/a_4 = 2\delta\omega_{\mathrm{f}}^2\;;\; \alpha_0 = a_0/a_4 = \omega_{\mathrm{f}}\omega_0^2\;.$$

Since the degree of the denominator polynomial $Q(p)$ or $q(p)$ is greater than that of the numerator polynomial $P(p)$ or $p_{\bullet}(p)$, e.g., $n = 4$ and $r = 1$, the transformed solution $\underline{s}\,(p)$ is a proper fraction, thus it vanishes as $p \to \infty$

$$\lim_{p\to\infty} \underline{s}(p) = 0 \quad \text{for} \quad n > r \tag{3.154}$$

Applying the inverse Laplace transform, Eq. (3.135a), to the response transform $\underline{s}\,(p)$ by substituting Eq. (3.142f) for $\underline{F}\,(p)$ the inversion formula of the

Laplace transform, Eq. (3.150a), yields the explicit form

$$s(t)=s_{\omega}(t)=\frac{\hat{F}\omega_0^2\cos\varphi_{0F}}{k}\frac{1}{2\pi j}\int_{\sigma_1-j\infty}^{\sigma_1+j\infty}\frac{p+\omega_f\tan\varphi_{0F}}{p^4+2\delta p^3+\left(\omega_0^2+\omega_f^2\right)p^2+2\delta\omega_f^2 p+\omega_f\omega_0^2}e^{tp}dp$$

for $t \geqq 0$

(3.150b).

To evaluate the Laplace integral being a complex integral the *theory of analytic functions* (complex variable functions) will be utilized by means of theorems for complex integration from the first. That means, there's no need for an analytic continuation as carried out in 3.2.4 in order to evaluate the real Fourier integral. The inversion formula, Eq. (3.135a), is defined in 3.2.7 as a complex integral with infinite limits of integration (improper integral) ranging from $\sigma_1 - j\infty$ to $\sigma_1 + j\infty$. The constant σ_1 is retained to indicate a straight line integration path paralleling the imaginary axis apart the positive real value $+\sigma_1$, thus lying within the region of convergence, Eq. (3.136b), here of $\underline{s}(p)$: $\sigma_1 > c$. The evaluation of the line integral along this vertical line, called the *Bromwich path* and sometimes being abbreviated by the letters Br, is in general complicated. However, by a suitable modification of the path to a simple closed curve C (contour) the difficulty can be overcome and results may be obtained with ease on account of *Cauchy's integral theorem*. The path of integration thus will be closed around a contour by tracing a line segment C_1 parallel with the imaginary axis and an arc of circle C_2 of radius R crossing the left-half plane. Thus, presupposing the transform $\underline{s}(p)$ being analytic (regular) on and inside the redrawed contour C except at a finite number of interior singularities (poles) p_k lying in the left-half p-plane the complex improper integral may be replaced by
the *corresponding contour integral*

$$\frac{1}{2\pi j}\int_{\sigma_1-j\infty}^{\sigma_1+j\infty}\underline{s}(p)\,e^{tp}dp=\frac{1}{2\pi j}\lim_{R\to\infty}\int_C \underline{s}(p)e^{tp}dp \quad \text{for } \sigma_1 > c \qquad (3.155a),$$

wherein the length of the path C increases without limit as R approaches infinity. The contour C being decomposed into two portions C_1 and C_2 permits the integration to be taken term by term, and in the counterclockwise sense

$$\int_C \underline{s}(p)e^{tp}dp=\int_{\sigma_1-jR}^{\sigma_1+jR}\underline{s}(p)\,e^{tp}dp+\int_{C_2}\underline{s}(p)e^{tp}dp \qquad (3.155b).$$

The transform $\underline{s}(p)$ tends to zero uniformly on C_2 as $R \to \infty$ in respect of *Jordan's lemma*

$$\lim_{R\to\infty}\int_{C_2}\underline{s}(p)e^{tp}dp \to 0 \quad \text{for } t \geqq 0 \qquad (3.155c),$$

that means

$$\underline{s}(p) \to 0 \quad \text{as } |p| \to \infty \quad \text{on } C_2,$$

because the transform $\underline{s}(p)$ is a rational function, additionally a proper fraction with $n > m$.

Thus, if R is large enough to contain all poles p_k and if R is greater than the maximum $|p_k|$, then the integral along C is independent of R. Hence, the value of the contour integral is given by 2πjtimes the sum of the residues of the integrand at those points. This result is known as
Cauchy's residue theorem

$$2\pi j\, s(t) = \lim_{R\to\infty} \int_C e^{tp}\, \underline{s}(p) dp = 2\pi j \sum_{k=1}^{n} \mathrm{Res}\left[\underline{s}(p)\, e^{tp}\right] \quad \text{for } t \geq 0 \quad (3.155d).$$

The proof of theorems on limiting contours, e.g., as $R \to \infty$, requires the knowledge of the *root locations of the integrand* in Eq. (3.150b), i.e., the interiour singularities (poles) of the displacement response transform $\underline{s}(p)$, Eq. (3.142d,f) must be defined. Taking up the p-plane geometry of the displacement transfer function $G(p)$ which characterizes the generalized frequency behaviour of a damped oscillator possessing significant behaviour (asymptotic stability), the characteristic root location is given by *system-poles* being generally simple and lying in the left half-plane, in the case of less-than-critical damping they are a pair of complex conjugates $p_{1,2}$, Eq. (3.18). This results from the characteristic polynomial $H(p)$ being part of the denominator of the integrand.

Now, however, the denominator consists of two factorized parts the second of which being due to the driving transform $\underline{F}_{\omega}(p)$. This factor vanishes, the response transform thus cannot remain finite at the "singularities of the excitation", being pure imaginary roots and denoting
the *excitation-poles*

$$p_{3,4} = \pm j\omega_f \quad (3.156a)$$

for a transient sinusoid with the forcing (angular) frequency ω_f. In addition an *excitation-zero*

$$p_{01} = -\omega_f \tan \varphi_{0_F} \quad (3.156b)$$

with the force-excitation initial phase φ_{0F} due to a non-zero-crossing excitation is located on the negative real axis, whereby it can be stated, that a zero does not affect the time response in so direct a way.

Considering the jω-axis as the boundary of convergence of $\underline{s}(p)$, i.e., if $c = 0$, the path of integration may be closed on the imaginary axis, consequently $\sigma_1 = c = 0$, provided that the integration through a singularity will be avoided by *indenting the contour*. The line segment C_1 thus will be traced with small semicircles of radius r around the excitation-poles $p_{3,4}$, Fig. 3.17.

Including in the proof of theorems on limiting contours that the integrand is bounded in absolute value as $r \to \infty$, [36], [37], [41], the value of the contour integral is given by Cauchy's residue theorem, Eq. (3.155d).

Before carrying out the residue calculus a representation of the response transform $\underline{s}(p)$ being an alternative to the ratio of polynomials, Eq. (3.142f), is given by expanding $\underline{s}(p)$ into linear factors, called

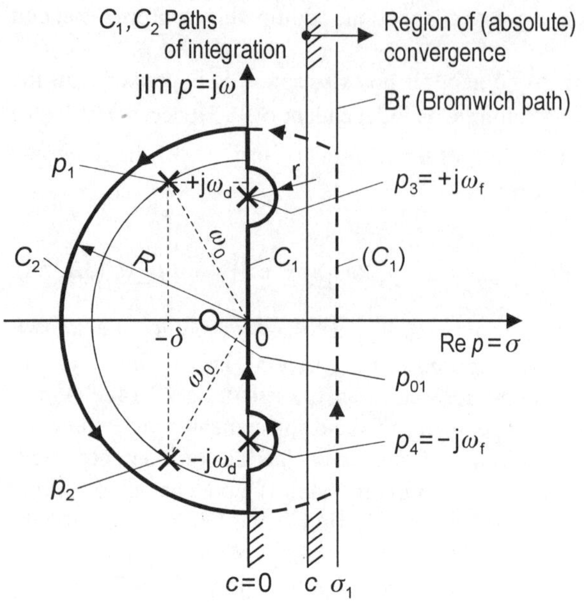

Fig. 3.17. Complex p-plane with intended closed curve surrounding either the system-poles $p_{1,2}$ and the excitation-poles $p_{3,4} = \pm j\omega_f$ in the left-half plane for an inversion by contour integration referred to transient harmonic response

the *factored form* of the rational function

$$\underline{s}_\omega(p) = K_s \frac{\prod_{i=1}^{r=1}(p - p_{0_i})}{\prod_{k=1}^{n=4}(p - p_k)} = K_s \frac{p - p_{0_1}}{(p - p_1)(p - p_2)(p - p_3)(p - p_4)}$$

$$= \frac{\hat{F}\omega_0^2 \cos\varphi_{0F}}{k} \frac{p - (-\omega_f \tan\varphi_0)}{[p - (-\delta + \alpha)][p - (-\delta - \alpha)](p - j\omega_f)(p + j\omega_f)} \tag{3.142h}$$

where K_s is the response factor, p_{0i}; $i = 1,2,\ldots,r$; are called the zeros, and p_k ; $k = 1,\ldots,n$; the poles of $\underline{s}(p)$.

Thus, a complete specification of the response transform is given by the r zeros p_{0i}, the n poles p_k, and the response factor K_s. Since $\underline{s}(p)$ is a rational function with real coefficients, zeros and poles must be real or occur in complex conjugate pairs.

A commonly used method of inverse transformation starting from the factored form of the rational algebraic fraction continues with the expansion into the *partial fractions*

$$\underline{s}_\omega(p) = \sum_{k=1}^{n=4} \frac{R_k}{p - p_k} = \frac{R_1}{p - p_1} + \frac{R_2}{p - p_2} + \frac{R_3}{p - p_3} + \frac{R_4}{p - p_4} \tag{3.142j}$$

Accordingly a complicated transform is reduced to the sum of simpler terms being in shape for the separate inversion by use of transform tables.

Contrary to the partial-fraction method the *method of residues* implies an effective expansion of the inversion formula approach being a direct method since it provides the evaluation of the Laplace integral directly. Consequently the unknown algebraic coefficients, usually denoted A, B, C, D and calculated by a comparison of coefficients, now are determined by the particular coefficients $A_{k,-1}$ of the corresponding Laurent series of the integrand $f(p) = \underline{s}(p)e^{tp}$ expanded near each of the singular points (poles) p_k. The coefficient $A_{k,-1}$ is associated with the only term $A_{k,-1}/(p - p_k)$ contributing to the contour integration of $f(p)$ by surrounding the interiour singularity (pole) $p = p_k$ in a closed curve. As already pointed out in 3.2.4, this coefficient $A_{k,-1}$ is called the residue of $f(p)$ at $p = p_k$, the value of which is denoted by $\text{Res}(p_k) = R_k$. The residue can be determined by evaluating the limit indicated in Eq. (3.99).

The outlined method of residues though being rigorous in view of mathematics will be modified for a practical carrying out of the calculus of residues. By cancelling the kernel of inverse transformation e^{tp} in the complex function $f(p)$, the *residue* will be calculated not of the integrand $f(p)$ of the Laplace integral but only of the response transform $\underline{s}(p)$, since *the limit is evaluated as* $p \to p_k$. In the case of poles of order one, called *simple poles*, hence it follows

$$\operatorname*{Res}_{p=p_k} \underline{s}(p) = R_k = \lim_{p \to p_k} (p - p_k)\underline{s}(p) = \left[(p - p_k)\underline{s}(p)\right]_{p=p_k} \qquad (3.157a).$$

However, operating with the residue calculus suchlike reduced the imperfect inverse transformation must be completed by use of an adapted transform pair for each term of the expansion given by Eq. (3.142j). With the minor modification the inversion formula approach adopts the approved partial-fraction method to a certain extent, and the following response calculation involves the use of an available transform table.

Thus, by use of factoring process, Eq. (3.142h), the residue of $\underline{s}(p)$ at the excitation-pole $p = p_3$, Eq. (3.155a), is easily obtained as

$$\begin{aligned} R_3 &= K_s\left[(p - j\omega_f)\frac{p_\bullet(p)}{(p - j\omega_f)(p + j\omega_f)h(p)}\right]_{p=j\omega_f} = K_s\frac{p_\bullet(j\omega_f)}{2j\omega_f h(j\omega_f)} \\ &= \frac{\hat{F}\omega_0^2\cos\varphi_{0F}}{k}\,\frac{j\omega_f - (-\omega_f\tan\varphi_{0F})}{2j\omega_f[j\omega_f - (-\delta + \alpha)][j\omega_f - (-\delta - \alpha)]} \qquad (3.158a). \\ &= \frac{\hat{F}\omega_0^2\cos\varphi_{0F}}{k}\,\frac{1 - j\tan\varphi_{0F}}{2\left[(\delta^2 - \alpha^2 - \omega_f^2) + 2j\delta\omega_f\right]} \end{aligned}$$

where the linear factors are combined into the reduced numerator polynomial of the response transform, Eqs. (3.142d,f):

$$p_\bullet(p) = \prod_{i=1}^{r=1}(p - p_{0_i}) = p - (-\omega_f\tan\varphi_0) \qquad (3.159a)$$

reduced characteristic polynomial

$$h(p) = \prod_{\nu=1}^{n=2}(p - p_\nu) = (p - p_1)(p - p_2) \qquad (3.159b).$$

Due to the fact that any rational algebraic fraction $\underline{s}(p) = P(p)/Q(p)$ with real coefficients b_i, a_k, Eq. (3.142e), has the property that $Q(p^*) = Q^*(p)$ for all p, the residues of $\underline{s}(p)$ at poles only occuring in complex conjugate pairs are also complex conjugates. Hence, at the excitation-poles $p_{3,4}$ being a pair of pure imaginary conjugates the appropriate residues $R_{3,4}$ are equally related:

$$R_4 = R_3^* \quad \text{because of } p_4 = p_3^* = -\mathrm{j}\omega_\mathrm{f} \qquad (3.158b),$$

and the value of R_4 results directly from converting Eq. (3.158a) to

$$R_4 = K_\mathrm{s}\frac{p_\bullet(-\mathrm{j}\omega_\mathrm{f})}{-2\mathrm{j}\omega_\mathrm{f}h(-\mathrm{j}\omega_\mathrm{f})} = \frac{\hat{F}\omega_0^2\cos\varphi_{0\mathrm{F}}}{k}\,\frac{1+\mathrm{j}\tan\varphi_{0\mathrm{F}}}{2\left[(\delta^2-\alpha^2-\omega_\mathrm{f}^2)-2\mathrm{j}\delta\omega_\mathrm{f}\right]} \qquad (3.158c).$$

By using L'Hospital's rule the limit indicated in Eq. (3.157a) can be evaluated alternatively by the *differential form*

$$\operatorname*{Res}_{p=p_\mathrm{k}} \underline{s}(p) = R_\mathrm{k} = \lim_{p\to p_\mathrm{k}}\frac{(p-p_\mathrm{k})P'(p)+P(p)}{Q'(p)} = \left[\frac{P(p)}{Q'(p)}\right]_{p=p_\mathrm{k}} = \frac{P(p_\mathrm{k})}{Q'(p_\mathrm{k})} \qquad (3.157b).$$

For the special case of a harmonic transient excitation represented by a pair of poles lying on the axis of imaginaries the denominator polynomial $Q(p)$ will be reduced to the characteristic polynomial $H(p)$, Eq. (3.147), and a separated factor $\left(p^2+\omega_\mathrm{f}^2\right)$ owing to the transient sinusoid

$$\underline{s}_\omega(p) = \frac{P(p)}{(p-\mathrm{j}\omega_\mathrm{f})(p+\mathrm{j}\omega_\mathrm{f})H(p)} = \frac{P(p)}{(p^2+\omega_\mathrm{f}^2)H(p)} = K_\mathrm{s}\frac{p_\bullet(p)}{(p^2+\omega_\mathrm{f}^2)h(p)}$$

(3.142g).

Thus, by use of the latter differential form, Eqs. (3.157b), (3.142g), the residue of $\underline{s}(p)$ at the system-pole $p = p_1$, Eq. (3.18), is favourably calculated by

$$R_1 = \lim_{p\to p_1} K_\mathrm{s}\left[\frac{p_\bullet(p)}{(p^2+\omega_\mathrm{f}^2)\frac{h(p)-h(p_1)}{p-p_1}}\right] = K_\mathrm{s}\left[\frac{p_\bullet(p)}{(p^2+\omega_\mathrm{f}^2)\frac{\mathrm{d}}{\mathrm{d}p}h(p)}\right]_{p=p_1} \qquad (3.160a).$$

$$= K_\mathrm{s}\frac{p_\bullet(p_1)}{(p_1^2+\omega_\mathrm{f}^2)h'(p_1)}$$

First derivative of the reduced characteristic polynomial $h(p)$ w.r.t. p using the product law:

$$\frac{\mathrm{d}h(p)}{\mathrm{d}p} = h'(p) = (p-p_2)+(p-p_1)$$
$$h'(p_1) = p_1 - p_2 \qquad (3.161)$$

$$R_1 = K_\mathrm{s}\frac{p_\bullet(p_1)}{(p_1^2+\omega_\mathrm{f}^2)(p_1-p_2)} = \frac{\hat{F}\omega_0^2\cos\varphi_{0\mathrm{F}}}{k}\,\frac{(-\delta+\alpha)-(-\omega_\mathrm{f}\tan\varphi_{0\mathrm{F}})}{\left[(-\delta+\alpha)^2+\omega_\mathrm{f}^2\right]2\alpha} \qquad (3.160b).$$

Case 1: *Less-than-critical damping* (underdamped system), $p_{1,2}$ referred to Eq. (3.18), $\alpha = \mathrm{j}\omega_\mathrm{d}$.

$$R_1 = K_\mathrm{s}\frac{(-\delta+\mathrm{j}\omega_\mathrm{d})-(-\omega_\mathrm{f}\tan\varphi_{0\mathrm{F}})}{\left[(-\delta+\mathrm{j}\omega_\mathrm{d})^2+\omega_\mathrm{f}^2\right]2\mathrm{j}\omega_\mathrm{d}} = \frac{\hat{F}\omega_0^2\cos\varphi_{0\mathrm{F}}}{k}\frac{-\delta+\mathrm{j}\omega_\mathrm{d}+\omega_\mathrm{f}\tan\varphi_{0\mathrm{F}}}{2\mathrm{j}\omega_\mathrm{d}\left[(\omega_\mathrm{f}^2-\omega_0^2)+2\delta^2-2\mathrm{j}\delta\omega_\mathrm{d}\right]} \quad (3.160c).$$

Hence, the system-poles $p_{1,2}$ being a pair of complex conjugates lying in the left half-plane the appropriate residues $R_{1,2}$ are equally related:

$$R_2 = R_1^* \quad \text{because of } p_2 = p_1^* = -\delta-\mathrm{j}\omega_\mathrm{d} \qquad (3.160d),$$

and the value of R_2 results directly from converting Eqs. (3.160b,c) to

$$R_2 = K_\mathrm{s}\frac{p_\bullet(p_2)}{(p_2^2+\omega_\mathrm{f}^2)h'(p_2)}$$

$$= K_\mathrm{s}\frac{(-\delta-\mathrm{j}\omega_\mathrm{d})-(-\omega_\mathrm{f}\tan\varphi_{0\mathrm{F}})}{\left[(-\delta-\mathrm{j}\omega_\mathrm{d})^2+\omega_\mathrm{f}^2\right](-2\mathrm{j}\omega_\mathrm{d})} = \frac{\hat{F}\omega_0^2\cos\varphi_{0\mathrm{F}}}{k}\frac{-\delta-\mathrm{j}\omega_\mathrm{d}-\omega_\mathrm{f}\tan\varphi_{0\mathrm{F}}}{(-2\mathrm{j}\omega_\mathrm{d})\left[(\omega_\mathrm{f}^2-\omega_0^2)+2\delta^2+2\mathrm{j}\delta\omega_\mathrm{d}\right]} \quad (3.160d).$$

The *modified partial-fraction expansion* (differential form) for simple poles results in

$$\underline{s_\omega}(p) = K_\mathrm{s}\left\{\frac{p_\bullet(\mathrm{j}\omega_\mathrm{f})}{2\mathrm{j}\omega_\mathrm{f}h(\mathrm{j}\omega_\mathrm{f})}\frac{1}{p-\mathrm{j}\omega_\mathrm{f}} + \frac{p_\bullet(-\mathrm{j}\omega_\mathrm{f})}{(-2\mathrm{j}\omega_\mathrm{f})h(-\mathrm{j}\omega_\mathrm{f})}\frac{1}{p+\mathrm{j}\omega_\mathrm{f}} + \sum_{\nu=1}^{n=2}\frac{p_\bullet(p_\nu)}{(p_\nu^2+\omega_\mathrm{f}^2)h'(p_\nu)}\frac{1}{p-p_\nu}\right\} \quad (3.142h).$$

The actual problem of inverse transformation is now a simple one by use of an *elementary transform pair* referred to exponential functions and denoted by the *correspondence*

$$\operatorname{Re} p \geqq c = 0\,; \quad \frac{R_\mathrm{k}}{p-p_\mathrm{k}} \;\bullet\!\!\overset{\mathrm{L}^{-1}}{\longrightarrow}\!\!\circ\; R_\mathrm{k}\mathrm{e}^{p_\mathrm{k}t}\,; \quad t \geqq 0 \qquad (3.162a).$$

Substituting the transform pair, Eq. (3.162a), into any term of the partial-fraction expansion, Eq. (3.142j), the inverse transformation of $\underline{s}(p)$ will be performed term by term.

The first two corresponding terms consisting in a complex (harmonic) exponential and its conjugate can be summarized in a single trigonometric function. However, it is preferable for response calculation to retain this function in complex notation. By use of the operator Re half the real part of either one of the conjugates must be taken. Thus, the general solution (complete response) is obtained as the joint result of the first two corresponding terms, and which is a sinusoidal modification of the joint step-response formula, known as
Heaviside's expansion theorem

$$s_\omega(t) = K_\mathrm{s}\Bigg\{\underbrace{\operatorname{Re}\left[\frac{p_\bullet(\mathrm{j}\omega_\mathrm{f})}{\mathrm{j}\omega_\mathrm{f}h(\mathrm{j}\omega_\mathrm{f})}\mathrm{e}^{\mathrm{j}\omega_\mathrm{f}t}\right]}_{=s_{\omega_\mathrm{s}}(t)} + \underbrace{\sum_{\nu=1\ldots}^{n=2}\frac{p_\bullet(p_\nu)}{(p_\nu^2+\omega_\mathrm{f}^2)h'(p_\nu)}\mathrm{e}^{p_\nu t}}_{=s_{\omega_\mathrm{t}}(t)}\Bigg\}\,; \quad t \geqq 0 \quad (3.163a),$$

in which the first term $s_{\omega s}(t)$ represents a forced vibration (steady-state response), and the second $s_{\omega t}(t)$ a free vibration (transient response).

Indeed, this theorem specifying the case of one excitation-pole lying at the origin (step response) has been generalized about a pair of pure imaginary poles on the present form (transient harmonic response), Eq. (3.163a).

Taking up the concept of complex excitations and responses (phasor method) of 3.1.3, the term of steady-state vibration $s_{\omega t}(t)$ can be summarized on account of the phasor of the excitation force $\underline{\hat{F}}$ and its conjugate

$$\underline{\hat{F}}^* = \hat{F}\,\mathrm{e}^{-\mathrm{j}\varphi_{0\mathrm{F}}} = \hat{F}(\cos\varphi_{0\mathrm{F}} - \mathrm{j}\sin\varphi_{0\mathrm{F}}) \tag{3.164a}$$

and the displacement response complexor, Eq. (3.54), converted to

$$G(\mathrm{j}\omega_\mathrm{f}) = \frac{\omega_0^2}{k}\frac{1}{h(\mathrm{j}\omega_\mathrm{f})} \tag{3.164b}$$

so that the particular actual response can be written in the real part of the complex response, Eqs. (3.57), (3.58), as

$$\begin{aligned} s_{\omega_\mathrm{s}}(t) &= \mathrm{Re}\left[G(\mathrm{j}\omega_\mathrm{f})\hat{F}\frac{\mathrm{j}\omega_\mathrm{f}\cos\varphi_{0\mathrm{F}} + \omega_\mathrm{f}\sin\varphi_{0\mathrm{F}}}{\mathrm{j}\omega_\mathrm{f}}\mathrm{e}^{\mathrm{j}\omega_\mathrm{f}t}\right] = \mathrm{Re}\left[G(\mathrm{j}\omega_\mathrm{f})\mathrm{j}\omega_\mathrm{f}\frac{\underline{\hat{F}}^*}{\mathrm{j}\omega_\mathrm{f}}\mathrm{e}^{\mathrm{j}\omega_\mathrm{f}t}\right] \\ &= \mathrm{Re}\left[G(\mathrm{j}\omega_\mathrm{f})\underline{\hat{F}}^*\mathrm{e}^{\mathrm{j}\omega_\mathrm{f}t}\right] = |G(\mathrm{j}\omega_\mathrm{f})|\hat{F}\,\mathrm{Re}\left\{\mathrm{e}^{\mathrm{j}[\omega_\mathrm{f}t - (-\mathrm{arc}\,G(\mathrm{j}\omega_\mathrm{f}) + \varphi_{0\mathrm{F}})]}\right\};\quad t \geqq 0 \end{aligned} \tag{3.164c}$$

Hence, it follows the original or t-domain solution from Eq. (3.163a) by substituting Eqs. (3.158a), (3.18), (3.160c), (3.164c), and the actual response will be rewritten at first in the real part of the complex notation:

Complete displacement response of the underdamped system

$$\begin{aligned} s_\omega(t) = \frac{\hat{F}\omega_0^2\cos\varphi_{0\mathrm{F}}}{k}\Bigg\langle &\mathrm{Re}\left\{\frac{1 - \mathrm{j}\tan\varphi_{0\mathrm{F}}}{\left[(\delta^2 + \omega_\mathrm{d}^2 - \omega_\mathrm{f}^2) + \mathrm{j}(2\delta\omega_\mathrm{f})\right]}\mathrm{e}^{\mathrm{j}\omega_\mathrm{f}t}\right\} \\ &+ \mathrm{e}^{-\delta t}\mathrm{Re}\left\{\frac{-\delta + \mathrm{j}\omega_\mathrm{d} + \omega_\mathrm{f}\tan\varphi_{0\mathrm{F}}}{\mathrm{j}\omega_\mathrm{d}\left[(\omega_\mathrm{f}^2 - \omega_0^2) + 2\delta^2 - 2\mathrm{j}\delta\omega_\mathrm{d}\right]}\mathrm{e}^{\mathrm{j}\omega_\mathrm{d}t}\right\}\Bigg\rangle;\quad t \geqq 0 \end{aligned} \tag{3.163b}$$

and finally in real notation:

Complete displacement response for the oscillator initially at rest

$$\begin{aligned} s_\omega(t) = \hat{F}\underbrace{\frac{1}{\left|\sqrt{(k - m\omega_\mathrm{f}^2)^2 + (c\omega_\mathrm{f})^2}\right|}}_{=|G(\mathrm{j}\omega_\mathrm{f})| = A(\omega_\mathrm{f})}&\Bigg\langle\Big\{\cos\Big[\omega_\mathrm{f}t - \underbrace{(\psi + \varphi_{0_\mathrm{F}})}_{=\varphi_{0_\mathrm{s}} = -\mathrm{arc}\,G(\mathrm{j}\omega_\mathrm{f}) + \varphi_{0_\mathrm{F}}}\Big]\Big\} \\ &- \frac{\omega_0}{\omega_\mathrm{d}}\left|\sqrt{1 - \delta\frac{\omega_\mathrm{f}}{\omega_0^2}\sin 2\varphi_{0\mathrm{F}} - \left[1 - \left(\frac{\omega_\mathrm{f}}{\omega_0^2}\right)^2\right]\sin^2\varphi_{0\mathrm{F}}}\right|\mathrm{e}^{-\delta t}\cos(\omega_\mathrm{d}t - \chi)\Bigg\rangle;\quad t \geqq 0 \end{aligned} \tag{3.163c}$$

where ψ equals Eq. (3.40a);

$$\varphi_{0_s} = \arctan\frac{2\delta\omega_f}{\omega_0^2-\omega_f^2} + \arctan(\tan\varphi_{0F}) = \psi + \varphi_{0F} \tag{3.164d}$$

according to Eq. (3.37b);

and
$$\chi = \arctan\frac{2\delta\omega_d}{-\delta^2+\omega_d^2-\omega_f^2} + \arctan\frac{-\delta+\omega_f\tan\varphi_{0F}}{\omega_d} \tag{3.164e}$$

which equals Eq. (3.44a) in case of a force-excitation initial phase being omitted (zero-crossing excitation) $\varphi_{0F} = 0$.

Case 2: *Greater-than-critical damping* (overdamped system), $p_{1,2}$ referred to Eq. (3.25), $\alpha = \lambda$.

The response calculation equally makes use of the expansion theorem for at least one pair of pure imaginary poles (of excitation), Eq. (3.163a), by substituting λ instead of $\mathrm{j}\omega_d$ into the second term. Thus, representing a non-oscillatory transient response the second term $s_{\omega t}(t)$ results from the appropriate residues $R_{1,2}$ at the system-poles $p_{1,2}$ being real and negative (real-roots-case).

Case 3: Considering the *limiting condition of critical damping*, $p_1 = p_2$ referred to Eq. (3.31), $\alpha = 0$, a remarkable pole configuration occurs as the system-poles coalesce on the real axis and δ reaches ω_0 (repeated-roots case). For this exceptional case of second-order system response the expansion theorem, Eq. (3.163a), being suited only for simple poles, is no more valid.

Generalizing the method of residues to *multiple-order poles* the *partial-fraction coefficients* (coefficients of the corresponding Laurent series expansion) of $\underline{s}(p)$ at a pole p_k can be evaluated by

$$A_{k,-\lambda} = \frac{1}{(m_k-\lambda)!}\left\{\frac{d^{m_k-\lambda}}{dp^{m_k-\lambda}}\left[(p-p_k)^{m_k}\,\underline{s}(p)\right]\right\}_{p=p_k} \tag{3.157c}$$

with the multiplicity (order) m_k of each pole p_k for n distinct poles where $k = 1,\ldots,n$ and $\lambda = 1,\ldots,m_k$.

The particular coefficients associate with the proper number $\lambda = 1$ are identical with the residues of $\underline{s}(p)$ at each distinct pole $p = p_k$

$$\operatorname*{Res}_{p=p_k}\underline{s}(p) = A_{k,-1} = R_k \qquad \text{for } k = 1,\ldots,n \tag{3.157d}$$

in conformity with Eq. (3.157a).

The inverse transformation starting from the *factored form* of the rational algebraic fraction

$$\underline{s_\omega}(p) = K_s\frac{\prod_{i=1}^{r=1}(p-p_{0i})}{\prod_{k=1}^{n=4}(p-p_k)^{m_k}} = K_s\frac{p-p_{01}}{(p-p_1)^2(p-p_2)(p-p_3)} \tag{3.142k}$$

continues with the expansion into *partial fractions*. In general, for each pole p_k of multiplicity m_k there are m_k terms

$$\underline{s_\omega}(p) = \sum_{k=1}^{n=4} \sum_{\lambda=1}^{m_k} \frac{A_{k,-\lambda}}{(p-p_k)^\lambda} = \frac{A_{1,-1} = R_1}{(p-p_1)} + \frac{A_{1,-2}}{(p-p_1)^2} + \frac{R_2}{p-p_2} + \frac{R_3}{p-p_3} \quad (3.142l).$$

In particular, the two coefficients $A_{1,-1} = R_1$; $A_{1,-2}$ at the system-pole $p_1 = -\delta = -\omega_0$ of 2nd order (double pole) follow from Eq. (3.157c) for $k = 1, m_k = 2$

$$\begin{aligned} A_{1,-2} &= K_s\left[(p-p_1)^2 \underline{s_\omega}(p)\right]_{p=p_1} \\ &= K_s\left[(p-p_1)^2 \frac{p_\bullet(p)}{(p-p_1)^2 (p^2+\omega_f^2)}\right]_{p=p_1} = K_s \frac{p_\bullet(p_1)}{(p_1^2+\omega_f^2)} \\ &= \frac{\hat{F}\omega_0^2 \cos\varphi_{0F}}{k} \frac{-\omega_0 - (-\omega_f \tan\varphi_{0F})}{\omega_0^2 + \omega_f^2} \end{aligned} \quad (3.160d)$$

$$\begin{aligned} A_{1,-1} &= R_1 = K_s \frac{1}{1!}\frac{d}{dp}\left[(p-p_1)^2 \underline{s_\omega}(p)\right] = K_s \frac{d}{dp}\left[\frac{p_\bullet(p)}{(p^2+\omega_f^2)}\right]_{p=p_1} \\ &= K_s \frac{d}{dp}\left[\frac{p-p_{01}}{(p^2+\omega_f^2)}\right]_{p=p_1} = K_s\left[\frac{(p^2+\omega_f^2) - 2p(p-p_{01})}{(p^2+\omega_f^2)^2}\right]_{p=p_1} \\ &= \frac{\hat{F}\omega_0^2 \cos\varphi_{0F}}{k} \frac{(\omega_0^2+\omega_f^2) + 2\omega_0(-\omega_0 + \omega_f \tan\varphi_{0F})}{(\omega_0^2+\omega_f^2)^2} \end{aligned} \quad (3.160e).$$

The already known residues R_2 and R_3 at the excitation-poles p_2 and p_3, respectively, are given by Eqs. (3.158a), (3.158c), therein denoted R_3 and R_4.

The actual problem of inverse transformation can be solved by use of an *elementary transform pair* referred to the product of positive powers of t and an exponential function being denoted by the *correspondence*:

$$\operatorname{Re} p \geqq c = 0 ; \quad \frac{A_{k,-\lambda}}{(p-p_k)^\lambda} \;\bullet\!\!\xrightarrow{\;L^{-1}\;}\!\!\circ\; A_{k,-\lambda} \frac{t^{\lambda-1}}{(\lambda-1)!} e^{p_k t} ; t \geqq 0 \quad (3.162b)$$

where $k = 1,\dots,n$ *and* $\lambda = 1,\dots,m_k$.

Taking pattern from the general solution form associated with at least one pair of pure imaginary poles (excitation-poles $p_{2,3}$), Eq. (3.163a), the joint result is

$$\begin{aligned} s_\omega(t) = K_s \Bigg\langle & \underbrace{\left\{ \operatorname{Re}\left[\frac{p_\bullet(j\omega_f)}{j\omega_f h(j\omega_f)} e^{j\omega_f t}\right]\right\}}_{=s_{\omega_s}(t)} \\ & \underbrace{+ \left[\frac{d}{dp}\left(\frac{p_\bullet(p_1)}{p_1^2+\omega_f^2}\right) + \frac{p_\bullet(p_1)}{p_1^2+\omega_f^2} t\right] e^{-\omega_0 t}}_{\dots\;=s_{\omega_t}(t)} \Bigg\rangle ; t \geqq 0 \end{aligned} \quad (3.163d)$$

and finally in real notation:

Complete displacement response of the critically damped system initially at rest

$$s_\omega(t)=\hat{F}\underbrace{\frac{1}{k+m\omega_f^2}}_{=|G(j\omega_f)|=A(\omega_f)}\left\langle\left\{\cos\left[\omega_f t-\underbrace{(\psi+\varphi_{0F})}_{\varphi_{0s}=\mathrm{arc}\,G(j\omega_f)+\varphi_{0F}}\right]\right\}\right.$$

$$\left.-\omega_0\left\{\left(1-\frac{\omega_f}{\omega_0}\tan\varphi_{0F}\right)\cos\varphi_0-\frac{\cos\varphi_0}{\omega_0}\left[1-\frac{2}{1+(\omega_f/\omega_0)^2}\left(1-\frac{\omega_f}{\omega_0}\tan\varphi_{0F}\right)\right]t\right\}e^{-\omega_0 t}\right\rangle$$

for $t \geqq 0$

(3.163e)

where $\psi=\arctan\dfrac{2\sqrt{km}\,\omega_f}{k-m\omega_f^2}=\arctan\dfrac{2\omega_0\omega_f}{\omega_0^2-\omega_f^2}$ (3.164e);

and $\varphi_{0s}=\psi+\varphi_{0F}$ according to Eq. (3.37b).

Contrary to the classical method treated in 3.1.2 the general solution obtained by the Laplace transform method must not be fitted to zero initial values, and the time-response calculation thus yields automatically the actual displacement as a *zero-state response*.

Convolution Approach

Due to a causal excitation function, per definition (one-sided transformation), Eq. (3.93), the convolution integral in the limits of Eq. (3.152a,b) holds true, whereat the incidental term vanishes as the initial value of the unit step response, Eq. (3.89c), for the proper case of $G(p)$ with $m < n$ (dominant inertia), Eq. (3.89b). In consequence of the convolution theorem reduced to the standard transform pair, Eq. (3.153), the convolution integral in detail is identical with that of Eq. (3.92b) in 3.2.3.

Writing the actual displacement response $s_\omega(t)$ in an explicit form by use of the actual transient harmonic excitation, Eq. (3.138a), rewritten as the real part of a complex notation

$$F_\omega(t)=\hat{F}\cos(\omega_f t-\varphi_0)=\frac{1}{2}\left[\underline{\hat{F}}^* e^{j\omega_f t}+\underline{\hat{F}}e^{-j\omega_f t}\right]=\mathrm{Re}\left[\underline{\hat{F}}^* e^{j\omega_f t}\right] \quad (3.165)$$

the evaluation of the *convolution integral* results in

$$\begin{aligned} s_\omega(t) &= \int_{\tau=0}^{t} g(\tau)F_\omega(t-\tau)\,d\tau=\int_0^t g(\tau)\,\mathrm{Re}\left[\underline{\hat{F}}^* e^{j\omega_f(t-\tau)}\right]d\tau \\ &= \mathrm{Re}\left[\underline{\hat{F}}^* e^{j\omega_f t}\int_0^t e^{-j\omega_f\tau}g(\tau)\,d\tau\right];\quad t\geqq 0 \end{aligned} \quad (3.163g).$$

For a damped oscillator of significant behaviour (asymptotic stability) the equality between frequency response function $G(p)$ for $p=j\omega$, Eq. (3.145), is valid, so the response characteristic $g(t)$ in the t-domain is absolutely integrable and defined as the inverse Fourier transform of $G(\omega)$,

$$G(j\omega)=\int_0^\infty e^{-j\omega\tau}g(\tau)\,d\tau=\mathrm{L}[g(t),j\omega]=\mathrm{F}[g(t)] \quad (3.166a)$$

and the decomposition of the integral in brackets, Eq. (3.163g), into the partial integrals

$$\int_0^t e^{-j\omega_f\tau} g(\tau)d\tau = \int_0^\infty e^{-j\omega_f\tau} g(\tau)d\tau - \int_t^\infty e^{-j\omega_f\tau} g(\tau)d\tau \qquad (3.167)$$

yields the general solution (complete response)

$$s_\omega(t) = \underbrace{\mathrm{Re}\left[\hat{\underline{F}}^* e^{j\omega_f t} \int_{\tau=0}^{\infty} e^{-j\omega_f\tau} g(\tau)d\tau\right]}_{=s_{\omega_s}(t)} - \underbrace{\mathrm{Re}\left[\hat{\underline{F}}^* e^{j\omega_f t} \int_{\tau=t}^{\infty} e^{-j\omega_f\tau} g(\tau)d\tau\right]}_{=s_{\omega_t}(t)}; \ t \geqq 0$$

(3.163h)

in which the first term $s_{\omega s}(t)$ represents a forced vibration (steady-state response), and the second $s_{\omega t}(t)$ a free vibration (transient response).

The term of steady-state vibration $s_{\omega s}(t)$ can be summarized on account of Eq. (3.166a) to the particular actual response identical with Eq. (3.164c).

Writing the unit pulse response $g(t)$ in an explicit form for the underdamped system, Eq. (3.83a), rewritten as the real part of the complex notation

$$g(t) = \frac{1}{m\omega_d} e^{-\delta t} \frac{1}{2j}\left(e^{j\omega_d t} - e^{-j\omega_d t}\right) = \frac{1}{m\omega_d} e^{-\delta t} \mathrm{Re}\left[j e^{-j\omega_d t}\right]; \ t \geqq 0 \qquad (3.166b)$$

the term of transient vibration $s_{\omega t}(t)$ also can be obtained in the complex notation

$$s_{\omega_t}(t) = \frac{1}{m\omega_d} \mathrm{Re}\left[\hat{\underline{F}}^* e^{j\omega_f t} j \int_t^\infty e^{[-\delta - j(\omega_f + \omega_d)]\tau} d\tau\right]; \quad t \geqq 0 \qquad (3.168)$$

which simplifies the response calculation. Certainly, the unit pulse response also can be used in real notation. However, the evaluation of the convolution integral then requires an integration by parts performed twice because of an integrand being a triple product of function of time.

Hence, it follows the actual response from the sum of the terms both evaluated, and in real notation finally designated as the complete response for the oscillator initially at rest. This result is identical with Eq. (3.163c), and also with Eq. (3.44a) by assumption of zero initial phase $\varphi_{0F} = 0$.

Response Data Plotting. The time history of the oscillatory displacement response (combined motion) in normalized form is given in Fig. A.4 of the Appendix A.

3.2.10
The Graphical Interpretation of the Transfer Function. *Conformal Mapping*

The Laplace transform method has proved its usefulness for graphical system representation in terms of block diagrams. The interrelation of functional blocks is connected with the *transfer function concept* being already introduced in 2.1.1.

In addition, the advantage of the Laplace transform to convert linear differential equations into algebraic equations provides an effective method of response calculation for systems affected especially by transient excitations (time-response analysis). This is carried out in 3.2.9 for evaluating the actual displacement response to a harmonic shock excitation by use of the known p-domain-response characteristic, the transfer function. However, the algebraic description of dynamic system's behaviour enables not only the portrayal of block diagram structures or the graphical representation of complete actual responses in the t-domain (time history curves).

In system (or vibration) analysis and design the key role is played by representing the dynamic behaviour of systems in the ω-domain (frequency-response analysis). The graphical interpretation of the transfer function leads to a concise graphical description of a system's frequency response characteristic as pointed out by the following sections.

Transfer-function Analysis by Special Types of Forcing Functions

Besides the great importance of sinusoidal excitation special types of aperiodic input functions are useful mainly dealt with in controls. Those types also present an approach for modelling vibrations and are classified as *singularity functions*. In electric circuitry such a kind of (whole) functions is called *typical waveforms*. It concerns functions continuously varying in time except for a certain initial time, usually occuring at zero time $t = 0$, and they can be obtained from oneanother by successive integration or differentiation. Singularity functions being reduced to abstract input variables designated as "non-dimensional quantities" are not suited to modelling until they are defined as dynamic system variables classified in 1.2. Thus, special types of inputs applied to a translational mechanical system are indicated as normalized external forces. A convenient normalization can be derived from the forcing function $F(t)$ by relating it to a pertinent reference quantity.

The normalized excitations most used in the analytical investigation of system transient response are the unit step or Heaviside function $u_0(t)$, and the unit pulse $\delta(t)$. In vibration analysis suchlike idealized step- or pulse-type excitations are often used as reference input and experimentally realized as a constant force excitation (simple step force), or with some investigation into impact-excitation technique, as a short-duration pulse force also applicable to large structures.

Unit Pulse. A rectangular shock pulse at $t = 0$ idealized by the impulse value of unity and zero pulse duration is called *unit pulse*, *Dirac* or *delta function* $\delta(t)$, Fig. 3.18a. This normalized shock excitation is defined as the inverse Laplace transform of unity denoted by the *correspondence*

$$\Delta(p) = 1 \quad \bullet\!\!\xrightarrow{\;L^{-1}\;}\!\!\circ \quad \delta(t) = \begin{cases} 0 & \text{for } t \neq 0 \\ \infty & \text{for } t = 0 \end{cases} \qquad \int_{-\infty}^{t} \delta(\tau)\,\mathrm{d}\tau = 1 \tag{3.169}$$

The delta function can be regarded as the result of a *limiting process* basing on a rectangular pulse of small pulse duration τ_0 and large amplitude equal to $1/\tau_0$. As $\tau_0 \to 0$, the amplitude tends to infinity, but in such a way that the impulse

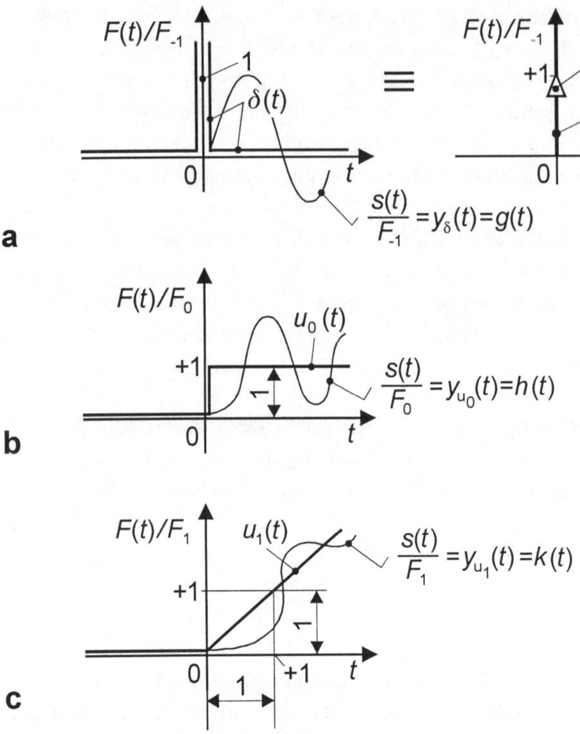

Fig. 3.18. Singularity functions and normalized responses of an underdamped system. **a** Unit pulse; **b** unit step; **c** unit ramp

(pulse area) remains constant and equal to unity. The singularity of a one-sided rectangular pulse can be depicted as a thick solid line and a single white arrow denoting the value of the impulse, Fig. 3.18a, showing the symbolic graph on the right side.

It should be stated that the Laplace integral, Eq. (3.135a), is only true as an integral in the analytical (Riemann) sense for functions not involving singularities. On the rigorous basis of generalized functions the delta function $\delta(t)$ is interpreted as a *delta functional* and the inverse Laplace transform as a distribution [36], [38], [39].

From the transformed zero-state response to an idealized rectangular pulse forcing function $F(t) = F_{-1}\delta(t)$

$$\underline{s}(p) = F_{-1}Y_\delta(p) = F_{-1}G(p)\cdot 1 = F_{-1}G(p) \qquad (3.170)$$

it follows the important property that the *transfer function* $G(p)$ *equals* the *pulse response transform* $\underline{s}(p)$ by relating it to the impulse (pulse force area) $F_{-1} = F_0\tau_0$:

$$G(p) = \frac{F_{-1}Y_\delta(p)}{F_{-1}} = Y_\delta(p) = \mathrm{L}[y_\delta(t)] = \mathrm{L}[g(t)] \qquad (3.171a).$$

Thus, the normalized pulse response $y_\delta(t)$ is identical with the *unit pulse response* (weighting function) $g(t)$.

Unit step. A suddenly rising constant excitation at $t=0$ idealized by the maximum height of unity and zero rise time is called *unit step* or *Heaviside function* $u_0(t)$, Fig. 3.18b. This normalized transient excitation constitutes an elementary transform pair denoted by the *correspondence*

$$U_0(p)=\frac{1}{p} \quad \bullet\!\!\xrightarrow{\mathrm{L}^{-1}}\!\!\circ \quad u_0(t)=\begin{cases}0 & \text{for } t<0\\ 1 & \text{for } t>0\end{cases} \qquad (3.172).$$

From the transformed zero-state response to a rectangular step forcing function $F(t)=F_0u_0(t)$

$$\underline{s}(p)=F_0Y_{u_0}(p)=F_0G(p)U_0(p)=F_0G(p)\frac{1}{p} \qquad (3.173)$$

it follows that the *transfer function* $G(p)$ *equals* the *step response transform* $\underline{s}(p)$ *multiplied by p* by relating it to the maximum height (constant force) F_0:

$$G(p)=\frac{F_0Y_{u_0}(p)p}{F_0}=pY_{u_0}(p)=p\mathrm{L}\left[y_{u_0}(t)\right]=p\mathrm{L}\left[h(t)\right] \qquad (3.171b).$$

Thus, the normalized step response $y_{u0}(t)$ is identical with the *unit step response* $h(t)$, finally on account of the differentiation theorem, Eq. (3.139a), the first derivative w.r.t. time of the unit step response $\dot{h}(t)$ equals the unit pulse response $g(t)$, Eq. (3.89a). Conversely, the unit step response is the integral of the unit pulse response corresponding to the integration of the singularity function

$$u_0(t)=\int_{-\infty}^{t}\delta(\tau)\,\mathrm{d}\tau \quad ; \quad h(t)=\int_{-\infty}^{t}g(\tau)\,\mathrm{d}\tau \qquad (3.147a).$$

Unit Ramp. A constant slope excitation at $t=0$ retaining finite rise time idealized by the maximum slope of unity is called *unit ramp* $u_1(t)$, Fig. 3.18c. This normalized transient excitation constitutes an elementary transform pair denoted by the *correspondence*

$$U_1(p)=\frac{1}{p^2} \quad \bullet\!\!\xrightarrow{\mathrm{L}^{-1}}\!\!\circ \quad u_1(t)=r(t)=\begin{cases}0 & \text{for } t<0\\ t & \text{for } t\geqq 0\end{cases} \qquad (3.175).$$

From the transformed zero-state response to a linearly increasing forcing function $F(t)=F_1u_1(t)$, written in a compact form by use of $u_1(t)=t\bullet u_0(t)$ (synthesizing property of unit step by annihilating any portion of $F(t)$ for $t<0$)

$$\underline{s}(p)=F_1Y_{u_1}(p)=F_1G(p)U_1(p)=F_1G(p)\frac{1}{p^2} \qquad (3.176)$$

it follows that the *transfer function* $G(p)$ equals the *ramp response transform* $\underline{s}(p)$ *multiplied by* p^2 by relating it to the maximum slope (rate of force rise)

$F_1 = F_0/\tau_0$:

$$G(p) = \frac{F_1 Y_{u_1}(p) p^2}{F_1} = p^2 Y_{u_1}(p) = p^2 \mathrm{L}\left[y_{u_1}(t)\right] = p^2 \mathrm{L}\left[k(t)\right] \quad (3.171c).$$

Thus, the normalized ramp response $y_{u1}(t)$ is identical with the *unit ramp response* $k(t)$, finally on account of the differentiation theorem for higher derivatives, Eq. (3.139a), the second derivative w.r.t.time of the unit ramp response $\ddot{k}(t)$ equals the unit pulse response $g(t)$. Conversely, the unit ramp response is the double integral of the unit pulse response corresponding to the successive integration of singularity functions

$$r(t) = \iint_{-\infty}^{t} \delta(\tau)\,\mathrm{d}^2\tau \quad ; \quad k(t) = \iint_{-\infty}^{t} g(\tau)\,\mathrm{d}^2\tau \quad (3.174b).$$

Exponentially Swelling Unit Sinusoid. A suddenly rising sinusoidal excitation at $t = 0$ the "amplitude" of which swells exponentially (increasing envelope) idealized by the initial amplitude of unity and zero rise time is called *exponentially swelled unit sinusoid (section)* $\mathrm{e}^{\sigma_\mathrm{f} t} u_\omega(t)$, Fig. 3.19a. This normalized transient excitation in complex notation of sectional complex exponentials constitutes a sum of elementary transform pairs denoted by the *correspondence*

$$U_{p_\mathrm{f}}(p) = \frac{1}{2}\left[\frac{1}{p - p_\mathrm{f}} + \frac{1}{p - p_\mathrm{f}^*}\right] \quad \bullet\!\!\overset{\mathrm{L}^{-1}}{\longrightarrow}\!\!\circ \quad \frac{1}{2}\left[\mathrm{e}^{p_\mathrm{f} t} + \mathrm{e}^{p_\mathrm{f}^* t}\right] u_0(t) = \mathrm{Re}\left[\mathrm{e}^{p_\mathrm{f} t}\right] u_0(t)$$

$$= \frac{p - \sigma_\mathrm{f}}{(p - \sigma_\mathrm{f})^2 + \omega_\mathrm{f}^2}; \ \mathrm{Re}\, p > \sigma_\mathrm{f} \qquad = \mathrm{e}^{\sigma_\mathrm{f} t} u_\omega(t) = \begin{cases} 0 & \text{for } t < 0 \\ \mathrm{e}^{\sigma_\mathrm{f} t} \cos\omega_\mathrm{f} t & \text{for } t \geq 0 \end{cases}$$

(3.177).

From the forced vibration (steady-state response)

$$s_{p_\mathrm{s}}(t) = \mathrm{Re}\left[G(p_\mathrm{f}) \hat{F} \mathrm{e}^{p_\mathrm{f} t}\right] = \mathrm{Re}\left[\underline{\hat{s}}\, \mathrm{e}^{p_\mathrm{f} t}\right]; \ \text{for } t > 0 \quad (3.178)$$

caused by a generalized transient sine forcing function $F(t)$ at the complex frequency ("of the drive") $p_\mathrm{f} = \sigma_\mathrm{f} + \mathrm{j}\omega_\mathrm{f}$, written in a compact form by use of $\mathrm{e}^{\sigma_\mathrm{f} t} u_\omega(t) = \left(\mathrm{e}^{\sigma_\mathrm{f} t} \cos\omega_\mathrm{f} t\right) \bullet u_0(t)$ (synthesizing property of unit step):

$$F(t) = \hat{F}\, \mathrm{Re}\left[\mathrm{e}^{p_\mathrm{f} t}\right] \bullet u_0(t) = \hat{F} \mathrm{e}^{\sigma_\mathrm{f} t} u_\omega(t) \quad (3.179)$$

it follows that the *transfer function* $G(p)$ at $p = p_\mathrm{f}$ equals the *complex ratio* of the displacement *response phasor at zero time* ("initial phasor") $\underline{\hat{s}}$ to the *phasor of the excitation* force $\underline{\hat{F}} = \hat{F}$:

$$G(p_\mathrm{f}) = \frac{\underline{\hat{s}}}{\underline{\hat{F}}} = \frac{\underline{\hat{s}}}{\hat{F}} \quad (3.171d).$$

To each complex frequency p_f a single response ratio of phasors is assigned being identical with a particular value of the transfer function $G(p) = G(p_\mathrm{f})$.

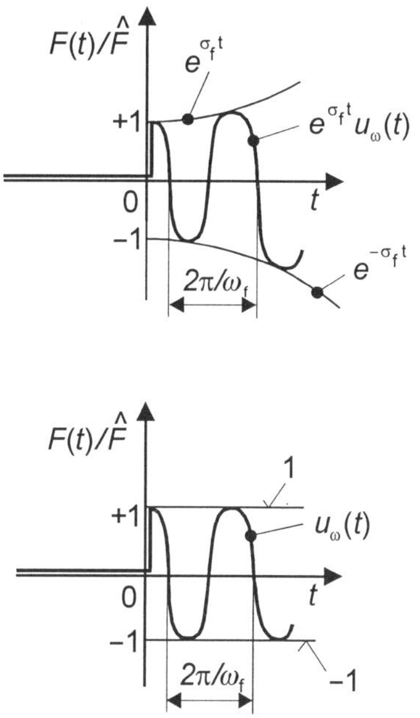

Fig. 3.19. Special types of forcing functions. **a** Exponentially swelling sinusoid; **b** sinusoid section

Indeed, an exponentially swelling sinusoid $(\sigma_f > 0)$ may be considered as a valid concept of interpreting the generalized complex frequency response $G(p)$ but with regard to growing up excitation functions without any practical purpose of vibration data analysis. To the contrary decaying-amplitude functions $(\sigma_f < 0)$, favourably the following constant-amplitude excitation $(\sigma_f = 0)$, hold both an interpreting concept and a practical purpose.

Unit Sinusoid Section. A suddenly rising sinusoidal excitation at $t = 0$ idealized by the amplitude of unity and zero rise time is called *sinusoid section* $u_\omega(t)$, Fig. 3.19b. This normalized transient excitation in complex notation of sectional "harmonic" exponentials constitutes a sum of elementary transform pairs denoted by the *correspondence*

$$U_\omega(p) = \frac{1}{2}\left[\frac{1}{p - j\omega_f} + \frac{1}{p + j\omega_f}\right] \bullet\!\!\xrightarrow{\;L^{-1}\;}\!\!\circ\ \frac{1}{2}\left[e^{j\omega_f t} + e^{-j\omega_f t}\right]u_0(t) = \mathrm{Re}\left[e^{j\omega_f t}\right]u_0(t)$$

$$= \frac{p}{p^2 + \omega_f^2}; \quad \mathrm{Re}\,p > 0_1 \qquad\qquad = u_\omega(t) = \begin{cases} 0 & \text{for } t < 0 \\ \cos\omega_f t & \text{for } t \geqq 0 \end{cases} \tag{3.180}$$

From the forced vibration (steady state response)

$$s_{\omega_s}(\mathrm{t}) = \mathrm{Re}\left[G(j\omega_f)\hat{F}e^{j\omega_f t}\right] = \mathrm{Re}\left[\hat{\underline{s}}\,e^{j\omega_f t}\right]; \quad \text{for } t > 0 \tag{3.181}$$

caused by a transient sine forcing function $F(t)$ at the forcing (angular) frequency ω_f, written in a compact form by use of $u_\omega(t) = (\cos\omega_f t) \bullet u_0(t)$ (synthesizing property of unit step):

$$F(t) = \hat{F}\,\mathrm{Re}\left[e^{j\omega_f t}\right] \bullet u_0(t) = \hat{F}u_\omega(t) \qquad (3.182)$$

it follows the important property that the transfer function $G(p)$ at $p = j\omega_f$, or in respect of p being replaced by $j\omega$, thus the *frequency transfer function* $G(j\omega)$ *at* $\omega = \omega_f$ equals the *complex ratio* of the displacement *response phasor* $\underline{\hat{s}}$ to the *phasor of the excitation* force $\underline{\hat{F}} = \hat{F}$:

$$G(p_f)\big|_{p_f = j\omega_f} = G(j\omega_f) = \frac{\underline{\hat{s}}}{\underline{\hat{F}}} = \frac{\underline{\hat{s}}}{\hat{F}} \qquad (3.171e).$$

To each forcing (angular) frequency ω_f corresponds a single response ratio of phasors assigning a particular value of the transfer function (termed a complexor): $G(j\omega) = G(j\omega_f)$.

Complex System Parameter

Vector Representation of Phasor Ratios (complexors). Returning to the vector representation of sinusoids, 3.1.3, the actual response in the steady state, (forced vibration) can be gathered solely from the one complex system parameter, the displacement-response phasor, Eq. (3.53), as demonstrated by use of phasor method. Thus, in diagram representation the steady-state interrelation is figured in terms of constant (resting) phasors, corresponding to the system variables, which form the resultant phasor defined as *complexor*. By reason of causal systems, especially for a passive mechanical system, the phasor of the displacement response $\underline{\hat{s}}$ lags the phasor of the excitation force $\underline{\hat{F}} = \hat{F}$ by the displacement phase angle φ_{0s}, Fig. 3.20a.

The *magnitude* $|\underline{\hat{s}}|$ *of the response phasor* $\underline{\hat{s}}$ denoting the displacement amplitude $\hat{s}$ of the forced vibration is obtained by *multiplying* the amplitude of the excitation force phasor by the modulus of the frequency transfer function, Eq. (3.38a)

$$|\underline{\hat{s}}| = \hat{s} = |G(j\omega_f)|\,|\underline{\hat{F}}| = A(\omega_f)\hat{F} \qquad (3.183a).$$

The *angle (or argument)* arc $\underline{\hat{s}}$ *of the response phasor* $\underline{\hat{s}}$ designating the phase angle φ_{0s} of the forced vibration, thus fixing the phasor direction, is determined by *adding* to the phase angle φ_{0F} of the excitation force phasor the argument of the frequency transfer function

$$\mathrm{arc}\ \underline{\hat{s}} = -\varphi_{0_s} = \mathrm{arc}\ G(j\omega_f) + \mathrm{arc}\ \underline{\hat{F}} = -\psi_1 - \varphi_{0_F} = -\psi_1 \ ; \ \varphi_{0_F} = 0 \qquad (3.183b)$$

in particular identical with the phase difference ψ_1, Eq. (3.40a,b).

By forming the *complex ratio* of the displacement-response phasor $\underline{\hat{s}}$ to the phasor of the excitation force $\underline{\hat{F}}$ the resulting complex system parameter may be

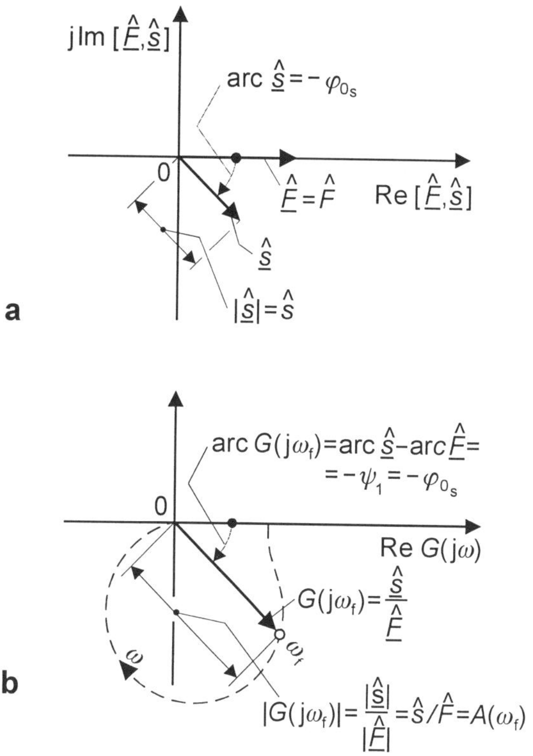

Fig. 3.20. Complex representation of sinusoidal steady-state response. **a** Phasor diagram of displacement response and excitation force; **b** diagram of displacement-response complexor (dynamic compliance)

identified with the *displacement response complexor* $G(j\omega_f)$, in conformity with Eqs. (3.54), (3.171e).

The complexor also can be thought of as a vector localized at the origin (radius vector) in a complex plane where the *magnitude of the complexor* is denoted the *displacement-response factor* $A(\omega_f)$, Eq. (3.55a)

$$|G(j\omega_f)| = A(\omega_f) = |\hat{\underline{s}}|/|\hat{\underline{F}}| = \hat{s}/\hat{F} \qquad (3.184a),$$

which in brevity is called the *amplitude ratio*, and the (angle) *argument of the complexor* fixing its direction is identical with the *phase difference* (phase shift) ψ_1, Eq. (3.56)

$$\text{arc } G(j\omega_f) = -\psi_1 = \text{arc } \hat{\underline{s}} - \text{arc } \hat{\underline{F}} = -\varphi_{0_s} + \varphi_{0_F} = -\varphi_{0_s}\,;\, \varphi_{0_F} = 0 \qquad (3.184b).$$

An evident definition with reference to the physical interpretation of complexors infers from the pair of dynamic system variables being measured or defined at the terminals of the model system. The underlying translational mechanical system is represented by the relation between an applied force and a resulting motion expressed as *displacement*. The ratio of the pertinent phasor quantities defines the *dynamic compliance* $\underline{C}(\mathrm{j}\omega_{\mathrm{f}})$

$$G(\mathrm{j}\omega_{\mathrm{f}}) = \hat{\underline{s}}/\hat{\underline{F}} = \underline{C}(\mathrm{j}\omega_{\mathrm{f}}) \tag{3.185}$$

It is also called the *receptance* α by several authors, see 4.2. It should be pointed out that the relation between the displacement phasor and the force phasor depends on the angular frequency ω, thus, the resultant phasor changes both, its length and its direction, as ω varies from 0 to ∞. For each frequency variable assigned to a particular forcing frequenc $\omega = \omega_{\mathrm{f}}$ the corresponding complexor $G(\mathrm{j}\omega_{\mathrm{f}})$ may be plotted as a single point in the complex plane, called, in this case, the *dynamic compliance plane*, Fig. 3.20b.

Polar Representation of the Frequency-response Characteristic

As ω varies the radius vector revolves clockwise about the origin and Eqs. (3.183a,b) constitute the polar equations of the frequency-response function (or frequency transfer function) $G(\mathrm{j}\omega)$, Eqs. (3.78a), (3.145), traced by the locus of the vector's tip. The frequency locus plot connecting all the points which represent the totality of forced vibrations (or sinusoidal steady-state responses) for varying forcing (angular) frequency is called the *polar frequency response locus*, in controls termed in short the *polar plot*, Fig. 3.20b, dashed curve.

Contrary to the complex ratio of two phasors (complexor) $G(\mathrm{j}\omega_{\mathrm{f}})$, introduced in 3.1.3 for sinusoidal quantities at the same frequency, the frequency-response function $G(\omega)$ has been generally defined for deterministic vibrations in 3.2.3, in addition involving random vibrations in 3.2.5, as the quotient of two Fourier transforms, Eqs. (3.79), (3.115). Thus, a more profitable interpretation of the polar plot tends to the actual response $s(t)$ related to the excitation function $F(t)$ by graphically representing the ratio of their Fourier spectral densities (continuous spectra) $\underline{s}(\omega)$, $\underline{F}(\omega)$. Respectively, the relation between the cross-correlation function at $F(t)$ and $s(t)$ and the autocorrelation of $F(t)$ can be figured by the ratio of their power spectral densities $S_{Fs}(\omega)$, $S_{FF}(\omega)$. Varying in ω over the definition range of the related Fourier transforms the frequency-response function $G(\omega)$ does not depend on the type of excitation function. Beyond the analysis of sinusoidal steady-state behaviour basing on the sweeping of a forcing frequency range the vibration data analysis can be performed by applying a transient function (shock excitation) or a stochastic process (random excitation) defined as reference input.

Just for the system identification of mechanical structures the requirements of data processing are more efficiently satisfied by virtue of frequency concepts basing on non-periodic and random vibrations (spectral analysis), 3.2.2, 3.2.4, 3.2.5.

Conformal Mapping. An equivalent polar plot of the transfer function $G(p)$ being a rational function of the complex variable p (or s) can be deduced from the *method of conformal mapping*. Starting from a vector diagram which presents the characteristic polynomial in factored form, Eq. (3.159b), the linear factors are interpreted by a vector subtraction. Basing on the complex-roots case the vectors have their tails respectively at one of the two system-poles p_1, p_2, and their heads both at the fixed point p. Thus, the transfer function, Eqs. (3.172), (3.83a), can be calculated at the single complex frequency p by use of vector properties, in particular of multiplying operations on complex numbers, Fig. 3.21a.

Though being valid for the total range of the complex frequency p the vector calculus may be done in practice for a frequency range of interest referred to the "complex frequency of the drive", [3], [27 to 29].

The point $G(p)$ corresponding to a fixed point p is called the image point with respect to the mapping defined by the *analytic function* (or corresponding function) $G(p)$. If the point p moves along a straight line $\sigma_1 = const$ (level curve) in the p-plane, the corresponding point $G(\sigma_1 + j\omega)$ will travel along a curve that

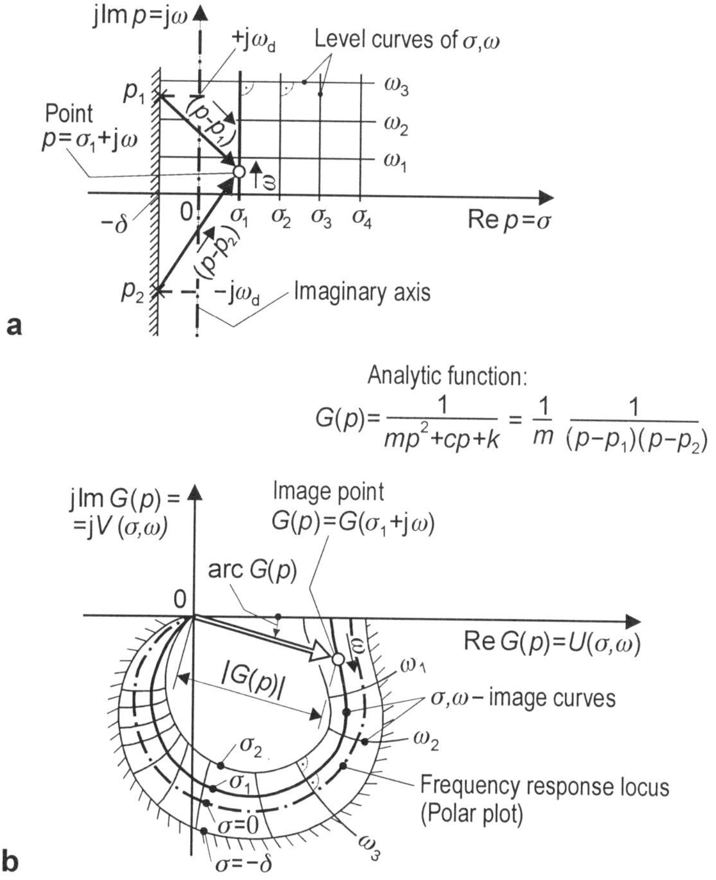

Fig. 3.21. Transfer function graphically illustrated by the σ,ω-field map. **a** Complex representation in the p-plane by the root loci vector diagram; **b** image curves in the $G(p)$-plane

defines the image of the level curve in the $G(p)$-plane (or transfer-function plane), called accordingly in this case the *generalized dynamic compliance plane*, Fig. 3.21b. Mapping the upper right half-plane $\sigma > -\delta$ onto the $G(p)$-plane the *image curves* form a set of corresponding field lines (loci of constant σ) and a conjugate set of "equipotentials" (loci of constant ω). Those curvilinear squares are interpreted as the *polar plot* (referred to $\sigma = 0$) *with its accompanying σ,ω-field map.*

The image of a small figure conforms to the original figure in the sense that it has approximately the same shape. So the images of the level curves intersecting at right angles (rectangular grid) make the same angle at each point of intersection (orthogonal grid). Thus, the mapping onto the $G(p)$-plane is said to be *angle-preserving* (conformal), [44].

3.2.11 The Graphical Interpretation of the Frequency-Response Function. *Frequency Response Plots*

The σ,ω-field map suited for the graphical interpretation of the transfer function $G(p)$ also can be used for illustrating frequency-response curves. For this purpose let the point p move along the line $\sigma = 0$ in the p-plane and the corresponding point $G(\mathrm{j}\omega)$ will travel along a curve that defines the *image of the imaginary axis* onto the $G(p)$-plane. Thus, the field line specified for $\sigma = 0$, indicated in Fig. 3.21b by the dot-dash line, is identical with the *polar plot* appertaining to the *frequency-response function* of the system $G(\mathrm{j}\omega)$.

Polar Frequency Response Locus (Nyquist plot)

Supposing a second-order linear system of significant behaviour the two characteristic roots (system-poles) p_1, p_2 constantly lie in the left-half plane. Thus, the region of convergence of $G(p)$ contains the $\mathrm{j}\omega$-axis in its interiour, i.e., if $c < 0$ the equality, Eq. (3.145a), holds true with reference to $G(\omega)$ being associated with $G(\mathrm{j}\omega)$.

The graphical representation of the frequency-response function by its polar plot in the complex *$G(\mathrm{j}\omega)$-plane (Nyquist-plane)* presupposes *specifications* of

- the *dynamic variable quantities* defined or measured as the excitation and the response function;
- the *pair of points in space* referred to excitation and response terminating the dynamic system model (input and output terminals).

Driving-point versus Transfer Response Characteristic. The frequency-response function is a complex system parameter which is referred to a point i or j in a linear time-invariant system where excitation and response may be taken at the same point or at different points in the same system. The considered translational mechanical system is represented by a motion response and an excitation force. The motion of the translational system occurs in response to *any forcing function of*

time which may be either a harmonic (sinusoidal) or perodic excitation, or else a transient or random excitation. Since the motion response and the excitation force are taken *at the same point i* or *j* in the same system the frequency-response function is defined with respect to *auto-spectral analysis* by the quotient of two Fourier transforms taken from the functions of time at the excitation point (input terminal), thus termed
the *direct (or driving-point) frequency-response function*

$$G_{ii}(j\omega) = G_{jj}(j\omega) = \frac{F[s_i(t)]}{F[F_i(t)]} = \frac{\underline{s}_i(\omega)}{\underline{F}_i(\omega)} = \frac{\underline{s}_j(\omega)}{\underline{F}_j(\omega)} \quad (3.186a).$$

In the specified application of a simple *harmonic excitation* the frequency-response function is formed by
the *response ratio of phasors (complexor)*

$$G_{ii}(j\omega_f) = G_{jj}(j\omega_f) = \frac{\underline{\hat{s}}_i}{\underline{\hat{F}}_i} = \frac{\underline{\hat{s}}_j}{\underline{\hat{F}}_j} = \underline{C}_{ii}(j\omega_f) \quad (3.186b).$$

The motion response being expressed in terms of a *displacement* the corresponding phasor ratio at the same point *i* or *j* is designated the *direct (or driving-point) dynamic compliance* (or direct receptance) of a structure $\underline{C}_{ii}(j\omega_f)$.

Since the motion response is taken *at one point* in a system and the excitation force is taken *at another point* $i \neq j$ in the same system the frequency-response function is defined with respect to *cross-spectral analysis* by the quotient of two Fourier transforms taken from the functions of time at different points, i.e., the response transform at point *j* (output terminal) to the excitation transform at point *i* (input terminal), thus termed
the *transfer frequency-response function*

$$G_{ij}(j\omega) = \frac{F[s_j(t)]}{F[F_i(t)]} = \frac{\underline{s}_j(\omega)}{\underline{F}_i(\omega)} \quad (3.187a).$$

Specifying a simple *harmonic excitation* the frequency-response function is formed by
the *response ratio of phasors (complexor)*

$$G_{ij}(j\omega_f) = \frac{\underline{\hat{s}}_i}{\underline{\hat{F}}_i} = \underline{C}_{ij}(j\omega_f) \quad (3.187b).$$

The motion response being expressed in terms of a *displacement* the corresponding phasor ratio at the different points *i* and *j* is designated the *transfer dynamic compliance* (or cross receptance) of a structure $\underline{C}_{ij}(j\omega_f)$.

Reciprocity. Operating on a linear elastic structure the excitation can be removed from *i* and placed at *j*, then the previous displacement at *j* will exist at *i*. Thus, equality between corresponding pairs of transfer dynamic compliances being measured or determined by definition is confirmed between *i* and *j*, and the *principle of dynamic reciprocity* is valid stating that

$$\underline{C}_{ij}(j\omega) = \underline{C}_{ji}(j\omega) \quad (3.187c).$$

It indicates that a mechanical structure can transmit energy equally well in both directions (two-way energy transmission systems).

The excitation may be, alternately, force applied to the system or motion of the foundation that supports the system. Since the significant response is given by the resultant force to an exciting motion if proves convenient to invert the quotient of Fourier transforms, in particular to form the inverse phasor ratio, defining the *reciprocal of the direct and the transfer frequency-response function*

$$G_{ii}^{-1}(j\omega) = \frac{F_i(\omega)}{\underline{s}_i(\omega)}; \quad G_{ii}^{-1}(j\omega_f) = \frac{\underline{\hat{F}_i}}{\underline{\hat{s}_i}} = \underline{K_{ii}}(j\omega_f) \tag{3.188a,b}$$

and

$$G_{ij}^{-1}(j\omega) = \frac{F_i(\omega)}{\underline{s}_j(\omega)}; \quad G_{ij}^{-1}(j\omega_f) = \frac{\underline{\hat{F}_i}}{\underline{\hat{s}_j}} = \underline{K_{ij}}(j\omega_f) \tag{3.189a,b}$$

The exciting motion being expressed in terms of a *displacement* the corresponding phasor ratios at the same point or at different points in the system are designated the *direct and the transfer dynamic stiffness* of a structure $\underline{K}_{ii}(j\omega_f)$ and $\underline{K}_{ij}(j\omega_f)$, respectively. Nevertheless, Eqs. (3.188a,b), (3.189a,b) hold equally true in case of a mobility measurement by exciting with a force at a single point (constant-force generator) and measuring the translational response motions on the structure.

The motion response being expressed in terms of a *velocity* the phasor ratios of the resulting velocity to the driving force at the same point or at different points in the system are, corresponding to Eqs. (3.186b), (3.187b), designated the *direct and the transfer (mechanical) mobility* of a structure $\underline{Y}_{ii}(j\omega_f)$ and $\underline{Y}_{ij}(j\omega_f)$, respectively

$$\frac{\underline{\hat{\upsilon}_i}}{\underline{\hat{F}_i}} = \underline{Y_{ii}}(j\omega_f) \quad (3.190) \quad ; \quad \frac{\underline{\hat{\upsilon}_j}}{\underline{\hat{F}_i}} = \underline{Y_{ij}}(j\omega_f) \tag{3.191}$$

The mobility, sometimes termed the mechanical admittance following the response ratio concept of electrical circuits, can be expressed alternately by its reciprocal, i.e., in terms of the *mechanical impedance* $\underline{Z}(j\omega_f)$. Taken at the same point or at different points the inverse phasor ratio will be specified as the *direct and the transfer mechanical impedance* of a structure $\underline{Z}_{ii}(j\omega_f)$ and $\underline{Z}_{ij}(j\omega_f)$, respectively

$$\frac{\underline{\hat{F}_i}}{\underline{\hat{\upsilon}_i}} = \underline{Z_{ii}}(j\omega_f) \quad (3.192) \quad ; \quad \frac{\underline{\hat{F}_i}}{\underline{\hat{\upsilon}_j}} = \underline{Z_{ij}}(j\omega_f) \tag{3.193}$$

The direct impedance indicates whether a structure *resists* (absorbs) the vibration, whereas the transfer impedance shows if the structure *transmits* the vibration or isolates.

Example 3.1: Forced Vibration due to Motion. The model system of a simple oscillator introduced in 3.1 can be used to illustrate the frequency-response function referred to alternative excitation at different points in space, Fig. 3.22.

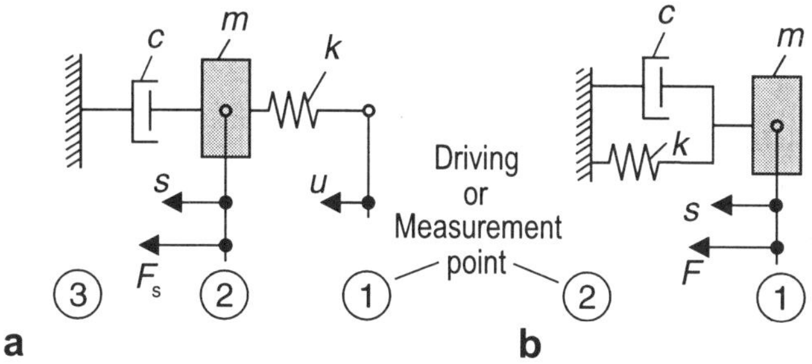

Fig. 3.22. Damped single degree-of-freedom system. **a** Harmonic excitation acting indirectly on the mass (spring-controlled); **b** acting directly on the mass (force-controlled)

In case of an excitation displacement $u(t)$ at point 1 acting indirectly on the mass m over a spring k (spring-controlled), the mechanical system is governed by the equation of motion

$$\begin{aligned} m\ddot{s} &= -c\dot{s} - k(s-u) \\ m\ddot{s} + c\dot{s} + ks &= ku = F_s(t) \end{aligned} \qquad (3.194a).$$

A harmonic excitation of motion $u(t) = \hat{u}\cos(\omega_f t - \varphi_{0u})$ causes a suchlike excitation of force $F_s(t) = ku(t)$ which is represented by the complex excitation $\underline{F}_s(t)$ according to Eq. (3.49), but with the force phasor $\underline{\hat{F}}_s = k\underline{\hat{u}}$

$$\underline{F}_s(t) = \underline{\hat{F}}_s e^{j\omega_f t} \qquad (3.195a).$$

Substituting Eq. (3.195a), into Eq. (3.194a) the phasor representation introduced in 3.1.3 results in the phasor equation

$$-m\omega_f^2\underline{\hat{s}} + jc\omega_f\underline{\hat{s}} + k\underline{\hat{s}} = \underline{\hat{F}}_s \qquad (3.196a).$$

The sinusoidal steady-state response is described by the displacement-response phasor $\underline{\hat{s}}$. Specifying a harmonic excitation and taking the motion response and excitation force at the same point 2 the *direct frequency-response function* $G_{22}(j\omega)$ at $\omega = \omega_f$ is defined as:

$$G_{22}(j\omega_f) = \frac{1}{k + jc\omega_f - m\omega_f^2} \qquad (3.197)$$

forming the *response ratio of phasors (complexor)*

$$\begin{aligned} G_{22}(j\omega_f) &= \frac{\underline{\hat{s}}}{\underline{\hat{F}}_s} = \underline{C_{22}}(j\omega_f) \\ &= \frac{1}{k}\frac{1}{1 + j\frac{c}{k}\omega_f - \frac{m}{k}\omega_f^2} = \frac{1}{k}\frac{1}{1 + j2\frac{\delta}{\omega_0}\frac{\omega_f}{\omega_0} - \left(\frac{\omega_f}{\omega_0}\right)^2} \end{aligned} \qquad (3.198a).$$

With respect to motion response in terms of a displacement the complexor, Eq. (3.198a), defines the *driving-point dynamic compliance* $\underline{C_{22}}(j\omega_f)$.

Example 3.2: Forced Vibration due to Force. In case of an excitation force $F(t)$ at point 1 acting directly on the mass m (force-controlled), Fig. 3.22b, it follows the equation of motion

$$\begin{aligned} m\ddot{s} &= -c\dot{s} - ks + F(t) \\ m\ddot{s} + c\dot{s} + ks &= F(t) \end{aligned} \qquad (3.194b).$$

The harmonic excitation of force $F(t) = \hat{F}\cos(\omega_f t - \varphi_{0F})$ is represented according to Eq. (3.49) by the complex excitation $\underline{F}(t)$ with the force phasor $\hat{\underline{F}}$

$$\underline{F}(t) = \hat{\underline{F}}\, e^{j\omega_f t} \qquad (3.195b).$$

So it follows by Eqs. (3.194b), (3.195b) the phasor equation

$$-m\omega_f^2 \hat{\underline{s}} + jc\omega_f \hat{\underline{s}} + k\hat{\underline{s}} = \hat{\underline{F}}.$$

Specifying a harmonic excitation and taking the motion response and excitation force at the same point 1 the direct frequency-response function $G_{11}(j\omega)$ at $\omega = \omega_f$ is identically defined by Eq. (3.198a), forming the *response ratio of phasors (complexor)*

$$G_{11}(j\omega_f) = \frac{\hat{\underline{s}}}{\hat{\underline{F}}} = \underline{C_{11}}(j\omega_f) \qquad (3.198b)$$

and being designated the *driving-point dynamic compliance* $\underline{C_{11}}(j\omega_f)$.

Frequency Response and Frequency Normalization. A tool of vibration data analysis being of considerable usefulness and convenience is *normalization* which means adjusting the scale of a function or of a variable to a new value. In the operation of *data processing* for obtaining frequency-response characteristics the frequency transfer function $G(j\omega)$ including the frequency variable ω will usually be changed in dimension, preferably it becomes a "non-dimensional quantity" (quantity of dimension one).

It is a fundamental property of linear systems that magnitude and angle of the response phasor $\hat{\underline{s}}$, Eqs. (3.183a,b), are independent of the magnitude of the excitation phasor $\hat{\underline{F}}$ (and the initial conditions s_0, $\dot{s}_0$). By reason of that the response ratio of phasors (complexor) $G(j\omega_f)$, Eqs. (3.186b), (3.187b), can be *adjusted in variable and scale* at each forcing frequency ω_f. Hence, this also holds true for the frequency-response function $G(\omega) = G(j\omega)$, Eqs. (3.186a), (3.187a), over its total range of frequency values ω to be performed in several steps. Separate steps are already introduced in 3.1.1, at first by defining the *system parameters* (vibratory specifications) δ, ζ, ω_0, Eqs. (3.3), (3.5), (3.6a). Moreover, it is convenient to normalize the variable "forcing (angular) frequency" ω_f by relating it to the undamped natural frequency ω_0, thus in effect by transforming to the *frequency ratio* η, Eq. (3.39), as the new variable. This transforming step has been previously applied in 3.1.2 to the "displacement-response factor" $A(\omega_f)$, Eq. (3.38a), for gaining the magnification factor R_d, Eq. (3.38b), as the new non-dimensional amplitude ratio. In quite a similar way the complex frequency-response function $G(j\omega)$ can be transformed into a function of the new variable η, [49].

The constant quantity $G(j\omega_f)$ at $\omega_f = 0$ defines the *static response factor* $G(0)$. This real particular parameter, to be adjusted as the (proportional) "gain of the system", is, in this case, identical with the *compliance of the spring*

$$C_k = 1/k \qquad (3.199a).$$

If the complex system parameter $G(j\omega_f/\omega_0)$ varying with η will be related to the multiplying constant $G(0)$ the transformation finally results in the *normalized response ratio of phasors (normalized complexor)*

$$\frac{\underline{C}_{ii}\left(j\frac{\omega_f}{\omega_0}\right)}{C_k} = \left(\frac{\hat{s}}{\underline{\hat{F}}_s}\right) \Bigg/ \left(\frac{1}{k}\right) = \frac{\hat{s}\,k}{\underline{\hat{F}}_s} = \frac{\hat{s}\,k}{\underline{\hat{F}}} = \underline{Y}_k(j\eta_1) = \frac{1}{1 + j2\zeta\eta_1 - \eta_1^2} \qquad (3.198c).$$

The motion response being originally expressed in terms of a displacement the corresponding normalized phasor ratio (at the same point i) is designated the *normalized (direct) dynamic compliance* (or direct receptance) of a structure $\underline{Y}_k(j\eta_1)$.

This new function is expressed in terms of only two quantities, the frequency ratio η and the damping ratio ζ. It turns out that by normalizing of frequency variable and of response ratio of phasors (complexor) the number of parameters can be diminished by relating element parameters to form system parameters, moreover the parameter values can be reduced to an order of unity, making otherwise long and tedious computations relatively simple. Thus, the normalizing transformation of frequency responses proves useful as a convenient *data reduction method* to the frequency domain.

Interpreting Remarks to Normalized Polar Plots. Using the vector representation of response phasor ratios the frequency response can be plotted in terms of polar coordinates by the locus of the tip of the displacement-response complexor (dynamic compliance), Eq. (3.198b), as the driving frequency ω_f (henceforth equivalent to ω) varies from 0 to ∞, Fig. 3.20b.

An alternative interpretation of polar plotting was given by the image of the imaginary axis of the complex frequency plane (p-plane) onto the transfer-function plane ($G(p)$- or generalized dynamic compliance plane), Fig. 3.21b.

By successively transforming from ω_f into the new frequency variable η_1 furthermore from $\underline{C}_{ii}(j\omega_f)$ into the new frequency-response function $\underline{Y}_k(j\eta_1)$ the displacement-response complexor will be normalized, Eq. (3.198c), and finally the adjusted (data-reduced) polar plot of $\underline{Y}_k(j\eta)$ can be traced out as the frequency ratio η_1 (henceforth equivalent to η) varies from 0 to ∞.

With respect to the Nyquist stability theorem applied to open-loop transfer functions in controls the polar plot being sketched on uniform scales is termed the *Nyquist plot* (after Harry Nyquist).

Response Data Plotting. The polar frequency-response loci (Nyquist plots) of *normalized dynamic compliance* $\underline{Y}_k(j\eta)$ are shown in Fig. B.1 of the Appendix B.

The new coordinates being *polar coordinates* of magnitude and phase are differently scaled. The polar radii of (vector) magnitudes are uniformly scaled in absolute numbers, and the (negatively) oriented angles of (vector) directions are uniformly scaled in the unit degree, thus generating the complex $\underline{Y}_k(j\eta)$ -plane (Nyquist plane) as a *circular curve chart.*

The effect of the system parameter *damping ratio* ζ on the *frequency-response characteristic* (direct dynamic compliance or direct receptance) $\underline{C}_{ii}(j\omega)$ is represented graphically in the normalized form $\underline{Y}_k(j\eta)$ by a set (or family) of polar plots for various amounts of damping. The damping effect is indicated by the *change in dynamic compliance* of a structure, i.e., by the change in displacement amplitude and phase related to the constant-amplitude force.

Specifications of Polar Plots. A polar plot (Nyquist plot) is specified by
- the *curve shape* (base of a range of loci);
- the *frequency scale* of the curve (index numbers of frequency);
- the *shape signature at high and low frequencies* (initial slope and termination asymptotes).

Respecting the polar representation of a set of normalized frequency responses the adjusted *curve shapes* are defined by the loci (hereat bicircular quartics) for selected damping ratio values $\zeta = const$,and the *frequency scales* are outlined as intersecting curves for selected frequency ratio values $\eta = const$. The portrayed loci and frequency lines thus generate the cellular-type structure of a meshwork. Nevertheless, it should be pointed out that the traced loci are parametric curves associated with the same specified field line referred to $\sigma = 0$ (image of the imaginary axis), though for different ζ-parameter values. Accordingly the presented parametric curves form a non-orthonormal meshwork contrary to the σ,ω-field map formed by image curves for different σ-values, Fig. 3.21b.

The *shape signature* is portrayed by a locus starting perpendicularly to the real axis from unity (normalized static response) at $\eta = 0$, and terminating asymptotically on the real axis in the origin at $\eta = \infty$. The terminal slope of asymptotic approach depends on the frequency response being a ratio of two polynomials such that the degree n of the denominator is greater than m of the numerator, Eq. (3.89b). For $n - m = 2$, related to the present function $\underline{Y}_k(j\eta)$, a sluggish dynamic behaviour dominates caused by a quadratic or second-order factor as denominator. Hence, a low-pass filtering property is indicated with a flat-tuned (hazy) bandwidth.

Significant Frequencies of Resonance. Resonance of a system exists when any change, however small, in the forcing frequency causes a decrease in the forced response. This phenomenon, sometimes called *resonance of amplitude*, is associated with the radius vector's greatest length coinciding with a rectangular angle of vector and tangent to the polar plot at point of tangency. The geometric locus of the maxima of magnitudes (greatest displacement-response factors) according to the greatest compliance values for various amounts of damping (dash-and-dot line) indicates the *resonance frequency ratio* referred to the *displacement amplitude*

$$\eta_r = \omega_r / \omega_0 = \sqrt{1 - 2\zeta^2} \tag{3.39a}$$

The *(displacement) resonance frequency* ω_r is lower by a small amount than the damped natural frequency ω_d, Eq. (3.19a).

The singular value of forcing frequency $\omega_f = \omega_0$, defined in 3.1.2 as the *condition of resonance*, pertains to a phenomenon sometimes called *resonance of phase*. This is a consequence of the fact that the phase angles are 90° for all amounts of damping. The *resonance frequency ratio* $\eta_0 = 1$, Eq. (3.46a), referred to the *(displacement) phase* is associated with the radius vector's orthonormal direction to the real axis. The geometric locus of rectangular arguments (phase lags in quadrature) coincides with the (negative) imaginary axis.

Q factor and Bandwidth. In resonance testing it is convenient to obtain a measure of damping by determining the amplitude of vibration at resonance. Conversely, if the amount of damping is known it is simple to make an estimate of the amplitude of vibration at resonance. Inserting Eq. (3.39a) back to normalized frequency response, Eq. (3.198c), the *maximum compliance magnitude* is obtained

$$\left|\underline{Y}_{\mathrm{k}}(\mathrm{j}\eta)\right|_{\max} = (R_{\mathrm{d}})_{\max} = \frac{1}{2\zeta\left|\sqrt{1-\zeta^2}\right|} \quad ; \quad \zeta < \frac{1}{2}\sqrt{2} \qquad (3.39b)$$

being in conformity with the *peak of the magnification factor* and which, for small ζ, can be approximated by

$$\left|\underline{Y}_{\mathrm{k}}(\mathrm{j}\eta)\right|_{\eta_0} = Q = \frac{1}{2\zeta} = \frac{1}{2c/c_{\mathrm{c}}} \qquad (3.39c).$$

The *compliance magnitude specified by the resonance frequency ratio* $\eta = \eta_0 = 1$ is known as the *Q factor* (quality factor). This term attributed to electrical circuitry is a measure of the sharpness of resonance. For a resonant oscillatory single-degree-of freedom system, either mechanical or electrical, the Q factor amounts to large numbers thus indicating small values of damping. This vibratory specification, Eq. (3.39c), is used for determining the equivalent viscous damping coefficient c_{eq}, Eq. (3.5b).

In practice, it may be difficult to measure the damping ratio by Eqs. (3.39b,c) because the static deflection being very small must be determined for plotting the normalized frequency-response function. Therefore it is customary to apply the *bandwidth method*, sometimes referred to as the *half-power method*, which utilizes frequency-response plots without normalization.

This technique requires very accurate measurement. Once the peak of the displacement response amplitude $\left|\hat{\underline{s}}\right|_{\max} = \hat{s}_{\max}$ and the resonance frequency ω_{r} have been located, the so called half-power points can be determined. Since the energy dissipated per cycle is proportional to $\hat{s}^2$, the energy dissipated will be reduced by 50% as for as the amplitude is reduced by a factor of $1/\sqrt{2}$. Thus, the half-power displacement amplitude is defined

$$\hat{s}_{\mathrm{p}} = (1/2)\sqrt{2}\hat{s}_{\max} = 0{,}707\ \hat{s}_{\max} \qquad (3.39d)$$

pertaining to the upper and lower cut-off frequencies ω_{U}, ω_{L} either side of ω_{r}. As for lightly damped systems the (displacement) resonance frequency ω_{r} is very nearly equal to the natural frequency ω_0 the frequency ω_{r} can be approximately replaced by ω_0.

For evaluating compliance test data it is also advantageous to use the polar plot in connection with the bandwidth method. For an experimental determination of Eq. (3.39c) the measured magnitude and phase of the (driving-point) dynamic compliance $\underline{C}_{\mathrm{ii}}(\mathrm{j}\omega)$ are plotted, that means by tracing localized vectors representing graphically the response ratios of phasors (complexors) $\hat{\underline{s}}_{\mathrm{i}}/\hat{\underline{F}}_{\mathrm{i}}$ in the Nyquist plane. A smooth curve can be drawn through the ends of the radius vectors for an accurate locating of the half-power points by polar plotting. Use of the bandwidth of the response locus involves accurately determining the frequencies ω_{U}, ω_{L}, sometimes called the *half-power frequencies*, for which

$$\tan\frac{c\omega_{\mathrm{L,U}}}{k - m\omega_{\mathrm{L,U}}^2} = \pm 1 \qquad (3.39e).$$

Those points thus occur when the phase angle amounts to the particular values

$$\begin{aligned}\operatorname{arc} \underline{C}_{\mathrm{ii}}(\mathrm{j}\omega_{\mathrm{L,U}}) = -\psi_{\mathrm{L,U}} &= -\arctan\frac{c\omega_{\mathrm{L,U}}}{k - m\omega_{\mathrm{L,U}}^2} \\ &= -45^\circ \text{ and } -135^\circ\end{aligned} \qquad (3.39f).$$

Relating the *interval between the upper and lower cut-offs*, equivalent to the *bandwidth* $\Delta\omega$, to the frequency of maximum response approximated by ω_0 the damping ratio ζ, and hence the Q factor associated with any mode of vibration can be found

$$\frac{\omega_U - \omega_L}{\omega_0} = \frac{\Delta\omega}{\omega_0} = 2\zeta = \left(2c/c_c\right) = \frac{1}{Q} \qquad (3.39g).$$

The method is also effective when the damping is *hysteretic* because the locus as ω increases is part of a circle [31].

Though being reserved to the interpretation of mobility test data an advantage of polar plots worth to be mentioned permits that the data may be enhanced by *circle-fitting procedures*. This becomes important for extracting modal damping coefficients from the test data, see [18]: Alternative plotting methods.

Logarithmic Frequency Response Locus (Nichols plot)

A graphical representation of frequency response locus being alternative to the polar plot is based on a plot in terms of parallel (Cartesian) coordinates by use of logarithmic transformation.

With respect to the proved manual technique of system design in controls for obtaining closed-loop response information from the open-loop data (Nichols chart) this graphical interpretation is termed the *Nichols plot* (after Nathaniel B. Nichols).

Starting from the polar form, Eqs. (3.79), (3.115), the frequency-response function can be transformed into a complex logarithmic function written in the natural logarithm of $G(j\omega)$

$$\begin{aligned}\ln G(j\omega) &= \ln\left[\left|G(j\omega)\right| e^{j \operatorname{arc} G(j\omega)}\right] \\ &= \ln\left|G(j\omega)\right| + j \operatorname{arc} G(j\omega)\end{aligned} \qquad (3.200).$$

The result permits to express the complex system parameter by its modulus and argument in terms of a real and an imaginary part. The real part is equal with the *logarithmic amplitude (frequency-)response*

$$\ln\left|G(j\omega)\right| = \ln A(\omega) \qquad (3.201a),$$

often referred to as the *logarithmic gain*, and the imaginary part equals the *phase (frequency-) response*, Eq. (3.80b)

$$\operatorname{arc} G(j\omega) = -\psi(\omega) \qquad (3.201b).$$

Remark on Logarithmic Quantities and Unities. Basing on the previously treated frequency-response normalization, Eqs. (3.198c), (3.199a), the logarithm of a non-dimensional displacement response factor (magnification factor R_d or normalized amplitude ratio A_2 / A_1), Eq. (3.38b), can be interpreted as a *response level of amplitudes* L_R of dimension one equal to unity.

The *neper*, N_p, being a special name for unity, is used as a unit for logarithmic quantities. In practice, the non-coherent unit *bel*, B, based on common (Briggs') logarithms lg (base 10), preferably the sub-multiple *decibel*, dB, is widely used.
Generally

$$L_R = \ln\left(A_2/A_1\right)\mathrm{Np} = 2\lg\left(A_2/A_1\right)\mathrm{B} = 20\lg\left(A_2/A_1\right)\mathrm{dB}$$
$$\text{with } 1\,\mathrm{dB} = \left[(\ln 10)/20\right]\mathrm{Np} = 0{,}1151293\,\mathrm{Np}.$$

Especially

$$\left|\frac{G(\mathrm{j}\omega)}{G(0)}\right|_{\mathrm{dB}} = 20\,\mathrm{lg}\left|\frac{G(\mathrm{j}\omega)}{G(0)}\right| \mathrm{dB} \tag{3.202}$$

accordingly normalized by the static response factor $G(0)$ as the proper reference magnitude [49].

Special Features of the Nichols Plot. Due to the simplicity of algebraic relations the advantages of the transfer function representation turned out in 2.1.1 also hold for a frequency-response function representation. In particular with regard to the cascade or tandem connection being the most important of all interconnections of components the overall frequency-response function is equal to the multiple product of the individual frequency responses. Using logarithmic amplitude responses the tedious vector multiplication of response complexors following polar plotting can be replaced by gain addition for each component which is easy to carry out graphically based on the relation:

$$\ln \prod_{\lambda=1}^{\ell} G_{\lambda}(\mathrm{j}\omega) = \sum_{\lambda=1}^{\ell} \ln\left|G_{\lambda}(\mathrm{j}\omega)\right| + \mathrm{j}\sum_{\lambda=1}^{\ell} \mathrm{arc}\, G_{\lambda}(\mathrm{j}\omega) \tag{3.203}$$

- The *resonance of amplitude* being associated with the greatest (logarithmic) gain is indicated by a horizontal tangent to the logarithmic frequency-response locus.
- A *change of (proportional) gain* of the system coinciding with a change of the static response factor $G(0)$ merely shifts the locus parallel to the abscissa.
- The *inverse of the locus*, in this case, the dynamic stiffness $\underline{K}(\mathrm{j}\omega)$ as the inverse of the dynamic compliance $\underline{C}(\mathrm{j}\omega)$, introduces a negative sign due to the relations

  $$1/\mathrm{lg}\, R_{\mathrm{d}} = -\,\mathrm{lg}\, R_{\mathrm{d}} \quad ; \quad \arctan(-\psi) = -\arctan\psi \tag{3.204}$$

  so that the curve shape remains invariant by inversion while the scales of coordinate axes must be mirrored only.

Response Data Plotting. The logarithmic frequency-response loci (Nichols plots) of the *normalized dynamic compliance* $\underline{Y}_{\mathrm{k}}(\mathrm{j}\eta)$ are shown in Fig. B.2 of the Appendix B.

The new coordinates being *rectangular axes* of magnitude and phase are differently scaled. The ordinate of magnitudes is logarithmically scaled by both, absolute numbers and numbers graduating the non-dimensional quantity in the special unit decibel, and the abscissa of phase angles is uniformly scaled in the unit degree, thus, generating the complex $\mathrm{lg}\,\underline{Y}_{\mathrm{k}}(\mathrm{j}\eta)$ -plane (Nichols plane) as a *semilogarithmic curve chart.*

Respecting the rectangular representation (Nichols plot) of a set of normalized frequency responses the curves are adjusted by the logarithmic transformation

$$\begin{aligned}\mathrm{lg}\frac{\underline{C}_{\mathrm{ii}}\left(\mathrm{j}\frac{\omega}{\omega_0}\right)}{C_{\mathrm{k}}} &= \mathrm{lg}\left|\underline{Y}_{\mathrm{k}}(\mathrm{j}\eta)\right| + \mathrm{j}\,\mathrm{arc}\,\underline{Y}_{\mathrm{k}}(\mathrm{j}\eta)\cdot \mathrm{lg}\,\mathrm{e} \\ &= -\,\mathrm{lg}\left|\sqrt{(1-\eta^2)^2 + (2\zeta\eta)^2}\right| - \mathrm{j}\arctan\frac{2\zeta\eta}{1-\eta^2}\cdot \mathrm{M}_{10}\end{aligned} \tag{3.205}$$

into common logarithms with the modulus of Briggs' logarithmic system $M_{10} = \lg e = 0{,}4343$ which is ordinarily omitted from the imaginary part of Eq. (3.205). Henceforth the adjusted *curve shapes* are defined by the loci (hereat higher parabolic plane curves) for selected damping ration values $\zeta = const$, and the *frequency scales* are outlined as intersecting curves for selected frequency ratio values $\eta = const$. The *shape signature* is portrayed by a locus starting horizontally from the amplitude ratio unity, $R_{\mathrm{d}} = 1$, respectively from the pertinent amplitude level in the special unit decibel, $L_{\mathrm{R}} = 0\,\mathrm{dB}$, versus $0°$ at $\eta = 0$, and terminating asymptotically versus the –180° phase line at $\eta = \infty$.

The geometric locus of the greatest (logarithmic) gains being associated with the *resonance of amplitude* (dash-and-dot line) indicates the resonance frequency ratios η_{r}, whereas the vertical intersecting line $\eta = 1$ coinciding with the –90° phase line is associated with the *resonance of phase* for all the damping amounts.

Logarithmic Frequency Plots (Bode plot)

A common alternative to the frequency response locus in the complex $\lg G(\mathrm{j}\omega)$-plane (Nichols plane) results from the graphical representation of rectangular frequency responses, i.e., by portraying the complex system parameter, Eq. (3.200), separately by its component parts versus the logarithmic frequency. The pair of real-valued frequency response curves is called the *logarithmic frequency plots.*

With respect to the development of design methods based on the open-loop frequency response, in particular by illustrating performance criteria of stability (phase and gain margins) for setting the parameters of controllers, this graphical interpretation is termed the *Bode plot* (after Henryk W. Bode).

Special Features of the Bode Plot. The advantages of logarithmic transformation already summarized with reference to the Nichols plot are largely preserved. This proves true for several comparative properties such as the gain shifting property. In case of a *change of gain constant* the total gain will be shifted parallel to the ordinate by that amount:

- Apart from being easier to construct the logarithmic frequency plots combine additional advantages the most significant of which are worth being mentioned.
- The *range of forcing frequency* on a logarithmically scaled axis of abscissas ($\lg\omega$-abscissa) is much larger.
- The *graphical addition of contributions* made by the various terms of a factored frequency-response function permits the plotting of composite curves using specified types of product terms (building blocks).
- The *straight-line approximation* reduces the frequency response plotting to a polygon (or broken line) sketching owing to the straight-line slopes and breakpoint (or corner) frequencies (angular) of the individual terms.
- More *accurate plots* can be sketched by tracing back to standard terms.

For a simplified straight-line approximation restricting the specified product terms only to first order factors (P-T_1-elements) with real roots the composite frequency function in the *factored form of a time-constant representation*

$$G(\mathrm{j}\omega) = \mathrm{K}_{\mathrm{G}_0} \frac{\prod\limits_{\mu=1}^{m}(1+T_{\mathrm{D}\mu}\mathrm{j}\omega)}{\prod\limits_{\nu=1}^{n}(1+T_{\mathrm{s}_\nu}\mathrm{j}\omega)} = \mathrm{K}_{\mathrm{G}_0} \frac{(1+T_{\mathrm{D}1}\mathrm{j}\omega)\cdots}{(1+T_{\mathrm{s}_1}\mathrm{j}\omega)(1+T_{\mathrm{s}_2}\mathrm{j}\omega)\cdots(1+T_{\mathrm{s}_n}\mathrm{j}\omega)} \quad (3.206\text{a})$$

describes in general the overall behaviour deduced from components most commonly used as *building blocks*.

The system parameters are given by

the *static response factor* (proportional gain): $K_{\mathrm{G}0} = K_\mathrm{p}$,

the *rate times* $T_{\mathrm{D}\mu}$,

the *delay times, time constants* $T_{\mathrm{s}\nu}$,

respectively by their reciprocals:

the *breakpoint frequencies*, also termed $\omega_{\mathrm{cf,D}\nu} = 1/T_{\mathrm{D}\mu}$

the *corner (angular) frequencies* $\omega_{\mathrm{cf,s}\nu} = 1/T_{\mathrm{s}\nu}$.

Expressed for logarithmic quantities using decibel units, Eq. (3.202), the rectangular frequency responses result in

$$\begin{aligned} |G(\mathrm{j}\omega)|_{\mathrm{dB}} &= 20\lg K_{\mathrm{G}0} + 20\sum_{\mu=1}^{m}\lg\left|1+T_{\mathrm{D}\mu}\mathrm{j}\omega\right| - 20\sum_{\nu=1}^{n}\lg\left|1+T_{\mathrm{s}\nu}\mathrm{j}\omega\right| \mathrm{dB} \\ \operatorname{arc} G(\mathrm{j}\omega) &= \operatorname{arc} K_{\mathrm{G}0} + \sum_{\mu=1}^{m}\operatorname{arc}\left(1+T_{\mathrm{D}\mu}\mathrm{j}\omega\right) - \sum_{\nu=1}^{n}\operatorname{arc}\left(1+T_{\mathrm{s}\nu}\mathrm{j}\omega\right) \end{aligned} \quad (3.206\text{b}),$$

so that the composite logarithmic frequency plots (Bode plot) can be constructed by summing up the contributions of 1st order factors. Both, the logarithmic gains and the phase angles in the numerator are added to the gain of the constant factor, while those in the denominator are subtracted. The gain constant $K_{\mathrm{G}0}$ is simply a horizontal straight line, whereas the 1st order gains are approximately represented by the asymptotes of two limiting cases. Approaching very small or else very high frequencies the component curves of each individual term $(1+\mathrm{j}T\omega)^{\pm 1}$ are approximated by a horizontal straight line broken upward or downward by a straight line of unity slope ±1, equalling ±20 dB/decade. The intersection of each of the *low-frequency* and the *high-frequency asymptotes* yields an appropriate *breakpoint (or corner) frequency* ω_{cf}. The composite curve is approximated by "blending" the component asymptote segments ranging from the pertinent breakpoint frequency to one of the limiting frequencies.

The approximative gain curve together with the phase curve may provide information enough, at any rate it facilitates the sketching of exact composite curves.

The composite frequency function without derivative action (absence of numerator dynamics due to missing lead elements: $T_{\mathrm{D}\mu} = 0$)

$$G(\mathrm{j}\omega) = \underline{C}(\mathrm{j}\omega) = C_\mathrm{k} \frac{1}{(1+T_{\mathrm{s}_1}\mathrm{j}\omega)(1+T_{\mathrm{s}_2}\mathrm{j}\omega)} \quad (3.207\text{a})$$

describes in particular the behaviour of a passive mechanical second-order system characterized by denominator factors (lag elements: $T_{s\nu} > 0, n = 2$) dominating with growing frequency.

The generic mechanical system the motion variable of which being expressed by a displacement has a complex frequency characteristic $G(\mathrm{j}\omega)$ designated as dynamic compliance $\underline{C}(\mathrm{j}\omega)$. This characteristic consists of a multiplying constant and of two 1st-order storage elements as building blocks.

The system parameters already defined, Eqs. (3.12), (3.28a), (3.28b), (3.199a), are

the compliance of the spring : $K_{G0} = K_p = C_k = 1/k$,

and the delay times : T_{s1}, T_{s2} .

Those time constants, Eqs. (3.28a), (3.28b), are real-valued only in case of *greater-than-critical damping*, thus, implying that Eq. (3.24) holds true.

On account of Eq. (3.206b) the *logarithmic dynamic compliance* of the *over-damped system* is represented by the composite logarithmic gain

$$\begin{aligned}|\underline{C}(\mathrm{j}\omega)|_{\mathrm{dB}} &= 20\,\lg C_k - 20\sum_{\nu}^{2} \lg|1 + T_{s\nu}\mathrm{j}\omega| = 20\lg\left|\frac{1}{k}\right| - 20\sum_{\nu}^{2}\lg\sqrt{1+(T_{s\nu}\omega)^2} \\ &= const - 10\,\lg\left(1 + T_{s1}^2\omega^2\right) - 10\,\lg\left(1 + T_{s2}^2\omega^2\right)\,\mathrm{dB}\end{aligned} \qquad (3.207\mathrm{b}),$$

and the composite phase angle

$$\begin{aligned}\operatorname{arc}\,\underline{C}(\mathrm{j}\omega) &= \operatorname{arc}\,C_k - \operatorname{arc}\,(1 + T_{s1}\mathrm{j}\omega) - \operatorname{arc}\,(1 + T_{s2}\mathrm{j}\omega) \\ &= 0 - \arctan(T_{s1}\omega) - \arctan(T_{s2}\omega)\end{aligned} \qquad (3.207\mathrm{c}).$$

Accordingly the Bode plot of dynamic compliance can be constructed by subtracting the gains from the compliance of the spring, and the phase angles from zero, respectively.

In the *critical damping case* the delay times coincide with the time constant T related to the natural period of oscillation T_0, Eq. (3.11), and the corner frequency ω_{cf} equals the natural frequency ω_0, Eq. (3.6a). It follows

$$T_{s1} = T_{s2} = T = \frac{1}{\omega_0} \quad ; \quad \omega_{cf} = \frac{1}{T} = \omega_0 \qquad (3.208\mathrm{a})$$

with the *normalized time constant* τ

$$\tau_{s1} = \tau_{s2} = \tau = T\omega_0 = 1 \quad ; \quad \eta_{cf} = \frac{1}{\tau} = \frac{1}{T\omega_0} = 1 \qquad (3.208\mathrm{b}),$$

and the *corner frequency ratio* η_{cf} , both of which are *unity*.

Response Data Plotting. The logarithmic frequency plots (Bode plots) of the *normalized dynamic compliance* $\underline{Y}_k(\mathrm{j}\eta)$ are shown in Fig. B.3 of the Appendix B.

The new coordinates being *rectangular axes* of either magnitude versus frequency or phase versus frequency are differently scaled. The ordinate of magnitudes is logarithmically scaled by both, absolute numbers and those in the special unit decibel, whereas the ordinate of phase is uniformly scaled by numbers in the units radian and degree. The abscissas of frequency are logarithmically scaled for both curve plottings. The rectangular coordinates thus generate coordinate frames for a pair of real-valued frequency plots at first as a *full-logarithmic*, at second as a *semilogarithmic curve chart* [49].

Respecting the rectangular representation (Bode plot) of a set of normalized frequency responses the *pair of adjusted curves* is defined by the amplitude response (logarithmic gain) and the phase response, both for selected damping ratio values $\zeta = const$ and for a logarithmic *range of frequency ratio* varying over two decades $\lg(\eta_2 / \eta_1) = \lg(10/0{,}1) = 2$.

The portrayed pair of logarithmic frequency plots shows clear *resemblance to the frequency-response curves* in terms of ordinary rectangular coordinates. Those uniformly scaled plots are proper for the common representation of frequency characteristics in vibrations known as the amplitude- and phase-frequency ratio responses. Both are graphical interpretations of the already introduced (dynamic) *magnification factor* (non-dimensional displacement response factor) R_d, Eq. (3.38b), and of the *phase difference* (phase lag) ψ, Eq. (3.40b). Uniformly scaled frequency response plots are commonly used to determine *vibratory specifications*. So the interval $\Delta\omega$ between cut-off frequencies serves the ascertainment of damping ratio ζ and Q factor basing on the half-power points, Eqs. (3.39d), (3.39f). The bandwidth $\Delta\omega$, already introduced by use of the Nyquist plot, in addition gives evidence in transmitting properties (filtering properties) of the system.

The determination of *frequency-domain specifications* takes advantage of logarithmically scaled plots (Bode plot). So the bandwidth $\Delta\omega$ defines a single transmission band extending from zero frequency up to a finite frequency, thus indicating a *low-pass filter*. The finite frequency then will be specified by the half-power frequency occuring when the power consum of the system reduces to one-half the maximum value at $\omega = 0$. By relating the dynamic compliance $\underline{C}_{ii}(j\omega)$ to the multiplying constant "compliance of the spring" C_k, Eq. (3.199a), the system frequency characteristic is adjusted in variable and scale, hence the logarithmic frequency response in new coordinates starts from the normalized dynamic compliance $\underline{Y}_k(j\eta)$ of the overdamped system

$$\frac{\underline{C}_{ii}(j\frac{\omega}{\omega_0})}{C_k} = \underline{Y}_k(j\eta) = \frac{1}{(1 + j\tau_{s1}\eta)(1 + j\tau_{s2}\eta)} \tag{3.209a}$$

normalized (logarithmic) gain of dynamic compliance

$$\begin{aligned} \left|\underline{Y}_k(j\eta)\right|_{dB} &= 20\lg 1 - 20\sum_{\nu}^{2}\sqrt{1 + (T_{s\nu}\eta)^2} \\ &= 0 - 10\lg\left(1 + T_{s1}^2\eta^2\right) - 10\lg\left(1 + T_{s2}^2\eta^2\right)\,\text{dB} \end{aligned} \tag{3.209b}$$

An amplitude response dropping to 0,707 being equivalent to the *logarithmic gain dropping of* 3 dB below the "peak value" of unity or 0 dB at $\eta = 0$ according to

$$\frac{\left|\underline{Y}_k(j\eta_U)\right|}{\underline{Y}_k(0)} = 0{,}707\,; \quad \left|\underline{Y}_k(j\eta_U)\right|_{dB} = 20\lg 1 + 20\lg 0{,}707\,\text{dB} = 0\,\text{dB} - 3{,}01\,\text{dB} = -3\,\text{dB} \tag{3.209c}$$

shows that the *pass-band* $\Delta\eta$ is thus defined by the
upper cut-off frequency ratio $\Delta\eta$

$$\frac{\Delta\omega}{\omega_0} = \Delta\eta = \eta_U - 0 = \eta_U \tag{3.210a}$$

Within this single pass-band $\Delta\eta$ the attenuation to oscillations is relatively small (–3dB), whereas the attenuation to oscillations of frequencies beyond the upper cut-off frequency is relatively large.

This holds true for magnitude plots without an increase at resonance ($\zeta > (1/2)\sqrt{2}$), even for those with sufficiently small increase in magnitude ($\zeta > 0{,}5$).

Regarding the underdamped case the time-constant representation basing on 1st order factors, Eq. (3.207a), is no more suited. Therefore, the frequency function of the system will be represented only by a 2nd-order term (*P-T₂*-element) implying complex roots

$$\frac{\underline{C}_{ii}(\mathrm{j}\frac{\omega}{\omega_0})}{C_k} = \underline{Y}_k(\mathrm{j}\eta) = \frac{1}{1 + \mathrm{j}2\zeta\eta - \eta^2} \qquad (3.211a)$$

normalized (logarithmic) gain of dynamic compliance

$$\left|\underline{Y}_k(\mathrm{j}\eta)\right|_{\mathrm{dB}} = -20\lg\sqrt{\left(1-\eta^2\right)^2 + (2\zeta\eta)^2} = -10\lg\left[\left(1-\eta^2\right)^2 + (2\zeta\eta)^2\right]\mathrm{dB} \qquad (3.211b).$$

For magnitude plots with sharp resonant amounts $(\zeta \ll 1)$ the bandwidth $\Delta\eta$, defines by use of 3 dB magnitude drop a single transmission band extending from a lower cut-off frequency greater than zero to a finite upper cut-off frequency, thus indicating a *band-pass filter*. The pass-band centre frequency coincides with the (displacement) resonance frequency ω_r being approximated by the natural frequency ω_0

$$\frac{\left|\underline{Y}_k(\mathrm{j}\eta_{L,U})\right|}{\underline{Y}_k(\mathrm{j}\eta_0)} = 0{,}707\,; \quad \left|\underline{Y}_k(\mathrm{j}\eta_{L,U})\right|_{\mathrm{dB}} = 20\lg\left|\frac{1}{2\zeta}\right| + 20\lg 0{,}707\ \mathrm{dB} = -20\lg|2\zeta|\ \mathrm{dB} - 3\ \mathrm{dB} \qquad (3.211c).$$

The *pass-band* is thus defined by the interval between the upper and the lower cut-offs

$$\Delta\eta = \eta_U - \eta_L \qquad (3.210b).$$

Since the pass-band width is relatively narrow, the frequency selectivity at resonance is a sharp one, the oscillatory system has the property of a *narrow-band filter* with a relatively large attenuation to oscillations at frequencies out of both of the cut-off frequencies.

Using the Bode plot a further understanding of system's behaviour in the regions of small and large frequency ratios can easily be visualized by *straight-line approximation*, see Fig. B.4 of the Appendix B.

Following Eq. (3.207b) and specifying the 2nd-order system as *critically damped* the two 1st-order factors of the denominator are identical, thus the only one system parameter given by the normalized time constant and its reciprocal is reduces to unity, Eq. (3.208b). The composite logarithmic gain and phase angle of the normalized dynamic compliance $\underline{Y}_k(\mathrm{j}\eta)$ are then

$$\begin{aligned}\left|\underline{Y}_k(\mathrm{j}\eta)\right|_{\mathrm{dB}} &= -20\lg\left(1+\tau^2\eta^2\right) = -20\lg\left(1+\eta^2\right)\mathrm{dB}\\ \operatorname{arc}\underline{Y}_k(\mathrm{j}\eta) &= -2\arctan(\tau\eta) = -2\arctan\eta\end{aligned} \qquad (3.212a).$$

Considering the limiting cases it follows a *low-frequency asymptote*

$$\left|\underline{Y}_k(\mathrm{j}\eta)\right|_{\mathrm{dB}} \approx -20\lg 1 = 0\ \mathrm{dB} \qquad \text{for } \eta \ll 1 \qquad (3.213a)$$

being horizontal, and a *high-frequency asymptote*

$$\left|\underline{Y}_k(\mathrm{j}\eta)\right|_{\mathrm{dB}} \approx -20\lg\eta^2 = -40\lg\eta\ \mathrm{dB} \qquad \text{for } \eta \gg 1 \qquad (3.213b)$$

having a slope of $-40\ \mathrm{dB/decade}$. It indicates that the high-frequency slope of a 2nd-order system is steeper than that of a 1st-order system $(-20\ \mathrm{dB/decade})$.

The high-frequency and low-frequency asymptotes intersect at the *corner frequency ratio* $\eta_{cf} = (1/\tau) = 1$. The component asymptote segments determined by Eqs. (3.213a,b) are

identical with the logarithmic gains of storage element compliances $|C_k|$, $|C_m|$. Relating them to the multiplying constant "compliance of the spring" C_k, Eq. (3.199a), the element frequency characteristics are adjusted in variable and scale to the
normalized compliance of the spring Y_{kk} (magnitude)

$$|Y_{kk}| = |C_k|/C_k = |1/k|/(1/k) = 1 \tag{3.214a}$$

normalized compliance of the mass Y_{km} (magnitude)

$$|Y_{km}| = |C_m|/C_k = |-1/(m\omega^2)|/(1/k) = k/(m\omega^2) = 1/\eta^2 \tag{3.214b}$$

leading to the
normalized (logarithmic) gains of storage compliances

$$|Y_{kk}|_{dB} = 0 \text{ dB} \quad ; \quad |Y_{km}|_{dB} = -40 \lg|\eta| \text{ dB} \tag{3.215a}$$

As a result of Eqs. (3.213a,b), (3.215a) it is obvious that the *composite frequency behaviour* of the structure is sufficiently interpreted by the *compliance of the spring* $|Y_{kk}|$ in the region of *small frequencies*, or else by the *compliance of the mass* $|Y_{km}|$ corresponding to *large frequencies*. Accordingly the element behaviour is illustrated by its frequency effect on the magnitude plot (covered frequency band). So the compliance of the mass causes the dynamic compliance curve of the composite structure to break down at the corner frequency ratio η_{cf}, thus indicating the system's filtering property of a *low-pass filter*.

Regarding the *underdamped case* the straight-line sketching ceases to be suitable for an accurate construction of the Bode plot. Indeed, for lightly damped systems the exact curves deviate substantially from the asymptotes in the vicinity of the corner frequency. The *correction to the asymptotic approximation* at the corner frequency being available in tabular form for distinct amounts of damping can be used to detail the region of resonance surrounding the corner frequency. The detail is marked in the magnitude plot of Fig. B.4 specified for $\zeta = 0{,}2$.

The correction amount can be evaluated by the difference of the logarithmic gain of dissipator element compliance $|C_c|$ (pointed tangent) at the corner frequency ratio $\eta_{cf} = (1/\tau) = 1$ to the low-frequency asymptote, Eq. (3.210a). Defining with respect to Eqs. (3.214a,b) the
normalized compliance of the damper Y_{kc} (magnitude)

$$|Y_{kc}| = |C_c|/C_k = |-j/(c\omega)|/(1/k) = k/(c\omega) = 1/(2\zeta\eta) \tag{3.214c}$$

normalized(logarithmic) gain of dissipator compliance

$$\lg|Y_{kc}|_{dB} = -20 \lg|2\zeta\eta| \text{ dB} \tag{3.215b}$$

the *correction gain* for $\zeta = 0{,}2$ follows from Eq. (3.210a)

$$0 + (-20 \lg 0{,}4) = -20 \lg 0{,}4 = 7{,}96 \text{ dB}. \tag{3.215c}$$

Though it is more accurate to extend the composite frequency function, Eq. (3.206a), by 2nd-order standard terms with complex roots there is no advantage with reference to straight-line approximation. The logarithmic gains being convergent at the limiting frequencies for all damping amounts are approximated by the same asymptotes. It is irrelevant to the goodness of approximation to deduce the asymptotes either from a quadratic or from linear denominator factors representing only the overdamped (or critically damped) system. Nevertheless, the construction of the exact composite curve requires the contribution of a 2nd-order standard term (building block), Eqs. (3.211a,b), for portraying the resonance phenomenon of the underdamped system.

Since the significant response is given by the resultant force to an exciting motion it proves convenient to represent graphically the reciprocal of the (direct) frequency-response function $G_{ii}^{-1}(j\omega_f)$ defining the (direct) dynamic stiffness of a structure $\underline{K}_{ii}^{-1}(j\omega_f)$, Eq. (3.188b). Accordingly the *reciprocal of the static response factor* $G^{-1}(0)$ is identical with
the *stiffness of the spring*

$$K_k = 1/C_k = k \qquad (3.199b).$$

If the inverse complex system parameter $G_{ii}^{-1}(j\omega_f/\omega_0)$ varying with the new variable η will be related to the multiplying constant $G^{-1}(0)$ the transformation in variable and scale results in a normalized response ratio of phasors (at the same point *i*), which being already defined by Eqs. (3.188b), (3.198c) is designated the *normalized (direct) dynamic stiffness* of a structure $\underline{X}_k(j\eta_1)$

$$\frac{\underline{K}_{ii}\left(j\frac{\omega_f}{\omega_0}\right)}{K_k} = \left(\frac{\hat{\underline{F}}_i}{\hat{\underline{s}}_i}\right)\bigg/k = \frac{\hat{\underline{F}}_i}{\hat{\underline{s}}_i k} = 1/\underline{Y}_k(j\eta_1) = \underline{X}_k(j\eta_1) = 1 + j2\zeta\eta_1 - \eta_1^2 \qquad (3.216).$$

The adjusted (data-reduced) polar plot of $\underline{X}_k(j\eta)$, Eq. (3.216), representing the inverse frequency response of $\underline{Y}_k(j\eta)$, Eq. (3.198c), which is graphically interpreted in detail and drawn in Fig. B.1 of the Appendix B, also can be plotted in terms of polar coordinates (Nyquist plot) as the frequency ratio η varies from 0 to ∞.

Response Data Plotting. The polar frequency-response loci (Nyquist plots) of *normalized dynamic stiffness* $\underline{X}_k(j\eta)$ are shown in Fig. B.5 of the Appendix B.

The effect of the system parameter *damping ratio* ζ on the *inverse frequency-response characteristic* (direct dynamic stiffness $\underline{K}_{ii}(j\omega)$) is represented graphically in the normalized form $\underline{X}_k(j\eta)$ by a set (or family) of polar plots for various amounts of damping. The damping effect is indicated by the *change in dynamic stiffness* of a structure, i.e., by the change in force amplitude and phase related to the constant-amplitude displacement.

The adjusted *curve shapes* are defined by the loci (hereat Runge parabolas) for selected damping ratio values $\zeta = const$, and the *frequency scales* are outlined as intersecting vertical straight lines for selected frequency ratio values $\eta = const$. The *shape signature* is portrayed by a locus starting perpendicularly to the real axis from unity (normalized static response) at $\eta = 0$, and terminating in fanwise spreading curves at $\eta = \infty$. The degree *n* of the denominator being smaller than *m* of the numerator, hereat $m - n = 2$, related to the present function $\underline{X}_k(j\eta)$ a fast dynamic behaviour dominates caused by the derivative action of a second-order factor as numerator (numerator dynamics). Hence, large magnitude response ratios of dynamic stiffness respecting major resonant vibrations indicate the high-pass filtering property.

Resonance of amplitude is associated with the radius vector's shortest length coinciding with a rectangular angle of vector and tangent to the polar plot at point of tangency. The geometric locus of the minima of magnitudes (smallest displacement response factors) according to the smallest stiffness values for various amounts of damping (dash-and-dot line forming a semiellipse) indicates the *resonance frequency ratio* η_r, Eq. (3.39a). The *frequency ratio* $\eta_0 = 1$ pertaining to the *resonance of phase* is associated with the geometric locus of rectangular arguments (phase leads in quadrature) that coincides with the (positive) imaginary axis.

In particular with regard to the cascade or tandem connection and the advantage of carrying out graphically with ease the vector multiplication of response complexors the logarithmic transformation will be applied likewise to the inverse complex system parameter $G_{ii}^{-1}(j\omega_f/\omega_0)$ related to the multiplying constant $G^{-1}(0)$.

Response Data Plotting. The logarithmic frequency-response loci (Nichols plots) of *normalized dynamic stiffness* $\underline{X}_k(j\eta)$ are shown in Fig. B.6 of the Appendix B.

One of the special features of the Nichols plot is that the curve shape remains invariant by inversion so that the dynamic stiffness loci, Fig. B.6, are *mirror images* of the dynamic compliance loci, Fig. B.2, being reflected in the unity- or 0dB-line coincident to the *axis of symmetry*.

The logarithmic frequency plots (bode plots) of *normalized dynamic stiffness* $\underline{X}_k(j\eta)$ are shown in Fig. B.7 of the Appendix B.

Considering that the advantages of logarithmic transformation are preserved by tracing the component parts separately versus the logarithmic frequency the magnitude and phase plots of dynamic stiffness, Fig. B.7, are also *mirror inversions* of dynamic compliance curves, Fig. B.3, by reflecting both in the unity- or 0dB-line and in the 0^0 or 0rad-line, respectively.

System's behaviour in the regions of small and large frequency ratios can be visualized by *straight-line approximation*, see Fig. B.8 of the Appendix B.

It is easy to show, that the composite logarithmic gain and phase angle of the normalized dynamic stiffness $\underline{X}_k(j\eta)$

$$|\underline{X}_k(j\eta)|_{dB} = 20\lg(1+\tau^2\eta^2) = 20\lg(1+\eta^2)\,\text{dB}$$

$$\text{arc}\,\underline{X}_k(j\eta) = 2\arctan(\tau\eta) = 2\arctan\eta \qquad (3.212b)$$

are the mirrored normalized dynamic compliance curves, Eq. (3.212a). Thus, the slopes of the high-frequency asymptotes do not differ in steepness but only in sign with reference to Eqs. (3.213a,b), and the *high-frequency* and *low-frequency asymptotes* intersect at the same *corner frequency ratio* $\eta_{cf} = (1/\tau) = 1$. The component asymptote segments determined by the inverse (positive-valued) Eqs. (3.213a,b) are identical with the logarithmic gains of storage element stiffnesses $|K_k|$, $|K_m|$. Relating them, and equally the dissipator element stiffness $|K_c|$, to the multiplying constant "stiffness of the spring" K_k, Eq. (3.199b), the element frequency characteristics are the

normalized stiffness of the spring X_{kk} (magnitude)

$$|X_{kk}| = |K_k|/K_k = k/k = 1 \qquad (3.216a),$$

normalized stiffness of the mass X_{km} (magnitude)

$$|X_{km}| = |K_m|/K_k = |-m\omega^2|/k = (m\omega^2)/k = \eta^2 \qquad (3.216b),$$

normalized stiffness of the damper X_{kc} (magnitude)

$$|X_{kc}| = |K_c|/K_k = |jc\omega|/k = (c\omega)/k = 2\zeta\eta \qquad (3.216c)$$

leading to the *normalized (logarithmic) gains of store stiffnesses*

$$|X_{kk}|_{dB} = 0\,dB \quad ; \qquad |X_{km}|_{dB} = 40\ lg\,|\eta|\,dB \qquad (3.217a),$$

and *of dissipator stiffness*, respectively

$$|X_{kc}|_{dB} = 20\ lg\,|2\zeta\eta|\,dB \qquad (3.217b).$$

As a result of Eqs. (3.216a,b), (3.217a) it is obvious that the *composite frequency behaviour* of the structure is sufficiently interpreted by the *stiffness of the spring* $|X_{kk}|$ in the region of *small frequencies*, or else by the *stiffness of the mass* $|X_{km}|$ corresponding to *large frequencies*. Accordingly the element behaviour is illustrated by its frequency effect on the magnitude plot (covered frequency band). So the stiffness of the mass causes the dynamic stiffness curve of the composite structure to break upward at the corner frequency ratio η_{cf} thus indicating the system's filtering property of a *high-pass filter*.

The *correction gain*, already determined for $\zeta = 0{,}2$ following Eq. (3.210a), is a deviation of the same value but of negative sign (of opposite direction).

3.3 Comparison of the Fourier and the Laplace Transform Methods (References to Applications)

Integral transforms being well suited to the analysis of model systems the behaviour of which is governed by linear differential equations are applied to a variety of engineering problems. Compared to the classical method treated in 3.1 *integral transform methods* are a mathematical technique that now as before is appropriate to solve and explain engineering problems in like manner, 3.2.

The *Laplace transform* widely considered as resulting from a broader and more basic method apparently has displaced the Fourier integral as the main tool of analysis. One reason seems to be the belief that the Laplace transform can handle a more general class of functions. But this is only apparent provided that the extension of integral transform theorems to *generalized functions* (distributions) comes into use in view of engineering problems, [36]. Above all, modern frequency concepts in vibration data analysis, shortly denoted *spectral analysis*, have brought anew *Fourier transforms* into focus though being obtained from digital data. Spectral analysis implies the data reduction of time histories to the frequency domain representing transient (shock) functions just as random functions, [42], [43], [111].

3.3.1 Fourier Transform Method. *Advantages and Disadvantages*

The Fourier transform is an essential tool of analysis of linear time-invariant systems implying features suited to particular problems in vibrations as follows.

Advantages

- On condition that the Fourier transform of a function exists this integral representation is more illustrative in the sense that the transform is a function of the real frequency ω instead of the complex frequency s (or p). For purposes of physical interpretation of actual measured data, the *Fourier spectrum representation* (continuous spectrum) has a greater intuitive appeal with respect to the frequency content of *non-periodic signals* (spectral density functions), 3.2.2.

- The Fourier transformation being primarily adapted for *time-limited signals* hence appears to be fitted to non-periodic vibrations caused by *forcing functions suddenly applied and then removed* (transient excitations of pulse-type such as shock pulses), 3.2.2. This integral representation can be visualized as a *logical extension* (passing to limit) *of approved Fourier techniques* fitted to periodic (multi-sinusoidal) signals (pulse-train excitations such as shock pulsatings) and their representation by the Fourier spectrum (line spectrum), 3.2.1.
- The Fourier transform is extensively used in connection with *statistical methods* for analysing vibrating systems. The integral transform representation appears as the spectral decomposition of a stationary (or slowly time-varying) *random process* specified by the transform of the autocorrelation function (power spectral density), 3.2.5.
- The frequency-domain analysis uses for the experimental determination of system parameters or mode shapes *data-processing methods* basing on the *discrete Fourier transform* (DFT) which is an estimate of the continuous Fourier transform (applied to mobility, respectively dynamic compliance measurements).

Disadvantages

- The *class of time functions* being Fourier-transformable is *restricted* thus covering only a part of time-varying functions occuring in practice, and so the steady (or continuously time-varying) sinusoidal signals (simple harmonic excitations) are excluded.
  This *difficulty* can be *overcome* basing on mathematically rigorous derivations by extending the transformation theorems to a suited class of *distributions*, (tempered distributions), 3.2.2.
  Another method of overcoming takes advantage of frequency-domain representations for the steady-state analysis of systems. The *phasor representation* of sinusoids is turned to account profiting by the striking simularity between the frequency-response function and the complex response ratio of phasors (complexor), 3.1.3.
- The evaluation of the inversion integral being a real integral for which the interval of integration is not finite (improper integral) turns out to be complicated.
  Performing the inversion formula approach this *difficulty* will be *by-passed* by *extending* the real ω-domain *to the complex s- (or p-) plane* (analytic continuation) and applying the method of residues, 3.2.4.
- *Initial or boundary conditions* at the origin must be *particularly brought in* either by use of the finite Fourier integral (one-sided transformation), or more generally with regard to the ordinary Fourier integral by applying some artifices in derivation (Cauchy's initial-value problem).

3.3.2 Laplace Transform Method. *Advantages and Disadvantages*

The Laplace transform is an essential tool of analysis of linear time-invariant systems appearing suited to more general cases of transient excitations in vibrations. Respecting control system analysis the transient behaviour frequently is of more interest than the steady-state behaviour. For that reason the Laplace transform tends to replace other forms of differential equation representation owing to the particular features as follows.

Advantages

- The restrictions of convergence are more temperate being due to the "convergence" factor in the Laplace kernel so that the class of *Laplace-transformable functions* covers *practically all functions of possible interest in engineering applications*, so the steady sinusoidal signals restricted to varying values of time which are positive (causal functions), 3.2.7.
- The Laplace transform being a complex analytic function of s (or p) enables to *evaluate the inversion integral* from the first on the *background of complex analysis* (theorems for complex integration). Thus, a preceding step in extending the real ω-domain to the complex s- (or p-) plane (analytic continuation) can be saved, and the inversion formula approach is preferably performed by combining the *partial fraction expansion* with the *method of residues*. In this way the last step of response calculation is taken with ease by using *elementary transform pairs* listened in available transform tables, 3.2.9.
- The Laplace transformation primarily adapted for *time-unlimited signals* though being bounded to positive values of time is an approved solution technique fitted to non-periodic vibrations caused by *forcing functions suddenly applied* (transient excitations of step-type such as a harmonic shock excitation), 3.2.8. The dynamic behaviour thus can be analysed by a source suddenly turned on and remaining stationary thereafter. *Starting-up or turn-on operations* generate in a simple way dynamic (starting) load conditions the responses to which are significant for structure design just as for system design.
- *Initial or boundary conditions* are *automatically incorporated* in the transformed solution for any arbitrary excitation. The complete response thus being gained at the first attempt is represented by a superposition of the system response to both the excitation (zero-state response) and the non-zero initial conditions (zero-driving response), 3.2.8.

Disadvantages

- The integral representation of functions in the Laplace- (s- or p-) domain is complicated respecting graphical interpretation for engineering applications, 3.2.10, because the transform is a complex analytic function (conformal mapping).

  For purposes of physical interpretation namely of the s- (or p-) domain response characteristic this *difficulty* may be *overcome* by substituting

$s\,(\text{or}\,p) = \mathrm{j}\omega$, thus profiting by the close relation of the transfer function to the frequency-response function (boundary function on the imaginary axis), 3.2.8, with regard to the well acquainted frequency-response analysis, 3.2.3.

- Time functions varying in negative values of time (non-causal functions) *cannot be uniquely represented* as an inverse L-transform. That is due to the definition range of time functions for this integral representation (one-sided transformation).
  The *difficulty* of cutting off signals for times prior to zero may be *by-passed* by introducing some artifice to simulate or approximate effects prior to zero. To handle this situation some authors use the *bilateral Laplace transform* (two-sided transformation) existing only in a vertical strip of the *s*- (or *p*-) domain, [36], [45], and not in a right-half plane of convergence, 3.2.7.
- Following the one-sided definition range of original (or time) functions the Laplace transform method is *unsuitable* for purposes of data processing *in random vibrations*, 3.2.6.

To summarize, that form of transformation should be used which conveniently handles the class of functions concerned (types of excitation and motion as well as model systems being of interest in vibrations) and simplifies the response calculation, whether the transformation is Fourier, Laplace, or even a more general type of advanced transform methods (Fourier-Bessel, Mellin, Hankel, *z* transform) which are treated by more specialized references, see [36] to [45].

4 Transform Analysis Methods of Vibratory Systems (Frequency-response Characteristics)

The representation of mechanical systems by integral transform methods basing on Fourier techniques just as on the Laplace transform enables to perform frequency concepts suited to linear model systems.

As an approved method in vibration data processing the frequency-response analysis of configurations of mechanical objects (elastic systems) will be treated more in detail. To define the dynamic characteristic of a structure the frequency-response function will be interpreted by both the phasor method (response phasor ratios) and the Fourier transform method (response transform quotients). Thus, covering steady-state analysis as well as spectral analysis with reference to transient and random vibrations, the frequency concept will be specified by *frequency-response (or dynamic) characteristics appropriate to structural configurations*. According to its graphical interpretation by response data plotting which includes phasor diagram representation and frequency response plots, respectively, the complex system parameter is defined by the (mechanical) mobility being wide-spread in vibration data analysis (mobility measurements), [18], in addition by the dynamic compliance (or receptance) being complementary to mobility and convenient for complex structures in machinery.

As an introductory section acts the reference to the model representation through a *mathematical system expressed in different formulations*. There a pointed out in summary the formulations of differential equations commonly used in vibrations in conjunction with diagram representations, regarding in particular the mechanical network. Fundamental relationships between network variables (interconnective relationships or mechanical circuit theorems) are presented for a direct derivation of overall dynamic compliances by viewing translational mechanical networks (circuits).

4.1 Formulation of Dynamic Equations (Equations of Motion)

The derivation of system equations basing on direct representation of lumped dynamic systems by use of mechanical model networks has been introduced in 2.3.4. However, a more general but concise survey of different formulations of dynamic equations now implying integral transforms (transform models) will be given in the following. Defined sets of relationships between abstract mathematical objects can be obtained by two principal classifications of mathematical systems. They are devided into *synthetical and analytical dynamics (or methods) in vibrations*, [53].

The first classification, also termed the *direct procedure*, [54], is a most convenient one for engineering. Subdividing systems into components (or elements) the dynamic equations are derived from element relations by formulating interconnection requirements. Furthermore, the elasticity involved in interactions between elements can be inserted into vibratory systems by several facilities. Finally, the visual inspection of network diagrams enables to bypass the difficulty of solving differential equations. The overall dynamic behaviour thus will be determined by use of transform models and their inherent topological connections between network elements.

4.1.1 Analytical Dynamics. *Mathematical System by Analytical Methods*

Newtonian Mechanics (vectorial mechanics)

The relationships between translational mechanical system variables specifying changes in the two distinct types of state, the kinetic and the static translational state, are identified and described by *laws enunciated by Newton* (Newton's second law, conservation of momentum and work). For the description of lumped parameter models the interaction of a finite number of isolated abstract objects will be identified by extending the fundamental laws of objects (element laws) to the mechanical model system (composite system of rigid bodies or simply of particles).

Basic Tool. Newtonian mechanics implies the application of *free-body diagrams* for each of the interacting passive elements being of the type "mass". Kinematical constraints cause internal forces being included as external forces in the free-body diagram. Physical coordinates and forces are used which are in general vector quantities. By reason of that Newtonian mechanics are also referred to as *vectorial mechanics*.

D'Alembert's Principle (dynamic equilibrium)

The principle of virtual work concerned with the static equilibrium of a system can be extended to dynamic systems. If there are some unbalanced forces acting upon a particle then according to Newton's second law the force resulting vector must be equal to the rate of change of the linear momentum vector.

Basic Tool. The case of static equilibrium is extended to the case occuring by a resultant force that is in equilibrium with the inertia force, thus describing the case of *dynamic equilibrium*. As a result of *variational principle* being fashioned from static into dynamic equilibrium the equations of motion can be derived without considering explicitly the interacting forces. D'Alembert's principle though giving a complete formulation of the problems of mechanics is still basically a vectorial approach. Accordingly this method remains closely related to Newtonian mechanics.

Lagrangian Mechanics (analytical mechanics)

This variational approach commonly referred to as analytical mechanics describes the motion in terms of more abstract *generalized coordinates*. The equations of motion are derived by means of scalar functions, namely, the kinetic and potential energy, and an infinitesimal expression known as the virtual work in the case of nonconservative systems. The equations of motion can be derived without the need of free-body diagrams.

Basic Tool. Lagrangian mechanics operates on the formulation of a set of general differential equations known as *Lagrange's equations of motion*. The *kinetic energy and the potential energy* can be expressed in terms of generalized displacements and velocities, respectively, and the *virtual work* is taken into account by generalized virtual displacements and forces, accordingly. It is remarkable that not all of the scalar quantities necessarily represent physical quantities. The advantages of Lagrange's equations become more and more evident as the number of degrees of freedom of the system increases.

Hamilton's Principle. A different formulation, based on d'Alembert's principle considers the total motion of the composite system between two instants. This leads to an *integral principle stated by Hamilton* that reduces problems of dynamics to the investigation of a scalar integral. Tracing the actual path (true path) in the configuration space the value of the integral will be rendered stationary being actually a minimum. This condition leads to the equations of motion also for non-conservative systems obtaining dissipator elements (dampers) with appreciable more ease than the Newtonian approach.

The Lagrangian formulation shares with d'Alembert's principle the advantage of not having to deal with constraint forces, but the advantages go beyond that. One remarkable advantage is that all the equations of motion can be derived from scalar quantities. By this it proves that the formulation is invariant with respect to the applied coordinate system. Moreover, the integral representation requires velocities only instead of accelerations. Nevertheless, the extended Hamilton's principle used for the Lagrangian approach is a formulation and not a solution of the problems of dynamics. It belongs to a broader class of principles, called *principles of least action*, [30], [40], [41], [54], [55], [56].

4.1.2 Synthetical Dynamics. *Mathematical System by Synthetical Methods*

Method of Influence Coefficients

In more complex mechanical systems (structures) it is of interest to express the relation between the translational displacement (deflection) at a point and unit forces (loads) acting at various points of the system attributed to the property of materials, termed elasticity. Thus, the static translational state of a given object configuration (elastic system) will be specified by its flexibility characteristic.

Basic Tool. The mechanical system variables (loads and deflections) are related by the *flexibility influence coefficients* being adopted extensively in structural engineering. For linear structures the superposition principle holds true and the total deflection at a point is obtained by summing up the contributions of all forces. Conversely, the total force at a point due to unit displacements at various points is given by defining the *stiffness influence coefficients* and summing up the contributions of all displacements. The various influence coefficients are to be considered as elements of the *flexibility matrix* or of its inverse, respectively, called the *stiffness matrix.*

In vibrations also a kinetic translational state must be specified being a type of state supplementary to the static one, and attributed to the property of matter, termed inertia. This is done by replacing the static loads by *inertia loads.* Such relations between the mechanical variables (forces and accelerations) are justified on account of the modelling condition of linearity. Oscillating in a normal mode all coordinates of the model system oscillate with the same frequency, and maintain fixed ratios with the other coordinates. Accordingly all inertia forces hold up fixed ratios with the other inertia forces. The result is a set of equations of motion in terms of *influence coefficients* where the *force* each particle (body) applies is *expressed by the product of its mass and acceleration.*

The force thus is defined in terms of changes in translational mechanical state which includes both the static and the kinetic type, 1.2.1. Accordingly the force-motion performance equations can be developed in matrix notation by replacing the matrix of influence coefficients by the *inverse of the system matrix* and by substituting a frequency modulus. Moreover, the inverse system matrix is defined as the product of multiplying the flexibility matrix and the mass matrix already introduced in 2.3.4, [30], [34], [41], [56].

Matrix Methods for Lumped Parameter Systems. The matrix formulation being a convenient way to handle simultaneous equations of motions is a particularly important method of vibration data analysis for multi-degree-of-freedom systems. The algebra of matrices forms the basis of many computer simulations to vibration problems. One of the most useful numerical methods of finding the characteristic values and mode shapes for a vibration system is the *method of matrix iteration.* This technique relies on the behaviour properties of a square matrix to converge on the characteristic vector through successive multiplications of the matrix by itsself. Knowing the characteristic vector (or mode shape) also determines the characteristic values, [56], [57].

Transfer Matrices (finite element method)

Matrix iteration using the behavioural properties of square matrices is an excellent technique for solving the characteristic value problem. However, complex systems require a large number of generalized coordinates to describe the total motion, so this method can be slow and cumbersome. In addition, entered errors are cumulative, and the total error can become quite large after successive iterations As an alternate, a finite number of elements may be used to reduce the size of matrices to the number of variables which are necessary to describe completely the motion of the system. This technique approximating the characteristic variables of the system piecewise over the elements is called the *finite element method.* It proceeds from one station of the discretized model to the other one in matrix notation by virtue of the transfer matrix.

Basic Tool. A large system is broken down into subsystems with simple elastic and dynamic properties. The formulation is in terms of the *state vector* which is a column matrix of the state variables representing physical variables such as internal force and displacement or velocity. The elements transfer the value of the state variables from one point to the next one. These points connecting the elements are called nodes not to misinterpret as zero vibration points. The dynamic properties caused by the mass of a subsystem are contained in the *point matrix*, whereas the elastic properties concentrated on the spring are described by the *field matrix*. Combining both elements to a section the state vector at one station is transferred to the following station by a square matrix, called the *transfer matrix for the section*. In terms of these state variables, previously numerical calculations are made to proceed from one end of complex systems to the other, the natural frequencies being established by satisfying the appropriate boundary conditions. For a complex system structure the transfer matrix can be obtained by repeatedly multiplying the transfer matrices for all the sections being successively connected, [57], [58], [59].

The *similarity* in transfer matrix procedure *to the two-port diagram representation* (the extension of which to multiports included) thus becomes obvious regarding the transfer matrix of the overall signal 4-pole as the result of a cascade connection, see 2.2.7, Eq. (2.65).

Mechanical Network Analysis (mechanical circuitry)

Concepts from linear system theory having to deal with the study of stability characteristics as with the derivation of frequency response relations (ω-domain models) prefer the approved network or circuit representation originating from electrical circuitry.

The *fundamental principle* on which network analysis is based is that of *conservation of energy*. For network analysis purposes this is expressed as:

- net power summed over all the network elements vanishes, see 2.5.3, Eq. (2.91).

For each element of the network the power may be conceptually determined by a simultaneous measurement of a pair of rate variables (intensity or power variables). To describe the interconnection pattern of a network in terms of mathematical relations (network equations) *general orientation conventions* are introduced into the related diagram representation (network diagram), 2.3.1. Allocating an arrow direction with respect to the “through-propagating“ nature of one element power variable (force transducer) and a plus or minus sign respecting the “across-acting“ nature of the other element power variable (velocity transducer), a polarity to both measuring instruments and thus an algebraic sign to the element power is allocated. Positive directions of the through power variable (P-variable rate) and the across power variable (T-variable rate), or rather in preferably used terms the flow and the effort variable, where both of them are positively directed, indicate a power being absorbed in a network element. The pair of power variables may thus be equivalently represented by a single *oriented linear segment.* Adopting the same polarity convention for every power-variable pair the consistently oriented set of completely connected segments for a network results in an *oriented linear graph* (directed graph or digraph), in controls termed a *signal flow graph*, 2.3.5, see Fig. 2.27.

The representation of power-variable measurement pairs positively oriented by an associated convention in diagram representation leads to interconnective relationships between power variables. Considering coherent sets of interconnection as the *fundamental laws of general network analysis* those relationships are independent on whether a mechanical, a fluid or an electrical system is represented, illustrated by the following *pairs of postulates* (interconnection requirements).

Incidence (or vertex) relationship for flow variables (through power variables):

- For *translational mechanical networks*, the algebraic sum of all *forces*, including inertia forces, incident at a point of connection (node or mechanical vertex) is identically zero (d'Alembert's principle).
- For *rotational mechanical networks*, the same interconnection relation holds true respecting all *torques*.
- For *fluid networks*, e.g., for a hydraulic piping actin in an incompressible fluid, the algebraic sum of all *volume flow rates* at a connection point (junction) is identically zero (principle of the continuity of flow rates).
- For *electrical networks*, the algebraic sum of the *currents* incident at any junction point (node or electrical vertex) is identically zero. This vertex postulate is known as *Kirchhoff's current law* (conveniently abbreviated to KCL).

Boundary (or circuit) relationship for effort variables (across power variables):

- For *translational mechanical networks*, the algebraic sum of the *relative velocities* between component terminals (junction points) is identically zero when taken around any closed boundary of the network (loop or mechanical circuit).
- For *rotational mechanical networks*, the same interconnective relation is valid concerning the *relative angular velocities.*
- For *fluid networks*, the algebraic sum of the *pressure differences* taken around any closed path (fluidal circuit) is identically zero.
- For *electrical networks*, the algebraic sum of the *voltages*, across the element terminal pairs is identically zero when taken around any closed boundary of the network (loop or electrical circuit). This circuit postulate is known as *Kirchhoff's voltage law* (conveniently abbreviated to KVL).

The two sets of interconnective constraints on power variables may thus be regarded as a natural *generalization of Kirchhoff's laws* originating in electrical network analysis. The pairs of postulates defining the spatial or interconnective relationships, and being appropriately called the *through-measurement principle* and the *across-measurement principle*, are necessary and sufficient for the conservation of energy in the network model.

Remark on Mechanical Duality. It should be noticed that mechanical network analysis exceeds the simple approach for the derivation of equations of motion basing on *dynamic equilibrium statement*. It is obvious that, e.g., for translational mechanical systems the summing of oriented forces at a common junction (node) is equivalent to summing forces acting on a rigid body (or particle), thus being coincident to *d'Alembert's principle*. Though saving free-body diagrams by use of this principle the constraints are reduced to the interconnective relations between forces only. As already noted by *H. M. Trent*, [14], in mechanics the fact had been apparently overlooked that the oriented sum of pertinent motions in terms of velocities, as well as of displacements, vanishes around a closed loop in mechanical network models, [62]. This restriction imposed by the geometric constraints (compatibility requirement) states the continuity of space law which could logically be called *velocity or displacement principle*, [52]. It bases on the elasticity or the property of resisting deformation which is involved in interactions between sets of objects as well as the other property of matter given by its inertia or the property of resisting change of motion. Thus, the interactions are caused by the two properties of matter defining the *mechanical duality* in consequence of which interconnective relations should be stated. Since the across power variables (velocities), or else the integrated across power variables (displacements), are connected in addition to the commonly performed combining of through power variables (forces) two sets of constraints can be formulated. Accordingly, they are termed *dual interconnective relations*, [15]. By this a second set of performance equations (or of equations of interconnection) is available being essential for the derivation of the mechanical network equations, [15], [51], [52], [54], [62].

Basic Tool. Using the topological relationships which depend on the connection of the elements in the network diagram, as pointed out in 2.3.5, the fundamental laws of general network analysis are utilized for the derivation of *sets of network equations*.

For any *translational mechanical system* the following sets of relationships may be stated:

Component relationships between power variables for each isolated element

- Material or *constitutive relationships* defined by the dual relations between momentum and velocity just as between force and displacement in linear form

$$p = m\upsilon \qquad (4.1\text{a}) \quad ; \quad F = ks \qquad (4.1\text{b}).$$

- Temporal or *dynamic relationships* expressed by the dual relations between force and momentum as between velocity and displacement

$$F = \mathrm{d}p/\mathrm{d}t \qquad (4.2\text{a}) \quad ; \quad \upsilon = \mathrm{d}s/\mathrm{d}t \qquad (4.2\text{b}).$$

Spatial or interconnective relationships between power variables including the dual relations corresponding to flow and effort variables (through and across variables)

- *Incidence (or vertex) relationship between forces* (d'Alembert's principle)

$$\sum_{\mathrm{i}} F_{\mathrm{i}} = 0 \qquad (4.3\text{a}).$$

- *Boundary (or circuit) relationship between velocities* (velocity principle)

$$\sum_{\mathrm{j}} \upsilon_{\mathrm{j}} = 0 \qquad (4.4\text{a}),$$

thus, with reference to the associated temporal or dynamic relationship, Eq. (4.2b), being also true for the *relation between displacements*

$$\sum_{j} s_j = 0 \qquad (4.4b).$$

In vibrations the stated dual interconnective relations are also termed *force interconnection requirement* and *motion interconnection requirement*, respectively, [17].

Example 4.1: Two-mass System. To demonstrate the procedure the two-mass example , already treated in 2.3.2 and 2.3.4 for illustrating the direct representation of simple systems, will be taken up again. For the network diagram (mechanical circuit), Fig. 2.26b, and its associated oriented linear graph (mechanical signal flow graph), Fig. 2.27b, the initial position of vibratory motion should be fixed by the introduced static equilibrium condition. Hence, the relations will be reduced to governing equations of the free vibration of an undamped two-degree-of-freedom system. Being not affected by gravitational sources the branches (or line segments) for the force generators (weight forces) are cancelled and the interconnections thus imply the equivalence relation to the two-mass system outlined in Fig. 2.25, since oscillating horizontally. Guided by the associated system interconnecting diagrams, Fig. 2.26b and Fig. 2.27b, the *system relationships* of the two-mass example may be summarized by inspection as follows:
Constitutive relationships for each component assumed to be linear

$$\begin{aligned} p_1 &= m_1 \upsilon_1 \\ p_2 &= m_2 \upsilon_2 \end{aligned} \quad ; \quad \begin{aligned} F_{s1} &= k_1 s_1 \\ F_{s2} &= k_2 s_{21} = k_2 (s_2 - s_1) \end{aligned} \qquad (4.5)$$

Dynamic relationships for each component

$$\begin{aligned} F_{m1} &= \mathrm{d}p_1/\mathrm{d}t \\ F_{m2} &= \mathrm{d}p_2/\mathrm{d}t \end{aligned} \quad ; \quad \begin{aligned} \upsilon_1 &= \mathrm{d}s_1/\mathrm{d}t \\ \upsilon_{21} &= \mathrm{d}s_{21}/\mathrm{d}t \end{aligned} \qquad (4.6).$$

The constitutive and dynamic relationships are combined to give
Component relationships between individual power variables

$$\begin{aligned} F_{m1} &= m_1\, \mathrm{d}\upsilon_1/\mathrm{d}t \\ F_{m2} &= m_2\, \mathrm{d}\upsilon_2/\mathrm{d}t \end{aligned} \quad ; \quad \begin{aligned} \mathrm{d}F_{s1}/\mathrm{d}t &= k_1 \upsilon_1 \\ \mathrm{d}F_{s2}/\mathrm{d}t &= k_2 \upsilon_{21} \end{aligned} \qquad (4.7)$$

Interconnective vertex and circuit relationships between through power variables, also termed *force requirement* at

node (or vertex) 1: $\quad F_{m1} + F_{s1} - F_{s2} = 0 \qquad (4.8a)$

node (or vertex) 2: $\quad F_{m2} + F_{s2} = 0 \qquad (4.8b)$,

respectively between across power variables, also termed *velocity requirement* for

loop (or circuit) 2102: $\quad \upsilon_{21} + \upsilon_1 - \upsilon_2 = 0 \quad ; \quad \upsilon_{21} = \upsilon_2 - \upsilon_1 \qquad (4.9a)$

loop (or circuit) 101: $\quad \upsilon_{10} - \upsilon_1 = 0 \quad ; \quad \upsilon_{10} = \upsilon_1 \qquad (4.9b)$.

The continuing step consists in selecting an *independent set* of power variables the derivatives of which occur in the component relation set. In this case it is obviously $F_{s1}, F_{s2}, \upsilon_1, \upsilon_2$.

Accordingly the component and the interconnective relationships are combined to

$$\left(\mathrm{d}F_{s1}/\mathrm{d}t\right) = k_1 \upsilon_1 \qquad (4.10a)$$

$$\left(\mathrm{d}F_{s2}/\mathrm{d}t\right) = k_2 \upsilon_{21} = k_2(-\upsilon_1 + \upsilon_2) \qquad (4.10b)$$

$$\left(\mathrm{d}\upsilon_1/\mathrm{d}t\right) = F_{m1}/m_1 = (-F_{s1} + F_{s2})/m_1 \qquad (4.11a)$$

$$\left(\mathrm{d}\upsilon_2/\mathrm{d}t\right) = F_{m2}/m_2 = (-F_{s2})/m_2 \qquad (4.11b)$$

to gain simultaneous *1st-order differential equations* in the set of independent power variables, rewritten in *matrix notation*

$$\frac{\mathrm{d}}{\mathrm{d}t}\begin{bmatrix} F_{s1} \\ F_{s2} \\ \upsilon_1 \\ \upsilon_2 \end{bmatrix} = \begin{bmatrix} 0 & 0 & k_1 & 0 \\ 0 & 0 & -k_2 & k_2 \\ -m_1^{-1} & m_1^{-1} & 0 & 0 \\ 0 & -m_2^{-1} & 0 & 0 \end{bmatrix} \begin{bmatrix} F_{s1} \\ F_{s2} \\ \upsilon_1 \\ \upsilon_2 \end{bmatrix} \qquad (4.12).$$

This set of equations completely defines the system behaviour when taken together with auxiliary equations relating displacement to velocity, etc. For known, fixed values of element parameters (time-invariant models) the future system behaviour is completely determined for any given initial set of values of υ_1, υ_2 and F_{s1}, F_{s2}.

Such a set of system variables is called a set of *state variables*, and the pertaining simultaneous differential equations are termed the *state equations of the network* being the basic tool of the state-space approach.

Fundamental Sets of Network Equations. In any method of network analysis an *independent* set of vertex equations and an *independent* set of circuit equations must be selected, since the total sets of these respective equations are obviously not independent.

To establish fundamental relationships the *concept of trees* of linear graphs must be entered. A tree of the linear graph is a connected subgraph which contains all the nodes but no circuits, 2.3.5. The constituent element of the tree are termed branches, and the remaining elements not belonging to the tree are termed chords. Supposing that the system linear graph has e branches (elements) and n nodes (vertices) it is obvious that the tree will have

i. $(n - 1)$ branches, and
ii. $(e - n + 1)$ chords.

A total set of $2e$ independent equations in the $2e$ power variables is obtained from the e constitutive relationships and an application of interconnective relationships after choice of a tree:

Independent set of vertex equations. Any set of $(n - 1)$ vertex equations corresponding to $(n - 1)$ distinct vertices is an independent set of vertex equations.

Independent set of circuit equations. Any set of $(e - n + 1)$ circuit equations corresponding to $(e - n + 1)$ distinct circuits is an independent set of circuit equations.

In *advanced theory of linear graphs* those statements are enlarged upon a more rigorous algebraic procedure using the *matrix method.* A set of independent fundamental loops equal in number to the number of chords is formed by inserting the chords into the tree one at a time. Thus, for a given tree independent sets of both, vertex and circuit equations can be defined in terms of a single matrix being called the *dynamic transformation matrix* for a given network, [9], [15], [16], [51], [52], [60].

4.2 Frequency-response Characteristics (Concepts of Mobility and Dynamic Compliance)

Experimental investigations of the *dynamic characteristics of structures* have gained importance with respect to vibration data analysis using frequency concepts for data reduction. The frequency-response function can be determined from *mobility measurements* or measurements of related frequency-response functions, known as *accelerance* and *dynamic compliance.* Accelerance and dynamic compliance differ from mobility only by the motion response which can be expressed in terms of acceleration of displacement, respectively, instead of in terms of velocity. For simplification in measurements only the term mobility is used, [18].

Some applications being typical for experimental investigations aim at

- predicting the dynamic response of structures to a known or assumed input excitation;
- determining the modal characteristics of a structure (natural frequencies, mode shapes and damping ratios);
- determining the dynamic properties of materials in pure or composite form subjected to cyclic load sequences (complex modulus of elasticity, fatigue, and crack propagation).

A complete description of the frequency-response characteristic requires measurements of translational forces and motions taken along three mutually perpendicular axes as well as measurements of moments and rotational motions about these axes. Although in most applications there is no need to determine the *overall mobility matrix* being of size 6 N x 6 N for N locations of interest on a structure. The requirements of vibration data analysis are often satisfied by measuring of *driving-point mobility* and *a few of transfer mobilities.* Accordingly a force at a single point in a single direction will be exerted and the translational response motions at key points on the structure are recorded.

4.2.1 Equivalent Definitions of Frequency-response Function

For a lot of applications in experimental investigations the dynamic responses of a vibratory system are specified by *sinusoidal steady-state responses.* Thus, the various kinds of measured output/input ratios resulting from simple harmonic motions are formed by the complex ratio of the motion-response phasor to the phasor of the excitation force. Basing on the concept of complex excitations and responses this ratio represents a *complex system parameter*, called the *response ratio of phasors* (complexor), as treated in 3.1.3 and 3.2.10.

Furthermore, the equivalent definitions of frequency-response function depend on the kind of motion being expressed by preference in terms of velocity or displacement. In the case of measuring of velocity the phasor ratio is designated the (mechanical) mobility, and since a displacement is recorded the corresponding ratio will be designated the dynamic compliance.

(Mechanical) Mobility (phasor mobility) $\underline{Y}(j\omega_f)$

The complex ratio of the *velocity-response phasor*, taken at a point in a mechanical system, to the *excitation force phasor* at the same or another point in the system, is designated the *direct (or driving-point) and the transfer (mechanical) mobility* of a structure $\underline{Y}_{ii}(j\omega_f)$, and $\underline{Y}_{ij}(j\omega)$, respectively, 3.2.11, Eqs. (3.190), (3.191).

The mobility is sometimes called the *mechanical admittance.*

Dynamic Compliance (phasor compliance) $\underline{C}(j\omega_f)$

The complex ratio of the *displacement-response phasor*, taken at a point in a mechanical system, to the *excitation force phasor* at the same or another point in the system is designated the *direct (or driving-point) and the transfer dynamic compliance* of a structure, $\underline{C}_{ii}(j\omega_f)$, and $\underline{C}_{ij}(j\omega_f)$, respectively, 3.2.11, Eqs. (3.186b), (3.187b).

The dynamic compliance is called the *receptance* by several authors.

Since the significant response is given by the resultant force to an exciting motion it proves convenient to form the *inverse phasor ratio*, defining the reciprocal of the frequency-response function.

Remark on Reciprocals of Dynamic Characteristics. Historically, frequency-response functions of structures have often been expressed in terms of the reciprocal of one of the two dynamic characteristics pointed out before. Though being called mechanical impedance it should be noted that the arithmetic reciprocal of mechanical mobility does not, in general, represent any of the elements of the impedance matrix of the structure.

Mechanical Impedance (phasor impedance) $\underline{Z}(j\omega)$

The complex ratio of the *force (-response) phasor* to the *(excitation) velocity phasor* where the force and velocity may be taken at the same or different points in the system, is designated the *direct (or driving-point) and the transfer impedance* of a structure $Z_{ii}(j\omega_f)$, and $Z_{ij}(j\omega_f)$, respectively, 3.2.11, Eqs. (3.192), (3.193).

The mechanical impedance is the *inverse* of the (mechanical) mobility

$$\underline{Z}(j\omega_f) = \underline{Y}^{-1}(j\omega_f) \tag{4.13}$$

Mobility test data cannot be used directly as part of an analytic impedance model of the structure. To achieve compatibility of the data and the model, the impedance matrix of the model must be inverted to a mobility matrix, or vice versa, [18]. The same conditions are to be noticed regarding dynamic compliance test data used as part of a dynamic stiffness model of the structure.

Dynamic Stiffness (phasor stiffness) $\underline{K}(j\omega_f)$

The complex ratio of the *force (-response) phasor* to the *(excitation) displacement phasor* where the force and displacement may be taken at the same or different points in the system, is designated the *direct (or driving-point) and the transfer dynamic stiffness* of a structure, $\underline{K}_{ii}(j\omega_f)$, and $\underline{K}_{ij}(j\omega_f)$, respectively, 3.2.11, Eqs. (3.188b), (3.189b).

The dynamic stiffness is the *inverse* of the dynamic compliance

$$\underline{K}(j\omega_f) = \underline{C}^{-1}(j\omega_f) \tag{4.14}$$

The conjunction with the transfer function analysis, 3.2.10, becomes obvious for a sinusoidal excitation as special type of forcing function. The defined kinds of measured output/input ratios are related to the frequency transfer function the utility of which is evident choosing a particular form with reference to the different problems in vibrations.

Review of Mobility and Mechanical Impedance Concepts. Historically, the "black-box" technique was developed in the early part of 20th century by electrical engineers for handling the analysis of linear circuits. It was extended in the early 1920's by acousticians to connect electrical, mechanical, and acoustical elements, and further extended since the late 1930's to the present control theory. Mechanical impedance methods become popular in vibration theory when problems were attacked by drawing an *analogous electric circuit* to take advantage of the highly developed techniques of electric circuitry. There came to the fore a problem of choice between two analogies. Force analogous to current results in a mechanical-mobility analogue whereas force analogous to voltage results in a mechanical-impedance analogue, see 2.2.3. The *mobility form of the analogy*, also called true-connected or force-to-current analogy, has been strongly advocated by *W. Hähnle* and *F. A. Firestone*, [12], [13], [61]. They pointed out that the impedance form of the analogy, also called dual or force-to-volute analogy, lacked completeness in the laws for combining series and parallel elements, as well as interconnective relationships between power variables (Kirchhoff's laws). A definitive treatment of analogues and dualogues concerning topological relationships as well as the formulation of vertex and circuit equations of relating networks was given by *H. M. Trent*, [62]. Notwithstanding lacking completeness *mechanical impedance* apparently has become the more popular of the two concepts.

Several endeavours being taken to remove *lacks in electromechanical analogies* should be mentioned. Mechanical analogues of electrical networks only exist in a direct sense when all the electrical network capacitors have a single terminal in common which may be taken as the reference terminal. This corresponds with the Newtonian reality of all the masses to be measured with respect to the inertial reference system. This difficulty may be overcome by adding ideal couplers, i.e., unity-ratio transformers in the electrical network to isolate the electrical capacitors and break direct connections between them. The transformers may then be replaced by their analogue given by the lever as a mechanical transformer. Consequently, various models of levered mass elements have been introduced for *extending electromechanical analogies* by *E. Lehr, G. Lander, K. Federn, L. Cremer*, and *K. Klotter*, [63] to [66]. Thus, using *two-terminal mass elements*, called the "new elements", analogous networks can be formed including the series connection of masses, [53]. A *general treatment of analogues and dualogues* for network models, as presented by *A. G. J. McFarlane*, [15], bases on the linear graph theory. Distinguishing between planar and nonplanar graphs it is turned out, that a nonplanar graph does not have a dual. Well developed circuit, vertex and mixed transform analysis is utilized for network modelling of particular systems by constructing the corresponding oriented linear graph on the base of analogues irrespective of the physical system in terms of the storage and the conversion of energy.

In view of *impedance measurements* fundamental aspects of the analogies have been traced by *G.J. O'Hara*, [67]. Impedance measurements using blocking force responses apply a single excitation velocity, and an array of ratios of forces responding to this excitation velocity is measured. The structure has been deliberately constrained by blocking forces that maintain the velocity at zero at all points scheduled for observation of their respective impedance elements (blocked impedance data). The impedance elements being observed will therefore depend upon the number of location of the blocking forces. Mobility measurements to the contrary apply a single excitation force, and an array of ratios of velocities responding to this excitation force is measured. The structure has not been artificially constrained. No other external forces must be exerted at the points of interest during the measurements run. Observations made anywhere on the system do not affect one another. Therefore, mobilities describe invariant characteristics of the whole structure, impedances generally concern themselves only with segments. Blocked impedance data are dependent upon the number of observation points (degrees of freedom) considered and, consequently, do not possess invariant characteristics of a structure. It is obvious, with reference to Eq. (4.13), that single impedance elements can be calculated from measurements which were obtained without using blocking forces (free impedance data). However, this is equivalent to *measuring mobilities* (experimental mobility matrix) and *calculating impedances*, not measuring impedances.

This fundamental knowledge from experimental investigations in vibration and shock has been adopted for basic definitions and requirements in international standards, [18].

In practice it is *much easier to measure mobility* than blocked mechanical impedance because the boundary conditions of zero velocity at all points being observed are very difficult or impossible to achieve in practical experimental procedures. Thus, it is generally not possible to determine *transfer mechanical impedances* by experimental means, whereas *transfer mobilities* are approachable by measurements. However, in the special case in which a single point on the structure is considered the impedance matrix and the mobility matrix only have a single term, and therefore, the impedance (*driving-point impedance*) being a free impedance is the arithmetic reciprocal of the mobility.

Comparing Impedance and Mobility Data. *Experimental investigations* of the dynamic characteristics of structures result in *mobility type data*. In *mathematical modelling*, however, it is generally easier to use mass and stiffness matrices. In the frequency domain, these result in *blocked impedance data*. As being not an invariant characteristic of a structure the elements of an impedance matrix can be compared with those of an inverted mobility matrix only if all degrees of freedom of both matrices (points and directions) are identical. If the mathematical model, equally its impedance matrix, has more degrees of freedom than the experimental mobility matrix, it is necessary to convert impedance to mobility to allow comparison with the corresponding elements of an experimentally-determined mobility matrix, rather than vice versa, [68], [18].

Mobility Data and Modal Analysis. The major advantage of impedance and mobility methods for structural dynamics lies in what is usually called the *building block approach*. In contrast to the classical approach one obtains the vibration at only those points in the system which are of interest or are required to investigate design changes and/or dynamic performance of a particular system. Specified forms of frequency-response function are obtained by sinusoidally exciting a structure at a point and simultaneously measuring its vibrational response at the same or another point in the system as the forcing frequency is varied within a range of interest. Having determined the necessary mobility functions for various components of a system, it is possible to obtain the overall mobility of the composite system by combining these partial response characteristics analytically, [69], [70].

Modal analysis being useful to link experimental analysis with mathematical modelling is convenient for predicting the dynamic interaction of interconnected sub-structures. When using *experimental mobility data*, modal analysis uses statistical methods to extract modal parameters, including natural frequencies, damping and modal mass (or stiffness), within the frequency range of interest. When using *mathematical models*, the modal parameters can be extracted from the computed mass, stiffness and damping matrices of the sub-structures by eigen-value/eigen-vector computation or other matrix reduction procedures. These procedures are often more efficient than the direct inversion of the entire impedance matrix, [71], [72].

Review of Dynamic Compliance and Stiffness Concepts. Historically, mechanical impedance concepts originate in acoustics for modelling mixed domain system structures. Subsequently impedance and mobility methods turned out to be an effective tool for analysing the dynamic characteristics of complex structures as demonstrated by *R. Plunkett*, [73], and *S. H. Crandall*, [17]. Nevertheless, investigations into the dynamic behaviour of machine structures in operations focus their attention to vibratory effects on deformation to gather from displacement instead of velocity. Thus, the corresponding output/input ratios relating displacement and force define the equivalent forms of frequency-response function and its inverse which are designated dynamic compliance, occasionally receptance, or dynamic stiffness, respectively.

The equivalent notion *receptance* has been introduced by *W. J. Duncan* and *M. A. Biot*, [74]. The concept of receptance (or "dynamic flexibility") which had been dealt with in detail by *R. E. D. Bishop* and *D. C. Johnson*, [75], provides a link between the treatment of simple systems, multi-degree-of-freedom systems, and continuous (distributed) systems. It also provides a tool for breaking down complex problems into simpler parts whose receptances are known or tabulated. The overall receptance of composite systems will be investigated by connecting constituent parts (subsystems) or by adding a remote system (component).

B. M. Wundt, [76], is to be credited with developing a dynamic stiffness method for predicting the actual critical behaviour of turbine rotors.Taking into consideration the support system the dynamic stiffness at the bearings includes the effects of bearing bracket masses and stiffnesses as well as the bearing oil film stiffness. For a given rotor configuration there will be a continuous variation in dynamic stiffness as the speed varies. At particular speeds the rotor system dynamic stiffness is equal in magnitude and opposite in phase to the support system dynamic stiffness. Those speeds are critical speeds of the combined rotor-support system which may be determined by predictions, factory test, and field data being correlated, [77].

Instead of mechanical impedance the term "dynamic stiffness" was suggested by *F. Eisele*, [78], for describing dynamic effects on the stiffness of machine tool structures. Contrary to impedance assigned to mechanical circuits the equivalent notion dynamic stiffness (or dynresistance) is associated with the vibration resistance against deformation of structures in operation. Giving evidence on frequency-depending effects caused by forced oscillation as well as by self-excited vibration the dynamic stiffness thus turned out to be a dynamic characteristic well suited to common imaginations in structural and vibration engineering. Various steps in engineering design have been taken to enlarge the dynamic stiffness of machine structures. The main attention is directed to the types of instability causing chatter vibrations of metal-cutting machine tools. Attempts to intensify damping effects on the system including flexible supports, frictional damping of restrained joints as well as dynamic vibration absorbers proved to be very promising. Various improvements on structural dynamic characteristics covering the frequency range of interest already have been realized by *Sadowy, Lysen, Korner, Loewenfeld, Corbach, Armbruster, Stefaniak*, a.o., [79] to [85].

Advances in vibration measurements enabled the determination of three-dimensional dynamic characteristics of complex machine tool structures, as pointed out by *M. Weck* and *K. Teipel*, [86]. Graphical interpretations involving tool-workpiece interrelations result in frequency-response curves of oriented dynamic compliances.

Basic Requirements for Mobility Measurements. Requirements for the selection of motion transducers, force transducers and impedance heads, as well as transducer attachment methods and operational calibration measures are standardized to a certain extent using single-point translation excitation with an attached exciter up to impact excitation with an exciter being not attached to the structure, [72], [46].

The *basic criteria* of all *measurement transducers* which are important in acquiring adequate mobility data are as follows:

- Transducers shall have sufficient sensivity and low noise in order to obtain an adequate signal-to-noise ratio of the measurement chain for covering the dynamic range of the mobility of the structure;
- the natural frequency of the response transducer shall be far enough below or above the frequency range of interest that no unacceptable phase shift will occur;
- transducer sensivity shall be stable with time and have negligible d.c. drift;
- transducers shall be insensitive to extraneous environmental effects, such as temperature, humidity, magnetic, electrical and acoustical fields, strain and cross-axis inputs;
- transducer mass and rotational inertia of exciters shall be small so as to avoid dynamic loading of the structure under test, [87], [18].

Basic Tool. Dynamic Stiffness and dynamic compliance have turned to be an effective tool for handling vibration problems that bases on synthetical methods and aims at frequency-domain representation (ω-domain modelling). Mathematical systems for lumped parameter models can be derived either from *mechanical network analysis* as demonstrated by *K. Federn* using the dual type of electromechanical analogies (founded on impedance analogue), [65], [88], or else from the fundamental equations of mechanics as pointed out by *H.-Th. Woernle*, [89], basing on the *method of influence coefficients*. For both of the outlined synthetical methods the relation to *transfer matrices* is obvious. This is verified by defining the dynamic stiffness a column matrix referred to a station limiting a section. All sections are connected to a complex system structure by repeated multiplication, [65]. In the latter case transfer matrix procedure is applied to oscillatory elastic chains using continued-fraction development for repeated structures, [89], [53].

4.2.2 Dynamic Characteristics of Mechanical Elements. *Component Mobilities and Dynamic Compliances*

The behaviour of the *elementary mechanical elements*, treated in 2.3.2, can be expressed by a defined or measured relationship between an applied force F (through power variable) and the resulting motion in terms of the velocity υ (across power variable) or of the displacement s (integrated across power variable) resulting in the *component relationships* between a pair of power variables, called the *T-storage element law* (Hooke's law)

$$F = ks \text{ or } F = k\int_0^t \upsilon \mathrm{d}t \qquad (4.15a),$$

the *dissipator element law* (viscous damping behaviour)

$$F = c\dot{s} \text{ or } F = c\upsilon \qquad (4.15b),$$

the *P-storage-element law* (Newton's second law)

$$F = m\ddot{s} \text{ or } F = m\dot{\upsilon} \qquad (4.15c),$$

where for the spring and the damper relative motions between the two connection points or terminals 1, 2

$$s = \Delta s = s_1 - s_2 \quad ; \quad \upsilon = \Delta\upsilon = \upsilon_1 - \upsilon_2 \quad ,$$

are taken in general, and for the mass its absolute motion between the upper connection point (object terminal) 1 and a fixed point in space (inertial reference system)

$$s = s_1 \quad ; \quad \upsilon = \upsilon_1$$

is measured, Fig. 4.1.

The component relationships for each isolated element involve constitutive and dynamic relationships, 4.1.2, Eqs. (4.1a) to (4.1b), which may be graphically interpreted by the characteristics defining a potential energy storage element, a viscous frictional dissipator element, and a kinetic storage element, Fig. 4.2.

Phasor Relations for Mechanical Elements

Applying a complex excitation (forcing function), 3.1.3, Eq. (3.48), the dynamic responses are represented by the complex responses expressed in terms of the dis-

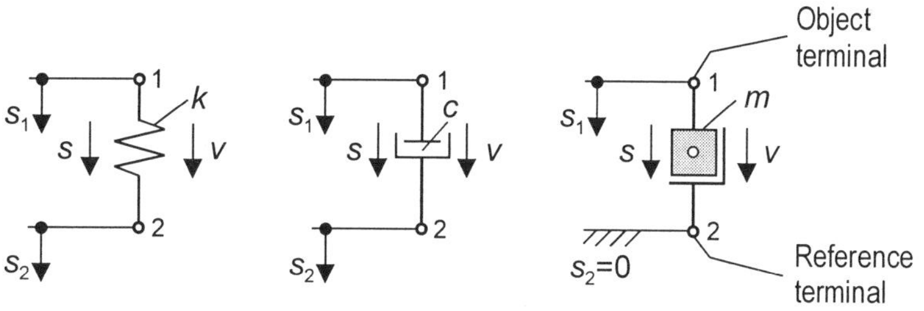

Fig. 4.1. Network symbols for basic linear time-invariant mechanical elements. **a** Spring; **b** damper; **c** mass

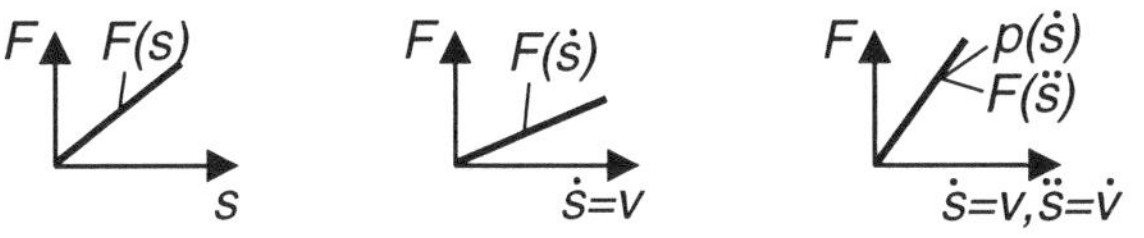

Fig.4.2. Linear element relationships. **a** Spring characteristic; **b** dissipator (damper) characteristic; **c** mass characteristic

placement, the velocity, and the acceleration

$$\begin{aligned} \underline{s}(t) &= \underline{\hat{s}}\,\mathrm{e}^{\mathrm{j}\omega_\mathrm{f} t} = (\underline{\hat{\upsilon}}/\mathrm{j}\omega_\mathrm{f})\,\mathrm{e}^{\mathrm{j}\omega_\mathrm{f} t} \\ \underline{\dot{s}}(t) &= \mathrm{j}\omega_\mathrm{f} \cdot \underline{\hat{s}}\,\mathrm{e}^{\mathrm{j}\omega_\mathrm{f} t} = \underline{\hat{\upsilon}} \cdot \mathrm{e}^{\mathrm{j}\omega_\mathrm{f} t} \\ \underline{\ddot{s}}(t) &= -\omega_\mathrm{f}^2 \cdot \underline{\hat{s}}\,\mathrm{e}^{\mathrm{j}\omega_\mathrm{f} t} = \mathrm{j}\omega_\mathrm{f} \cdot \underline{\hat{\upsilon}} \cdot \mathrm{e}^{\mathrm{j}\omega_\mathrm{f} t} \end{aligned} \tag{4.16}$$

By adopting complex sinusoids only sinusoidal steady-state responses (forced vibrations) are specified. Thus, using the phasor representation of power variables (phasor method), each mechanical element can be characterized by a *phasor relation* between the applied force $\underline{\hat{F}}$ and either the displacement $\underline{\hat{s}}$, or equivalently the velocity $\underline{\hat{\upsilon}}$

$$\underline{\hat{F}} = K_\mathrm{i}\,\underline{\hat{s}} \quad (4.17\mathrm{a}) \;; \qquad \underline{\hat{F}} = Z_\mathrm{i}\,\underline{\hat{\upsilon}} \quad (4.18\mathrm{a}).$$

Hence, the *complex ratios* (complexors) defined by the pertinent phasor quantities are termed

the *phasor stiffnesses* K_i *of the elementary (or 1-port) components k, c, m*

$$\begin{aligned} \underline{\hat{F}} &= k\underline{\hat{s}} = K_\mathrm{k}\,\underline{\hat{s}} \\ \underline{\hat{F}} &= \mathrm{j}c\omega_\mathrm{f}\,\underline{\hat{s}} = K_\mathrm{c}\,\underline{\hat{s}} \\ \underline{\hat{F}} &= -m\omega_\mathrm{f}^2\,\underline{\hat{s}} = K_\mathrm{m}\,\underline{\hat{s}} \end{aligned} \tag{4.17b}$$

or, respectively,

the *phasor impedances* Z_i *of the elementary (or 1-port) components k, c, m*

$$\begin{aligned} \underline{\hat{F}} &= (-\mathrm{j}k/\omega_\mathrm{f})\underline{\hat{\upsilon}} = Z_\mathrm{k}\,\underline{\hat{\upsilon}} \\ \underline{\hat{F}} &= c\underline{\hat{\upsilon}} = Z_\mathrm{c}\,\underline{\hat{\upsilon}} \\ \underline{\hat{F}} &= \mathrm{j}m\omega_\mathrm{f}\,\underline{\hat{\upsilon}} = Z_\mathrm{m}\,\underline{\hat{\upsilon}} \end{aligned} \tag{4.18b}$$

as pointed out in 3.2.10, (phasor ratios).

Costumarily, the symbol K is used for dynamic stiffness and Z for mechanical impedance referring electrical circuit elements, with a subscript i as required to indicate the specific element.

Transform Relations for Mechanical Elements

In a more generalized sense, taking transient and random vibrations into consideration, the mechanical elements will be characterized by a corresponding *relation of the Fourier transforms* (spectral densities) between the applied force $\underline{F}(\omega)$ and the pertinent motion variable $\underline{s}(\omega)$, or $\underline{\upsilon}(\omega)$

$$\underline{F}(\omega) = K_\mathrm{i}\,\underline{s}(\omega) \quad (4.17\mathrm{c}) \;; \qquad \underline{F}(\omega) = Z_\mathrm{i}\,\underline{\upsilon}(\omega) \quad (4.18\mathrm{c}).$$

The constitutive and dynamic relationships, Eq. (4.1b), result with reference to the differentiation theorem (Fourier transform method), 3.2.3, Eq. (3.74b), in the *transformed component relationships*

associated with the *displacement spectrum* $\underline{s}(\omega)$

$$\begin{aligned} \underline{F}(\omega) &= (\mathrm{j}\omega)^0 k\,\underline{s}(\omega) = k\underline{s}(\omega) = K_\mathrm{k}(\mathrm{j}\omega)\,\underline{s}(\omega) \\ \underline{F}(\omega) &= (\mathrm{j}\omega)^1 c\,\underline{s}(\omega) = \mathrm{j}c\omega\underline{s}(\omega) = K_\mathrm{c}(\mathrm{j}\omega)\,\underline{s}(\omega) \\ \underline{F}(\omega) &= (\mathrm{j}\omega)^2 m\underline{s}(\omega) = -m\omega^2\,\underline{s}(\omega) = K_\mathrm{m}(\mathrm{j}\omega)\underline{s}(\omega) \end{aligned} \tag{4.17d}$$

or, respectively, with the *velocity spectrum* $\underline{\upsilon}(\omega)$

$$\underline{F}(\omega) = (\mathrm{j}\omega)^{-1} k\underline{\upsilon}(\omega) = -\mathrm{j}(\mathrm{k}/\omega)\ \underline{\upsilon}(\omega) = Z_\mathrm{k}(\mathrm{j}\omega)\underline{\upsilon}(\omega)$$

$$\underline{F}(\omega) = (\mathrm{j}\omega)^{0} c\ \underline{\upsilon}(\omega) = c\underline{\upsilon}(\omega) = Z_\mathrm{c}(\mathrm{j}\omega)\ \underline{\upsilon}(\omega) \qquad (4.18\mathrm{d}).$$

$$\underline{F}(\omega) = (\mathrm{j}\omega)^{1} m\underline{\upsilon}(\omega) = \mathrm{j}m\omega\underline{\upsilon}(\omega) = Z_\mathrm{m}(\mathrm{j}\omega)\underline{\upsilon}(\omega)$$

The dynamic characteristics of the elementary (or 1-port) elements thus can be interpreted by the *transform quotients* of the pertinent Fourier transforms being designated

the *dynamic stiffness of the spring*

$$K_\mathrm{k}(\mathrm{j}\omega) = k \qquad (4.19\mathrm{a}),$$

the *dynamic stiffness of the damper*

$$K_\mathrm{c}(\mathrm{j}\omega) = \mathrm{j}c\,\omega \qquad (4.19\mathrm{b}),$$

the *dynamic stiffness of the mass*

$$K_\mathrm{m}(\mathrm{j}\omega) = -m\omega^2 \qquad (4.19\mathrm{c}),$$

or, respectively,

the *mechanical impedance of the spring*

$$Z_\mathrm{k}(\mathrm{j}\omega) = -\mathrm{j}k/\omega \qquad (4.20\mathrm{a}),$$

the *mechanical impedance of the damper*

$$Z_\mathrm{c}(\mathrm{j}\omega) = c \qquad (4.20\mathrm{b}),$$

the *mechanical impedance of the mass*

$$Z_\mathrm{m}(\mathrm{j}\omega) = \mathrm{j}m\omega \qquad (4.20\mathrm{c}).$$

The dynamic characteristics of the mechanical elements are defined in terms of *phasor relations* or of *transform relations*. The phasor relations are graphically represented by *constant (resting) phasors* (complexors), and the transform relations by *logarithmic frequency plots* (magnitude plots), Fig. 4.3.

The vector representation in the complex plane illustrates the relation of the dynamic characteristics of the different elements to oneanother by a counter-clockwise shift in phase around the angle $+\pi/2$ corresponding to the order of repeated multiplying with the differential factor $\mathrm{j}\omega_\mathrm{f}$.

The magnitude plots (gains) illustrate the dynamic characteristics for each element by straight lines with different slopes. The slopes are accordant to the degree of the frequency factor $(\mathrm{j}\omega)^{\ell}$ including sign and value of its power exponent ℓ. Regarding the component stiffnesses, Fig. 4.3a, the damper and, first of all, the mass are more operative at high frequencies, whereas the dynamic stiffness of the spring is independent of frequency. Following the mechanical impedance concept, Fig. 4.3b, the magnitude plot of the spring is effective at low frequencies, that one of the mass at high frequencies, whilst the mechanical impedance of the damper behaves constantly with frequency.

Mechanical elements can also be characterized by a transform relation being the reciprocal of the defined dynamic characteristics, Eqs. (4.19), (4.20).

The inverse response transform quotients of the pertinent Fourier transforms are designated

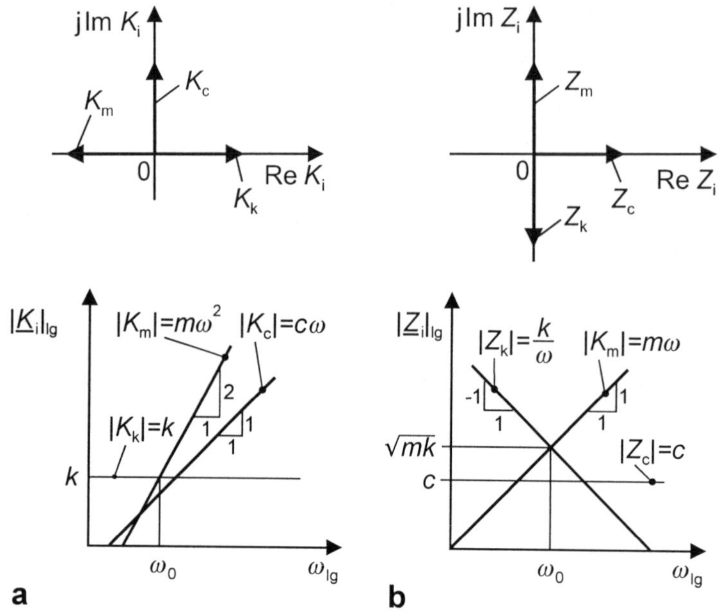

Fig. 4.3. Phasor diagrams and logarithmic frequency plots (magnitude plots). **a** Component dynamic stiffnesses; **b** component mechanical impedances

the *dynamic compliance of the spring*

$$C_k(j\omega) = 1/K_k(j\omega) = 1/k \quad (4.21a),$$

the *dynamic compliance of the damper*

$$C_c(j\omega) = 1/K_c(j\omega) = -j/(c\omega) \quad (4.21b),$$

the *dynamic compliance of the mass*

$$C_m(j\omega) = 1/K_m(j\omega) = -1/(m\omega^2) \quad (4.21c),$$

or, respectively,

the *mobility of the spring*

$$Y_k(j\omega) = 1/Z_k(j\omega) = j\omega/k \quad (4.22a),$$

the *mobility of the damper*

$$Y_c(j\omega) = 1/Z_c(j\omega) = 1/c \quad (4.22b),$$

the *mobility of the mass*

$$Y_m(j\omega) = 1/Z_m(j\omega) = -j/(m\omega) \quad (4.22c).$$

The graphical representation of the corresponding inverse dynamic characteristics is given by Fig. 4.4.

The logarithmic gains of the reciprocals are simply the negatives of the corresponding originals. As a result, the magnitude plots of the inverse component characteristics are the mirror images of the original ones, see Fig. 4.3.

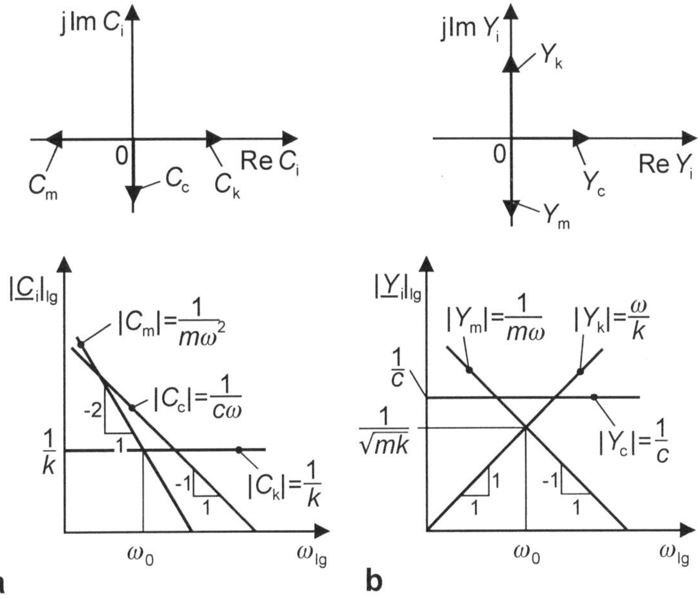

Fig. 4.4. Phasor diagrams and logarithmic frequency plots (magnitude plots). **a** Component dynamic compliances; **b** component mobilities

4.2.3 Dynamic Characteristics of Composite Systems. *Overall Mobility and Dynamic Compliance*

The overall behaviour of the *composite mechanical system*, treated in 2.3.5, can be expressed in terms of mathematical relations using the interconnection pattern of a network (topological relationships). Including the dual relations corresponding to flow and effort variables (through and across variables) the network diagram leads to the spatial or *interconnective relationships* between the power variables.

Force Interconnection Requirement. For translational mechanical networks a *junction* (node or mechanical vertex) usually combining the terminals of several components can be equivalently illustrated by a rod (shaft) or a bar (traverse), both symbolizing a rigid, inertia free, linearly guided point of connection, Fig. 4.5.

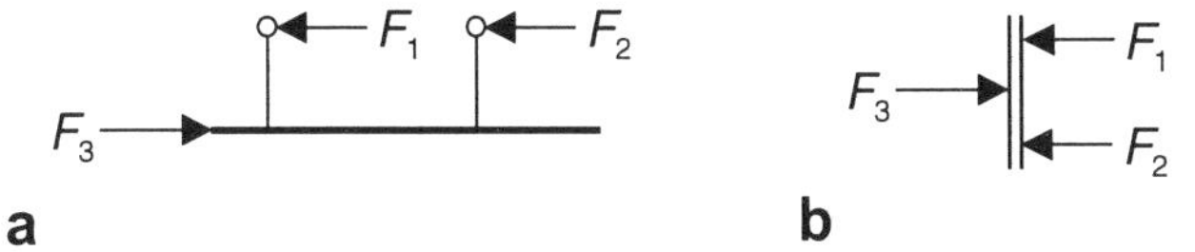

Fig. 4.5. Network symbols of a junction. **a** Rod (shaft) after K. Federn [65]; **b** bar (traverse) after S.H. Crandall [17], [54], and H.Th. Woernle [89]

Basing on dynamic equilibrium statement (d'Alembert's principle) the interconnective constraint on the pertinent flow variables at a common node (or vertex) results in
the *incidence (or vertex) relationship between forces*

$$\sum_{\mathrm{i}} F_{\mathrm{i}} \equiv 0 \qquad \left(\text{or} \quad \sum_{\mathrm{i}} T_{\mathrm{i}} \equiv 0\right) \tag{4.3b}$$

For rotational mechanical networks the same interconnective constraint holds true expressed in parenthesis between torques around a common axis.

Motion Interconnection Requirement. For translational mechanical networks the connection of two junctions (nodes or mechanical vertices) forms a *branch* (line segment) which can be illustrated by the dynamic model "2-terminal element", a rigid link, finally a "black" box, thus symbolizing either an elementary (or 1-port) mechanical element, or a rigid (bridged) connection, or else a component of unspecified or unknown relations, respectively, Fig. 4.6.

The inertial reference system is labelled rod or bar 0 (reference framework).

By a connected subnetwork having only two branches incident with each of the two junctions (rods or bars) a closed path will be formed. Any such closed path is termed a *boundary* (loop) of a network, Fig. 4.7.

A boundary or loop is also called a circuit (or a mesh). A *mechanical circuit* is defined as a closed path in space which includes one point on the inertial reference frame, [15], [51]. It is obvious that the motion around a circuit will be considered as not restricted (compatibility requirement).

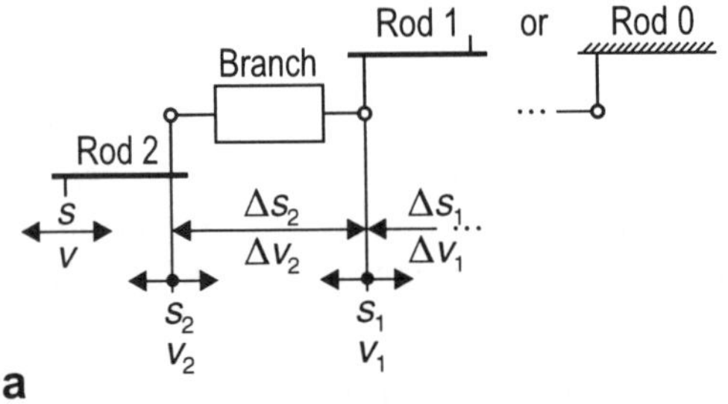

a

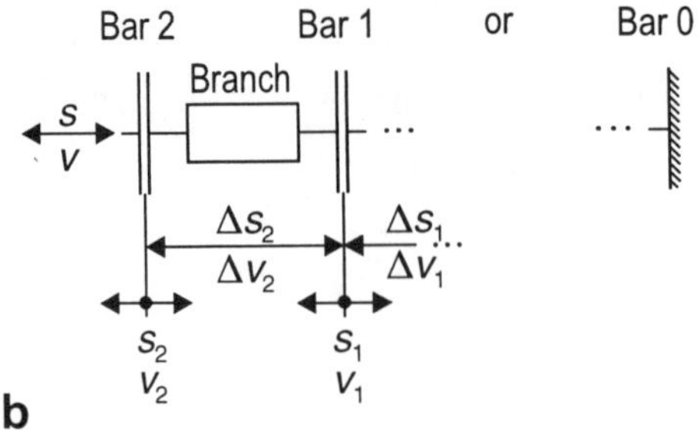

b

Fig. 4.6. Network symbol of an unspecified branch. **a** Junctions as rods; **b** junctions as bars

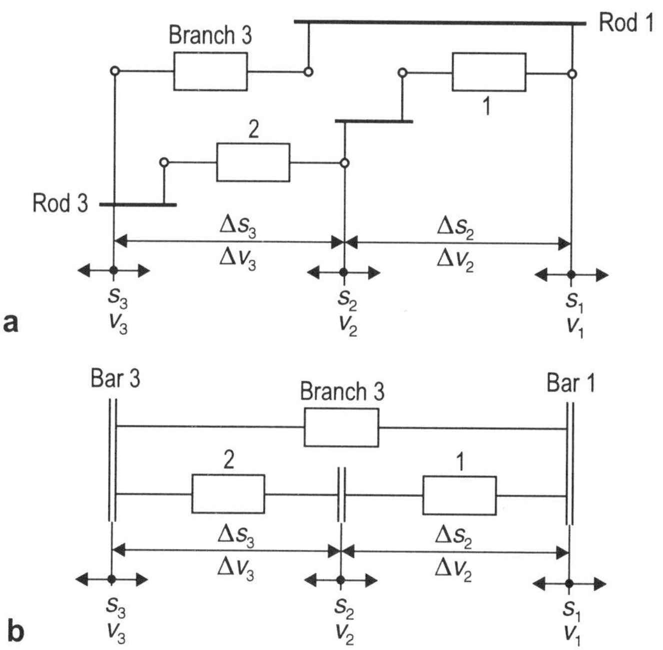

Fig. 4.7. Subnetwork of an unspecified boundary (loop). **a** Junctions as rods; **b** junctions as bars

Basing on the statement of geometric constraints (velocity or displacement principle) the interconnective constraint on the pertinent effort variables around any closed path (loop) results in
the *boundary (or circuit) relationship between relative displacements*

$$\sum_i \Delta s_i \equiv 0 \qquad \left(\text{or} \quad \sum_i \Delta\varphi_i \equiv 0\right) \tag{4.4c}$$,

being also true for the *relation between relative velocities*

$$\sum_i \Delta \upsilon_i \equiv 0 \qquad \left(\text{or} \quad \sum_i \Delta\dot{\varphi}_i \equiv 0\right) \tag{4.4d}$$.

For rotational mechanical networks the same interconnective constraints are valid expressed in parenthesis between relative angular displacements or velocities around a common axis.

Fundamental Configurations of Mechanical Elements (parallel-series connections)

The two sets of interconnective constraints on power variables, Eqs. (4.3b), (4.4c), (4.4d), may be regarded as a generalization of Kirchhoff's laws about mechanical circuits. The pair of postulates being necessary and sufficient for the conservation of energy in the mechanical network model is basic for an overall theoretical analysis of vibration systems.

A direct derivation of composite dynamic characteristics by viewing will be rendered possible for simple circuits. For this the dual interconnective relations are applied to special types of subsystem configurations known as the *fundamental configurations* connecting elements either in parallel or in series.

Elements in Parallel. Components connected in parallel are positively actuated by a coupler imposing on the structure a guided motion, so that the *displacement or velocity* at the connection rod or bar is *common across all the elements*, Fig. 4.8.

Parallel connection is indicated by the *special feature of total blocking of the structure* (with all the other connection points of the system "blocked") since any of the elements is constrained to have zero velocity.

Considering the sinusoidal steady state the force and motion variables can be written in terms of phasor quantities, furthermore the dynamic characteristics in terms of phasor relations.

Force interconnection requirement (vertex postulate):

$$\underline{\hat{F}} = \underline{\hat{F}_1} + \underline{\hat{F}_2} + \underline{\hat{F}_3} \tag{4.23}$$

implies that the force phasors sum up through the elements to the total force phasor.

Motion interconnection requirement (circuit postulate):

$$\underline{\hat{s}} = \underline{\hat{s}_1} = \underline{\hat{s}_2} = \underline{\hat{s}_3} \tag{4.24a}$$

is simply that the displacement phasor across all the elements is the same.

Forming the corresponding phasor ratios (force-displacement relations)

$$\frac{\underline{\hat{F}}}{\underline{\hat{s}}} = \frac{\underline{\hat{F}_1}}{\underline{\hat{s}_1}} + \frac{\underline{\hat{F}_2}}{\underline{\hat{s}_2}} + \frac{\underline{\hat{F}_3}}{\underline{\hat{s}_3}} \tag{4.24b}$$

the *overall dynamic stiffness of a parallel-connected structure*

$$\underline{K}(\mathrm{j}\omega_\mathrm{f}) = K_1 + K_2 + K_3 = K_\mathrm{k} + K_\mathrm{c} + K_\mathrm{m} = k + \mathrm{j}c\omega_\mathrm{f} - m\omega_\mathrm{f}^2 \tag{4.25a}$$

makes evident that the *component dynamic stiffnesses add.*

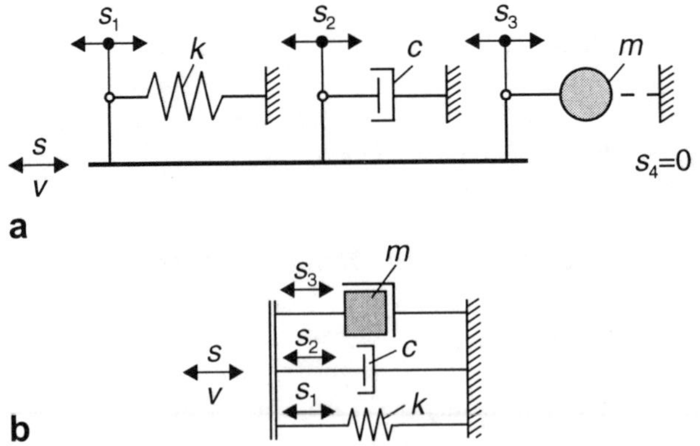

Fig. 4.8. Parallel-connected basic elements. **a** Junction as a rod; **b** junction as a bar

The same interconnective constraints are valid if motion will be expressed by the velocity phasor across all the elements

$$\hat{\underline{\upsilon}} = \hat{\underline{\upsilon}}_1 = \hat{\underline{\upsilon}}_2 = \hat{\underline{\upsilon}}_3 \qquad (4.26a).$$

Forming the corresponding phasor ratios (force-velocity relations)

$$\frac{\hat{\underline{F}}}{\hat{\underline{\upsilon}}} = \frac{\hat{\underline{F}}_1}{\hat{\underline{\upsilon}}_1} + \frac{\hat{\underline{F}}_2}{\hat{\underline{\upsilon}}_2} + \frac{\hat{\underline{F}}_3}{\hat{\underline{\upsilon}}_3} \qquad (4.26b)$$

the *overall mechanical impedance of a parallel-connected structure*

$$\underline{Z}(\mathrm{j}\omega_\mathrm{f}) = Z_1 + Z_2 + Z_3 = Z_\mathrm{k} + Z_\mathrm{c} + Z_\mathrm{m} = -\mathrm{j}\frac{k}{\omega_\mathrm{f}} + c + \mathrm{j}m\omega_\mathrm{f} \qquad (4.27a)$$

makes evident that the *component mechanical impedances add.*

Sinusoidal steady-state response (forced vibration) occurs in a mechanical circuit being subjected to a simple harmonic excitation. In the case of a motion being applied to the structure the significant response is given by the resultant force. Expressing the exciting motion in terms of a displacement the response ratio of phasors taken at the same point in the system is designated
the *driving-point (or direct) dynamic stiffness of the structure*

$$\underline{K}(\mathrm{j}\omega_\mathrm{f}) = \frac{\hat{\underline{F}}}{\hat{\underline{s}}} = (k - m\omega_\mathrm{f}^2) + \mathrm{j}c\omega_\mathrm{f} = K_\mathrm{Re} + \mathrm{j}K_\mathrm{Im} \qquad (4.25b),$$

a complex ratio (complexor) with
the *real part* (active dynamic stiffness)

$$K_\mathrm{Re} = k - m\omega_\mathrm{f}^2 \qquad (4.28a),$$

and the *imaginary part* (reactive dynamic stiffness)

$$K_\mathrm{Im} = c\omega_\mathrm{f} \qquad (4.28b).$$

The significant response is also given by expressing the exciting motion in terms of velocity. The response ratio of phasors at the same point defines
the *driving-point (or direct) mechanical impedance of the structure*

$$\underline{Z}(\mathrm{j}\omega_\mathrm{f}) = \frac{\hat{\underline{F}}}{\hat{\underline{\upsilon}}} = c + \mathrm{j}(m\omega_\mathrm{f} - \frac{k}{\omega_\mathrm{f}}) = R + \mathrm{j}X \qquad (4.27b),$$

a complex ratio (complexor) with
the *real part* (mechanical resistance)

$$R = c \qquad (4.29a),$$

and the *imaginary part* (mechanical reactance)

$$X = m\omega_\mathrm{f} - (k/\omega_\mathrm{f}) \qquad (4.29b).$$

For the present network configuration the resultant force phasor $\hat{\underline{F}}$ can be rewritten as an algebraic product of the exciting motion phasor $\hat{\underline{s}}$, or $\hat{\underline{\upsilon}}$, and the dynamic characteristic of the composite system expressed by
the *phasor stiffness of the parallel-connected structure* $\underline{K}(\mathrm{j}\omega_\mathrm{f})$

$$\hat{\underline{F}} = \left[\left(k - m\omega_\mathrm{f}^2\right) + \mathrm{j}c\omega_\mathrm{f}\right]\hat{\underline{s}} = \underline{K}(\mathrm{j}\omega_\mathrm{f})\hat{\underline{s}} \qquad (4.30),$$

or, respectively, by
the *phasor impedance of the parallel-connected structure* $\underline{Z}(\mathrm{j}\omega_\mathrm{f})$

$$\underline{\hat{F}} = \left[c + \mathrm{j}\left(m\omega_\mathrm{f} - \frac{k}{\omega_\mathrm{f}} \right) \right] \underline{\hat{\upsilon}} = \underline{Z}(\mathrm{j}\omega_\mathrm{f})\underline{\hat{\upsilon}} \qquad (4.31).$$

The related characteristics, Eqs. (4.25a,b), (4.27a,b), can be converted into one another with ease by multiplying the phasor stiffness, Eq. (4.25a,b), and $\mathrm{j}\omega_\mathrm{f}$ together, or respectively, the phasor impedance, Eq. (4.27a,b), and $1/(\mathrm{j}\omega_\mathrm{f})$ together, whereat $\mathrm{j}\omega_\mathrm{f}$ is representing the 1st derivative of phasors, and $1/(\mathrm{j}\omega_\mathrm{f})$ is denoting the integral of phasors

$$\underline{K}(\mathrm{j}\omega_\mathrm{f}) = \mathrm{j}\omega_\mathrm{f}\,\underline{Z}(\mathrm{j}\omega_\mathrm{f}) \quad \Leftrightarrow \quad \underline{Z}(\mathrm{j}\omega_\mathrm{f}) = \frac{1}{\mathrm{j}\omega_\mathrm{f}}\,\underline{K}(\mathrm{j}\omega_\mathrm{f}) \qquad (4.32).$$

The overall dynamic behaviour of mechanical elements connected in parallel is characterized by adding the component mechanical impedances, Eq. (4.27a). Though being *analogous* to the combination of electrical system components the *summing up of component impedances* nevertheless describes a *series-connected electric circuit* with reference to the “classical“ type of electromechanical analogies (dual or force-to-voltage analogy), 2.2.3.

Following the concept of related characteristics by adding the component dynamic stiffnesses, Eq. (4.25a), the presented analogy has been modified to the “practical“ type by *K. Federn*, [65], 4.2.1.

Using the vector representation of response phasor ratios the overall frequency response can be plotted in terms of polar coordinates by the *locus of* $\underline{K}(\mathrm{j}\omega)$, *and of* $\underline{Z}(\mathrm{j}\omega)$, Eqs. (4.25a,b), (4.27a,b), in the complex plane called, in this case, the *dynamic stiffness plane*, or the *mechanical impedance plane*, respectively, Fig. 4.9.

The composite polar frequency-response loci (Nyquist plots), see 3.2.11, of dynamic stiffness and mechanical impedance can be constructed by graphical vector addition of contributions made by the individual phasor ratios. The steady-state interrelation of mechanical components is illustrated by the *phasor ratio polygon* combining the individual characteristics with the composite dynamic characteristic of the *parallel-connected structure*.

Response Data Plotting. Polar plots (Nyquist plots) of *normalized dynamic stiffness* are shown in Fig. B.5 of the Appendix B.

The composite logarithmic frequency plots (Bode plots) also can be constructed by summing up the contributions of component curves, Fig. 4.10.

Beyond resonance the dynamic stiffness approaches straight lines appertaining to one of the storage element characteristics as shown in Fig. 4.3. This property is used for reducing the frequency response plotting to broken line sketching by a *straight-line approximation*, Fig. 4.10a.

Following the mechanical impedance concept, Fig. 4.10b, the symmetric composite curve is approximated by a low-frequency and a high-frequency asymptote which intersect at the corner frequency equalling natural frequency ω_0.

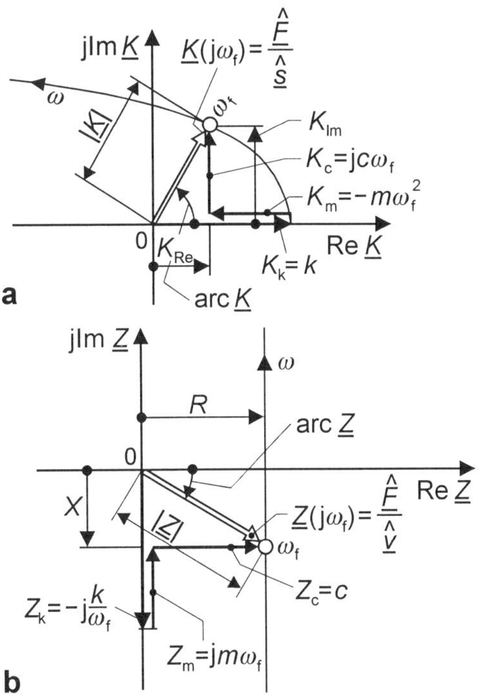

Fig. 4.9. Phasor diagrams for parallel-connected basic elements. **a** In the dynamic stiffness plane; **b** in the mechanical impedance plane

For lightly damped systems the exact curves deviate substantially from the a-symptotes. The corrections of the logarithmic frequency curves to the asymptotic approximation at the corner frequency are marked by the correction gains.

Response Data Plotting. Logarithmic frequency plots (Bode plots) of *normalized dynamic stiffness* are shown in Fig. B.7, the composite logarithmic gain with the asymptotes inclusive are visualized in Fig. B.8 of the Appendix B.

For most of the practical applications an external force acts on the structure, and the significant response is given by a pertinent motion variable. Expressing phasor ratios in terms of the inverse input-output relation, i.e., by the dynamic stiffness, or the mechanical impedance, those composite characteristics only have *historical significance* in vibration data analysis. All commonly used test procedures and requirements result in the determination of the reciprocal of one of the above-named dynamic characteristics, [18], [72], being designated as follows.

Forming the inverse phasor ratio (displacement-force relation) corresponding to Eq. (4.25b) taken at the same point in the system,
the *driving-point dynamic compliance of the structure*

$$\underline{C}(\mathrm{j}\omega_\mathrm{f}) = \frac{1}{\underline{K}(\mathrm{j}\omega_\mathrm{f})} = \frac{\hat{\underline{s}}}{\hat{\underline{F}}}$$

$$= \frac{1}{(k - m\omega_\mathrm{f}^2) + \mathrm{j}c\omega_\mathrm{f}} = \frac{k - m\omega_\mathrm{f}^2 - \mathrm{j}c\omega_\mathrm{f}}{(k - m\omega_\mathrm{f}^2) + (c\omega_\mathrm{f})^2} = \frac{K_\mathrm{Re} - \mathrm{j}K_\mathrm{Im}}{K_\mathrm{Re}^2 + K_\mathrm{Im}^2} = C_\mathrm{Re} + \mathrm{j}C_\mathrm{Im} \qquad (4.33a)$$

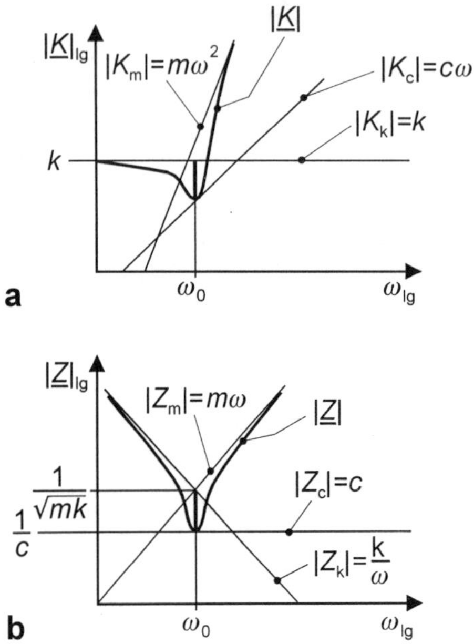

Fig. 4.10. Logarithmic frequency plots (magnitude plots) for parallel-connected basic elements. **a** Dynamic stiffness plots; **b** mechanical impedance plots

is defined being a complex ratio (complexor), which is sometimes called the direct receptance, with the *real part* (active dynamic compliance)

$$C_{\mathrm{Re}} = \frac{k - m\omega_{\mathrm{f}}^2}{(k - m\omega_{\mathrm{f}}^2)^2 + (c\omega_{\mathrm{f}})^2} = \frac{K_{\mathrm{Re}}}{K_{\mathrm{Re}}^2 + K_{\mathrm{Im}}^2} \qquad (4.34a),$$

and the *imaginary part* (reactive dynamic compliance)

$$C_{\mathrm{Im}} = -\frac{c\omega_{\mathrm{f}}}{(k - m\omega_{\mathrm{f}}^2)^2 + (c\omega_{\mathrm{f}}^2)} = -\frac{K_{\mathrm{Im}}}{K_{\mathrm{Re}}^2 + K_{\mathrm{Im}}^2} \qquad (4.34b).$$

Forming the inverse phasor ratio (velocity-force relation) corresponding to Eq. (4.27b) taken at the same point in the system,
the *driving-point (mechanical) mobility of the structure*

$$\underline{Y}(\mathrm{j}\omega_{\mathrm{f}}) = \frac{1}{\underline{Z}(\mathrm{j}\omega_{\mathrm{f}})} = \frac{\hat{\underline{\upsilon}}}{\hat{\underline{F}}}$$

$$= \frac{1}{c + \mathrm{j}(m\omega_{\mathrm{f}} - \frac{k}{\omega_{\mathrm{f}}})} = \frac{c - \mathrm{j}(m\omega_{\mathrm{f}} - \frac{k}{\omega_{\mathrm{f}}})}{c^2 + (m\omega_{\mathrm{f}} - \frac{k}{\omega_{\mathrm{f}}})^2} = \frac{R - \mathrm{j}X}{R^2 + X^2} = G + \mathrm{j}B \qquad (4.35a)$$

is defined being a complex ratio (complexor), which is sometimes called the direct mechanical admittance, with the *real part* (mechanical conductance)

$$G = \frac{c}{c^2 + (m\omega_{\mathrm{f}} - \frac{k}{\omega_{\mathrm{f}}})^2} = \frac{R}{R^2 + X^2} \qquad (4.36a),$$

and the *imaginary part* (mechanical susceptance)

$$B = -\frac{m\omega_f - \frac{k}{\omega}}{c^2 + (m\omega_f - \frac{k}{\omega_f})^2} = -\frac{X}{R^2 + X^2} \qquad (4.36b).$$

The related characteristics, Eqs. (4.33a), (4.35a), can be converted into one another with ease by multiplying the phasor compliance, Eq. (4.33a), and $j\omega_f$ together, or respectively, the phasor mobility, Eq. (4.35a), and $1/j\omega_f$ together

$$\underline{C}(j\omega_f) = \frac{1}{j\omega_f}\underline{Y}(j\omega_f) \quad \Leftrightarrow \quad \underline{Y}(j\omega_f) = j\omega_f\,\underline{C}(j\omega_f) \qquad (4.37).$$

For a parallel connected structure the elements have the same displacement phasor, respectively velocity phasor, across them, so that
the *component dynamic compliances*

$$\frac{1}{\underline{C}(j\omega_f)} = \frac{1}{C_k} + \frac{1}{C_c} + \frac{1}{C_m} \qquad (4.38),$$

and the *component mobilities*

$$\frac{1}{\underline{Y}(j\omega_f)} = \frac{1}{Y_k} + \frac{1}{Y_c} + \frac{1}{Y_m} \qquad (4.39)$$

combine by the reciprocal rule corresponding to Eqs. (4.25a), (4.27a).

For visualizing the graphical vector addition of contributions made by the reciprocals of the individual phasor ratios the multiplication by
the *conversion factor*

$$\underline{C}\,C_k \qquad (4.40)$$

yields the reduction of phasor compliance, Eq. (4.33a), to the term

$$C_k = \underline{C}(j\omega_f) + \frac{C_k}{C_c}\underline{C}(j\omega_f) + \frac{C_k}{C_m}\underline{C}(j\omega_f) \qquad (4.41),$$

and, respectively, the multiplication by the *conversion factor*

$$\underline{Y}\,Y_k \qquad (4.42)$$

results in the reduction of phasor mobility, Eq. (4.35a), to the term

$$Y_k = \underline{Y}(j\omega_f) + \frac{Y_k}{Y_c}\underline{Y}(j\omega_f) + \frac{Y_k}{Y_m}\underline{Y}(j\omega_f) \qquad (4.43).$$

Thus, the above-named inverse phasor ratios, Eqs. (4.33a), (4.35a), are fitted to construct the composite polar frequency-response loci of the dynamic compliance of the structure

$$\underline{C}(j\omega_f) = \frac{1}{\frac{1}{C_k} + \frac{1}{C_c} + \frac{1}{C_m}} = \frac{1}{(k - m\omega_f^2) + jc\omega_f} = \frac{1}{\underline{K}(j\omega_f)} = \frac{\hat{\underline{s}}}{\hat{\underline{F}}} \qquad (4.33b),$$

as well as of the (mechanical) mobility of the structure

$$\underline{Y}(j\omega_f) = \frac{1}{\frac{1}{Y_k} + \frac{1}{Y_c} + \frac{1}{Y_m}} = \frac{1}{c + j(m\omega_f - \frac{k}{\omega_f})} = \frac{1}{\underline{Z}(j\omega_f)} = \frac{\hat{\underline{\upsilon}}}{\hat{\underline{F}}} \qquad (4.35b).$$

The *inverse phasor ratio polygon* illustrates the combination of the reduced individual characteristics with the composite dynamic characteristic of the parallel-connected structure by use of the reciprocal rule, Fig. 4.11.

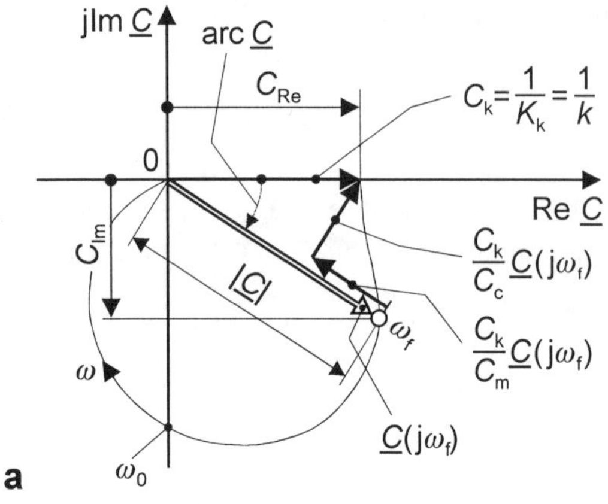

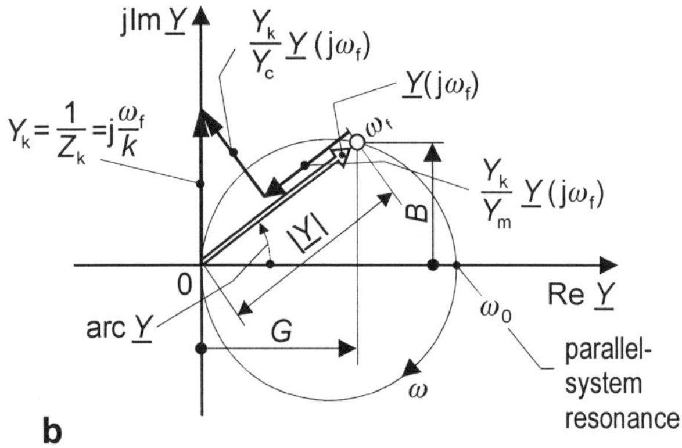

Fig. 4.11. Phasor diagrams and polar plots for parallel-connected basic elements. **a** In the dynamic compliance plane; **b** in the mobility plane

In the mobility plane the parallel mechanical system is represented by a *circle* with the maximum magnitude at the natural (or resonance) frequency ω_0, Fig. 4.11b.

Response Data Plotting. Polar plots (Nyquist plots) of *normalized dynamic compliance* are shown in Fig. B.1 of the Appendix B.

The composite logarithmic frequency plots (Bode plots) also can be constructed by summing up the contributions of component curves, Fig. 4.12.

The logarithmic gains of the reciprocals are only the negatives of the corresponding originals. As a result, the magnitude plots of the inverse composite characteristics are the *mirror images* of the original ones as shown in Fig. 4.10. Just as illustrated before logarithmic frequency plotting may be reduced to straight-line approximation using the magnitude plots of the inverse component characteristics represented in Fig. 4.4.

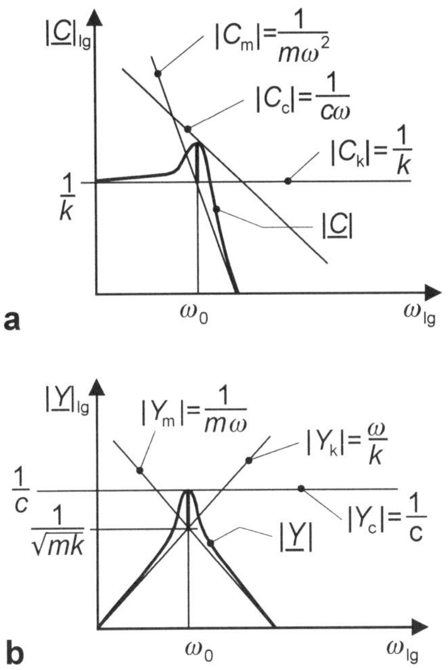

Fig. 4.12. Logarithmic frequency plots (magnitude plots) for parallel-connected basic elements. **a** Dynamic compliance plots; **b** mobility plots

At low frequencies the dynamic compliance is governed almost entirely by the spring. With regard to experimental investigations the *parallel mechanical system* is said to be *stiffness "controlled" below resonance*. At high frequencies the mass dominates. Thus, the parallel system is said to be *mass "controlled" above resonance*, Fig. 4.12a.

Following the (mechanical) mobility concept, Fig. 4.12b, the symmetric composite curve is approximated by a low-frequency and a high-frequency asymptote intersecting at natural (or resonance) frequency ω_0. At resonance the mobility is equal to the reciprocal of the (viscous) damping coefficient c, and the correction gain is marking out the gain difference between approximative and exact composite curves.

The parallel system requires a small excitation-force input to the structure for generating a resulting response motion at resonance frequency.

Response Data Plotting. Logarithmic frequency plots (Bode plots) of *normalized dynamic compliance* are shown in Fig. B.3, the composite logarithmic gain with the asymptotes inclusive are visualized in Fig. B.4 of the Appendix B.

Elements in Series. Components connected in series are separately actuated, each of them between two yielding attachments transmitting through the structure an excitation force, so that the *force* at the connecting rods or bars is *common through all the elements*, Fig. 4.13.

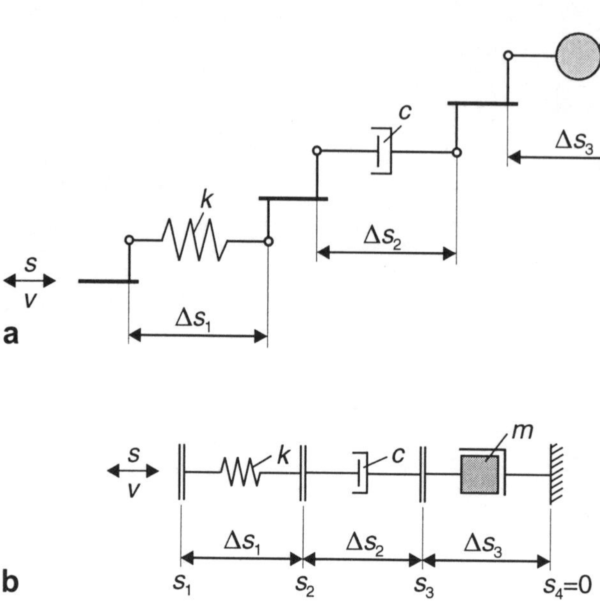

Fig. 4.13. Series-connected basic elements. **a** Junctions as rods; **b** junctions as bars

Series connection is indicated by the *special feature of zero restraining of the structure* (with all the other connection points of the system "free") since any of the elements is constrained to have zero velocity.

Considering the sinusoidal steady state the system variables and dynamic characteristics can be written in terms of phasors.

Force interconnection requirement (vertex postulate):

$$\underline{\hat{F}} = \underline{\hat{F}_1} = \underline{\hat{F}_2} = \underline{\hat{F}_3} \tag{4.44}$$

is simply that the force phasor through all of the elements is the same.

Motion interconnection requirement (circuit postulate):

$$\underline{\hat{s}} = \underline{\Delta\hat{s}_1} + \underline{\Delta\hat{s}_2} + \underline{\Delta\hat{s}_3} \tag{4.45a}$$

implies that the (relative) displacement phasors sum up across the elements to the total displacement phasor.

Forming the corresponding phasor ratios (displacement-force relations)

$$\frac{\underline{\hat{s}}}{\underline{\hat{F}}} = \frac{\underline{\Delta\hat{s}_1}}{\underline{\hat{F}_1}} + \frac{\underline{\Delta\hat{s}_2}}{\underline{\hat{F}_2}} + \frac{\underline{\Delta\hat{s}_3}}{\underline{\hat{F}_3}} \tag{4.45b}$$

the *overall dynamic compliance of a series-connected structure*

$$\underline{C}(\mathrm{j}\omega_\mathrm{f}) = C_1 + C_2 + C_3 = C_\mathrm{k} + C_\mathrm{c} + C_\mathrm{m} = \frac{1}{k} - \mathrm{j}\frac{1}{c\omega_\mathrm{f}} - \frac{1}{m\omega_\mathrm{f}^2} \tag{4.46a}$$

makes evident that the *component dynamic compliances add.*

The same interconnective constraints are valid if motion will be expressed between the (relative) velocity phasors across all the elements

$$\underline{\hat{\upsilon}} = \underline{\Delta \hat{\upsilon}_1} + \underline{\Delta \hat{\upsilon}_2} + \underline{\Delta \hat{\upsilon}_3} \qquad (4.47a).$$

Forming the corresponding phasor ratios (velocity-force relations)

$$\frac{\underline{\hat{\upsilon}}}{\underline{\hat{F}}} = \frac{\underline{\hat{\upsilon}_1}}{\underline{\hat{F}_1}} + \frac{\underline{\hat{\upsilon}_2}}{\underline{\hat{F}_2}} + \frac{\underline{\hat{\upsilon}_3}}{\underline{\hat{F}_3}} \qquad (4.47b)$$

the *overall (mechanical) mobility of a series-connected structure*

$$\underline{Y}(\mathrm{j}\omega_\mathrm{f}) = Y_1 + Y_2 + Y_3 = Y_\mathrm{k} + Y_\mathrm{c} + Y_\mathrm{m} = \mathrm{j}\frac{\omega_\mathrm{f}}{k} + \frac{1}{c} - \mathrm{j}\frac{1}{m\omega_\mathrm{f}} \qquad (4.48a)$$

makes evident that the *component mobilities add.*

Expressing the response motion in terms of a displacement the response ratio of phasors taken at the same point in the system is designated
the *driving-point (or direct) dynamic compliance of the structure*

$$\underline{C}(\mathrm{j}\omega_\mathrm{f}) = \left(\frac{1}{k} - \frac{1}{m\omega_\mathrm{f}^2}\right) - \mathrm{j}\frac{1}{c\omega_\mathrm{f}} = \frac{\underline{\hat{s}}}{\underline{\hat{F}}} = C_\mathrm{Re} + \mathrm{j}C_\mathrm{Im} \qquad (4.46b),$$

being a complex ratio (complexor), which is sometimes called the direct receptance, with the *real part* (active dynamic compliance)

$$C_\mathrm{Re} = \frac{1}{k} - \frac{1}{c\omega_\mathrm{f}^2} \qquad (4.49a),$$

and the *imaginary part* (reactive dynamic compliance)

$$C_\mathrm{Im} = -\frac{1}{c\omega_\mathrm{f}} \qquad (4.49b).$$

The significant response is also given by expressing the response motion in terms of velocity. The response ratio of phasors at the same point defines
the *driving-point (or direct) (mechanical) mobility of the structure*

$$\underline{Y}(\mathrm{j}\omega_\mathrm{f}) = \frac{1}{c} + \mathrm{j}\left(\frac{\omega_\mathrm{f}}{k} - \frac{1}{m\omega_\mathrm{f}}\right) = \frac{\underline{\hat{\upsilon}}}{\underline{\hat{F}}} = G + \mathrm{j}B \qquad (4.48b),$$

being a complex ratio (complexor), which is sometimes called the direct mechanical admittance, with the *real part* (mechanical conductance)

$$G = \frac{1}{c} \qquad (4.50a),$$

and the *imaginary part* (mechanical susceptance)

$$B = \frac{\omega_\mathrm{f}}{k} - \frac{1}{\omega_\mathrm{f} m} \qquad (4.50b).$$

For the present network configuration the response motion phasor $\underline{\hat{s}}$, or $\underline{\hat{\upsilon}}$, can be rewritten as an algebraic product of the exciting force phasor $\underline{\hat{F}}$ and the dynamic characteristic of the composite system expressed by
the *phasor compliance of the series-connected structure* $\underline{C}(\mathrm{j}\omega_\mathrm{f})$

$$\underline{\hat{s}} = \left[\left(\frac{1}{k} - \frac{1}{m\omega_\mathrm{f}^2}\right) - \mathrm{j}\frac{1}{c\omega_\mathrm{f}}\right]\underline{\hat{F}} = \underline{C}(\mathrm{j}\omega_\mathrm{f})\underline{\hat{F}} \qquad (4.51),$$

or, respectively, by
the *phasor mobility of the series-connected structure* $\underline{Y}(\mathrm{j}\omega_\mathrm{f})$

$$\hat{\underline{\upsilon}} = \left[\frac{1}{c} + \mathrm{j}\left(\frac{\omega_\mathrm{f}}{k} - \frac{1}{m\omega_\mathrm{f}}\right)\right]\hat{\underline{F}} = \underline{Y}(\mathrm{j}\omega_\mathrm{f})\hat{\underline{F}} \qquad (4.52).$$

The related characteristics, Eqs. (4.46a,b), (4.48a,b), can be converted into one another with ease by multiplying the phasor compliance, Eq. (4.46a,b), and $\mathrm{j}\omega_\mathrm{f}$ together, or respectively, the phasor mobility, Eq. (4.48a,b), and $1/(\mathrm{j}\omega_\mathrm{f})$ together

$$\underline{C}(\mathrm{j}\omega_\mathrm{f}) = \frac{1}{\mathrm{j}\omega_\mathrm{f}}\underline{Y}(\mathrm{j}\omega_\mathrm{f}) \quad \Leftrightarrow \quad \underline{Y}(\mathrm{j}\omega_\mathrm{f}) = \mathrm{j}\omega_\mathrm{f}\,\underline{C}(\mathrm{j}\omega_\mathrm{f}) \qquad (4.53).$$

The overall dynamic behaviour of mechanical elements connected in series is characterized by adding the component mobilities, which are sometimes called the component mechanical admittances, Eq. (4.48a). Though being *analogous* to the combination of electrical system components the *summing up of component admittances* nevertheless describes a *parallel-connected electric circuit* with reference to the "classical" type of electromechanical analogies. Following the concept of related characteristics by adding the component dynamic compliances, Eq. (4.46a), the presented analogy has been called the "practical" type, [65].

Using the vector representation of response phasor ratios the overall frequency response can be plotted in terms of polar coordinates by the *locus of* $\underline{C}(\mathrm{j}\omega_\mathrm{f})$ *, and of* $\underline{Y}(\mathrm{j}\omega_\mathrm{f})$, Eqs. (4.46a,b), (4.48a,b), in the complex plane called, in this case, the *dynamic compliance plane*, or the *mobility plane*, respectively, Fig. 4.14.

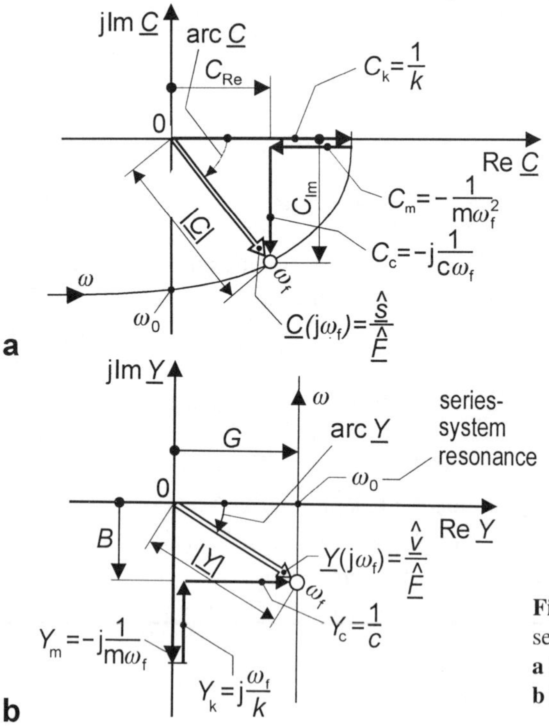

Fig. 4.14. Phasor diagrams for series-connected basic elements. **a** In the dynamic compliance plane; **b** in the mobility plane

The composite polar frequency-response loci (Nyquist plots) of dynamic compliance and (mechanical) mobility can be constructed by graphical vector addition of individual phasor ratios. The combination of individual characteristics with the composite dynamic characteristic of the *series-connected structure* is illustrated by the *phasor ratio polygon.*

In the mobility plane the series mechanical system is represented by a *straight line* crossing the real axis at natural (or resonance) frequency ω_0 because the imaginary part (mechanical susceptance) is zero at this frequency, Fig. 4.14b.

The composite logarithmic frequency plots (Bode plots) also can be constructed by summing up the contributions of component curves, Fig. 4.15.

At low frequencies the dynamic compliance is governed almost entirely by the mass, whereas at high frequencies the spring dominates. Regarding test requirements the *series mechanical system* is thus said to be *mass "controlled" below resonance*, and respectively *stiffness "controlled" above resonance*, Fig. 4.15a.

Following the (mechanical) mobility concept, Fig. 4.15b, logarithmic frequency plotting evidently illustrates that the dynamic characteristic of the series system is *opposite* to that of a *parallel system* as shown in Fig. 4.12b. At resonance the mobility is equal to the reciprocal of the damping coefficient *c*, and the correction gain is marked respecting a *straight-line approximation* by the component characteristics represented in Fig. 4.4.

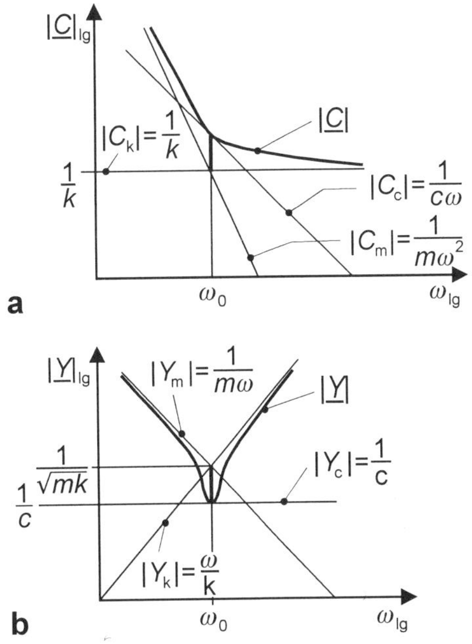

Fig. 4.15. Logarithmic frequency plots (magnitude plots) for series-connected basic elements. **a** Dynamic compliance plots; **b** mobility plots

The resonance of a series mechanical system is often called an "*antiresonance*" because a large excitation force is required to cause a motion response at natural (or resonance) frequency. Considering test equipments for generating vibration the series connection typifies a test structure mounted on the vibration exciter in the environmental type of test rather than a parallel type specimen which would require the mass to be resiliently supported by a spring (suspension system), with the mass directly attached to the exciter.

4.2.4 General Transform Analysis Principles. *Mechanical Circuit Theorems*

Structures are composed of several series and parallel combinations. The frequency-response characteristic of an arbitrary mechanical system can be analysed by subdividing a corresponding dynamic model system into the phasor or transform relations of the elementary (or 1-port) components. Applying the *procedure of simplification* an equivalent model system even for a complicated structure (lumped parameter system with large degrees of freedom) may be defined, and its response when combined with a remote system of known dynamic characteristic can be predicted.

Network Diagram Reduction

Forming interconnective relationships by phasor ratios has proved an useful instrument of direct derivation of system relationships. Previously, the deriving of interconnective relations by inspection has been confined to fundamental configurations of elements only (parallel-series connections), 4.2.3.

Nevertheless, a directly deriving procedure also can be obtained for mechanical circuits having passive elements both in parallel and in series. By use of devices for *simplification of networks* a base for introducing a *general lumped-system analysis* may be presented being appropriate to cover a lot of interconnection problems. Forming equivalent characteristics of *repeated structures (or sections) in parallel and in series* a reduction of network diagrams can be realized appearing as a counterpart to the reduction of block diagrams and signal flow graphs by fundamental configurations, 2.1.4.

Basic Tool. For lumped-system analysis the *reduction rules of mechanical circuits* are formulated as follows:

- for mechanical *subsystems (or sections) in parallel* the component dynamic *stiffness* (respectively the component mechanical *impedances*) have to be *added*;
- for mechanical *subsystems in series* the component dynamic *compliances* (respectively the component *mobilities*) have to be *added*.

Combining all the sections by successive reduction the final result will be the dynamic characteristic of the composite system represented either by an overall dynamic stiffness or by its inverse, an overall dynamic compliance (respectively by an overall mechanical impedance or mobility).

Maxwell's Reciprocity Theorem

Reciprocity, a transmission property related to *transfer frequency-response characteristics*, presupposes systems being composed of linear and bilateral elements. A *bilateral element* is one through which forces are transmitted equally well in either direction passing its connection.

If a mechanical generator operates on a system of linear bilateral elements by an excitation force (constant-force generator) at the point *i*, the exciter can be removed from *i* and placed at the point *j* on the structure; then the former motion response at *j* will exist at *i*, provided the frequency characteristics at all points are unchanged. Expressing the motion responses at *j*, respectively at *i*, in terms of a displacement the reciprocity theorem states equality for
the *transfer dynamic compliances* (or cross receptance)

$$\underline{C}_{ij}(j\omega) = \underline{C}_{ji}(j\omega) \qquad (4.54a).$$

Measuring response velocities at *j* and *i* the reciprocity theorem holds true for the related characteristics at different points, i.e., for
the *transfer (mechanical) mobilities*

$$\underline{Y}_{ij}(j\omega) = \underline{Y}_{ji}(j\omega) \qquad (4.55a).$$

This theorem also will be stated by use of a generator exerting a motion on the structure at the point *i* (constant-displacement or constant-velocity generator), and the resulting force is measured at the point *j*. Replacing the exciter from *i* to *j* the reciprocals of the related characteristics at different points are equalling in terms of
the *transfer dynamic stiffness*

$$\underline{K}_{ij}(j\omega) = \underline{K}_{ji}(j\omega) \qquad (4.54b),$$

or, respectively, of
the *transfer mechanical impedance*

$$\underline{Z}_{ij}(j\omega) = \underline{Z}_{ji}(j\omega) \qquad (4.55b).$$

Reciprocity simplifies the analysis of two-way energy transmission systems since the complex ratios of system relationships need be determined for only one direction.

Superposition Theorem

If more than one mechanical generator acts on a system of linear bilateral elements, the force or motion response at a point in the system can be determined by *adding the response to each of the generators*, taken at any time. Consequently, the other generators are substituted for their restrained-state characteristic being measured or defined, i.e., for their internal impedances. This theorem is useful for analysing systems having *several mechanical generators*. Furthermore, the superposition of sinusoidal terms simplifies the response calculation for mechanical structures being subjected to *periodic excitation functions*.

Foster's Reactance Theorem

Mechanical circuits connecting only storage elements (springs and masses) without including any dissipator element (damper) the interconnective relationships result in a frequency-response characteristic taken at the same point on a structure which is pure imaginary.

For that reason the driving-point impedance, or, respectively, the driving-point stiffness and their reciprocals, have additional properties constituting *Foster's theorem*. Denoting the overall dynamic characteristic of a *network configuration without transmission loss* by

$$G(\mathrm{j}\omega) = \mathrm{j}\,\mathrm{Im}G(\mathrm{j}\omega) = \mathrm{j}G(\omega) \tag{4.56}$$

then the following properties hold in general:

i. The system function $G(\omega)$ is *real* for all ω;
ii. The slope $\mathrm{d}G/\mathrm{d}\omega$ is always *positive*;
iii. At the origin $\omega = 0$ the function $G(\omega)$ has either a *pole* or a *zero*;
iv. All poles and zeros are *simple*; i.e., there are no multiple-order poles;
v. The *characteristic roots and the zeros alternate*, i.e., there is always a zero between two poles;
vi. The system function $G(\omega)$ is defined by the *location of its characteristic roots* (system-poles) and its zeros except for a multiplying constant (static response factor) $G(0)$.

The transfer impedance and related transfer characteristics do not involve completely the above mentioned function properties. Mostly, the conditions ii and v are not satisfied with regard to transfer characteristics of lossless systems.

Thévenin's Theorem

Mechanical circuits which contain one or more vibration sources (exciters or active elements) having an output terminal to a load for the energy supply can be represented by an *ideal constant-force generator* in parallel with an equivalent mechanical impedance connected to the load, Fig. 4.16a.

The *Thévenin equivalent network* may be determined by applying output-terminal constraints, introduced in 2.2.3. Referring the behaviour of 2-terminal sources, in particular the complete mechanical source derived from the 2-port parameter method in 2.2.5, first the output point will be restrained (no motion permitted at the output). The output force being transmitted by the attachment point of the equipment becomes the *blocked force* $\underline{F}_{\mathrm{oc}}$ (the subscript "oc" emphasizes open-circuit condition for measurement, i.e., $\underline{F}_{\mathrm{oc}}$ is an open-circuit force phasor). Secondly, the output point is released from load connection to move freely (no force exerted at the output), and the output velocity becomes the *free velocity* $\underline{v}_{\mathrm{sc}}$

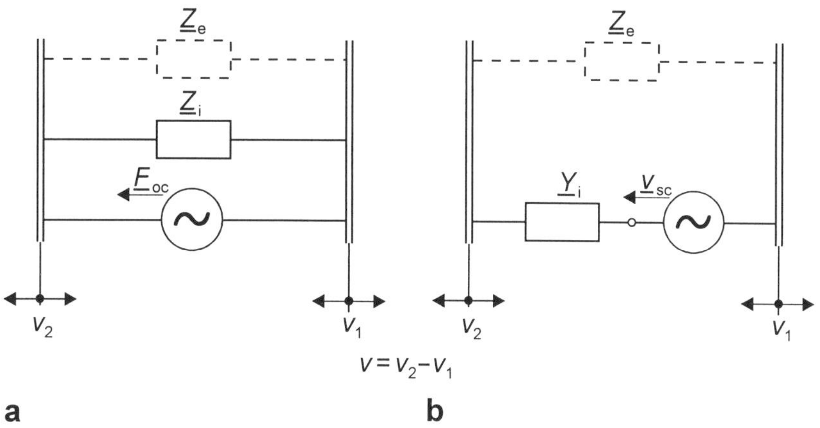

Fig. 4.16. Equivalent arrangements of mechanical circuits. **a** Thévenin's equivalent network; **b** Norton's equivalent network

(the subscript "sc" emphasizes short-circuit condition, i.e., $\underline{v}_{sc}$ is a closed-circuit velocity phasor). Then the *internal impedance* (parallel impedance) $\underline{Z}_i$ can be determined by the phasor ratio of the measured quantities $\underline{F}_{oc}$ and $\underline{v}_{sc}$.

$$\underline{Z}_i = \underline{F}_{oc} / \underline{v}_{sc} \qquad (4.57a).$$

A great advantage is derived from this equivalent system in that attention is focused on the characteristics of a system at its output point and not on its individual components. By this the system response can be predicted with ease when the load attached to the output is varied.

Norton's Theorem

Mechanical circuits which contain vibration sources having an output connection to components can be represented alternatively by an *ideal constant-velocity generator* in series with an equivalent (mechanical) mobility connected to the components, Fig. 4.16b.

The *Norton equivalent network* is corollary (dual) to Thévenin's equivalent system expressed in terms of the measured quantities defined above as well as in terms of the inverse phasor ratio defining the *internal mobility* (series mobility) $\underline{Y}_i$

$$\underline{Y}_i = \underline{v}_{sc} / \underline{F}_{oc} \qquad (4.57b).$$

Thus, the identity proves true

$$\underline{Z}_i \underline{Y}_i = 1 \qquad (4.57c).$$

The same advantages in lumped-system analysis exist as with Thévenin's parallel representation. The equivalent system to be preferred depends upon the type of structure. For experimental investigations of the equivalent dynamic characteristics it is usually easier to measure the free velocity than the blocked force on large heavy structures, while the converse is advisable for light structures. In any case, one representation is easily derived from the other corresponding with the identity of Eq. (4.57c).

The various mechanical circuit theorems treated above in short can be used as an aid in network modelling of mechanical systems and in response calculation. In particular, the problem of *adding a remote system* to an already complex system will be simplified. Thévenin's equivalent is useful when the separate subsystem is to be inserted in a branch where the force is already known. Norton's equivalent is advantageous when the subsystem is to be bridged between two points whose relative velocity is already known.

The mechanical circuit theorems are the analogues of the well-known theorems employed in the analysis of electric circuits, see 2.2.5, Fig. 2.17.

Generalized Kirchhoff's Laws

The interconnective relationships between the power variables at common nodes (or vertices), and around closed paths (loops) are the fundamental laws of general network analysis. A large number of sets of network equations can be derived by use of *force interconnection requirement* and of *motion interconnection requirement*. If the system is not too large ingenious elimination among these equations will result in the desired system relationships, 4.1.2.

The combination of *generalized Kirchhoff's laws* with the concept of *mechanical mobility*, or *dynamic compliance*, respectively, provides an effective method for lumped-system analysis also including complicated mechanical circuits. The dynamic characteristic of a composite system is commonly defined as a *response phasor ratio* (phasor mobility or phasor dynamic compliance) respecting sinusoidal steady-state analysis only. Furthermore, the transform analysis method replaces response phasor ratios by *response transform quotients*. Relating the Fourier transform pairs of power variables structural configurations are represented by a frequency-response characteristic in generalized form adapted to vibration data analysis implying transient and random vibrations, 4.2.1.

In lumped-system analysis the velocities throughout a mechanical circuit are commonly evaluated by applying the statement of dynamic equilibrium (d'Alembert's principle) at each junction (node) where the velocity is unknown. Once the velocities are determined, any desired forces can be evaluated. On the base of this principle (force interconnection requirement) the *equations of motions* as well as the *system transform relationships* in terms of response phasor ratios (or, possibly in terms of response transform quotients) may be determined by inspection of the network diagram, 2.3.4.

Basic Tool. According to the interconnective force relationships at any node (or vertex) phasor force equations can be stated. Then the phasor forces are expressed in terms of relative phasor velocities and phasor impedances of the components. Finally, the desired system relationship will be described by a *set of vertex (or node) equations in phasor notation*. When the phasor equations are written so that the unknown velocities form columns, the equations are in the proper form for a determinant solution for any of the unknowns.

By elimination the unknown velocity referred to a single point in a system the phasor ratio of force and velocity at the same or different points in the system is formed resulting in an *overall mechanical impedance of the structure*. Since the dynamic behaviour will be represented by a force-displacement relationship the same directly deriving procedure can be performed expressing motion in terms of relative phasor displacements. Thus, the related characteristic given by an *overall dynamic stiffness of the structure* will be deduced being more convenient in machinery.

Suited to a lot of problems in vibrations the above-outlined *vertex transform analysis* (method of node forces) will be sufficient for deducing dynamic characteristics of any structure consisting of mechanical (phasor) sources and basic elements. Singly it may be indispensable for deriving the desired system relationship that phasor velocity equations around enough closed loops are stated which include each element at least once. Then the phasor velocities are expressed in terms of phasor forces and phasor mobilities of the components. Finally, the desired system relationship will be described by a *set of circuit (or loop) equations in phasor notation*. The phasor equations are solved for the unknown forces. Thus, the inverse phasor ratio of force and velocity, likewise of displacement, referred to the same or different points in the system result in an *overall (mechanical) mobility*, or, respectively, in an *overall dynamic compliance of the structure*.

The above-sketched *circuit transform analysis* (method of loop velocities) provides a second set of phasor equations. Together with the first one the two constraints cover dual interconnective relations.

For *large-scale systems* the process of deriving as sets of network phasor equations as overall dynamic characteristics will be further *systematized* by use of matrix methods to establish *fundamental transform relationships*, entering the component characteristics of different types of elements in other matrices, and finally obtaining independent sets of vertex and circuit equations by an automatic sequence of matrix operations, 4.1.2.

For developing the techniques of large-scale network analysis in detail one should be referred to more specialized literature, [11], [16], [17], [52], [60], [75], [90].

4.2.5
Graphical Methods to Mechanical System Design. *Selecting Vibratory Specifications by Polar Diagrams*

The presentation of mobility type data makes use of different graphs to suit this presentation to various applications. The advantages of logarithmic transformation for portraying frequency responses separately by their magnitude (gain) and phase versus log frequency have been marked out in 3.2.11. In addition to the favoured logarithmic plotting (Bode design) there is sometimes advantage in using alternative plotting methods.

The *polar representation of the frequency-response characteristic* plots the real and imaginary components as functions of frequency. It is also desirable to plot the data in polar coordinates (Nyquist plot) as illustrated in 4.2.3. The polar diagram the measured data of which may be enhanced by a circle-fitting procedure plays a part in extracting modal parameters from test data, [18].

In the following it will be shown that polar diagrams are also suited to judge the vibratory effects on machine structures in operations. For the investigation of the dynamic characteristic of structural members or parts the *concept of mechanical impedance* and that of *dynamic stiffness* will equally be applied. The concepts express equivalent definitions for different kinds of inverse frequency response functions, as treated in 4.2.1. Which of both concepts claims actual usefulness depends on the nature of problems to be fixed by the investigator. Whereas in vibration analysis impedance is well acquainted with yielding the *principal modes of mechanical circuits*, dynamic stiffness gives evidence on frequency-depending effects associated with the *vibratory resistance against deformation.*

For that reason the polar diagram of dynamic stiffness is appropriate for extracting *performance criteria* which are fundamental to *mechanical system design*, in that case for proportioning of main structures like bedplates or frames of the machine to permissible values of deformation owing to forced oscillation or self-excited vibration. Performance criteria in terms of calculated or estimated parameter values related to composite dynamic characteristics are called *frequency-response specifications*.

Vibratory Specifications by Varying Forcing Frequency

Using the vector representation of response phasor ratios the related overall frequency response is plotted in terms of polar coordinates by the locus of the composite characteristics $\underline{K}(\mathrm{j}\omega)$, Eq. (4.25a), in the dynamic stiffness plane, or $\underline{Z}(\mathrm{j}\omega)$, Eq. (4.27a), in the mechanical impedance plane, respectively, Fig. 4.17.

The steady-state interrelation of mechanical components is illustrated by the contributions made by the individual phasor ratios combining
the *phasor stiffnesses* K_i *of the elementary components*

$$\begin{aligned} \underline{K}(\mathrm{j}\omega = 0) &= K_k = k = K_k X_{kk} = k \cdot 1 \\ \underline{K}(\mathrm{j}\omega = \mathrm{j}\omega_0) &= K_c = \mathrm{j}c\omega = K_k X_{kc} = k \cdot \mathrm{j}2\zeta\eta \\ \underline{K}(\mathrm{j}\omega = \infty) &= K_m = -m\omega^2 = K_k X_{km} = k \cdot (-\eta^2) \end{aligned} \quad (4.58a,b,c),$$

or, respectively,
the *phasor impedances* Z_i *of the elementary components*

$$\begin{aligned} \underline{Z}(\mathrm{j}\omega = 0) &= Z_k = -\mathrm{j}k/\omega = Z_0 \tilde{X}_{kk} = \sqrt{mk} \cdot (-\mathrm{j}/\eta) \\ \underline{Z}(\mathrm{j}\omega = \mathrm{j}\omega_0) &= Z_c = c = Z_0 \tilde{X}_{kc} = \sqrt{mk} \cdot 2\zeta \\ \underline{Z}(\mathrm{j}\omega = \infty) &= Z_m = \mathrm{j}m\omega = Z_0 \tilde{X}_{km} = \sqrt{mk} \cdot \mathrm{j}\eta \end{aligned} \quad (4.59a,b,c)$$

with the composite characteristic of the parallel-connected structure defined by the complex ratio of phasors taken at the same point in the system.

The response ratio of phasors is associated to a single forcing (angular) frequency, and thus assigned to one point of frequency locus plot. The magnitude of the phasor ratio (amplitude ratio) being identical with the length of the localized vector (radius vector) is related to the excitation force by multiplying constant

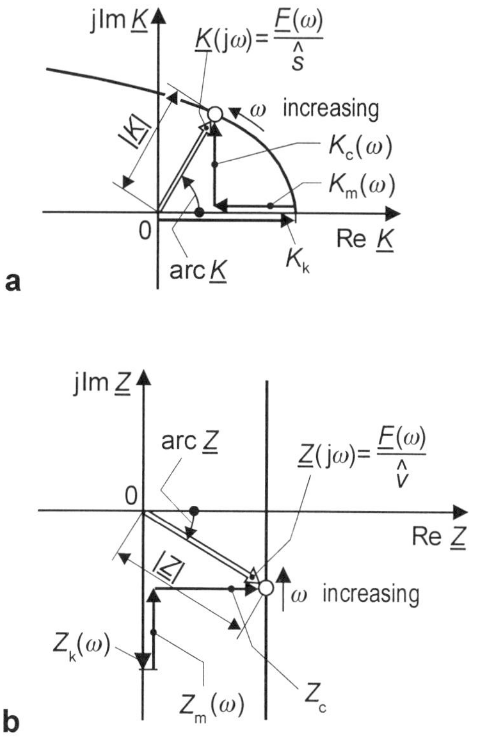

Fig. 4.17. Polar plots for parallel-connected basic elements. **a** Polar diagram of dynamic stiffness $\underline{K}(j\omega)$; **b** polar diagram of mechanical impedance $\underline{Z}(j\omega)$

phasors. Thereupon, the amplitude of the external force to be applied to the system follows from the phasor relation combining
the *modulus of dynamic stiffness* and *displacement amplitude*

$$\left|\hat{\underline{F}}\right|_\omega = \left|\hat{\underline{F}}_E\right|_\omega = |\underline{K}(j\omega)||\hat{\underline{s}}| \qquad (4.30a),$$

or, respectively,
the *modulus of mechanical impedance* and *velocity amplitude*

$$\left|\hat{\underline{F}}\right|_\omega = \left|\hat{\underline{F}}_E\right|_\omega = |\underline{Z}(j\omega)||\hat{\underline{v}}| \qquad (4.31a).$$

Specifying Driving Force Amplitude. If forcing frequency is allowed to vary within the frequency range of interest the change of excitation force in amplitude can be gathered by the frequency locus plot following the change of radius vector in length.

One problem of system design in the frequency domain arises from predicting *performance criteria for a vibration generator* imparting its vibrations whether to a fatigue testing machine or to oscillatory devices in process engineering. Supposing a simple harmonic vibration the phasor relations, Eqs. (4.30a), (4.31a), permit to determine "dependent on frequency" the *driving force amplitude* $\hat{F}$ which is to be required for generating or maintaining a presupposed constant operating motion $\hat{s}$, or $\hat{\upsilon}$ (constant-amplitude motion).

The question gets an extension for estimating the driving force amplitude $\hat{F}$ being necessary to produce *forces transmitted through the individual components*, $\underline{\hat{F}}_{\mathrm{s}}$, $\underline{\hat{F}}_{\mathrm{d}}$, $\underline{\hat{F}}_{\mathrm{m}}$. For this purpose the phasor ratio polygon proves useful, so that the component ratios are combined with the known amplitude of operating displacement $\hat{s}$, or velocity $\hat{\upsilon}$, respectively, both of them being constant multipliers.

Specifying Significant System Parameters. For illustrating frequency effects on the dynamic characteristic of a structure the *normalizing transformation of frequency responses* is used being a convenient data reduction method suited to graphical interpretation in the frequency domain. For that, the response ratios of phasors are adjusted in variable and scale to be performed in the following steps as shown in 3.2.11. The static response factors are identical with
the *stiffness of the spring*

$$|\underline{K}(0)| = K_{\mathrm{k}} = k \qquad (4.60),$$

or, respectively,
the *indicial (or characteristic) mechanical impedance*

$$|Z_{\mathrm{k}}(\omega_0)| = |Z_{\mathrm{m}}(\omega_0)| = Z_0 = |\sqrt{mk}| \qquad (4.61).$$

Then, the forcing (angular) frequency is transformed into the new variable *frequency ratio*, Eq. (3.39). Finally, the frequency responses varying with the new variable will be related to the multiplying constants, Eqs. (4.60), (4.61), and result in "nondimensional response ratios" of phasors designated
the *normalized dynamic stiffness of the structure*

$$\frac{\underline{K}\left(\mathrm{j}\frac{\omega}{\omega_0}\right)}{K_{\mathrm{k}}} = \underline{X_{\mathrm{k}}}(\mathrm{j}\eta) = X_{\mathrm{kk}} + X_{\mathrm{kc}} + X_{\mathrm{km}} = 1 + \mathrm{j}2\zeta\eta - \eta^2 = \frac{\underline{F}(\mathrm{j}\eta)}{\underline{\hat{s}}\,k} \qquad (4.62),$$

or, respectively,
the *normalized mechanical impedance of the structure*

$$\frac{\underline{Z}\left(\mathrm{j}\frac{\omega}{\omega_0}\right)}{Z_0} = \underline{\tilde{X}_{\mathrm{k}}}(\mathrm{j}\eta) = \tilde{X}_{\mathrm{kk}} + \tilde{X}_{\mathrm{kc}} + \tilde{X}_{\mathrm{km}} = -\mathrm{j}\frac{1}{\eta} + 2\zeta + \mathrm{j}\eta = \frac{\underline{F}(\mathrm{j}\eta)}{\underline{\hat{s}}\sqrt{mk}} \qquad (4.63).$$

The number of parameters is diminished by combining element parameters to system parameters, so that the new functions, Eqs. (4.62), (4.63), are expressed in terms of only two quantities, the frequency ratio η and the damping ratio ζ. Thus, the effect of the *performance criterion for related energy dissipation* ζ on the frequency-response characteristics will be represented graphically in the normalized form by a set (or family) of *adjusted (data-reduced) polar plots* for various amounts of damping as η varies over some driving frequency range, Fig. 4.18.

The steady-state interrelation of mechanical components is illustrated by the polygon of "non-dimensional phasor ratios" resulting from

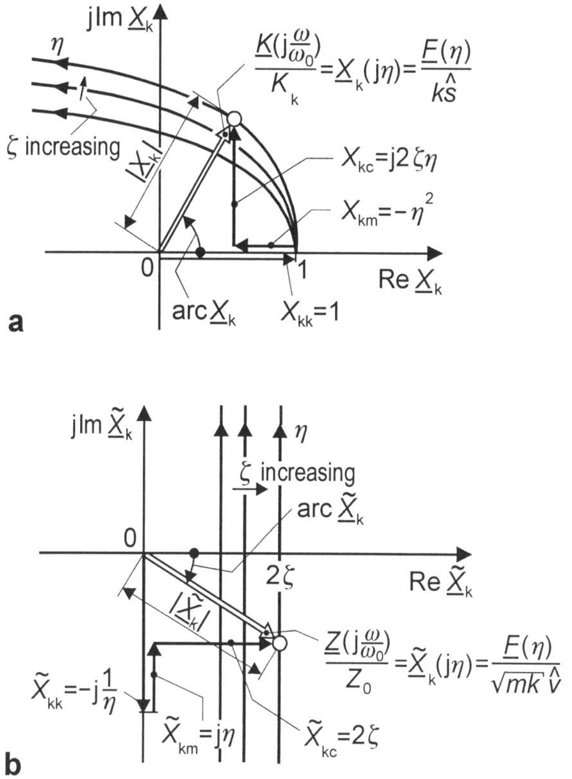

Fig. 4.18. Adjusted polar plots for parallel-connected basic elements. **a** Polar diagram of normalized dynamic stiffness $\underline{X}_{\mathrm{k}}(\mathrm{j}\eta)$; **b** polar diagram of normalized dynamic mechanical impedance $\underline{\widetilde{X}}_{\mathrm{k}}(\mathrm{j}\eta)$

the *normalized element stiffnesses*

$$
\begin{aligned}
K_{\mathrm{k}}/K_{\mathrm{k}} &= X_{\mathrm{kk}} &&= 1\\
K_{\mathrm{c}}/K_{\mathrm{k}} &= X_{\mathrm{kc}} &&= \mathrm{j}2\zeta\eta\\
K_{\mathrm{m}}/K_{\mathrm{k}} &= X_{\mathrm{km}} &&= -\eta^2
\end{aligned} \qquad (4.64\text{a,b,c}),
$$

or, respectively,
the *normalized element impedances*

$$
\begin{aligned}
Z_{\mathrm{k}}/Z_0 &= \widetilde{X}_{\mathrm{kk}} &&= -\mathrm{j}/\eta\\
Z_{\mathrm{c}}/Z_0 &= \widetilde{X}_{\mathrm{kc}} &&= 2\zeta\\
Z_{\mathrm{m}}/Z_0 &= \widetilde{X}_{\mathrm{km}} &&= \mathrm{j}\eta
\end{aligned} \qquad (4.65\text{a,b,c}).
$$

Response Data Plotting. Polar plots (Nyquist plots) of *normalized dynamic stiffness* are shown in Fig. B.5 of the Appendix B.

The requirement to be satisfied by vibration machines may be reverse to that one just treated. For this frequently applied driving operations the external force acting on the structure remains constant with frequency. The significant response to be expressed by a pertinent motion variable commonly results in the displacement at the same point in the structure, on occasion its velocity. Thus, forming the inverse phasor ratios corresponding to Eqs. (4.25a), (4.27a), displacement and velocity response phasors are related to the locus of the composite characteristic $\underline{C}(\mathrm{j}\omega)$, Eq. (4.33a), in the dynamic compliance plane, or $\underline{Y}(\mathrm{j}\omega)$, Eq. (4.35a), in the (mechanical) mobility plane, respectively, Fig. 4.19.

The steady-state interrelation of mechanical components is illustrated by the reciprocals of individual phasor ratios combining

the *phasor compliances* C_i *of the elementary components*

$$\begin{aligned} \underline{C}(\mathrm{j}\omega = 0) &= C_k = 1/k = C_k Y_{kk} = 1/k \\ \underline{C}(\mathrm{j}\omega = \omega_0) &= C_c = -\mathrm{j}/(c\,\omega) = C_k Y_{kc} = \left(-\mathrm{j}/(2\zeta\eta)\right)/k \\ \underline{C}(\mathrm{j}\omega = \infty) &= C_m = -1/\left(m\omega^2\right) = C_k Y_{km} = \left(-1/\eta^2\right)/k \end{aligned} \quad (4.66\text{a,b,c}),$$

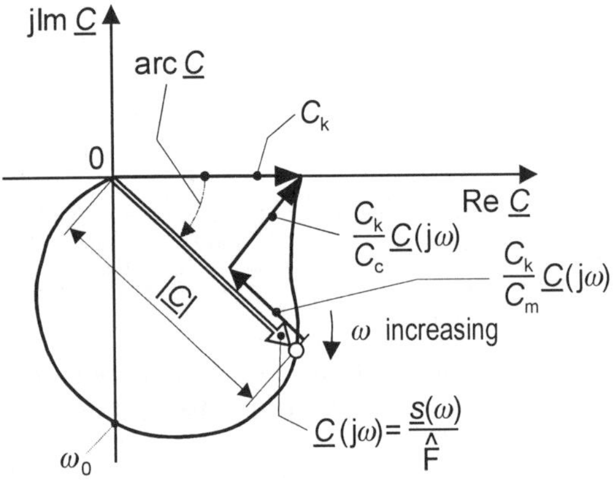

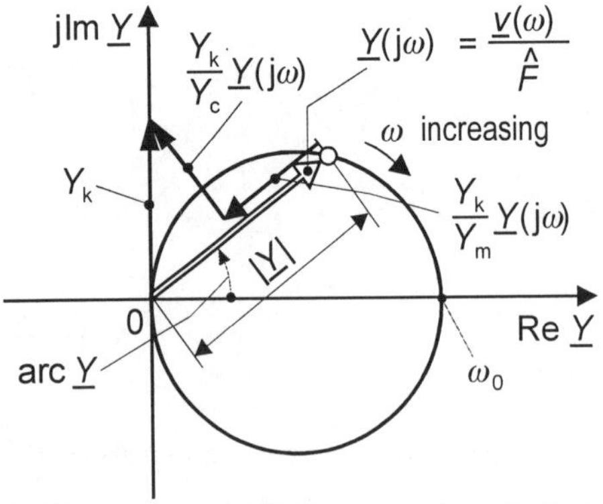

Fig. 4.19. Polar plots for parallel-connected basic elements. **a** Polar diagram of dynamic compliance $\underline{C}(\mathrm{j}\omega)$; **b** polar diagram of (mechanical) mobility $\underline{Y}(\mathrm{j}\omega)$

or, respectively,
the *phasor mobilities* Y_i *of the elementary components*

$$\begin{aligned} \underline{Y}(j\omega=0) &= Y_k = j\omega/k = Y_0 Y'_{kk} = j\eta/\sqrt{mk} \\ \underline{Y}(j\omega=j\omega_0) &= Y_c = 1/c = Y_0 Y'_{kc} = \left(1/(2\zeta)\right)/\sqrt{mk} \\ \underline{Y}(j\omega=\infty) &= Y_m = -j/(m\omega) = Y_0 Y'_{km} = \left(-j/\eta\right)/\sqrt{mk} \end{aligned} \qquad (4.67a,b,c)$$

with the composite characteristic of the parallel-connected structure due to Eqs. (4.41), (4.43). The magnitude of the inverse phasor ratio (inverse amplitude ratio) being identical with the length of the localized vector is related to the motion response by multiplying constant phasors. Thereupon, the amplitude of the resultant motion response of the system follows from the phasor relation combining the *modulus of dynamic compliance* and *force amplitude*

$$|\hat{\underline{s}}|_\omega = |\underline{C}(j\omega)||\hat{\underline{F}}| \qquad (4.33c),$$

or, respectively,
the *modulus of mechanical mobility* and *force amplitude*

$$|\hat{\underline{v}}|_\omega = |\underline{Y}(j\omega)||\hat{\underline{F}}| \qquad (4.34c).$$

Specifying Operating Motion Amplitude. The change of motion response in amplitude can be gathered by the frequency response locus plot following the change of radius vector in length.

In the following the problem of system design in the frequency domain arises from predicting *performance criteria for a vibration machine* being the reverse of the preceding driving characteristics. Supposing a simple harmonic vibration the phasor relations, Eq. (4.33c), (4.34c), permit to determine "dependent on frequency" the *operating motion amplitude* $\hat{s}$, or $\hat{v}$, which can be realized by applying a presupposed constant driving force $\hat{F}$ (constant-amplitude force). The question can be extended for estimating the amplitudes of *occuring forces transmitted through the individual components* $\hat{\underline{F}}_s$, $\hat{\underline{F}}_d$, $\hat{\underline{F}}_m$, due to an acting driving force $\hat{F}$. For this purpose the inverse phasor ratio polygon proves useful, so that the reduced component ratios are combined with the known driving force amplitude related to the dynamic compliance of the spring $\hat{F}/C_k$, or related to the impedance of the spring $\hat{F}/Z_k$, respectively, both of them being constant multipliers.

Response Data Plotting. Polar plots (Nyquist plots) of normalized dynamic compliance are shown in Fig. B.1 of the Appendix B.

Vibratory Specifications by Element Parameter Estimation

For illustrating the varying effect of lumped parameters (discrete elements spring, mass, and damper) on the dynamic behaviour of structures in the frequency domain it is not suitable to carry out a normalizing transformation of frequency responses. Instead of using a polar representation adjusted in variable and scale the polar plotting of frequency-response functions will be realized without any data reduction (direct response locus).

For mechanical engineering applications machine structures have to meet the demand that deformations caused by oscillations must be limited. Diminishing of translational amplitude is an improvement of mechanical system's forced response which corresponds in the frequency domain to an increase in dynamic stiffness magnitude. Thus, the problem of vibrations-reduced structural design is focused on the proper *performance criterion* formed by the *modulus of dynamic stiffness* of machine structures.

Adopting the fundamental configuration of basic elements in parallel to a first approach the overall frequency response is plotted by the locus of the composite characteristic $\underline{K}(\mathrm{j}\omega)$, Eq. (4.25a). Supposing a sinusoidal vibration the varying parameter effect on the overall dynamic stiffness can be estimated for any fixed forcing (angular) frequency ω_f. Assigning definite lumped parameter values to frequency locus plots a correspondence will be set up between points and curves in the Nyquist plane, Fig. 4.20.

An enlargement of both the spring constant k and the damping coefficient c causes an increase in dynamic stiffness, whereas a growth of mass m involves a decrease in dynamic stiffness. This varying parameter tendency proves true with reference to forcing frequencies ranging below resonance.

Since the driving frequency passes the region of resonance an estimation of dynamic stiffness values must imply the complete frequency range of interest. Related to the fundamental configuration being subjected directly to an external force or excited indirectly via a spring the figured frequency locus plot is a proper tool for estimating lumped parameters in the sense of an increase in dynamic stiffness, Fig. 4.21.

Thus, a respective change of storage element parameters k, m made above resonance causes an effect which acts on the frequency characteristic of the structure yet in contrary sense to that below resonance. It should be noticed that a *change of dissipator element parameter c* merely effects *without reversing its tendency* when passing through resonance.

Furthermore, it is obvious that resonance is associated with the radius vector's smallest length, 3.2.11. The minimum of modulus coincident with the smallest dynamic stiffness magnitude indicates a significant frequency. Being referred to the deformation amplitude this singular frequency is designated *displacement resonance frequency* ω_r, Eq. (3.39a), the value of which is some what lower than natural frequency ω_0, Eq. (3.6a).

Design Specifications for Vibration-reduced Structures

Summarizing frequency and parameter effects on the frequency characteristic of the structure it can be stated that the required *increase in dynamic stiffness* will be obtained under conditions of *minor resonant vibration* by

- slowing up forcing frequency ω_f;
- enlarging the spring constant (stiffness) k;
- diminishing the mass m;

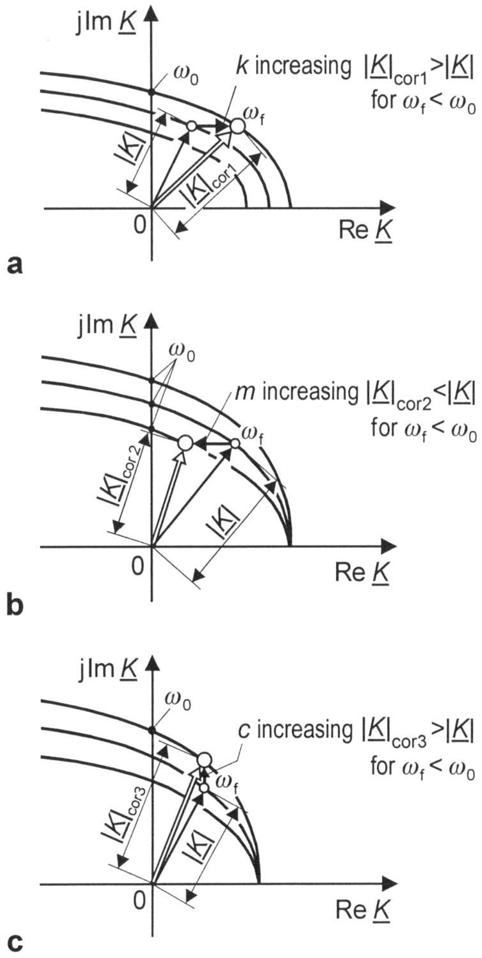

Fig. 4.20. Polar diagrams of dynamic stiffness $\underline{K}(\mathrm{j}\omega)$ due to varied lumped parameters of basic elements in parallel. **a** Change of spring constant (stiffness) *k*; **b** change of mass *m*; **c** change of linear damping coefficient *c*

or, under conditions of *major resonant vibration* by

- speeding up forcing frequency ω_f;
- decreasing the spring constant *k*;
- enlarging the mass *m*;

and within *all regions including resonance* by

- enlarging the damping coefficient *c*.

The demand for reducing vibrations in structural design can be satisfied by a variation of element parameters suited for limiting vibrational deformations, thus

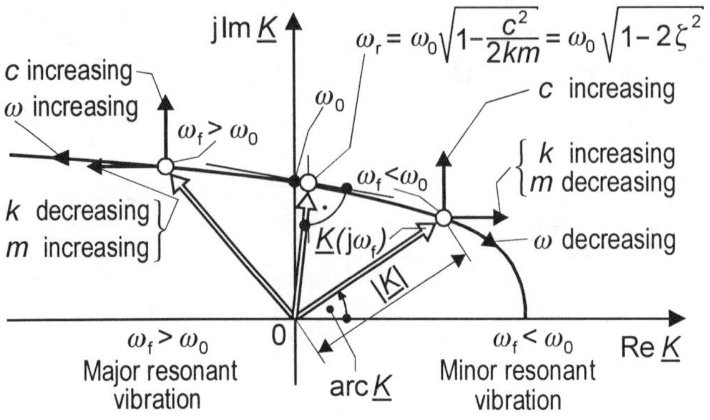

Fig. 4.21. Polar diagram of dynamic stiffness $\underline{K}(j\omega)$ with improvement of frequency response by a lumped parameter estimation tending towards reduced vibrations below and above resonance

resulting in a *stiff lightweight construction* since operating below resonance, or in a *resilient heavyweight construction* above resonance. Stiffening of structures is often mistaken for an increase in dynamic stiffness in a general sense, though this holds true only for operational regions below resonance. On the other hand an intensifying of energy dissipation takes effect on all the regions especially around resonance range.

The stated *performance criteria for vibration-reduced structures* can be interpreted physically as a systematic approach based on utilizing the *tuning property of mechanical systems*. The recommended variation of forcing frequency and/or estimation of storage element parameters is equivalent to separating driving and natural frequency from one another, an equalizing procedure being called *detuning* (or tuning off resonance) by analogy to electrical communication systems. Compared to detuning a strengthening of the damping effect signifies a smoothing down procedure against sharpness of resonance, termed *flat tuning*.

Various attempts to modify characteristic modes and to intensify energy dissipation have been started in structural engineering. The improvements achieved in mechanical system's frequency response include auxiliary mass attachments, flexible foundations (or supports), frictional damping of joints restrained by force as well as dynamic vibration absorbers and detuners (auxiliary vibratory systems), [79] to [86].

Review of Vibration-reduced Design. Historically, the requirements for a *stiff lightweight construction* have been satisfied by developing weight-saving steelframe structures of machines. Attending at first to *plate frames* sectional areas could be enlarged by use of sets of plates. Due to an increase in resistance to elongation mainly longitudinal vibrations have been reduced. Thereupon, design of machine structures passed from plate frame into *hollow-frame constructions*. Exemplified by a ribbed girder forming a zig-zag framework of web plates high flexural stiffness could be realized. This box-type construction was introduced by *E. Peters*, [90], for machine tool bed plates, Fig. 4.22.

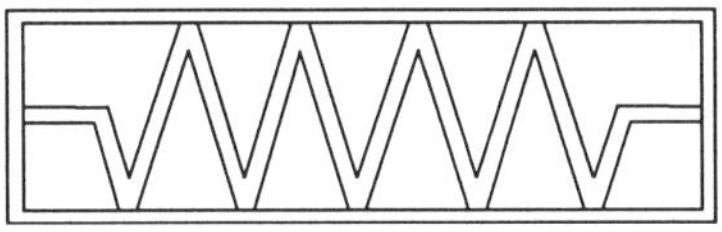

Fig. 4.22. Welded hollow-girder zigzag rib-strengthened after E. Peters

To enlarge resistance to torsion lathe beds in *tubular frame construction* after *W. Möbius* exemplified by various embodiments have been well tried in working, [95].

Forming *cellular structures* by use of strutbraced sections after *C. Krug*, [91], [92], a considerable increase in resistance to bending and twisting was realized which permitted to reduce lateral and torsional vibrations, e.g., on grinding machine structures.

Subassemblies of steel parts are joint together thus forming body frameworks composed by screw connections. Using the *frictional damping effect* in force-locking (bolted) joints a wide-band increase in dynamic stiffness could be obtained as pointed out by *O. Kienzle*, and extensively summarized by *K. Bobek, A. Heiß, and F. Schmidt*, [93]. Comparing steel frame structures being welded and integral cast pieces the typifying damping ratios could be specified by measuring: welded steel structures $\zeta = 0{,}07$;
cast-iron structures $\zeta = 0{,}03$.

Changing from solid to liquid friction stiffness and damping characteristics of joints have been varied utilizing the squeeze film effect. Although, dependency of stiffness as of roughness between friction surfaces behave in a reverse sense. A strengthening of restraining force as well as a refinement of surface finish involve indeed an increase in stiffness bat a decrease in damping as being traced out by *M. Weck*, and *G. Petuelli*, [94).

The requirement for a *resilient heavy-weight construction* can be satisfied preferably by cast-iron structures. However, it is not expedient to enlarge wall thickness of delicate castings for a growth in weight. It will rather be appropriate to couple auxiliary masses on machine structures or to support them by a flexible foundation.

At any rate, cast-iron structures involve a considerable variability in box-type design. Bracings or ribbings can be arranged along the path of lines of force (or flux of force), so that an increase in form stability efficiently can be realized. To intensify dissipation of energy the core sand can be leaved in the cavities of castings. Furthermore, the filling up of hollow-frame constructions with polymer concrete, resin-bound and of low shrinkage, proved to be very effective, [96], [97].

Due to either rubbing or squeezing action between fully supporting solid or liquid friction surfaces whether appertaining to bolted joints or to slideways of moved traverses the frequency response characteristic will be improved by dissipative energy effects. For that reason the rubbing effect has been intensified by use of additional frictional dampers. Damped absorbers may consist of bundles of sheet-metal laminations with intermediate plastic layers after *K. Loewenfeld*, [82], or of high-damping polymers with low functional relationship of temperature after *H.W. Lysen*, [80], [98].

Control System Design for Vibration-reduced Structures

In the last several decades the performance of structures has been improved by the use of closed-loop systems, 2.1.2. By means of control system design the desired performance of a dynamic system can be achieved through a feedback compensation whereby actuators apply forces to a structure based on the structure response as measured by sensors. In contrast, the above-mentioned attempts improve the performance of *passive element configurations* (mechanical plants) through the use of materials or devices enhancing the damping and stiffness characteristics of a structure. This has turned out to be suited for *reducing* both *emissions of oscillation* (active vibration isolation) and *imissions of oscillation and shock* (passive vibration and shock isolation), as well as for diminishing undue strain in structural members, [99]. However, the demands made on machinery, structures, and dynamic systems are continuously increasing, so that the dynamic performance requirements are always rising. Consequently, the *control of structures* has been introduced to complete the methods of improving system responses.

Motion Control. For industrial applications engineering systems are frequently designed to perform, with high precision, a specified task or follow a desired motion. Controlled hydraulic actuators are used for the automatic position control of machine tools and robots at which the damping effect will be adjusted by feedback compensation instead of intensifying energy dissipation.

Active Damping. Reducing vibrations by control has been realized at first for weight-saving constructions in aircraft and power craft, such as rotor blades of helicopters or road vehicle suspension systems, [100], [101]. High-speed hydraulic actuators basing on electrodynamic moving coil principle have been applied to milling spindle mountings for reducing self-excited vibration, [102]. By use of solid-state actuators on the base of piezoelectric and magnetostrictive effects microactuators of high dynamics have been developed for active damping purposes. They already have proved in operation, e.g., at fast moving, lightweight robotic arms, or at the center point, i.e., close by the area of tool action at plain grinding machines, [103], [104].

Control of Structures. This topic blending together various disciplines, namely analytical dynamics, structural dynamics and modern control theory for a dynamic system design, combined, may be referred to more specialized literature for advanced studies, [30], [31].

Vibratory Specification by System Parameter Estimation

Taking advantage of normalization by use of adjusted (data-reduced) polar plots the improvement of passive system's response can be achieved by varying the performance criterion for dissipation as given by the significant system parameter *damping ratio* ζ.

Example 4.2: Reduced Vibration due to Frictional Damping. The frame of a machine (engine bed) has the following measured parameters:

spring constant (stiffness) $k = 220\ N/\mu m$

damping ratio $\zeta = 0{,}008$.

Following Eq. (4.62) the dynamic stiffness magnitude

$$\left|\underline{K}(\mathrm{j}\omega_\mathrm{f})\right| = k \cdot \left|\underline{X}_\mathrm{k}(\mathrm{j}\eta_1)\right| = k \cdot \left|\sqrt{(1-\eta_1^2)^2 + (2\zeta\eta_1)^2}\right| \tag{4.68a}$$

is defined, resulting for $\eta = 1$ in

$$\left|\underline{K}\right|_1 = k \cdot 2\zeta = 1{,}6 \cdot 2{,}2 \cdot 10^{-1} = 3{,}52\,\mathrm{N/\mu m} \tag{4.68b}$$

thus being at resonance only 1/63th of (static) stiffness!

Some additional form of damping may be introduced into the structure. By utilizing rubbing action between mounted friction surfaces dissipative performance will be varied resulting in a considerable improvement on the system parameter

damping ratio $\zeta = 0{,}05$.

By that, the dynamic stiffness magnitude

$$|\underline{K}|_2 = k \cdot 2\zeta = 1 \cdot 2{,}2 \cdot 10 = 22\,\mathrm{N/\mu m} \tag{4.68c}$$

now amounts at resonance to 1/10th of stiffness. Especially around resonance range any intensifying of damping effect becomes effective for an increase in dynamic stiffness.

4.2.6 Some Exercises in Transform Analysis Methods. *Applying Dynamic Compliance Techniques*

For proving their suitability the previously defined frequency-response characteristics will be applied to systems or structures with two or more than two degrees of freedom. Though references are usually made to impedance and mobility analysis techniques the related characteristics basing on the *concept of dynamic stiffness and dynamic compliance* will be favoured as proving a powerful tool in *general lumped-system analysis*. Stiffness and compliance is more convenient for problems the phasor relations of which between forces and relative motion variables should be expressed in terms of displacements instead of velocities.

Otherwise, the equivalent definitions of various kinds of response ratios of phasors can be converted into one another with ease as outlined in 4.2.3.

Forming equivalent characteristics of repeated structures (or sections) in parallel as in series a directly deriving procedure for *simplification of networks* (basic tool of network diagram reduction) will be applied as treated in 4.2.4.

Taking pattern from linear graph theory a basic form of *cut-set analysis* will be additionally introduced being suited to mechanical network diagrams with junctions illustrated by rods (shafts). This procedure may be considered as an advanced device for simplification of networks. To gain composite dynamic characteristics the interconnective vertex relationship between forces will be modified by the *concept of supernodes*. Defining the rod as a complex node, termed a supernode after *R.E. Scott*, [104], and dividing it into parts, equations for any set of nodes can be found by simply adding together the vertex equations for the appropriate nodes. These equations are called *cutset equations*, [52], [60].

Example 4.3: Force Transducer in Fatigue Testing. One of the basic requirements for force- and motion-measurement transducers is that transducer mass should be small so as to avoid dynamic loading of the structure under test. In addition, the requirement for force transducers implies selecting stiffness of the transducer and its components so that no resonances involving this stiffness occur within the frequency range of interest.

Though covering mobility measurements, [18], the above-mentioned basic definitions also come true in materials testing measurements, [105].

The model system of an uniaxial fatigue testing machine, reduced to a two-degree-of-freedom-system undergoing a harmonic motion that provides excitation at the point 2, consists of a mechanical source acting by a displacement of the amplitude $\hat{u}$ on a configuration of two subsystems, Fig. 4.23.

The mechanical source being modelled by an ideal constant-displacement generator is approximately embodied by the equipment of a crank wheel drive which belongs to *direct-drive mechanical vibration generator systems*. Hydraulic vibration generator systems are also appropriate for generating vibration in terms of an exciting displacement since the actuator operates

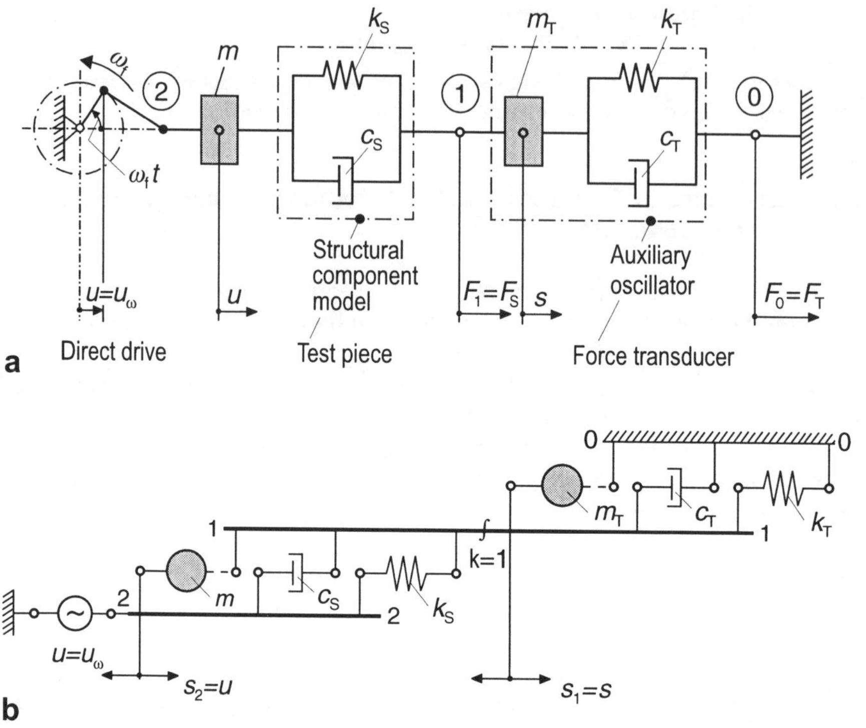

Fig. 4.23. Fatigue testing system. **a** Schematic diagram of test equipment and force-measuring system; **b** mechanical network with junctions as rods

in a stroke-controlled system. Undergoing this condition of testing the pressure driven piston is assumed to be controlled by its position relative to the cylinder.

The two-element subsystem connecting a spring and a damper in parallel describes the test piece (or specimen) by the *model of a structural component* in conformity with the Voigt-Kelvin model characterizing visco-elastic behaviour of materials.

Furthermore, the three-element subsystem of basic elements in parallel forms an *auxiliary oscillator* representing the force transducer (load cell). Its location within the test equipment is important for the avoidance of errors caused by mass loading. Frequently, the force transducer is fitted between the test piece (at resting grip) and the load frame with a maximum stiffness, thus implying the requirement for a rigid frame (braced girder or trussing) of a high lowest natural frequency (fundamental frequency).

The subsystems are specified by the component parameters:

k_S	Test piece stiffness	k_T	Force transducer stiffness
c_S	Damping coefficient of test piece	c_T	Damping coefficient of force transducer (perhaps to be neglected)
m_S	Mass of the test piece and partly of the equipment (moving gripping device, drive mechanism)	m_T	Mass of the force transducer and load train components (load cell and loading fittings, i.e., effective end mass)

u_ω Sinusoidal displacement applied to the test piece

s Sinusoidal displacement of transducer attachment (at resting grip)

F_S Load response of the test piece (at resting grip); in short: *true force*

F_T Load indicated by the force measuring-system (at force indicator); in short: *indicated force*.

The problem of force measurement reduces to gathering the unknown true force F_S from the known indicated force F_T. A conclusion from the output measurement point 0 back to the input measurement point 1 can be drawn only by knowledge of the related frequency response of the force transducer. Contrary to problems of the type analysis or synthesis the problems arising in *instrumentation, measurement, or calibration* are of that type, that a transmission system is used to record the response, from which the excitation can be deduced.

Taking the resulting forces at the different points 1 and 2 in the same measuring system the desired *transfer frequency-response characteristic* at $\omega = \omega_f$ is defined as

$$G_{01}(j\omega_f) = \frac{\underline{\hat{F}_1}}{\underline{\hat{F}_0}} = \frac{\underline{\hat{F}_S}}{\underline{\hat{F}_T}} \tag{4.69a}$$

forming a non-dimensional phasor ratio which is designated the *force transmissibility*.

Rod 0: Loading frame as reference frame; subsystem in parallel yields the phasor relation between *indicated force* and common (absolute) displacement

$$\underline{\hat{F}_T} = \left(K_{k_T} + K_{c_T}\right)\underline{\hat{s}} \tag{4.70}$$

by adding the dynamic stiffnesses of the force transducer components incident to the connecting rod 0.

Rod 1: Being considered as a supernode rod 1 if cut at station $k = 1$ would separate the mechanical network into two parts. Forming a set of two nodes phasor relations can be rewritten for appropriate sets of element branches which terminate in one of the different sections of the rod.

Cut station $k = 1$:

Left section in parallel gives a phasor relation between forces summing at the left face of the cross section, identical with true force, and common relative displacement

$$\underline{\hat{F}_S} = \left(K_{k_S} + K_{c_S}\right)\left(\underline{\hat{u}} - \underline{\hat{s}}\right) \tag{4.71}$$

by adding the dynamic stiffnesses of the test piece components incident at the connecting left part of rod 1.

Right section in parallel leads to an equivalent phasor relation between forces summing at the right face of the cross section and (absolute) displacement common to all of the force transducer components.

The two parts of the rod will be joint due to the equilibrium condition relating internal forces at the opposite faces of a cross section

$$\underline{\hat{F}_k} + \underline{\hat{F}_{k+1}} = 0 \tag{4.72}$$

Thus, by inspecting the elements that join the rod sections the circuit relationship between relative displacements follows from a *cutset equation*

$$\underline{\hat{F}_k} = \left(K_{k_S} + K_{c_S}\right)\left(\underline{\hat{u}} - \underline{\hat{s}}\right) = \left(K_{k_T} + K_{c_T} + K_{m_T}\right)\underline{\hat{s}} = -\underline{\hat{F}_{k+1}} \tag{4.73a}$$

$$\underline{\hat{u}} - \underline{\hat{s}} = \frac{K_{k_T} + K_{c_T} + K_{m_T}}{K_{k_S} + K_{c_S}}\,\underline{\hat{s}} \tag{4.73b}$$

Replacing the force phasors in Eq. (4.69a) by Eqs. (4.70), (4.71)), and substituting Eq. (4.73b) for the phasor relation between relative displacements
the *force transmissibility*

$$G_{01}(\mathrm{j}\omega_\mathrm{f}) = \frac{\underline{\hat{F}_\mathrm{S}}}{\underline{\hat{F}_\mathrm{T}}} = \frac{K_{\mathrm{k_T}} + K_{\mathrm{c_T}} + K_{\mathrm{m_T}}}{K_{\mathrm{k_T}} + K_{\mathrm{c_T}}} = \frac{k_\mathrm{T} + \mathrm{j}c_\mathrm{T}\omega_\mathrm{f} - m_\mathrm{T}\omega_\mathrm{f}^2}{k_\mathrm{T} + \mathrm{j}c_\mathrm{T}\omega_\mathrm{f}} \tag{4.69b}$$

is specified by the complex ratio of concerned component dynamic stiffnesses. For simplification the dynamic stiffness of the damper associated with the force transducer may be neglected as $c_\mathrm{T} \approx 0$, and the amplitude ratio of indicated and true force is reduced to
the *magnitude of undamped force transmissibility*

$$|G_{01}(\mathrm{j}\omega_\mathrm{f})| = \frac{|\underline{\hat{F}_\mathrm{S}}|}{|\underline{\hat{F}_\mathrm{T}}|} = \frac{k_\mathrm{T} - m_\mathrm{T}\omega_\mathrm{f}^2}{k_\mathrm{T}} = 1 - \eta_1^2 \tag{4.74}$$

The *relative dynamic accuracy error*

$$\varepsilon = \frac{|\underline{\hat{F}_\mathrm{T}}| - |\underline{\hat{F}_\mathrm{S}}|}{|\underline{\hat{F}_\mathrm{S}}|} = \left|G_{01}^{-1}(\mathrm{j}\omega_\mathrm{f})\right| - 1 = \frac{\eta_1^2}{1 - \eta_1^2} \leqq 1\% \tag{4.75a}$$

being expressed as a percentage of the true force amplitude $\hat{F}_\mathrm{S}$ represents a performance criterion which is used for selecting a force transducer well suited to equipment characteristics. To smother the effect of (uncompensated) inertia laoadings on the force-measuring system the estimation of storage element parameters k_T, m_T relevant to a transducer selection must be suited to meet the requirement for a tuning off resonance. Admitting a dynamic accuracy error that amounts at most 1 percent, Eq. (4.75a), the detuning is sufficiently performed due to the parameter varying effect since

$$\eta_1 \leqq 0{,}1 \qquad \text{or} \qquad \omega_0 \geqq 10\,\omega_\mathrm{f} \tag{4.75b}$$

That means, the practicable range of test frequency ω_f is limited by a force transducer the natural frequency ω_0 of which is at least as high as the 10-fold of that test frequency ω_f being applied as maximum frequency.

Mass Cancellation. Though transducer selection and designing of the attachment hardware have a high priority in force measurements the detuning criterion, Eq. (4.75b), may become unrealizable. Then, a *processing of transducer signals* will be applied that includes a compensation procedure, commonly known as *mass cancellation*. The acceleration signal at the point of load response (resting grip) is obtained and multiplied, either in an analogue circuit or digitally, by the total effective mass to be compensated. This force signal is subtracted from the force transducer signal to obtain the net load response (true force) acting on the force-measuring system under test.

Electronic mass cancellation cannot compensate for rotational inertia loadings; it can only compensate for translational inertia loadings at the driving point and in the direction of the excitation. All other spurious forces can only be minimized by choosing transducers of low inertia, [106], [107], [108].

Example 4.4: Torsional Vibrations of Rotors on a Flexible Shaft. To analyse the torsional vibration of crankshafts of steam engines, combustion engines, rotor systems of turbines and generators a considerable body of literature shows solutions to those problems, [53] to [57]. Supplementary to well known matrix methods, [58], [59], dynamic stiffness and compliance analysis techniques are applied. It will be demonstrated that frequency-response characteristics can be directly derived be reducing a translational mechanical network which is equivalent to a lumped torsional system.

The model system of a single shaft rotating in frictionless bearings with three discs rigidly attached to it, reduced to a three-degree-of-freedom torsional system undergoing a sinusoidal steady-state vibration, consists of a mechanical source acting by a harmonic torque of the amplitude $\hat{T}_3$ on the point 3. It marks the end of a configuration of repeated sections of massless flexible shafts and rigid masses, Fig. 4.24.

The mechanical source being modelled by an ideal constant-torque generator is approximately embodied by the drive shaft transmitting a forcing torque from a single engine or motor. The subsystems are specified by the component parameters:

J_i Moments of inertia of the discs

k_i Torsional stiffnesses of the shafts

φ_i Angular displacements of the discs (torsional oscillations).

Taking the torque excitation and the resulting angular displacement at the same point 3 in the torsional system the desired *direct frequency-response characteristic* at $\omega = \omega_f$ is

$$\underline{C}_{33}(\mathrm{j}\omega_f) = \frac{\hat{\underline{\varphi}}_3}{\hat{\underline{T}}_3} \tag{4.76a}$$

forming a phasor ratio which is designated the *driving-point dynamic compliance.*

To gain the dynamic characteristic of the composite system the *cut-set simplification* of networks proves useful again. Considering the rods as supernodes the mechanical network will be subdivided into different sections each of them joining an appropriate set of element branches. Thus, phasor relations of fundamental configurations can be applied to find the overall dynamic compliance of the structure by successive reduction. For that reason, all the sec-

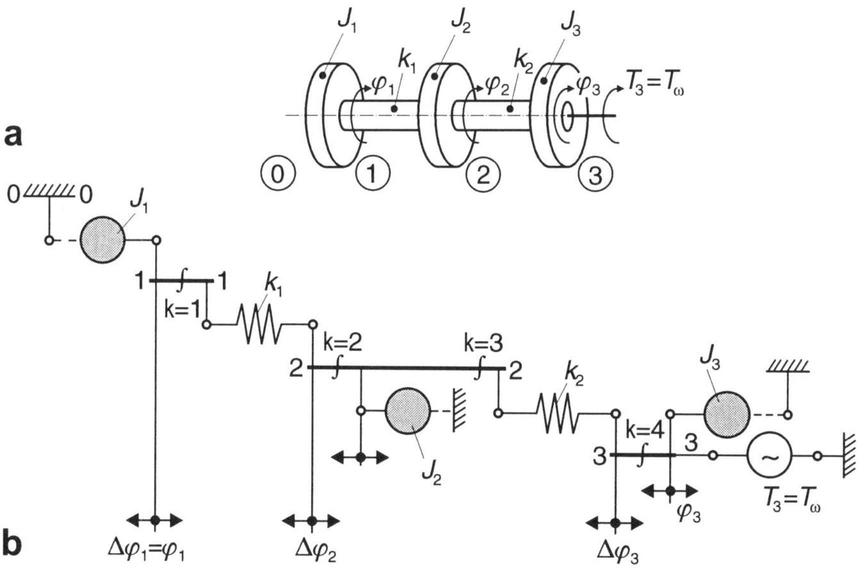

Fig. 4.24. Torsional system. **a** Schematic diagram of a three-rotor system; **b** mechanical network with junctions as rods

tions are numbered in an increasing order from left to right with the dynamic characteristic at each cut station also progressing to the right.

Rod 0: Reference frame (inertial reference system)

Rod 1: Cut station $k = 1$

The left section (incident at rod 1) the torque through the rotating mass J_1 is related to the component dynamic stiffness

$$\underline{\hat{T}_1} = K_1 \underline{\hat{\varphi}_1} = K_{J_1} \underline{\hat{\varphi}_1} \qquad (4.77a).$$

Forming the corresponding phasor ratio (torque-displacement relation) the component dynamic stiffness of the rotating mass J_1 equalling that one at the starting cut station $k = 1$

$$\frac{\underline{\hat{T}_1}}{\underline{\hat{\varphi}_1}} = K_1 = K_{J_1} = -J_1\omega_f^2 \qquad (4.77b).$$

Rod 2: Cut station $k = 2$

At the connecting sections in series the torque is common through the torsional spring k_1 incident at rod 1 and 2 (vertex postulate)

$$\underline{\hat{T}_2}_{(k=2)} = \underline{\hat{T}_1} = K_{1(k=1)} \underline{\hat{\varphi}_1} = K_{k_1} \underline{\hat{\varphi}_2} \qquad (4.78a),$$

and the angular displacements sum up across the element k_1 (circuit postulate)

$$\underline{\hat{\varphi}}_2 = \underline{\hat{\varphi}}_1 + \Delta\underline{\hat{\varphi}}_2 \qquad (4.78b).$$

The two cutset equations, Eqs. (4.78a,b), constituting a pair of postulates is used to form the corresponding phasor ratios (displacement-torque relations), thus illustrating that the dynamic compliances assigned to the jointed sections add

$$\frac{\underline{\hat{\varphi}_2}}{\underline{\hat{T}_2}_{(k=2)}} = \frac{\underline{\hat{\varphi}_1}}{\underline{\hat{T}_2}_{(k=2)}} + \frac{\Delta\underline{\hat{\varphi}_2}}{\underline{\hat{T}_2}_{(k=2)}} = \frac{1}{\underline{K_2}_{(k=2)}} = \frac{1}{K_{1(k=1)}} + \frac{1}{K_{k_1}} = \frac{1}{K_{J_1}} + \frac{1}{K_{k_1}} \qquad (4.78c)$$

resulting by inversion in the dynamic stiffness at the cut station $k = 2$

$$\underline{K_2}_{(k=2)} = -\frac{k_1 J_1 \omega_f^2}{k_1 - J_1\omega_f^2} \qquad (4.78d).$$

Rod 2: Cut station $k = 3$

At the connecting sections in parallel the angular displacement is common across the torsional spring k_1 and the rotating mass J_2 incident at rod 2 (circuit postulate)

$$\underline{\hat{\varphi}_2}_{(k=3)} = \underline{\hat{\varphi}_2}_{(k=2)} \qquad (4.79a),$$

and the torques sum up through the elements k_1, J_2 (vertex postulate)

$$\underline{\hat{T}_2} = \underline{\hat{T}_2}_{(k=2)} + \underline{\hat{T}_2'} = \underline{\hat{T}_2}_{(k=2)} + K_{J2} \underline{\hat{\varphi}_2} \qquad (4.79b).$$

Using the deduced cutset equations (pair of postulates) to form corresponding phasor ratios (torque-displacement relations) the dynamic stiffnesses assigned to the jointed sections add

$$\frac{\underline{\hat{T}_2}}{\underline{\hat{\varphi}_2}} = \frac{\underline{\hat{T}_2}_{(k=2)}}{\underline{\hat{\varphi}_2}} + \frac{\underline{T_2'}}{\underline{\hat{\varphi}_2}} = \underline{K_2} = \underline{K_2}_{(k=2)} + K_{J_2} = K_{J_2} + \frac{1}{\frac{1}{K_{J_1}} + \frac{1}{K_{k_1}}} \qquad (4.79c)$$

resulting by conversion in the dynamic stiffness at the cut station $k = 3$

$$\underline{K_2} = -\frac{k_1 J_1 \omega_f^2 + J_2\omega_f^2(k_1 - J_1\omega_f^2)}{k_1 - J_1\omega_f^2} \qquad (4.79d).$$

Rod 3: Cut station $k = 4$

At the connecting sections in series the torque is common through the torsional spring k_2 incident at rod 3 and 4 (vertex postulate)

$$\underline{\hat{T}_3}_{(k=4)} = \underline{\hat{T}_2} = K_2\,\underline{\hat{\varphi}_2} = K_{k_2}\,\underline{\Delta\hat{\varphi}_3} \tag{4.80a}$$

and the angular displacements sum up across the element k_2 (circuit postulate)

$$\underline{\hat{\varphi}}_3 = \underline{\hat{\varphi}}_2 + \underline{\Delta\hat{\varphi}}_3 \tag{4.80b}$$

Using the lately deduced cutset equations to form corresponding phasor ratios (displacement-torque relations) the dynamic compliances assigned to the jointed sections add

$$\frac{\underline{\hat{\varphi}_3}}{\underline{\hat{T}_3}_{(k=4)}} = \frac{\underline{\hat{\varphi}_2}}{\underline{\hat{T}_3}_{(k=4)}} + \frac{\underline{\Delta\hat{\varphi}_3}}{\underline{\hat{T}_3}_{(k=4)}} = \frac{1}{\underline{K_3}_{(k=4)}} = \frac{1}{\underline{K_2}} + \frac{1}{K_{k_2}} = \frac{1}{K_{k_2}} + \frac{1}{K_{J_2} + \dfrac{1}{\dfrac{1}{K_{J_1}} + \dfrac{1}{K_{k_1}}}} \tag{4.80c}$$

resulting by inversion in the dynamic stiffness at the cut station $k = 4$

$$\underline{K_3}_{(k=4)} = \frac{k_2\left[k_1 J_1 \omega_f^2 + J_2\omega_f^2(k_1 - J_1\omega_f^2)\right]}{\left[k_1 J_1\omega_f^2 + J_2\omega_f^2(k_1 - J_1\omega_f^2)\right] - (k_1 - J_1\omega_f^2)k_2} \tag{4.80d}$$

Rod 3: Driving point 3

At the connection sections in parallel the angular displacement is common across the torsional spring k_4 and the rotating mass J_3 incident at rod 3 (circuit postulate)

$$\underline{\hat{\varphi}_3}_{(k=4)} = \underline{\hat{\varphi}_3} \tag{4.81a}$$

and the torques sum up through the elements k_4, J_3 (vertex postulate)

$$\underline{\hat{T}_3} = \underline{\hat{T}_3}_{(k=4)} + \underline{\hat{T}_3'} = \underline{\hat{T}_3}_{(k=4)} + K_{J_3}\,\underline{\hat{\varphi}_3} \tag{4.81b}$$

Using the lately deduced cutset equations to form corresponding phasor ratios (torque-displacement relations) the dynamic stiffnesses assigned to the jointed sections add

$$\frac{\underline{\hat{T}_3}}{\underline{\hat{\varphi}_3}} = \frac{\underline{\hat{T}_3}_{(k=4)}}{\underline{\hat{\varphi}_3}} + \frac{\underline{\hat{T}_3'}}{\underline{\hat{\varphi}_3}} = \underline{K_3} = \underline{K_3}_{(k=4)} + K_{J_3} = K_{J_3} + \frac{1}{\dfrac{1}{K_{k_2}} + \dfrac{1}{K_{j_2} + \dfrac{1}{\dfrac{1}{K_{k_1}} + \dfrac{1}{K_{J_2}}}}} \tag{4.81c}$$

By joining at the final cut station the two parts of rod 3 all sections are connected. The consecutive set of cutset equations yields the overall dynamic characteristic. A structure of regularly alternating parallel and series connections is represented in the frequency domain by a *continued-fraction expansion* composed of element stiffnesses and compliances

$$\underline{C_{33}}(j\omega_f) = \frac{\underline{\hat{\varphi}_3}}{\underline{\hat{T}_3}} = \cfrac{1}{K_{J_3} + \cfrac{1}{C_{k_2} + \cfrac{1}{K_{J_2} + \cfrac{1}{C_{k_1} + 1/K_{J_1}}}}} \tag{4.76b}$$

$$\equiv \frac{1\,|}{|K_{J_3}} + \frac{1\,|}{|C_{k_2}} + \frac{1\,|}{|K_{J_2}} + \frac{1\,|}{|C_{k_1}} + \frac{1\,|}{|K_{J_1}}$$

The consecutive quotients of the division and reciprocation process will be resolved to an uniform fraction, finally resulting by inversion in
the *driving-point compliance* of the torsional system

$$\underline{C_{33}}(\mathrm{j}\omega_\mathrm{f})=\frac{k_1J_1\omega_\mathrm{f}^2-(k_1-J_1\omega_\mathrm{f}^2)(k_2-J_2\omega_\mathrm{f}^2)}{\left[(k_1-J_1\omega_\mathrm{f}^2)(k_2-J_2\omega_\mathrm{f}^2)-k_1J_1\omega_\mathrm{f}^2\right]J_3\omega_\mathrm{f}^2+k_2\left[(k_1-J_1\omega_\mathrm{f}^2)J_2\omega_\mathrm{f}^2+k_1J_1\omega_\mathrm{f}^2\right]}$$

(4.76c).

Taking resulting response and applied excitation at different points on a structure *transfer compliances* can be deduced by successive reduction multiplying continued fractions of subsystems. Assuming two-way energy transmission as a rule Maxwell's reciprocity theorem may be applied. Lumped-system analysis thus will be simplified by removing the exciter from *i* to *j*, or vice versa, 4.2.4.

5 The Flow of Power and Energy in Systems (Energy Transactions)

To power and energy interactions is not given one's full attention in many applications of systems modelling though being an essential matter of systematic approach to the design of complex systems.

Energy and power alone are the fundamental dynamic variables, the ultimate currency of all physical interaction and transaction. This statement being already emphasized by *Henry M. Paynter* gave a decisive impulse to describe energy bonding as a particular type of functional connectedness, and finally to develop the bond graph approach as a complete and formal discipline, 2.5, [22], [23].

To impart the understanding and insight of the transformation and flow of energy an introduction in brevity to energetics is given following some of the fundamental conclusions on this matter performed by Paynter, [22].

Thereupon, it will be propitious to bring into focus two specific types of relations representing the dynamic behaviour on an energetic basis.

Variables and Parameters of Energetic Systems

Energetic Systems. Within the scope of a general treatment of system-to-system interactions it is of particular interest for systems engineering to focus the attention on a class of physical systems called *energetic systems.* Several attempts have been made to describe the energy transactions which can occur across the boundaries of dynamic systems.

Review of Energy Concept. Historically, *J.C. Maxwell's* use of field concepts for light and electricity resulted in a theoretical structure appropriate not only for analysing electromagnetic radiation phenomena but also for revealing the energetic aspect of fields of all types. *J.H. Poynting* extended this basic conception in physics to the consideration of energy transport in an electrical network. This was continued by *C.P. Steinmetz* for developing a regimented theory for practical analysis and design of electrical circuits, and also by *O. Heaviside* for generalizing the statement of energy continuity. Concerned with the identity of energy *O. Lodge* implanted the concept of energy-matter conservation in the structure of the theory of energetics.

In parallel with the outlined development was that of the classical theory of thermodynamics.

Poynting, Heaviside, and Lodge stated that the transformation of energy is always accompanied by a transfer, thus presupposing the ether as a substance to permeate all space. For pre-sent purposes energy exists in and of itsself, requiring no material vessel in which to be stored or transported, [22].

Noncausal Description. Energy transactions may be defined by a pair of variables which together are a measure of the flow rate of energy equally termed the power across the interface between system and environment.

One of the variables is an *extensive factor* in that its magnitude depends on the extent of the portion of the system entering into energy transaction. The other variable is an *intensive factor* being a function only of the field in which the system resides. If the two variables are properly chosen, that means, if they are a pair of rate variables (intensity or power variables) in the sense of 1.2.2, their product will yield the instantaneous power exchanged. The factoring of power into two components is fundamental in mechanics (power = force · velocity) and thermodynamics (power = pressure · rate of volume change).

For the modelling process it is convenient to think of the intensive variable as an *effort*, $e(t)$, and the extensive variable as a *flow*, $f(t)$, so that their product will yield the *instantaneous power*, $P(t)$, Eq. (2.91), 2.5.

Being associated with the overall-energy state E the system enters into energy transactions with its environment at a number of localized regions on its boundary surface. Thus, in the case of the noncausal energy interconnection, the quantities e_i and f_i are the *external variables* of the systems.

An interconnection diagram for energetic systems, which may be a reticulated bond diagram, imparts an understanding of the transformation and flow of energy within the system and assists in the isolation of the essential energy interactions with the environment.

Causal Description. However, for a more detailed synthesis and preliminary analysis the noncausal energetic interconnection is usually not adequate and must be transformed into the *bilateral signal flow reticulation*. The description of the system is then completed by conceptually substituting for each element a black box for which the input-output functional dependency is specified.

Reticulation can be defined as the process by which a system is endowed with structure. Viewing the elements of a system simply as systems of lower order, called subsystems, the interactions between two elements are considered to be of the same class as those which may occur between two systems. Summarily, all possible interactions occuring within the dynamic system or across its boundaries may, in fact, be looked upon as *system-to-system interactions*.

A segregation of n inputs X_i from the n outputs Y_i can be imagined involving a conceptual deformation of the system such that all the inputs enter at one face while all the outputs leave from the opposite face.

The quantities X_i are referred to as the components of the *input vector* $\boldsymbol{x}$, and the quantities Y_i as the components of the *output vector* $\mathbf{y}$, where in matrix notation the vectors are denoted by small (lower case) bold-faced letters. Though being convenient to arrange the components in columns there is much advantage in spacing by writing the column vectors horizontally (in braces) $\boldsymbol{x} = \{X_1, \ldots, X_n\}$, and $\boldsymbol{y} = \{Y_1, \ldots, Y_n\}$. For each input-output pair (X_i, Y_i) there corresponds a power component $P = X_i\ Y_i$.

Passing on to algebraic operations for matrices one of the two column vectors must be rewritten in terms of a row vector (in brackets), i.e., $\boldsymbol{x}^T = [X_1, \ldots, X_n]$, thus performing the transposition of the input vector $\boldsymbol{x}$ denoted with the additional

mark (T as letter superscript) $\boldsymbol{x}^{\mathrm{T}}$, frequently reduced to the symbol with a typographical sign (prime) $\boldsymbol{x}'$. The matrix multiplication of the two real vectors yields the *scalar product (dot product)* summing up to a (1x1) matrix, i.e., to a real number being identical with
the *total power exchanged*

$$P = [X_1, \dots, X_n] \begin{bmatrix} Y_1 \\ \vdots \\ Y_n \end{bmatrix} = \boldsymbol{x}'\boldsymbol{y} = \sum_{i=1}^{n} X_i Y_i \qquad (5.1a).$$

Performing the transposition of a product $(\boldsymbol{x}'\boldsymbol{y})' = \boldsymbol{y}'\boldsymbol{x}$ an alternative form of the scalar product is found. Hence, the total power can be rewritten in two equivalent forms of matrix notation

$$P = \boldsymbol{x}'\boldsymbol{y} = \boldsymbol{y}'\boldsymbol{x} \qquad (5.1b).$$

Being a matrix of only one element the total power is a *scalar* defined by the property to be equal with its transpose, Eq. (5.1b).

In conformity with the principle of the *continuity of energy* the energy flow rate or power is *invariant* under a *coordinate transformation*. Supposing a linear transformation (or mapping) of $\boldsymbol{x}$ into $\boldsymbol{y}$ caused by the system it its interesting to note, that the pair of vectors ($\boldsymbol{x}$ and its image $\boldsymbol{y}$) does not go over into new coordinates $\bar{\boldsymbol{x}}, \bar{\boldsymbol{y}}$ by corresponding relations (by a cogredient transformation), but according to the opposite relations, i.e., by a *contragredient transformation*

$$\boldsymbol{x} = \boldsymbol{T}\bar{\boldsymbol{x}} \quad ; \quad \bar{\boldsymbol{y}} = \boldsymbol{T}'\boldsymbol{y} \qquad (5.2a),$$

where $\boldsymbol{T}$ is the transform matrix assumed to be nonsingular. Applying the transposition of a product, and the inverse transformation premultiplying by $(\boldsymbol{T}')^{-1}$, respectively, to the equations of vector transformation, Eq. (5.2a)

$$\boldsymbol{x}' = (\boldsymbol{T}\bar{\boldsymbol{x}})' = \bar{\boldsymbol{x}}'\boldsymbol{T}' \quad ; \quad (\boldsymbol{T}')^{-1}\bar{\boldsymbol{y}} = \boldsymbol{y} \qquad (5.2b),$$

the insertion of the obtained transpose and inverse, Eq. (5.2b), into Eq. (5.1a) yields the power in the new coordinate frame

$$P = \boldsymbol{x}'\boldsymbol{y} = \bar{\boldsymbol{x}}'\boldsymbol{T}'(\boldsymbol{T}')^{-1}\bar{\boldsymbol{y}} = \bar{\boldsymbol{x}}'\bar{\boldsymbol{y}} \qquad (5.1c).$$

The system is represented in particular by coefficient matrices (m x n matrices) which perform the linear transformation (mapping) of vector spaces either in original or in new coordinates. Those two assumed coefficient matrices related to the nonsingular $\boldsymbol{T}$ by Eq. (5.2a) are said to be *congruent*. The congruence transformation is put to good use in the transformation of *quadratic forms*, 5.1.2.

The *functional dependency*

$$\boldsymbol{y} = \psi(\boldsymbol{x}) \qquad (5.3a)$$

is of a most general form such that the entire history of $\boldsymbol{x}$ is scanned to yield a present value of $\boldsymbol{y}$. Therefore, it is applicable to the analysis of all processes in which the system might be involved, and in particular to transition processes from one steady-state condition to another.

Finally the system is conceptually replaced by a *module* which performs the operation on the input vector $\boldsymbol{x}$ being required to yield up the output vector $\boldsymbol{y}$.

The *generalized functional* ψ contains the intrinsic properties of the system in the form of a *set of parameters*.

A verification of the functional dependency, Eq. (5.3a), can be carried out by scanning the input vector $x(t-\tau)$ for $0 < \tau < \infty$ and yielding up a value for the output vector $Y(t)$. A particular form frequently used is the *vector-to-scalar transformation*

$$Y(t) = \psi[x(t)] \qquad (5.3b).$$

Considering an ideal element the energy stored in this element, $E(t)$, is yielded by the *integral formula* from 0 to ∞

$$\begin{aligned} Y(t) = E(t) &= \int_0^\infty P(t-\tau)\,\mathrm{d}\tau \\ &= \int_0^\infty e(t-\tau) \cdot f(t-\tau)\,\mathrm{d}\tau \end{aligned} \qquad (5.4),$$

wherein it might be identified

$$\boldsymbol{x} = [X_1, X_2] = [e, f].$$

For the *steady-state analysis*, a simpler *static functional* ϕ is applicable, yielding the present value of $\boldsymbol{y}$ corresponding to a present value of $\boldsymbol{x}$, thus performing a *static functional transformation*

$$\boldsymbol{y} = \phi(\boldsymbol{x}) \qquad (5.3c).$$

Hence, for a bilateral signal flow reticulation it becomes obvious that the variables of the system are the vectors $\boldsymbol{x}$ and $\boldsymbol{y}$, or, more precisely, their respective components.

Equilibrium Power Transfer. Sign for power as causality for effort and flow variables has already been outlined in the context of conventions for interconnected multiports, 2.5.2.

Conceiving one of two participating systems as the supplier and the other as the recipient of power the conventions above should be completed by the presumption of a *definite intersection between demand and supply effort-flow characteristics.* This implies the existence of a point of stable equilibrium operation. Modifications in characteristics to be performed by load or source matching for achieving a stable power transfer thus will be omitted.

The Steady State of Energetic Systems.

The analysis of steady state plays a dominant role in providing an overall understanding of the behaviour of energetic systems. An insight into steady-state behaviour of a system forms the basis upon which the analysis of its stability and transient behaviour may be founded. The stability of a system may be evaluated by observing the result of small excursions about a steady operating point. Moreover, a transient condition in a stable system is the means by which its operating state alters from one steady condition to another.

The two *types of steady state* are:

i. The *static case*, wherein the power flux is identically zero and all that is required is a statement of the distribution of internally stored energy.

ii. The *stationary case*, wherein the power flux is a constant, at least in the mean, and the system is in dynamic equilibrium with the internally stored energy. In the case of *strict stationarity* the time derivates of the flows both internal and external, vanish identically. A weaker condition is that of *quasi-stationarity* wherein the time averages of the flows are zero, i.e., each of the flows is fluctuating about some steady mean value.

Power Transactions in Vibrations

The analysis of physical systems on an energy basis is well approved in applications most of them being attached to circuits in electric power engineering.

The study of vibration is favourably engaged as in *converting shock and vibratory motion* into proportional parameters of the experienced motion, as in preventing or *reducing unwanted vibrations of structures*. Those two main subjects of vibrations are associated with either the *problem of measuring instruments* (or systems) or the *problem of system analysis* (or system identification). Consequently, in vibration theory scarcely arises the need to deal with energy flux and power flow.

However, vibration engineering also faces the claim of system design which includes at least the want of power flow. In particular, this proves necessary for generating *desired vibrations* being applied by way of the "vibration method" to a variety of industrial processes or to testing procedures. For this purpose, *(utility) vibration machines* engaged in various working operations are developed, as well *vibration testing equipments* are in use, the power supply of which requires an appropriate vibratory drive, [88], [123], to [125).

Thus, from the engineering viewpoint vibration theory should be occupied likewise with the *problem of system synthesis* covering the demand for rating and optimizing the driving power flow in machinery. In the following *two attempts* of an *energetic system approach* to the mechanical system subjected to vibrations are made. Both of the approaches related to power interactions between the power supply and the engineering component base on the diagram representation by a significant type of systematic diagram (interconnection diagram):

- One of them is concerned with the causal description of the *generalized transport process of energy* in the *two-port diagram* (mechanical 2-port), treated in 5.1;
- the other one deals with the *phasor power concept* basing on the acausal description of the *network diagram* (mechanical circuit), treated in 5.2.

By transforming time-dependent power transactions into the frequency domain the spectral decomposition of power combined with the algebraic function of complex power permits the insight into the steady-state behaviour of power and energy flow in reticular systems.

Finally, the relationship of *phasor power concept* to the *concepts of mechanical mobility* or *dynamic compliance* (linear one-port dynamic characteristics) and its usefulness for energetic system design will be demonstrated in 5.3.

5.1 Power Transmission through Linear Two Ports (Generalized Transport Process)

A *2-port* may be conceived as a *generalized transport process*, i.e., a process by which energy is transformed, transmitted, or transduced.

Besides the components of a communication system many common engineering components can be considered as a 2-port, and a power transmission device of complex system structures can be looked upon as a *string of 2-ports*, 2.2. and 2.6.

Two-port (or four-pole) representation combines the four terminals of a single entity to terminal pairs which shall be designated generally as the upstream and the downstream port, 1 and 2, being the only ones for which the device is to be represented and investigated.

Description of Linear Two Ports

The behaviour will be characterized in terms of the *input power* and the *output power*

$$P_1(t) = e_1(t) \cdot f_1(t) \quad ; \quad P_2(t) = e_2(t) \cdot f_2(t) \qquad (5.1d)$$

and therefore in terms of four variables:

$$e_1, f_1, e_2, f_2 .$$

Since at each port only one of these variables may be taken as an input, X_1, X_2, the existing interaction indicates that the output variable at each port, Y_1 as Y_2, is functionally determined by the input variables at both ports, X_1 and X_2. Using the *general functional operators for the 2-port* $\psi_a\ \psi_b$, the functional dependency in the time domain is:

$$\begin{aligned} Y_1(t) &= \psi_a\left[X_1(t), X_2(t)\right] \\ Y_2(t) &= \psi_b\left[X_1(t), X_2(t)\right] \end{aligned} \qquad (5.3d).$$

Standard Forms of Matrices. A linear 2-port can be described by 4-pole equations which may be summarized to a single causal statement in matrix form. The physically realizable particularizations favourably associated with the electrical field are termed the four *causal matrices* ***Z, Y, H, G***. Their relations to the possible connections of two ports have been outlined in 2.2.2.

In the following the most significant standard form will be taken up which relates the power states at the opposite terminals through the *transmission matrix* ***A***, respectively, Eqs. (2.24), (2.27), 2.2.1. While the ***A*** and ***B*** matrices are clearly noncausal they have the peculiar advantage that the overall coupled transmission matrix for two 2-ports in cascade (or tandem) may be obtained by direct matrix multiplication. By this, the transmission of power variables is described for the resultant two-terminal-pair network.

Transfer Characteristics of 2-port Networks. In systems modelling using matrix technique to describe the overall behaviour of two-port networks the arising problems are commonly placed in the three categories, [22]:

i. The *transfer problem*: The effort or flow at the output terminal is required in response to effort and flow at the input terminal, with ideal terminations generally assumed at the output terminal.
ii. The *transmission problem*: The power state at one port is looked for in terms of the power state at the second port with:
   - unrestricted terminal conditions, or
   - terminal impedance specified.
iii. The *insertion problem*: The effect of inserting a two port into a system in place of an ideal coupler, e.g., a rigid link in mechanics, is sought. Typically these problems are "filtering" and "protection" situations, where performance is measured in terms of the change in power, effort, or flow after insertion from that occuring before insertion.

The transmission characteristics of 2-port networks are specified for the general case by *four independent 2-port parameters*. These may be determined in terms of the elements of the transmission (or chain) matrix being tabulated in terms of converted causal matrices, [10], [22].

5.1.1 The Transmission Problem of Two-port Networks. *Unrestricted Terminal Conditions*

Causally Reticulated Two Port (signal four pole)

Using a transfer function block diagram with two inputs and two outputs the interpretation of a multivariable system can be placed on the two-port diagram, 2.2.6. By this, the generic *2-port in matrix form* implying conventional reference directions will be replaced by the *causal form of a signal 4-pole*, Fig. 5.1.

The signal 4-pole being a subsystem configuration of considerable usefulness in system engineering consists of 1-port components each of which represents one of the four 2-port parameters.

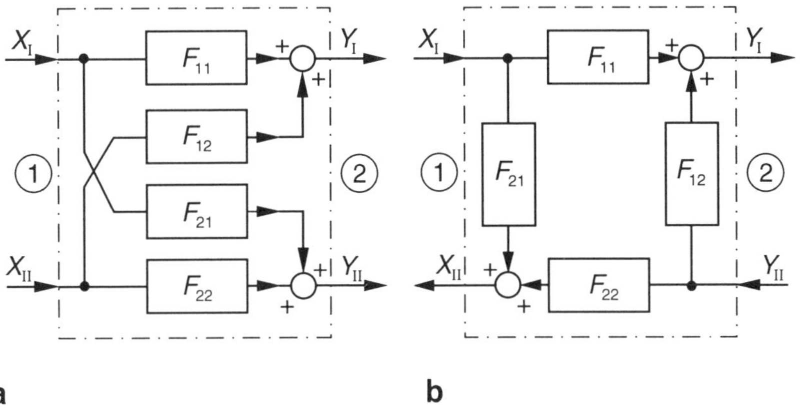

Fig. 5.1. Basic structures of a signal 4-pole. **a** Block diagram in the canonical configuration; **b** in the field configuration

In electrical network theory an equivalent circuit arrangement is given by the lattice network (or "bridge-net"). The frequency-response behaviour of this 4-element structure is described by 4 impedances, 2 of which are referred to as the "series arms", the other ones as the "cross arms". The corresponding 2-port parameters, also called the *general circuit parameters*, are expressed in terms of related impedances of the "arms", and which are determined by subjecting the lattice equivalent circuit of the 2-port to constraints. The equivalent definitions of the 2-port parameters pertaining to the kind of related phasor quantities are then short-circuit current ratio, open-circuit transfer impedance, short-circuit transfer admittance, and open-circuit voltage ratio, [16], [52], [60].

Though two equivalent block diagram configurations of the signal 4-pole in terms of direct and transfer frequency characteristics are realizable, Fig. 5.1a,b, it is more convenient to represent the causally reticulated 2-port by the *field configuration*, Fig. 5.1b. The segregation of the two inputs from the two outputs will be maintained by a conceptual deformation such that the two inputs X_I, X_{II} at port 1 are conceived as entering in opposite direction, accordingly the two outputs Y_I, Y_{II} as leaving from port 2 in opposite direction. This emphasizes *energetic interactions with causality* by an effort-flow couple, said a *pair of conjugate variables.*

Linear Transformation of Vector Spaces

In the following the power state will be expressed separately by the output quantities in response to both of the input quantities, thus being appropriate to the transfer problem, 5.1.

If the general functional operators for the 2-port, ψ_a and ψ_b, Eqs. (5.3d), can be assumed linear over the practical range of operation, then a most important and powerful signification subsists in the frequency domain. This is manifested in the reduction of the functional relationship between the (transforms of the) variables to a *pair of linear algebraic equations*

$$\begin{aligned} Y_I &= \psi_a[X_I X_{II}] = F_{11}X_I + F_{12}X_{II} \\ Y_{II} &= \psi_b[X_I X_{II}] = F_{21}X_I + F_{22}X_{II} \end{aligned} \qquad (5.5a),$$

or, in matrix notation

$$\begin{bmatrix} Y_I \\ Y_{II} \end{bmatrix} = \begin{bmatrix} F_{11} & F_{12} \\ F_{21} & F_{22} \end{bmatrix} \begin{bmatrix} X_I \\ X_{II} \end{bmatrix} \qquad (5.5b),$$

wherein one might identify the *input vector* and the *output vector*

$$\boldsymbol{x} = [X_I, X_{II}] \quad ; \quad \boldsymbol{y} = [Y_I, Y_{II}] \qquad (5.6).$$

Arranging them as column vectors the relationships will be reduced to a single *vector equation*

$$\boldsymbol{y} = \boldsymbol{F}\boldsymbol{x} \qquad (5.7a)$$

where $\boldsymbol{F}$ is the *inverse transmission matrix of the signal 4-pole.*

By this standard form of matrix a direct *spatial correspondence* to the ports themselves is established relating the power states, P_2 and P_1, at the opposite terminals of the linear 2-port.

Considering the *stationary case of steady-state* the forcing functions applied are restricted to the special type of sinusoidal excitations.

Variables and 2-port parameters may be pointed out in complex quantities, so that in general the 2-port parameters are complex system parameters (functional operators) formed by the ratios of two phasors. In particular, at any given driving frequency 4 suitable measurements will suffice to determine the parameter set of 4 phasor ratios (or complexors), and thus describe the sinusoidal steady-state responses of the output variable pair.

If the two-terminal network is *passive reciprocal* one constraining relation implies that there are only *three independent 2-port parameters*; now, three measurements will suffice to define the system.

The 4 "general circuit parameters" thus signify

- two *direct (or driving-point) frequency-response functions*: F_{11}, F_{22};
- two *transfer frequency-response functions*: F_{12}, F_{21}.

Implying boundary conditions and specifying output/input phasor ratios the 4 frequency characteristics are defined with the corresponding designations as force transmissibility, velocity transmissibility, furthermore as free impedance, blocked admittance (or mobility), respectively, [11], [17], [18].

5.1.2 The Transmission Problem related to Complex Power. *Generalized Quadratic Forms*

By a linear transformation of the vector space $\boldsymbol{x}$ and $\boldsymbol{y}$, performed by Eq. (5.7a), the power transaction is treated on the base of factoring power into two components. Indeed, the input/output relation of separate components represents the *effort-flow characteristic* yet being an *implicit form* of the proper *power state*.

The power is carried across the system boundary by transmission links (shafts, ducts, electrical conductors, waveguides, etc.). Respecting the power flow through an area of restricted extent the power state at any energy port is rather indicated in terms of an *effort-flow product* than in terms of the corresponding effort-flow couple, i.e., the energy coupling by a transmission link is physically represented by the scalar quantity *power* replacing the vectorial quantities of the individual conjugate variables, such as force and velocity. Hence, problems of the previously mentioned implicit type are involved with finding $\boldsymbol{y}$ as a function of $\boldsymbol{x}$, whereas those in the sequel deal with finding P_2 as a function of P_1. Properly speaking, the requirement of effort or flow at the leaving port in response to those at the entering port is appropriate to the category of transfer problems, but the demand for power representation at both ends applies especially to the *category of transmission problems*, as defined in 5.1, [22].

Instead of the customary effort-flow characteristic a proper *energy flow characteristic* will be presented in the following which expresses the *power state* in terms of an *explicit form* appropriate to the transmission problem, 5.1.

For this reason respecting sinusoidal steady state the power transaction of the 2-port may be defined by the *transmission ratio of phasor products*

$$\underline{A}_s = \frac{\underline{S}_Y}{\underline{S}_X} = \frac{\underline{\widetilde{Y}}_{II}^* \, \underline{\widetilde{Y}}_I}{\underline{\widetilde{X}}_{II}^* \, \underline{\widetilde{X}}_I} \tag{5.8}$$

expressing the relation of output power to input power in terms of effort-flow products. It is convenient to define the root-mean-square (r.m.s. value) as the modulus of phasors (additionally marked by a tilde as an overscript) specifying the individual variables as complex quantities representing sinusoidal quantities of the same frequency. Taking at each energy port the product of an effort phasor $\underline{\widetilde{X}}_{\mathrm{I}}$, or $\underline{\widetilde{Y}}_{\mathrm{I}}$, and the conjugate of a flow phasor $\underline{\widetilde{X}}^{*}_{\mathrm{II}}$, or $\underline{\widetilde{Y}}^{*}_{\mathrm{II}}$, respectively, Fig. 5.1b, the energy state at both ends is represented by the *complex input power* $\underline{S}_{\mathrm{X}}$, and by the *complex output power* $\underline{S}_{\mathrm{Y}}$.

Scalar Transformation of Power States

In the following it will be shown that the complex output power $\underline{S}_{\mathrm{Y}}$ can be determined directly from the input vector $\boldsymbol{x}$ via element connections of the (inverse) transmission matrix $\boldsymbol{F}$. Since indeed this standard form of a noncausal matrix can be assumed to be known, an explicit relationship will be derived following the transformation of the vector space $\boldsymbol{x}$ into $\boldsymbol{y}$, Eq. (5.7a), yet being adequate to perform a scalar transformation of the power state $\underline{S}_{\mathrm{X}}$ into $\underline{S}_{\mathrm{Y}}$, [108]. The two elements (r.m.s.-value components) assigned to each port are complex numbers thus forming *complex vectors*, called the input vector $\underline{\widetilde{\boldsymbol{x}}}$, and the output vector $\underline{\widetilde{\boldsymbol{y}}}$. Inserting those vectors into Eq. (5.7a), the corresponding equation is rewritten as

$$\underline{\widetilde{\boldsymbol{y}}} = \boldsymbol{F}\,\underline{\widetilde{\boldsymbol{x}}} \tag{5.7b}$$

This vector equation can be expanded as a linear combination of the components $\underline{\widetilde{X}}_{\mathrm{i}}$ of the input vector by use of a matrix notation introducing the (2x1) *submatrices* or column vectors $\boldsymbol{f}_{\mathrm{m}}$ of $\boldsymbol{F}$

$$\begin{bmatrix} \underline{\widetilde{Y}}_{\mathrm{I}} \\ \underline{\widetilde{Y}}_{\mathrm{II}} \end{bmatrix} = \begin{bmatrix} \boldsymbol{f}_1, \boldsymbol{f}_2 \end{bmatrix} \begin{bmatrix} \underline{\widetilde{X}}_{\mathrm{I}} \\ \underline{\widetilde{X}}_{\mathrm{II}} \end{bmatrix} = \boldsymbol{f}_1\,\underline{\widetilde{X}}_{\mathrm{I}} + \boldsymbol{f}_2\,\underline{\widetilde{X}}_{\mathrm{II}} \tag{5.6c}$$

Being represented by a (2-rowed) square matrix the transmission matrix $\boldsymbol{F}$ of a signal 4-pole can be written in terms of its (1x2) *submatrices* or row vectors $\boldsymbol{f}^{\ell}$

$$\boldsymbol{F} = \begin{bmatrix} F_{\ell\mathrm{m}} \end{bmatrix} = \begin{bmatrix} F_{11} & F_{12} \\ F_{21} & F_{22} \end{bmatrix} = \begin{bmatrix} \boldsymbol{f}^1 \\ \boldsymbol{f}^2 \end{bmatrix} \tag{5.9}$$

Consequently, the components $\underline{\widetilde{Y}}_{\mathrm{i}}$ of the output vector are obtained from the *scalar product* (dot product) of the ℓth row vector $\boldsymbol{f}^{\ell}$ and the column vector $\underline{\widetilde{\boldsymbol{x}}}$ at the input

$$\begin{aligned} \underline{\widetilde{Y}}_{\mathrm{I}} &= \boldsymbol{f}^1\,\underline{\widetilde{\boldsymbol{x}}} = \begin{bmatrix} F_{11}, F_{12} \end{bmatrix} \begin{bmatrix} \underline{\widetilde{X}}_{\mathrm{I}} \\ \underline{\widetilde{X}}_{\mathrm{II}} \end{bmatrix} \\ \underline{\widetilde{Y}}_{\mathrm{II}} &= \boldsymbol{f}^2\,\underline{\widetilde{\boldsymbol{x}}} = \begin{bmatrix} F_{21}, F_{22} \end{bmatrix} \begin{bmatrix} \underline{\widetilde{X}}_{\mathrm{I}} \\ \underline{\widetilde{X}}_{\mathrm{II}} \end{bmatrix} \end{aligned} \tag{5.10a}$$

To obtain the complex conjugate $\underline{\widetilde{Y}}_{\mathrm{II}}^*$ of one of the output components, e.g., $\underline{\widetilde{Y}}_{\mathrm{II}}$, the conjugate and transpose of the appertaining dot product $\boldsymbol{f}^2\underline{\widetilde{x}}$, must be formed

$$\underline{\widetilde{Y}}_{\mathrm{II}}^* = \overline{\left(\boldsymbol{f}^2\,\underline{\widetilde{x}}\right)'} = \left(\boldsymbol{f}^2\,\underline{\widetilde{x}}\right)^* = \underline{\widetilde{x}}^*\boldsymbol{f}_2^* = \left[\underline{\widetilde{X}}_{\mathrm{I}}^*\ \underline{\widetilde{X}}_{\mathrm{II}}^*\right]\begin{bmatrix} F_{21}^* \\ F_{22}^* \end{bmatrix} \quad (5.10b).$$

Modified Hermitian Form. Thus, the complex output power results from multiplying the conjugate dot product, Eq. (5.10b), and the corresponding dot product, Eq. (5.10a), together. By this "composed transformation" the input vector $\underline{\widetilde{x}}$ is transformed into a scalar form being similar, not equal to an expression designated in the theory of matrices as the Hermitian form. This standard form can be interpreted as the complex generalization from the real quadratic form. Compared to the above-mentioned standard specification the given expression denoted by $\underline{S}_{\mathrm{Y}}$ represents

the *modified Hermitian form* in the two complex variables $\underline{\widetilde{X}}_{\mathrm{I}}$, $\underline{\widetilde{X}}_{\mathrm{II}}$

$$\underline{S}_{\mathrm{Y}} = \underline{\widetilde{Y}}_{\mathrm{II}}^*\,\underline{\widetilde{Y}}_{\mathrm{I}} = \underline{\widetilde{x}}^*\boldsymbol{f}_2^*\boldsymbol{f}^1\underline{\widetilde{x}} = \underline{\widetilde{x}}^*L\,\underline{\widetilde{x}} \quad (5.11a).$$

Though being a scalar the value of the form $\underline{S}_{\mathrm{Y}}$ is not a real but a complex number. This implies a corresponding square matrix omitting in general symmetric properties.

Matrix of the Form L. The so-called dyadic product of the conjugate transpose of the row vector $\overline{\boldsymbol{f}^{2'}}$, being equal to the column vector $\boldsymbol{f}_2^*$, and the row vector $\boldsymbol{f}^1$ yields a (2x2) matrix of the rank 2, called the *coefficient matrix of the form*

$$\boldsymbol{L} = \boldsymbol{f}_2^*\boldsymbol{f}^1 = \begin{bmatrix} F_{21}^* \\ F_{22}^* \end{bmatrix}\left[F_{11},\ F_{12}\right] = \begin{bmatrix} F_{21}^*F_{11} & F_{21}^*F_{12} \\ F_{22}^*F_{11} & F_{22}^*F_{12} \end{bmatrix} = \left[L_{\ell\mathrm{m}}\right] \quad (5.12a).$$

This square matrix is complex but not Hermitian providing all its elements, $L_{\ell\mathrm{m}}$, are complex numbers (phasor ratio connectives). Thus, symmetric properties of $\boldsymbol{L}$ holding true for complex coefficients can not be assumed.

Standard Form of the Matrix L. Except for the trivial case of equal row vectors

$$\boldsymbol{f}^2 = \boldsymbol{f}^1$$

equivalent to

$$F_{21} = F_{11} \quad ; \quad F_{12} = F_{22} \quad (5.13a)$$

the square matrix, Eq. (5.12a), is a *Hermitian matrix*. This corresponds to a signal 4-pole characterized by a transmission matrix $\boldsymbol{F}$ said to be singular because the determinant of the square matrix is zero

$$\det F = F_{11}F_{22} - F_{21}F_{11} = 0 \quad (5.13b).$$

Then the matrix of the form, Eq. (5.12a), has the Hermitian property

$$\boldsymbol{L}^* = \boldsymbol{L} \quad (5.13c)$$

being the complex generalization from a real symmetric matrix. This matrix is in addition positive (semi-) definite since the output complex power holds in particular a *Hermitian form* being a real number.

Following from

$$\underline{\tilde{x}}^* f_1^* f^1 \underline{\tilde{x}} = \underline{\tilde{Y}_{\mathrm{I}}}^* \underline{\tilde{Y}_{\mathrm{I}}} = \left|\underline{\tilde{Y}_{\mathrm{I}}}\right|^2 \geqq 0 \qquad (5.13d)$$

for all $\underline{\tilde{X}}_{\mathrm{i}}$ or due to the decomposition of the Hermitian matrix into the sum of dyadic vector products

$$f_1^* f^1 + f_2^* f^2 = F^* F \qquad (5.14a),$$

and regarding the square of the norm

$$\begin{aligned} \underline{\tilde{x}}^* F^* F \underline{\tilde{x}} = \underline{\tilde{y}}^* \underline{\tilde{y}} = \left|\underline{\tilde{y}}\right|^2 &= \left|\underline{\tilde{Y}_{\mathrm{I}}}\right|^2 + \left|\underline{\tilde{Y}_{\mathrm{II}}}\right|^2 = 2\left|\underline{\tilde{Y}_{\mathrm{I}}}\right|^2 \\ \left|\underline{\tilde{Y}_{\mathrm{I}}}\right|^2 &= \frac{1}{2}\left|\underline{\tilde{y}}\right|^2 \geqq 0 \end{aligned} \qquad (5.14b)$$

for all $\underline{\tilde{X}}_{\mathrm{i}}$, $\underline{\tilde{Y}}_{\mathrm{i}}$, the power state is represented in this exceptional case by an active output power which consists in the square of the magnitude of the output component equalling half the square norm of the output vector, and which must be non-negative for passive 2-port networks.

The Hermitian matrix $\boldsymbol{L}$ as being singular a positive (semi-) definite form is existent which may become zero also for $\underline{\tilde{x}} \neq 0$.

Generalized Form of the Matrix L. For the general signal 4-pole the complex matrix of the form, Eq. (5.12a), can be written as the sum of a *Hermitian matrix* ***U*** and a *skew-Hermitian matrix* ***V***

$$L = U + V \qquad (5.12b)$$

where

$$\begin{aligned} U &= \frac{1}{2}\left(L + L^*\right) \\ V &= \frac{1}{2}\left(L - L^*\right) \end{aligned} \qquad (5.15a).$$

With respect to the dyadic product of the conjugate transpose of Eq. (5.12a)

$$L^* = \left(f_2^* f^1\right)^* = f_1^* f^2 = \begin{bmatrix} F_{11}^* \\ F_{12}^* \end{bmatrix} \left[F_{21},\, F_{22}\right] = \begin{bmatrix} F_{11}^* F_{21} & F_{11}^* F_{22} \\ F_{12}^* F_{21} & F_{12}^* F_{22} \end{bmatrix} \qquad (5.12c)$$

the following relations summing up or subtracting complex elements, hold true

$$\begin{aligned} U &= \frac{1}{2}\begin{bmatrix} F_{21}^* F_{11} + F_{11}^* F_{21} & F_{21}^* F_{12} + F_{11}^* F_{22} \\ F_{22}^* F_{11} + F_{12}^* F_{21} & F_{22}^* F_{12} + F_{12}^* F_{22} \end{bmatrix} = \left[U_{\ell \mathrm{m}}\right] \\ V &= \frac{1}{2}\begin{bmatrix} F_{21}^* F_{11} - F_{11}^* F_{21} & F_{21}^* F_{12} - F_{11}^* F_{22} \\ F_{22}^* F_{11} - F_{12}^* F_{21} & F_{22}^* F_{12} - F_{12}^* F_{22} \end{bmatrix} = \left[V_{\ell \mathrm{m}}\right] \end{aligned} \qquad (5.15b).$$

Complex Power at the Downstream Port

By matrix operations including quadratic forms it has been demonstrated that the complex output power $\underline{S}_{\mathrm{Y}}$ can be expressed in terms of connected elements, $F_{\ell \mathrm{m}}$, of the transmission matrix $\boldsymbol{F}$ combined with the r.m.s.value components $\underline{\tilde{X}}_i$ of the input vector $\underline{\tilde{\boldsymbol{x}}}$, where the energetic connectedness of the causally reticulated 2-port is described by the *modified Hermitian form* rewritten in full notation

$$\begin{aligned}
\underline{S}_Y &= \underline{\tilde{Y}}_{II}^* \underline{\tilde{Y}}_I \\
&= \underline{\tilde{\boldsymbol{x}}}^* \boldsymbol{L} \underline{\tilde{\boldsymbol{x}}} = \left[\underline{\tilde{X}}_I^* , \underline{\tilde{X}}_{II}^*\right] \begin{bmatrix} F_{21}^* F_{11} & F_{21}^* F_{12} \\ F_{22}^* F_{11} & F_{22}^* F_{12} \end{bmatrix} \begin{bmatrix} \underline{\tilde{X}}_I \\ \underline{\tilde{X}}_{II} \end{bmatrix} \\
&= \left[\underline{\tilde{X}}_I^* , \underline{\tilde{X}}_{II}^*\right] \begin{bmatrix} F_{21}^* F_{11} \underline{\tilde{X}}_I + F_{21}^* F_{12} \underline{\tilde{X}}_{II} \\ F_{22}^* F_{11} \underline{\tilde{X}}_I + F_{22}^* F_{12} \underline{\tilde{X}}_{II} \end{bmatrix} \\
&= \underline{\tilde{X}}_I^* \left(F_{21}^* F_{11} \underline{\tilde{X}}_I + F_{21}^* F_{12} \underline{\tilde{X}}_{II} \right) \\
&\quad + \underline{\tilde{X}}_{II}^* \left(F_{22}^* F_{11} \underline{\tilde{X}}_I + F_{22}^* F_{12} \underline{\tilde{X}}_{II} \right) = \sum_{\ell=1}^{n=2} \sum_{m=1}^{n=2} L_{\ell m} \underline{\tilde{X}}_\ell^* \underline{\tilde{X}}_m
\end{aligned} \qquad (5.11b).$$

Real and imaginary parts of the complex power are transacting components being significant for the insight into stationary flow of energy through a 2-port. It will be shown that the *decomposition of the complex power* reduces the introduced modified Hermitian form to corresponding standard forms involving either even or odd properties of symmetry.

Active Power (average power)

The real part of the complex power which describes its active (or in-phase) component, called the *active power*, leaving port 2

$$P_Y = \mathrm{Re}\left[\underline{S}_Y\right] = \frac{1}{2}\left(\underline{S}_Y + \underline{S}_Y^*\right) \qquad (5.16a)$$

results due to

$$\begin{aligned}
\mathrm{Re}\left[\underline{S}_Y\right] &= \frac{1}{2}\left(\underline{\tilde{\boldsymbol{x}}}^* \boldsymbol{L} \underline{\tilde{\boldsymbol{x}}} + \underline{\tilde{\boldsymbol{x}}}^* \boldsymbol{L}^* \underline{\tilde{\boldsymbol{x}}}\right) \\
&= \frac{1}{2}\left\{\underline{\tilde{\boldsymbol{x}}}^* (\boldsymbol{U} + \boldsymbol{V}) \underline{\tilde{\boldsymbol{x}}} + \underline{\tilde{\boldsymbol{x}}}^* (\boldsymbol{U} - \boldsymbol{V}) \underline{\tilde{\boldsymbol{x}}}\right\} = \underline{\tilde{\boldsymbol{x}}}^* \boldsymbol{U} \underline{\tilde{\boldsymbol{x}}}
\end{aligned} \qquad (5.16b)$$

in a standard form, termed the *Hermitian form*

$$P_Y = \mathrm{Re}\left[\underline{\tilde{Y}}_{II}^* \underline{\tilde{Y}}_I\right] = \underline{\tilde{\boldsymbol{x}}}^* \boldsymbol{U} \underline{\tilde{\boldsymbol{x}}} = \sum_{\ell=1}^{n=2} \sum_{m=1}^{n=2} U_{\ell m} \underline{\tilde{X}}_\ell^* \underline{\tilde{X}}_m \qquad (5.16c).$$

The *matrix of the form* $\boldsymbol{U}$ is the *Hermitian term* of the (complex) quadratic matrix $\boldsymbol{L}$, Eqs. (5.15a,b,c), thus the elements of the principal diagonal are real, whereas the elements of the secondary diagonal are complex conjugate.

In addition $\boldsymbol{U}$ is positive definite, that means non-zero for every choice of $\underline{\tilde{\boldsymbol{x}}}$ as the active output power P_Y is a *Hermitian form* which has the value of a *real number* being non-negative for passive 2-port networks. $\boldsymbol{U}$ is nonsingular, thus the corresponding Hermitian form is properly positive definite, the value of which only becomes zero for $\underline{\tilde{\boldsymbol{x}}} = 0$.

Reactive Power

The imaginary part of the complex power which describes its reactive (or quadrature) component, called
the *reactive power*, leaving port 2

$$Q_Y = \mathrm{Im}\left[\underline{S}_Y\right] = \frac{1}{2j}\left(\underline{S}_Y - \underline{S}_Y^*\right) \tag{5.17a}$$

accordingly results due to

$$\mathrm{Im}\left[\underline{S}_Y\right] = \frac{1}{2j}\left(\underline{\tilde{x}}^* \boldsymbol{L}\, \underline{\tilde{x}} - \underline{\tilde{x}}^* \boldsymbol{L}^* \underline{\tilde{x}}\right) \tag{5.17b}$$

in a complementary standard form, termed
the *skew-Hermitian form*

$$jQ_Y = j\mathrm{Im}\left[\underline{\tilde{Y}}_{II}^* \,\underline{\tilde{Y}}_I\right] = \underline{\tilde{x}}^* \boldsymbol{V}\, \underline{\tilde{x}} = \sum_{\ell=1}^{n=2}\sum_{m=1}^{n=2} V_{\ell m}\, \underline{\tilde{X}}_\ell^* \,\underline{\tilde{X}}_m \tag{5.17c}.$$

The *matrix of the form* $\boldsymbol{V}$ is the *skew-Hermitian term* of the (complex) quadratic matrix $\boldsymbol{L}$, Eqs. (5.15a,b,c), thus the elements of the principal diagonal are pure imaginary, whereas the elements of the secondary diagonal are equal except for the real parts.

$\boldsymbol{V}$ is not positive definite, the imaginary reactive output power jQ_Y is a *skew-Hermitian form* having the value of a *pure imaginary number* or zero for every choice of $\underline{\tilde{x}}$.

Transformation of power States in Summary

By using Eq. (5.11b) an efficient relationship is available to perform the transformation of the scalar power state $\underline{S}_X$ into $\underline{S}_Y$ with respect to the stationary case within the types of steady state.

Finally, the energetic system is conceptually replaced by a *module* which performs the operation on the complex input power $\underline{S}_X$ being required to yield up the complex output power $\underline{S}_Y$. This module previously conceived as a static functional can be explicitely defined by the matrix of the form $\boldsymbol{L}$ being a *modified Hermitian matrix*, a generalized expression holding for an energy storing and dissipating system subjected to a harmonic excitation.

The matrix $\boldsymbol{L}$ contains the *intrinsic properties of the system* concerning energetic interactions at the ports in form of a *set of composite parameters*, $L_{\ell m}$, being related to the 2-port parameters, $F_{\ell m}$, of the known transmission matrix $\boldsymbol{F}$, Eq. (5.9). The composite parameters, $L_{\ell m}$, are complex elements (functional operators) each of them representing a phasor ratio (or frequency characteristic) multiplied by a respective conjugate. Thus, the complex elements, $L_{\ell m}$, shortly termed *phasor ratio connectives*, are significant for the frequency behaviour of stationary energy flow through a linear 2-port.

5.1.3
The Power Transmission Factor. *Generalized Transmission Ratio*

Previously it has been demonstrated how to find the output power P_2 as a function of the input power P_1 due to matrix operations using the complex generalization of quadratic forms with reference to the sinusoidal steady state. Finally, the system could be conceptually replaced by the known matrix of the form $\boldsymbol{L}$ (corresponding to the transmission matrix $\boldsymbol{F}$) which performs the operation on the given input vector $\underline{\tilde{\boldsymbol{x}}}$ being required to yield up the *complex output power* $\underline{S}_{\mathrm{Y}}$. Though being of considerable interest the performed scalar transformation of the power state $\underline{S}_{\mathrm{X}}$ into $\underline{S}_{\mathrm{Y}}$ covers only one aspect of the transmission problem engaged in the flow of energy.

In particular with respect to system synthesis another aspect of the problem arises which looks for valuing energetic systems by a transmission characteristic which makes evident the varying effects on power being transmitted through. For this purpose an explicit relationship must be found relating the complex power which leaves to that one which enters the causally reticulated 2-port. As a result the *transmission factor of the complex power* can be defined involving the frequency characteristic of power transmission. This relationship is suited to illustrate frequency and 2-port parameter effects on the dynamic power flow behaviour (frequency-domain specifications).

Starting from the transmission ratio of phasor products, Eq. (5.8), 5.1.2, matrix operations will be applied returning to quadratic forms.

Scalar Transformation of Power State at the Input Port

In comparison with the complex output power $\underline{S}_{\mathrm{Y}}$ expressed in terms of a scalar form, Eqs. (5.11a,b), an adequate form must be found to represent the complex input power $\underline{S}_{\mathrm{X}}$ in matrix notation.

A (2-rowed) diagonal matrix with elements in the principal diagonal being all unity is called a unit matrix $\boldsymbol{I}$ and can be written in terms of its (1x2) *submatrices* or row vectors (unit vectors) $\boldsymbol{e}^{\ell}$

$$\boldsymbol{I} = [I_{\ell m}] = \begin{bmatrix} \mathbf{e}^1 \\ \mathbf{e}^2 \end{bmatrix} \qquad (5.18).$$

Consequently, the components $\underline{\tilde{X}}_{\mathrm{i}}$ of the input vector are obtained from the scalar product (dot product) of the ℓth row vector $\boldsymbol{e}^{\ell}$ and the column vector $\underline{\tilde{\boldsymbol{x}}}$ of the input

$$\begin{aligned} \underline{\tilde{X}}_{\mathrm{I}} &= \mathbf{e}^1\,\underline{\tilde{\boldsymbol{x}}} = [1\,,0]\begin{bmatrix} \underline{\tilde{X}}_{\mathrm{I}} \\ \underline{\tilde{X}}_{\mathrm{II}} \end{bmatrix} \\ \underline{\tilde{X}}_{\mathrm{II}} &= \mathbf{e}^2\,\underline{\tilde{\boldsymbol{x}}} = [0\,,1]\begin{bmatrix} \underline{\tilde{X}}_{\mathrm{I}} \\ \underline{\tilde{X}}_{\mathrm{II}} \end{bmatrix} \end{aligned} \qquad (5.19a).$$

The complex conjugate $\underline{\tilde{X}}^*_{\mathrm{II}}$ is formed by

$$\underline{\tilde{X}}^*_{\mathrm{II}} = \overline{(\mathbf{e}^2\,\underline{\tilde{\boldsymbol{x}}})'} = (\mathbf{e}^2\,\underline{\tilde{\boldsymbol{x}}})^* = \underline{\tilde{\boldsymbol{x}}}^*\mathbf{e}_2 = [\underline{\tilde{X}}^*_{\mathrm{I}}\ \underline{\tilde{X}}^*_{\mathrm{II}}]\begin{bmatrix} 0 \\ 1 \end{bmatrix} \qquad (5.19b).$$

Modified Quadratic Form. Thus, the complex input power results from multiplying the conjugate dot product, Eq. (5.19b), and the corresponding dot product, Eq. (5.19a), together. By this, the input vector $\underline{\tilde{x}}$ is transformed into a scalar form being a generalization from the real quadratic form. Compared to this standard specification the given expression denoted by $\underline{S}_X$ represents

the *modified quadratic form* in the two complex variables $\underline{\tilde{X}}_I$, $\underline{\tilde{X}}_{II}$

$$\underline{S}_X = \underline{\tilde{X}}_{II}^* \underline{\tilde{X}}_I = \underline{\tilde{x}}^* \mathbf{e}_2 \mathbf{e}^1 \underline{\tilde{x}} = \underline{\tilde{x}}^* \boldsymbol{E} \underline{\tilde{x}} \qquad (5.20a).$$

Though being a scalar the value of the form $\underline{S}_X$ is a complex number, and the correspondent coefficient matrix omits symmetric properties.

Matrix of the Form E. The dyadic product of the transpose of the row vector $\boldsymbol{e}^{2'}$, being equal to the column vector $\boldsymbol{e}_2$, and the row vector $\boldsymbol{e}^1$ yields a (2x2) matrix of the rank 1, called
the *coefficient matrix of the form*

$$\boldsymbol{E} = \mathbf{e}_2 \mathbf{e}^1 = \begin{bmatrix} 0 \\ 1 \end{bmatrix} [1, 0] = \begin{bmatrix} 0 & 0 \\ 1 & 0 \end{bmatrix} = [E_{\ell m}] \qquad (5.21a).$$

This square matrix is real but not symmetric.

Standard Form of the Matrix E. Except for the trivial case of equal row vectors (unit vectors)

$$\mathbf{e}^2 = \mathbf{e}^1$$

equivalent to

$$\boldsymbol{E} = \begin{bmatrix} 1 & 0 \\ 0 & 0 \end{bmatrix} \qquad (5.22a)$$

the square matrix, Eq. (5.21a), is a *symmetric matrix* with the property

$$\boldsymbol{E} = \boldsymbol{E}' \qquad (5.22b)$$

being in addition positive (semi-) definite since the input complex power holds in particular a real *quadratic form*.

Following from

$$\underline{\tilde{x}}^* \mathbf{e}_1 \mathbf{e}^1 \underline{\tilde{x}} = \underline{\tilde{X}}_I^* \underline{\tilde{X}}_I = \left|\underline{\tilde{X}}_I\right|^2 \geqq 0 \qquad (5.22c)$$

for all $\underline{\tilde{X}}_i$ or due to the decomposition of the symmetric unit matrix into the sum of dyadic unit-vector products

$$\mathbf{e}_1 \mathbf{e}^1 + \mathbf{e}_2 \mathbf{e}^2 = \boldsymbol{I}'\boldsymbol{I} = \boldsymbol{I} \qquad (5.23a),$$

in long hand

$$\begin{bmatrix} 1 & 0 \\ 0 & 0 \end{bmatrix} + \begin{bmatrix} 0 & 0 \\ 0 & 1 \end{bmatrix} = \begin{bmatrix} 1 & 0 \\ 0 & 1 \end{bmatrix} \qquad (5.23b),$$

and regarding the square of the norm

$$\underline{\tilde{x}}^* \boldsymbol{I}'\boldsymbol{I} \underline{\tilde{x}} = \underline{\tilde{x}}^* \underline{\tilde{x}} = |\underline{\tilde{x}}|^2 = \left|\underline{\tilde{X}}_I\right|^2 + \left|\underline{\tilde{X}}_{II}\right|^2 = 2\left|\underline{\tilde{X}}_I\right|^2$$
$$\left|\underline{\tilde{X}}_I\right|^2 = \frac{1}{2}|\underline{\tilde{x}}|^2 \geqq 0 \qquad (5.23c)$$

for all $\underline{\widetilde{X}}_{\mathrm{i}}$, the power state is represented in this exceptional case by an active input power which consists in the square of the magnitude of the input component equalling half the square norm of the input vector, and which entering the 2-port network cannot be negative.

Generalized Form of the Matrix E. For the general signal 4-pole the real matrix of the form, Eq. (5.21a), can be written as the sum of a real *symmetric matrix* $\boldsymbol{E}_{\mathrm{s}}$ and a real *skew-symmetric matrix* $\boldsymbol{E}_{\mathrm{a}}$

$$\boldsymbol{E} = \boldsymbol{E}_{\mathrm{s}} + \boldsymbol{E}_{\mathrm{a}}$$

$$\boldsymbol{E}_{\mathrm{s}} = \frac{1}{2}(\boldsymbol{E} + \boldsymbol{E}') \tag{5.21b}$$

$$\boldsymbol{E}_{\mathrm{a}} = \frac{1}{2}(\boldsymbol{E} - \boldsymbol{E}') \tag{5.24a}$$

With respect to the dyadic product of the conjugate transpose of Eq. (5.21a)

$$\boldsymbol{E}' = (\mathbf{e}_2 \mathbf{e}^1)' = \mathbf{e}_1 \mathbf{e}^2 = \begin{bmatrix} 1 \\ 0 \end{bmatrix} [0\,,1] = \begin{bmatrix} 0 & 1 \\ 0 & 0 \end{bmatrix} \tag{5.21c}$$

the following relations summing up or subtracting unit elements, hold true

$$\boldsymbol{E}_{s} = \frac{1}{2}\begin{bmatrix} 0 & 1 \\ 1 & 0 \end{bmatrix} = \left[E_{\mathrm{s}\,\ell\mathrm{m}} \right]$$

$$\boldsymbol{E}_{\mathrm{a}} = \frac{1}{2}\begin{bmatrix} 0 & -1 \\ 1 & 0 \end{bmatrix} = \left[E_{\mathrm{a}\,\ell\mathrm{m}} \right] \tag{5.24b}$$

Complex Power at the Upstream Port

It has been demonstrated that the complex input power $\underline{S}_{\mathrm{X}}$ can be expressed in terms of a unit element 1 combined with the r.m.s. value components $\underline{\widetilde{X}}_{\mathrm{i}}$ of the input vector $\underline{\widetilde{\boldsymbol{x}}}$, where the energetic connectedness is described by the *modified quadratic form* rewritten in full notation

$$\begin{aligned}
\underline{S}_{\mathrm{X}} &= \underline{\widetilde{Y}}_{\mathrm{II}}^{*}\, \underline{\widetilde{Y}}_{\mathrm{I}} \\
&= \underline{\widetilde{\boldsymbol{x}}}^{*}\, \boldsymbol{E}\, \underline{\widetilde{\boldsymbol{x}}} = \left[\underline{\widetilde{X}}_{\mathrm{I}}^{*},\ \underline{\widetilde{X}}_{\mathrm{II}}^{*} \right] \begin{bmatrix} 0 & 0 \\ 1 & 0 \end{bmatrix} \begin{bmatrix} \underline{\widetilde{X}}_{\mathrm{I}} \\ \underline{\widetilde{X}}_{\mathrm{II}} \end{bmatrix} \\
&= \left[\underline{\widetilde{X}}_{\mathrm{I}}^{*},\ \underline{\widetilde{X}}_{\mathrm{II}}^{*} \right] \begin{bmatrix} 0 \\ \underline{\widetilde{X}}_{\mathrm{I}} \end{bmatrix} \\
&= \underline{\widetilde{X}}_{\mathrm{II}}^{*}\, \underline{\widetilde{X}}_{\mathrm{I}} = \sum_{\ell=1}^{n=2} \sum_{\mathrm{m}=1}^{n=2} E_{\ell\mathrm{m}}\, \underline{\widetilde{X}}_{\ell}^{*}\, \underline{\widetilde{X}}_{\mathrm{m}}
\end{aligned} \tag{5.20b}$$

Active Power (average power)

The real part of the complex power, called the active power, entering port 1

$$P_{\mathrm{X}} = \mathrm{Re}\left[\underline{S}_{\mathrm{X}} \right] = \frac{1}{2}\left(\underline{S}_{\mathrm{X}} + \underline{S}_{\mathrm{X}}^{*} \right) \tag{5.25a}$$

results due to

$$\mathrm{Re}\left[\underline{S}_{\mathrm{x}}\right] = \frac{1}{2}\left(\underline{\tilde{x}}^{*} E\, \underline{\tilde{x}} + \underline{\tilde{x}}^{*} E'\, \underline{\tilde{x}}\right)$$

$$= \frac{1}{2}\left\{\underline{\tilde{x}}^{*}(E_{\mathrm{s}} + E_{\mathrm{a}})\underline{\tilde{x}} + \underline{\tilde{x}}^{*}(E_{\mathrm{s}} - E_{\mathrm{a}})\underline{\tilde{x}}\right\} = \underline{\tilde{x}}^{*} E_{\mathrm{s}}\, \underline{\tilde{x}} \tag{5.25b}$$

in a standard form termed the *real quadratic form*

$$P_{\mathrm{x}} = \mathrm{Re}\left[\underline{\tilde{Y}}_{\mathrm{II}}^{*}\, \underline{\tilde{Y}}_{\mathrm{I}}\right] = \underline{\tilde{x}}^{*} \boldsymbol{E}_{\mathrm{s}}\, \underline{\tilde{x}} = \sum_{\ell=1}^{\mathrm{n}=2} \sum_{\mathrm{m}=1}^{\mathrm{n}=2} E_{\mathrm{s}\ell\mathrm{m}}\, \underline{\tilde{X}}_{\ell}^{*}\, \underline{\tilde{X}}_{\mathrm{m}} \tag{5.25c}$$

The *matrix of the form* $\boldsymbol{E}_{\mathrm{s}}$ is the *symmetric term* of the (real) quadratic matrix $\boldsymbol{E}$, Eqs. (5.24a,b), thus the principal diagonal elements are real whereas the secondary diagonal elements are equal.

Using real matrices and bases for the real and imaginary vectors P_{X} can be expressed in the alternative form of the trace of a product of real matrices. Introducing a real vector composed of the geometric mean of the input components P_{X} furthermore can be expressed by the square of the appertaining vector norm, [120].

Reactive Power

The imaginary part of the complex power, called the reactive power, entering port 1

$$Q_{\mathrm{x}} = \mathrm{Im}\left[\underline{S}_{\mathrm{x}}\right] = \frac{1}{2\mathrm{j}}\left(\underline{S}_{\mathrm{x}} - \underline{S}_{\mathrm{x}}^{*}\right) \tag{5.26a}$$

accordingly results due to

$$\mathrm{Im}\left[\underline{S}_{\mathrm{x}}\right] = \frac{1}{2\mathrm{j}}\left(\underline{\tilde{x}}^{*} \boldsymbol{E}\, \underline{\tilde{x}} - \underline{\tilde{x}}^{*} \boldsymbol{E}'\, \underline{\tilde{x}}\right) \tag{5.26b}$$

in a complementary standard form, termed the *pure imaginary skew-quadratic form*

$$\mathrm{j}Q_{\mathrm{x}} = \mathrm{j}\,\mathrm{Im}\left[\underline{\tilde{X}}_{\mathrm{II}}^{*}\, \underline{\tilde{X}}_{\mathrm{I}}\right] = \underline{\tilde{x}}^{*} \boldsymbol{E}_{\mathrm{a}}\, \underline{\tilde{x}} = \sum_{\ell=1}^{\mathrm{n}=2} \sum_{\mathrm{m}=1}^{\mathrm{n}=2} E_{\mathrm{a}\ell\mathrm{m}}\, \underline{\tilde{X}}_{\ell}^{*}\, \underline{\tilde{X}}_{\mathrm{m}} \tag{5.26c}$$

The matrix of the form $\boldsymbol{E}_{\mathrm{a}}$ is the *skew-symmetric term* of the (real) quadratic matrix $\boldsymbol{E}$, Eqs. (5.24a,b), thus the principal diagonal elements are equal except for the opposite sign.

Introducing a real vector composed of the geometric mean of the input components Q_{X} furthermore can be expressed by the square of the appertaining vector norm, [120].

Generalized Transmission Ratio

The scalar transformation of power states representing the effort-flow product by the complex power at both of the energy ports has been carried out. This now permits to specify the explicit relationship covering the transmission problem of *valuing energetic systems* as follows.

Modified Rayleigh Quotient. Thus, the explicit relationship results from replacing the phasor products in the transmission ratio, Eq. (5.8), by the generalized quadratic forms, Eqs. (5.11a), (5.20a). By this, the input vector $\underline{\tilde{x}}$ is transformed into scalar forms the quotient of which is similar, not equal to an expression designated in the theory of matrices as the Rayleigh quotient. This standard form is fit for the determination of eigenvalues, e.g., for estimating the lowest natural frequency of lumped parameter systems (Rayleigh's principle). The quotient of quadratic forms including their generalizations even in complex notation is commonly written in the particular form where the denominator is reduced to the square of the vector norm of $\underline{\tilde{x}}$.

Compared to this standard specification the given expression denoted by $\underline{A}_S$ represents

the *modified Rayleigh quotient*

$$\underline{A}_S = \frac{\underline{S}_Y}{\underline{S}_X} = \frac{\underline{\tilde{Y}}_{II}^* \underline{\tilde{Y}}_I}{\underline{\tilde{X}}_{II}^* \underline{\tilde{X}}_I} = \frac{\underline{\tilde{x}}^* \underline{\underline{L}} \underline{\tilde{x}}}{\underline{\tilde{x}}^* \underline{\underline{E}} \underline{\tilde{x}}} = \mathrm{R_S}[\underline{\tilde{x}}] \qquad (5.27a),$$

and in terms of connected elements $F_{\ell m}$ of the transmission matrix $\boldsymbol{F}$ rewritten in full notation

$$\underline{A}_S = \frac{\left[\underline{\tilde{X}}_I^*, \underline{\tilde{X}}_{II}^*\right]\begin{bmatrix} F_{21}^* F_{11} & F_{21}^* F_{12} \\ F_{22}^* F_{11} & F_{22}^* F_{12} \end{bmatrix}\begin{bmatrix} \underline{\tilde{X}}_I \\ \underline{\tilde{X}}_{II} \end{bmatrix}}{\left[\underline{\tilde{X}}_I^*, \underline{\tilde{X}}_{II}^*\right]\begin{bmatrix} 0 & 0 \\ 1 & 0 \end{bmatrix}\begin{bmatrix} \underline{\tilde{X}}_I \\ \underline{\tilde{X}}_{II} \end{bmatrix}} = \frac{\sum_{\ell=1}^{n=2}\sum_{m=1}^{n=2} L_{\ell m} \underline{\tilde{X}}_\ell^* \underline{\tilde{X}}_m}{\sum_{\ell=1}^{n=2}\sum_{m=1}^{n=2} E_{\ell m} \underline{\tilde{X}}_\ell^* \underline{\tilde{X}}_m} \qquad (5.27b)$$

the energetic connectedness of the causally reticulated 2-port is indicated.

This quotient of generalized quadratic forms may be designated the *transmission factor of the complex power* of a 2-port.

Properties of the Rayleigh Quotient. Though being a quotient of scalars the value of $\underline{A}_S$ is not a real but a *complex number*. Except for the trivial case referred to both of the coefficient matrices of the form, $\boldsymbol{L}$ and $\boldsymbol{E}$, being Hermitian and real symmetric, respectively, $\underline{A}_S$ is reduced to a real number.

Furthermore, $\underline{A}_S$ is *bounded*, i.e., the denominator does not vanish by oneself, and the modified quadratic form $\underline{\tilde{x}}^* \boldsymbol{E} \underline{\tilde{x}}$ must be *positive definite*.

Hence, in the complex plane a proper region $[\boldsymbol{L}]$ is assigned to the matrix of the form $\boldsymbol{L}$, the so-called *field of values*. The boundary of this field can be marked out by the maximal values of the composite elements, $L_{\ell m}$, of the matrix of the form, i.e., by the least upper bound of estimated 2-port parameters, $F_{\ell m}$, presupposing an input vector $\underline{\tilde{x}}$ being normalized. This corresponds to a limited amount of power transmission performed by the signal 4-pole.

Finally, it may be taken profit from the Rayleigh quotient by use of the inherent relations to the eigenvector $\boldsymbol{x}$ corresponding to its components λ_ℓ of $\boldsymbol{L}$, and from its extreme value properties for solving *eigenvalue problems*, respectively, *optimal value problems* with reference to power transmission by complex systems.

Real and imaginary parts of the transmission factor of the complex power are significant for valuing the stationary flow of energy through a 2-port. It will be shown that the *decomposition of the power transmission factor* reduces the introduced generalized quadratic forms to corresponding standard forms involving properties of even or odd symmetry.

Active-power Transmission Factor

The real part of the complex power transmission factor which describes its active (or in-phase) component, called
the *active-power transmission factor*

$$A_P = \frac{P_Y}{P_X} = \frac{\mathrm{Re}\left[\underline{\widetilde{Y}}_{II}^{*}\,\underline{\widetilde{Y}}_{I}\right]}{\mathrm{Re}\left[\underline{\widetilde{X}}_{II}^{*}\,\underline{\widetilde{X}}_{I}\right]} = \frac{\underline{\widetilde{x}}^{*}\boldsymbol{U}\underline{\widetilde{x}}}{\underline{\widetilde{x}}^{*}\boldsymbol{E}_{s}\underline{\widetilde{x}}} = \mathrm{R_P}\left[\underline{\widetilde{x}}\right] \qquad (5.28a),$$

rewritten in full notation

$$A_P = \frac{\left[\underline{\widetilde{X}}_{I}^{*},\ \underline{\widetilde{X}}_{II}^{*}\right]\begin{bmatrix} F_{21}^{*}F_{11}+F_{11}^{*}F_{21} & F_{21}^{*}F_{12}+F_{11}^{*}F_{22} \\ F_{22}^{*}F_{11}+F_{12}^{*}F_{21} & F_{22}^{*}F_{12}+F_{12}^{*}F_{22}\end{bmatrix}\begin{bmatrix}\underline{\widetilde{X}}_{I} \\ \underline{\widetilde{X}}_{II}\end{bmatrix}}{\left[\underline{\widetilde{X}}_{I}^{*},\ \underline{\widetilde{X}}_{II}^{*}\right]\begin{bmatrix}0 & 1\\ 1 & 0\end{bmatrix}\begin{bmatrix}\underline{\widetilde{X}}_{I} \\ \underline{\widetilde{X}}_{II}\end{bmatrix}} = \frac{\sum_{\ell=1}^{n=2}\sum_{m=1}^{n=2} U_{\ell m}\underline{\widetilde{X}}_{\ell}^{*}\underline{\widetilde{X}}_{m}}{\sum_{\ell=1}^{n=2}\sum_{m=1}^{n=2} E_{s_{\ell m}}\underline{\widetilde{X}}_{\ell}^{*}\underline{\widetilde{X}}_{m}}$$

(5.28b)

corresponds to the Rayleigh quotient $\mathrm{R_P}\left[\underline{\widetilde{x}}\right]$.

This ratio relating the Hermitian form, Eq. (5.16c), to the real quadratic form, Eq. (5.25c), is a real positive number, and bounded, at which the matrices of the form $\boldsymbol{U}$ and $\boldsymbol{E}_s$, Eqs. (5.15a,b), (5.24a,b), are positive definite, furthermore, $\boldsymbol{E}_s$ is nonsingular.

The components $\underline{\widetilde{X}}_i$ of the complex input vector $\underline{\widetilde{x}}$ can be represented by the alternative of real matrices

$$\underline{\widetilde{X}}_{I} = \mathbf{e}\,u_I + \mathbf{j}\,\upsilon_I = \begin{bmatrix} u_I & -\upsilon_I \\ \upsilon_I & u_I\end{bmatrix} = \widetilde{\boldsymbol{X}}_{I}$$

$$\underline{\widetilde{X}}_{II}^{*} = \mathbf{e}\,u_{II} - \mathbf{j}\,\upsilon_{II} = \begin{bmatrix} u_{II} & \upsilon_{II} \\ -\upsilon_{II} & u_{II}\end{bmatrix} = \widetilde{\boldsymbol{X}}_{II}' \qquad (5.29a),$$

provided that the real and the imaginary unit vector are defined by the elementary matrices

$$\mathbf{e} = \begin{bmatrix}1 & 0\\ 0 & 1\end{bmatrix} \qquad \mathbf{j} = \begin{bmatrix}0 & -1\\ 1 & 0\end{bmatrix} \qquad (5.29b).$$

Introducing now the real vector $\boldsymbol{x}_m$ the components of which are formed by the geometric mean of the real and the imaginary part of $\underline{\widetilde{X}}_i$

$$\boldsymbol{x}_m = \begin{bmatrix} X_{m_1} \\ X_{m_2}\end{bmatrix} = \begin{bmatrix} \left|\sqrt{u_I u_{II}}\right| \\ \left|\sqrt{\upsilon_I \upsilon_{II}}\right|\end{bmatrix} \qquad (5.30a)$$

it becomes obvious that the real part of the product $\underline{\widetilde{X}}_{II}^{*}\underline{\widetilde{X}}_{I}$ corresponds to the scalar product

$$\boldsymbol{x}_m'\,\boldsymbol{x}_m = \left[\left|\sqrt{u_I u_{II}}\right|, \left|\sqrt{\upsilon_I \upsilon_{II}}\right|\right]\begin{bmatrix} \left|\sqrt{u_I u_{II}}\right| \\ \left|\sqrt{\upsilon_I \upsilon_{II}}\right|\end{bmatrix} = u_I u_{II} + \upsilon_I \upsilon_{II} = X_{m_1}^2 + X_{m_2}^2 = |\boldsymbol{x}_m|^2 \qquad (5.30b)$$

being equal to the square of the vector norm of $\boldsymbol{x}_m$

$$\mathrm{Re}\left[\underline{\widetilde{X}}_{II}^{*}\,\underline{\widetilde{X}}_{I}\right] = |\boldsymbol{x}_m|^2 = \left(u_I u_{II} + \upsilon_I \upsilon_{II}\right) \qquad (5.30c).$$

By use of the square norm of $\boldsymbol{x}_{\mathrm{m}}$ the equality to the standard form in the denominator, Eq. (5.25c), holds

$$\underline{\tilde{\boldsymbol{x}}}^{*} \boldsymbol{E}_{\mathrm{s}} \underline{\tilde{\boldsymbol{x}}} = \boldsymbol{x}'_{\mathrm{m}} \, \boldsymbol{x}_{\mathrm{m}} = |\boldsymbol{x}_{\mathrm{m}}|^{2} \qquad (5.31),$$

and a normalized input vector may be introduced

$$\underline{\tilde{z}} = \frac{\underline{\tilde{\boldsymbol{x}}}}{|\boldsymbol{x}_{\mathrm{m}}|} = \frac{1}{\left|\sqrt{u_{\mathrm{I}} u_{\mathrm{II}} + \upsilon_{\mathrm{I}} \upsilon_{\mathrm{II}}}\right|} \begin{bmatrix} \underline{\tilde{X}}_{\mathrm{I}} \\ \underline{\tilde{X}}_{\mathrm{II}} \end{bmatrix} \qquad (5.32a)$$

$$\underline{\tilde{z}}^{*} = \frac{\underline{\tilde{\boldsymbol{x}}}^{*}}{|\boldsymbol{x}_{\mathrm{m}}|} = \frac{1}{\left|\sqrt{u_{\mathrm{I}} u_{\mathrm{II}} + \upsilon_{\mathrm{I}} \upsilon_{\mathrm{II}}}\right|} \left[\underline{\tilde{X}}_{\mathrm{I}}^{*}, \underline{\tilde{X}}_{\mathrm{II}}^{*}\right] \qquad (5.32b).$$

Thus, a standard specification for the Rayleigh quotient can be found by transforming it via Eqs. (5.31), (5.32) into
the *Hermitian form*

$$A_{\mathrm{P}} = \underline{\tilde{z}}^{*} \boldsymbol{U} \, \underline{\tilde{z}} = R_{\mathrm{P}}\left[\underline{\tilde{z}}\right] \qquad (5.33).$$

As a result the active-power transmission factor A_{P} covers the *field of values* $[\boldsymbol{U}]$ of the *Hermitian matrix of the form* ***U****, Eq. (5.15a,b), ranging along a segment of the real axis (real-axis intercept).*

Reactive-power Transmission Factor

The imaginary part of the complex power transmission factor which describes its reactive (or quadrature) component, called
the *reactive-power transmission factor*

$$\underline{A_{\mathrm{Q}}} = \frac{Q_{\mathrm{Y}}}{Q_{\mathrm{X}}} = \frac{\mathrm{Im}\left[\underline{\tilde{Y}}_{\mathrm{II}}^{*} \underline{\tilde{Y}}_{\mathrm{I}}\right]}{\mathrm{Im}\left[\underline{\tilde{X}}_{\mathrm{II}}^{*} \underline{\tilde{X}}_{\mathrm{I}}\right]} = \frac{\underline{\tilde{\boldsymbol{x}}}^{*} \boldsymbol{V} \underline{\tilde{\boldsymbol{x}}}}{\underline{\tilde{\boldsymbol{x}}}^{*} \boldsymbol{E}_{\mathrm{a}} \underline{\tilde{\boldsymbol{x}}}} = \mathrm{R}_{\mathrm{Q}}\left[\underline{\tilde{\boldsymbol{x}}}\right] \qquad (5.34a),$$

rewritten in full notation

$$A_{\mathrm{Q}} = \frac{\left[\underline{\tilde{X}}_{\mathrm{I}}^{*}, \underline{\tilde{X}}_{\mathrm{II}}^{*}\right] \begin{bmatrix} F_{21}^{*}F_{11} - F_{11}^{*}F_{21} & F_{21}^{*}F_{12} - F_{11}^{*}F_{22} \\ F_{22}^{*}F_{11} - F_{12}^{*}F_{21} & F_{22}^{*}F_{12} - F_{12}^{*}F_{22} \end{bmatrix} \begin{bmatrix} \underline{\tilde{X}}_{\mathrm{I}} \\ \underline{\tilde{X}}_{\mathrm{II}} \end{bmatrix}}{\left[\underline{\tilde{X}}_{\mathrm{I}}^{*}, \underline{\tilde{X}}_{\mathrm{II}}^{*}\right] \begin{bmatrix} 0 & -1 \\ 1 & 0 \end{bmatrix} \begin{bmatrix} \underline{\tilde{X}}_{\mathrm{I}} \\ \underline{\tilde{X}}_{\mathrm{II}} \end{bmatrix}} = \frac{\sum_{\ell=1}^{n=2} \sum_{m=1}^{n=2} V_{\ell\mathrm{m}} \underline{\tilde{X}}_{\ell}^{*} \underline{\tilde{X}}_{\mathrm{m}}}{\sum_{\ell=1}^{n=2} \sum_{m=1}^{n=2} E_{\mathrm{a}\,\ell\mathrm{m}} \underline{\tilde{X}}_{\ell}^{*} \underline{\tilde{X}}_{\mathrm{m}}} \qquad (5.34b)$$

corresponds to the Rayleigh quotient $\mathrm{R}_{\mathrm{Q}}\left[\underline{\tilde{\boldsymbol{x}}}\right]$.

This ratio relating the skew-Hermitian form, Eq. (5.17c), to the pure imaginary skew-quadratic form, Eq. (5.26c), is a real number, and bounded, at which the matrix of the form $\boldsymbol{E}_{\mathrm{a}}$, Eqs. (5.24a,b), is nonsingular.

Introducing by use of Eqs. (5.29a), (5.29b) the real vector $\boldsymbol{x}_{\mathrm{g}}$ the components of which are formed by the geometric mean of the real and the imaginary part of $\underline{\tilde{X}}_{\mathrm{i}}$

$$\boldsymbol{x}_{\mathrm{g}} = \begin{bmatrix} |X_{\mathrm{g}_1}| \\ |X_{\mathrm{g}_2}| \end{bmatrix} = \begin{bmatrix} \left|\sqrt{\upsilon_{\mathrm{I}} u_{\mathrm{II}}}\right| \\ \mathrm{j}\left|\sqrt{\upsilon_{\mathrm{II}} u_{\mathrm{I}}}\right| \end{bmatrix} \qquad (5.35a)$$

it becomes evident that the imaginary part of the product $\underline{\tilde{X}}_{\mathrm{II}}^{*}\underline{\tilde{X}}_{\mathrm{I}}$ corresponds to the scalar product

$$\boldsymbol{x}_{\mathrm{g}}'\,\boldsymbol{x}_{\mathrm{g}} = \left[\left|\sqrt{\upsilon_{\mathrm{I}} u_{\mathrm{II}}}\right|, \mathrm{j}\left|\sqrt{\upsilon_{\mathrm{II}} u_{\mathrm{I}}}\right|\right]\begin{bmatrix}\left|\sqrt{\upsilon_{\mathrm{I}} u_{\mathrm{II}}}\right| \\ \mathrm{j}\left|\sqrt{\upsilon_{\mathrm{II}} u_{\mathrm{I}}}\right|\end{bmatrix} = \upsilon_{\mathrm{I}} u_{\mathrm{II}} - \upsilon_{\mathrm{II}} u_{\mathrm{I}} = X_{\mathrm{g}_1}^{2} + X_{\mathrm{g}_2}^{2} = \left|\boldsymbol{x}_{\mathrm{g}}\right|^{2} \quad (5.35\mathrm{b})$$

being equal to the square of the vector norm of $\boldsymbol{x}_{\mathrm{g}}$

$$Q_{\mathrm{x}} = \mathrm{Im}\left[\underline{\tilde{X}}_{\mathrm{II}}^{*}\,\underline{\tilde{X}}_{\mathrm{I}}\right] = \left|\underline{\tilde{x}}_{\mathrm{g}}\right|^{2} \quad (5.35\mathrm{c}).$$

By use of the square norm of $\boldsymbol{x}_{\mathrm{g}}$ the equality to the standard form in the denominator, Eq. (5.26c), holds

$$\underline{\tilde{\boldsymbol{x}}}^{*}\boldsymbol{E}_{\mathrm{a}}\,\underline{\tilde{\boldsymbol{x}}} = \boldsymbol{x}_{\mathrm{g}}'\,\boldsymbol{x}_{\mathrm{g}} = \left|\boldsymbol{x}_{\mathrm{g}}\right|^{2} \quad (5.36),$$

and a normalized input vector may be introduced

$$\underline{\tilde{\boldsymbol{r}}} = \frac{\underline{\tilde{\boldsymbol{x}}}}{\left|\boldsymbol{x}_{\mathrm{g}}\right|} = \frac{1}{\left|\sqrt{\upsilon_{\mathrm{I}} u_{\mathrm{II}} - \upsilon_{\mathrm{II}} u_{\mathrm{I}}}\right|}\begin{bmatrix}\underline{\tilde{X}}_{\mathrm{I}} \\ \underline{\tilde{X}}_{\mathrm{II}}\end{bmatrix} \quad (5.37\mathrm{a})$$

$$\underline{\tilde{\boldsymbol{r}}}^{*} = \frac{\underline{\tilde{\boldsymbol{x}}}^{*}}{\left|\boldsymbol{x}_{\mathrm{g}}\right|} = \frac{1}{\left|\sqrt{\upsilon_{\mathrm{I}} u_{\mathrm{II}} - \upsilon_{\mathrm{II}} u_{\mathrm{I}}}\right|}\left[\underline{\tilde{X}}_{\mathrm{I}}^{*},\ \underline{\tilde{X}}_{\mathrm{II}}^{*}\right] \quad (5.37\mathrm{b}).$$

Thus, a complementary standard specification for the Rayleigh quotient can be found by transforming it via Eqs. (5.36), (5.37) into the *skew-Hermitian form*

$$\mathrm{j}A_{\mathrm{Q}} = \underline{\tilde{\boldsymbol{r}}}^{*}\boldsymbol{V}\,\underline{\tilde{\boldsymbol{r}}} = \mathrm{R}_{\mathrm{Q}}\left[\underline{\tilde{\boldsymbol{r}}}\right] \quad (5.38).$$

As a result the reactive-power transmission factor jA_{Q} covers the *field of values* [***V***] of the *skew-Hermitian matrix of the form* ***V***, Eqs. (5.15a,b), ranging along a segment of the imaginary axis (imaginary-axis intercept).

The 2-port Power Transmission in Summary

Standard Forms Coincident to Power States and Efficiency. It has been demonstrated in view of both aspects arising from the transmission problem of power, i.e., the *transformation of power states* as well as the *efficiency of power flow*, that steady-state energetic interactions of a causally reticulated 2-port result in the complex generalization of quadratic forms. By decomposition of the complex power into an active (or in-phase) and a reactive (or quadrature) component the modified Hermitian form, Eqs. (5.11a,b), as the modified Rayleigh quotient, Eqs. (5.27a,b), can be reduced to the *corresponding standard forms* being more available for computing and significant for the insight into stationary flow of energy through a 2-port.

Representing the complex input vector $\underline{\tilde{\boldsymbol{x}}}$ by real matrices and applying normalizing transformation, moreover the Rayleigh quotient can be reduced to the explicit form of an integral notation. Thus, the power states in terms of active and reactive power, *P* and *Q*, just as the corresponding real and imaginary ports of the power transmission factor, A_{P} and A_{Q}, are both specified in the two complementary standard forms, the *Hermitian form*, Eqs. (5.16c), (5.33), and the *skew-Hermitian form*, Eqs. (5.17c), (5.38).

The corresponding standard matrices of the form, $\boldsymbol{U}$ and $\boldsymbol{V}$, contain the intrinsic properties of the energetic system concerning the *transmission problem*, i.e., the scalar transformation of power states which represents the *energy flow characteristic* (power). The corresponding sets of composite parameters, $U_{\ell m}$ and $V_{\ell m}$, being terms of phasor ratio connectives, are significant for the frequency behaviour of the *dissipative*, respectively the *idle component* of stationary energy flow. The real or pure imaginary elements, $U_{\ell m}$, $V_{\ell m}$, are furthermore related to the 2-port parameters (general circuit parameters), $F_{\ell m}$, of the transmission (or chain) matrix $\boldsymbol{F}$ assumed to be known. This coefficient matrix of the signal 4-pole contains the intrinsic properties of the energetic system concerning the *transfer problem*, i.e., the linear transformation of vector spaces which represents the *effort-flow characteristic* by segregation of vector components (individual conjugate variables), [120], [121].

Physical Interpretation of Power Efficiency. For *ideal 2-port elements* the equation of energy continuity holds in its simplified form

$$\sum_{i=1}^{n} P_i = \sum_{i=1}^{2} e_i f_i = 0 \tag{5.39}$$

where the sum is carried over the two ports. This presupposes that there cannot be internal dissipation or storage of energy.

Thus, for an *ideal transformer* being said static or lossless, the power balance requires equality of the active powers at both ends, and the active-power transmission factor A_P must be unity. This coincides with the *optimum efficiency* η_{opt}

$$P_1 = P_2 \quad ; \quad A_{Popt} = \eta_{opt} = P_2 / P_1 = 1 \tag{5.40a},$$

since efficiency η is considered as a conception common for qualifying the energy transport process in engineering devices.

In general, the real process of energy conversion cannot be represented without respecting internal energy losses. Thus, for a *real transformer* an inequality of the active powers at both ends must be ascertained which corresponds to an *active-power transmission factor* A_P, respectively an efficiency, being less than unity

$$P_1 > P_2 \quad ; \quad A_P = \eta = P_2 / P_1 < 1 \tag{5.40b}.$$

However, the classical conception of efficiency η conceives the power flux as a constant. This defines the stationary type of steady state but under the condition of *strict stationarity*. On the contrary, steady state of mechanical systems in vibrations implies the condition of *quasi-stationarity*, outlined in 5. Compared to the efficiency η the active-power transmission factor A_P cannot be considered as a constant except for the special case of ideal behaviour. In general, A_P represents the effectiveness of power flow only at a single frequency, because energy dissipation depends on frequency.

Nevertheless, the real process is not yet accomplished unless internal energy storage is taken into account. As a result, the effectiveness denoted A_P is merely referred to the dissipative component of power flow, whereas for qualifying its idle component the *reactive-power transmission factor* A_Q must be considered. A_Q represents the wattless power flow only at a single frequency. Regarding a spring-mass-damper system the energy storage interaction between the spring and the mass also depends on frequency.

For certain types of real elements, e.g., actual energy transformers, systems modelling takes advantage of embedding the losses and dynamics in the transmission elements adjoining to the terminals of an ideal transformer.

For *real 2-port elements* losses of effort and flow, as well as dynamic effects, such as inertance and capacitance, must be taken into consideration.

Thus, for a *real transmission element* the nonideal behaviour manifests itsself in dissipation, scattering, and storage of energy.

Overall Power States and Efficiency. As already turned out steady state plays a dominant role in providing an overall understanding of the behaviour of energetic systems. Looking upon complex systems as a power transmission represented by a string of 2-ports, the 4-pole cascade connection, 2.2.2 and 2.2.7, may be applied to the transmission problem for determining the power states in the explicit form of a *total energy flow characteristic*.

Relating the complex power at the output port h to the complex power at the input port h-1 of the hth 2-port

$$\underline{S_{\mathrm{Y}}}_{\mathrm{h}} \,/\, \underline{S_{\mathrm{Y}}}_{\mathrm{h-1}} \qquad (5.41),$$

the generalized transmission ratio may be computed by applying the associate connection rule at the ports, Eqs. (2.29), (2.65). Thus, the multiple product of all the individual quotients of generalized quadratic forms results in
the *total (or over-all) transmission factor of the complex power*

$$\underline{A_{\mathrm{S}}}_{1}^{\mathrm{n}} = \frac{\underline{S_{\mathrm{Y}}}_{\mathrm{n}}}{\underline{S_{\mathrm{X}}}_{1}} = \underline{A_{\mathrm{S}}}_{\mathrm{n-1}}^{\mathrm{n}} \cdot \underline{A_{\mathrm{S}}}_{\mathrm{n-2}}^{\mathrm{n-1}} \cdots \underline{A_{\mathrm{S}}}_{\mathrm{h-1}}^{\mathrm{h}} \cdots \underline{A_{\mathrm{S}}}_{1}^{2} = \prod_{\mathrm{h=1}}^{\mathrm{n}} \underline{A_{\mathrm{S}}}_{\mathrm{h-1}}^{\mathrm{h}} = \frac{\underline{\tilde{\boldsymbol{x}}}^{*} \boldsymbol{L}_{\mathrm{n}} \underline{\tilde{\boldsymbol{x}}}}{\underline{\tilde{\boldsymbol{x}}}^{*} \boldsymbol{E} \underline{\tilde{\boldsymbol{x}}}} \qquad (5.42a)$$

being equal to the modified over-all Rayleigh-quotient with the total field of values $[\boldsymbol{L}_{\mathrm{n}}]$ marked out by the maximum values of the composite matrix elements, $L_{\mathrm{n},\ell\mathrm{m}}$.

Though being analogous with successively multiplying individual efficiencies η_{h} together along the transmission line, following the classical conception of *total efficiency* η_{t}

$$\eta_{\mathrm{t}} = \prod_{\mathrm{h=1}}^{\mathrm{n}} \eta_{\mathrm{h}} \qquad (5.42b),$$

a successive transformation of vector spaces is more convenient in favour of a reducible way of calculating with quadratic forms. This preceding step results in the *overall transmission matrix* $\boldsymbol{F}_{\mathrm{n}}$ of the composite signal 4-pole, Eq. (2.65), at first. Then the dyadic product of the conjugate transpose of the row vector $\overline{\boldsymbol{f}_{\mathrm{n}}^{2'}}$, being equal to the column vector $\boldsymbol{f}_{2}^{\mathrm{n}*}$, and the row vector $\boldsymbol{f}_{\mathrm{n}}^{1}$ yields
the *overall coefficient matrix of the form*

$$\boldsymbol{L}_{\mathrm{n}} = \boldsymbol{f}_{2}^{\mathrm{n}*} \boldsymbol{f}_{\mathrm{n}}^{1} \qquad (5.43).$$

The real part of the modified over-all Rayleigh-quotient, Eq. (5.42a), which describes its active (or in-phase) component, may be called the *total active-power transmission factor.* This factor $A_{\mathrm{P}_1}^{\mathrm{n}}$ corresponds to the total efficiency η_{t} , Eq. (5.43a), for the special case of constant power flux.

5.2 Power Transmission through Mechanical Networks (Generalized Impedance)

A *1-port* may be thought of as a *generalized impedance*, some specific examples being resistance, capacitance, and inertance elements, together with all one-ported networks composed of such elements.

Classical dynamics has been primarily concerned with reticular systems and processes which can be effectively looked upon as a *plurality of 1-ports* suitably interconnected through *ideal energy functions*, 2.3 and 2.6.

Network (or circuit) representation generally assumes storage and dissipation of energy at a finite number of localized regions, "lumps", or "points", e.g., mass particles in mechanics, circuit elements in electric network theory. Such substitutes for the actual underlying continuous (or distributed) system have often been useful and productive.

Description of Networks

In electrical circuitry the networks are portrayed graphically by a meshwork of lines, each one of which represents a component or subsystem (circuit element) storing or dissipating, resp. converting energy. By use of a *transformed component model*, e.g., a $G(p)$-plane (transfer-function plane) model, each component may be considered as a *general one-port impedance element.*

The junctions are usually called nodes (or vertices). The line joining any two nodes is called a branch (or path), and any closed path made up of branches is called a loop (or circuit).

The topological properties of such networks are also usable for mechanical networks, 2.3.4.

5.2.1 The Transmission Problem of One-port Networks. *Functional Relationships*

One-port Relations (impedance relations)

The general functional operators ψ, introduced in 5.1, could represent extremely complex fields, processes, or networks, but one may always speak of them as *impedance relations*, so long as but one port of the two interacting systems were involved. Assuming both of the systems being isolated from the environment the restriction to a deterministic system is true capable of exchanging power only at a single port. For such elements the overall behaviour is defined by specifying the functional relationship between effort and flow at the single port of entry.

Dynamic Impedances

Starting out from causality the power transmission must depend upon the product of one input variable, $X(t)$, and one output variable, $Y(t)$. There are two general forms which exist for the nonequilibrium or transient case the *impedance functionals*

$$e(t) = \psi_{\text{ef}} * f(t) \tag{5.44a}$$

where $X(t) = f(t)$; $Y(t) = e(t)$, and

the *admittance functionals*

$$f(t) = \psi_{\text{fe}} * e(t) \tag{5.44b}$$

where $X(t) = e(t)$; $Y(t) = f(t)$.

Though being inverse functionals it must be distinguished between these two forms, since in view of the nonlinear property in general a well defined converse of a given functional relationship may not exist.

Review of Impedance Concept. Historically, the impedance concept grew out of the desire to generalize *Ohm's law* and the notion resistance to make certain elementary constant or *direct current (dc-) concepts* applicable to problems involving periodically varying or *alternating current (ac)*. This need arose in the course of growing generation of electricity and communication networks in the later part of 19th century. In dynamic analysis physical scientists, notable among them Helmholtz, Kelvin, Maxwell, and Heaviside, saw the analogous structure of electrodynamics and classical dynamics. For these early writers the natural analogy of the electrical impedance, relating voltage to current, was the relation of force to velocity. However, largely due to the historical precedence of static elastic analysis in mechanical problems, the principal variables in mechanics were taken to be force and displacement. In this sense, the *mechanical impedance* (respectively, *mobility concept*), finally being fitted to a force-to-displacement ratio by the *dynamic stiffness* (respectively, *concept of dynamic compliance*), may be considered as the attempt to generalize Hooke's law and the notion of a spring constant to problems in dynamics, as already treated in 4.2.1.

In the following the term *impedance* will be loosely used to describe all general effort-flow dynamic functional relationships. This renders unnecessary such distinction between varying definitions.

Linear One-port Impedances

If the general functional operators are assumed to be linear in the form

$$\psi_{\text{ef}} = Z \qquad ; \qquad \psi_{\text{fe}} = Y \tag{5.45a,b}$$

more customary definition of impedance are obtained based upon linear system behaviour.

Conventionally, these linear operators, themselves, have been associated with the concept of a linear impedance, **Z**, and its reciprocal, the linear admittance, ***Y***, since now follows

$$e = \psi_{\text{ef}} * f = \boldsymbol{Z} \bullet f \tag{5.46a}$$

$$f = \psi_{\text{fe}} * e = \boldsymbol{Z} \bullet e \tag{5.46b}$$

thus $\boldsymbol{Z} \bullet \boldsymbol{Y} = 1$; or $\boldsymbol{Y} = 1/\boldsymbol{Z}$.

However, for *linear systems with constant parameters* (time-invariant or stationary models) these operators may be expressed in the linear differential term (of derivatives with respect to time)

$$\boldsymbol{Z} = Z(\mathrm{D}) \quad ; \quad \boldsymbol{Y} = Y(\mathrm{D}) \tag{5.47}$$

where $\mathrm{D} = \mathrm{d}/\mathrm{d}t$.

By use of *transform models* (ω-domain models) and their inherent topological connections between the network elements the functional relationship can be expressed in the explicit form of the *frequency-response (or dynamic) characteristics*, being designated the *overall mobility* and its *related characteristics*, 4.2.3.

5.2.2 The Transmission Problem related to Complex Power. *Phasor Power*

By use of the introduced basic tools the flow of power and energy can be described over the extent of the system. Though it might be possible to consider the transient power flow the main interest turns upon the *steady state of energetic systems*.

This process may be viewed either in the time domain or in the frequency domain. At first, the instantaneous power $S(t)$ at the node (vertex) of the parallel-connected structure, a fundamental configuration of mechanical networks, will be considered. Furthermore, $S(t)$ is spectrally decomposed into $\underline{S}(\omega_i)$ but restricted to the single frequency component at driving frequency ω_f. This explicit form of power state, termed the complex power $\underline{S}(j\omega_f)$, represents the steady-state power relation by an algebraic function. With reference to the conventional description for ac-power systems, the components of real or active power flow $P(\omega_f)$ and wattless or reactive power flow $Q(\omega_f)$ are presented. Sometimes it proves convenient to reticulate further each subsystem configuration into its spectral components $P_k(\omega_f)$, $Q_k(\omega_f)$.

Harmonic Responses (phasor method)

Applying a periodic forcing function of arbitrary shape this periodic excitation will be approximated by a Fourier series, $F(t)$, in real or complex notation, Eqs. (3.60a-c), treated in 3.2.1.

Then, if the system is stationary and linear, the response, $s(t)$, must be that due to the superposition of responses, $s_n(t)$, caused by each simple harmonic of nth order, $F_n(t)$, acting alone. These harmonic responses, $s_n(t)$, can be derived by examining the behaviour of a linear operation excited by a sinusoid at the integral multiple of the fundamental frequency $\omega_n = n\omega_1$.

Specifying a simple harmonic excitation and taking the velocity response and excitation force at the same point in the system the power state is defined by the individual conjugate variables designated
the *actual excitation*

$$F_\omega(t) = \hat{F}\cos(\omega_f t + \varphi_F) = \tilde{F}\sqrt{2}\cos(\omega_f t + \varphi_F) \qquad (5.48a),$$

and the *actual response*

$$\upsilon_\omega(t) = \hat{\upsilon}\cos(\omega_f t + \varphi_\upsilon) = \tilde{\upsilon}\sqrt{2}\cos(\omega_f t + \varphi_\upsilon) \qquad (5.48b).$$

Representing the sinusoidal quantities of the effort-flow couple as phasors, 3.1.3, the power state is rewritten in complex notation designated the *complex excitation*

$$\underline{F_{\omega}}(t) = \hat{F}\,e^{j\varphi_F}\,e^{j\omega_f t} = \hat{\underline{F}}\,e^{j\omega_f t} = \underline{\tilde{F}}\sqrt{2}\,e^{j\omega_f t} \qquad (5.49a),$$

together with the *complex response*

$$\underline{\upsilon_{\omega}}(t) = \hat{\upsilon}\,e^{j\varphi_\upsilon}\,e^{j\omega_f t} = \hat{\underline{\upsilon}}\,e^{j\omega_f t} = \underline{\tilde{\upsilon}}\sqrt{2}\,e^{j\omega_f t} \qquad (5.49b)$$

where the corresponding complex constituents are determined by the characteristic parameters, i.e., by

the *phasor of the excitation force* (force phasor)

$$\hat{\underline{F}} = \hat{F}\,e^{j\varphi_F} = \sqrt{2}\,\tilde{F}\,e^{j\varphi_F} \qquad (5.50a),$$

the *phasor of the velocity response* (velocity phasor)

$$\hat{\underline{\upsilon}} = \hat{\upsilon}\,e^{j\varphi_\upsilon} = \sqrt{2}\,\tilde{\upsilon}\,e^{j\varphi_\upsilon} \qquad (5.50b);$$

in which the real constants are the amplitudes $\hat{F}$; $\hat{\upsilon}$, alternatively the r.m.s. values $\tilde{F} = \sqrt{2}\,\hat{F}/2$; $\tilde{\upsilon} = \sqrt{2}\,\hat{\upsilon}/2$, the initial phases φ_F ; φ_υ, and the *forcing angular frequency* (pulsatance) ω_f .

Review of Phasor Method and Remarks on Restrictions. Historically, the concept of complex excitation and response was introduced by *Ch.P. Steinmetz* for simplifying the response calculation in electric circuitry. This classical approach to sinusoidal steady state of linear systems is known in electrical engineering as the so-called *symbolic method.* Although the complex notation proves useful in dealing with many aspects of system analysis the use of phasors underlies restrictions concerning vectorial operations. In particular, the consideration of complex quantities as complex constituents alone, Eqs. (5.49a,b), and subsequently the taking up of only the real parts

$$F_{\omega}(t) = \mathrm{Re}\left[\underline{F_{\omega}}(t)\right] \quad ; \quad \upsilon_{\omega}(t) = \mathrm{Re}\left[\underline{\upsilon_{\omega}}(t)\right] \qquad (5.51a,b)$$

are steps marking a somewhat artificial procedure which is restricted to linear operations. Presupposing that the susperposition principle holds, the above mentioned simplification, Eqs. (5.51a,b), is thus restricted to operations such as addition, subtraction, differentiation, and integration which are the operations encountered in the solution of linear differential equations with constant coefficients.

In dealing with power and energy relationships for electric circuits, *E.A. Guillemin*, [16], pointed out, that for relationships which involve *quadratic expressions*, a similar procedure is not possible since it is not true that the real part of the square of a complex expression is equal to the square of its real part. Similar remarks on non-commutativity concerning squaring operation apply to the product of two complex quantities. That is to say, it is not correct to assume in connection with Eqs. (5.49a,b), that the product of two sinusoidal quantities expressed in complex notation

$$\mathrm{Re}\left[\underline{F_{\omega}}(t) \cdot \underline{\upsilon_{\omega}}(t)\right] = \mathrm{Re}\left[\hat{\underline{F}}\,\hat{\underline{\upsilon}}\,e^{j2\omega_f t}\right] \qquad (5.52a)$$

may be written in place of the product of two sinusoidal quantities in real notations, Eqs. 5.48a,b)

$$\mathrm{Re}\left[\underline{F_{\omega}}(t)\right] \cdot \mathrm{Re}\left[\underline{\upsilon_{\omega}}(t)\right] = \mathrm{Re}\left[\hat{\underline{F}}\,e^{j\omega_f t}\right] \cdot \mathrm{Re}\left[\hat{\underline{\upsilon}}\,e^{j\omega_f t}\right] \qquad (5.52b).$$

For multiplying sinusoids together the customary complex notation should be replaced by the alternative form given by one-half the sum of the exponential factor $e^{j\omega_f t}$ and its conjugate $e^{-j\omega_f t}$ with respect to *Euler's formula* (identity), Eqs. (3.20a,b). Implying an equivalent instead of only a representation of a sinusoidal quantity this form serves as a *correct substitute* for the actual excitation and response in the following, [16], [120].

In contrast to the costumary complex representation the alternative form, Eqs. (5.48c,d), substitutes for the actual excitation and response, Eqs. (5.48a,b)

$$F_{\omega}(t)=\frac{1}{2}\left[\underline{F_{\omega}}(t)+\underline{F_{\omega}}^{*}(t)\right]=\frac{1}{2}\left[\hat{\underline{F}}\,e^{j\omega_{f}t}+\hat{\underline{F}}^{*}e^{-j\omega_{f}t}\right]=\frac{1}{2}\sqrt{2}\left[\tilde{\underline{F}}\,e^{j\omega_{f}t}+\tilde{\underline{F}}^{*}e^{-j\omega_{f}t}\right] \tag{5.48c}$$

$$\upsilon_{\omega}(t)=\frac{1}{2}\left[\underline{\upsilon_{\omega}}(t)+\underline{\upsilon_{\omega}}^{*}(t)\right]=\frac{1}{2}\left[\hat{\underline{\upsilon}}\,e^{j\omega_{f}t}+\hat{\underline{\upsilon}}^{*}e^{-j\omega_{f}t}\right]=\frac{1}{2}\sqrt{2}\left[\tilde{\underline{\upsilon}}\,e^{j\omega_{f}t}+\tilde{\underline{\upsilon}}^{*}e^{-j\omega_{f}t}\right] \tag{5.48d}$$

where the complex quantities are paired off and determined by the force and velocity phasors, Eqs. (5.50a,b), together with
the corresponding *phasor conjugates*

$$\hat{\underline{F}}^{*}=\hat{F}e^{-j\varphi_{F}}=\sqrt{2}\,\tilde{F}e^{-j\varphi_{F}} \tag{5.50c}$$

$$\hat{\underline{\upsilon}}^{*}=\hat{\upsilon}e^{-j\varphi_{\upsilon}}=\sqrt{2}\,\tilde{\upsilon}e^{-j\varphi_{\upsilon}} \tag{5.50d}$$

Actual Energy Flow (modified phasor method)

In the following the behaviour of energetic systems will be characterized in the time domain describing the power state by the scalar quantity termed
the *instantaneous power* in real notation

$$\begin{aligned} S(t) &= F_{\omega}(t)\upsilon_{\omega}(t)=\hat{F}\hat{\upsilon}\cos(\omega_{f}t+\varphi_{F})\cos(\omega_{f}t+\varphi_{\upsilon}) \\ &= 2\hat{F}\hat{\upsilon}\cos(\omega_{f}t+\varphi_{F})\cos(\omega_{f}t+\varphi_{\upsilon}) \end{aligned} \tag{5.53a}$$

thus defining the product of the instantaneous values of the sinusoidal quantities force and velocity at a port.

Substituting the complex constituents and their conjugates for the sinusoidal quantities of the effort-flow product the power state is rewritten

$$\begin{aligned} S(t) &= \frac{1}{2}\left[\tilde{\underline{F}}\,e^{j\omega_{f}t}+\tilde{\underline{F}}^{*}e^{-j\omega_{f}t}\right]\left[\tilde{\underline{\upsilon}}\,e^{j\omega_{f}t}+\tilde{\underline{\upsilon}}^{*}e^{-j\omega_{f}t}\right] \\ &= \left[\frac{\tilde{\underline{F}}\tilde{\underline{\upsilon}}^{*}+\tilde{\underline{F}}^{*}\tilde{\underline{\upsilon}}}{2}+\frac{\tilde{\underline{F}}\tilde{\underline{\upsilon}}\,e^{2j\omega_{f}t}+\tilde{\underline{F}}^{*}\tilde{\underline{\upsilon}}^{*}e^{-j2\omega_{f}t}}{2}\right] \end{aligned} \tag{5.53b}$$

Multiplying the constituents together a term of partial products is obtained. Those products being complex numbers, shortly termed *phasor connectives*, are significant for the time behaviour of stationary energy flow through a linear one-port impedance.

Being rearranged in complex pairs and their conjugates it becomes obvious that this composite quadratic term including a constant and a time-varying component of power is a correct substitute for the instantaneous power. Considering now the pairs of phasor connectives as the real parts of a complex quantity, which may be defined in point as the *phasor of the instantaneous power* $\underline{S}(t)$, the actual scalar quantity $S(t)$, Eq. (5.53a), can be alternatively expressed in terms of the real part of the complex notation

$$S(t) = \mathrm{Re}\left[\underline{\tilde{F}}\,\underline{\tilde{\upsilon}}^* + \underline{\tilde{F}}\,\underline{\tilde{\upsilon}}\,\mathrm{e}^{\mathrm{j}2\omega_f t}\right] = \mathrm{Re}\left[\underline{S}(t)\right] \qquad (5.53c).$$

Thus, the concepts of complex quantities and *phasor representation*, 3.1.3., will be *reutilized* as a rule of operation in *power transactions*, though not for abridging the multiplication of sinusoids, but the more applied to the correct substitute for multiplying operations which yields

- the geometric interpretation of the effort-flow product by use of *vector representation* in the complex plane (*phasor of instantaneous power*);
- the derivation of the *energy flow characteristic* in the frequency domain by defining the complex power (*phasor power*).

Vector Representation of Power (instantaneous power). Complex quantities representing the actual quantity of a simple harmonic as well as of harmonics are figured in the complex plane by rotating phasors, likewise by counterrotating phasor pairs, localized at the origin as shown in Fig. 3.3, 3.1.3, as in Fig. 3.6, 3.2.1.

Contrary to sinusoids and harmonics both being symmetrical quantities the scalar quantity of *instantaneous power* differs from an alternating dependence on time thus requiring some modifications in vector representation.

Viewing the composite quadratic form of power, Eqs. (5.53a-c), the actual energy flow appears as a pulsating quantity associated to a periodic quantity of non-zero mean value. For the special case of the underlying energy flow the *phasor* suited to represent the pulsating property of energy is a *total vector* resulting from *component phasors* in two equivalent forms of localized vectors. Both forms of decomposition base on the linear operation of vector addition including *constant* (resting) and *time-varying* (rotating) *phasors*, the latter of which being in general localized outside the origin and rotating at double the forcing angular frequency. Thus, taking the corresponding projections on the real axis as customary the geometric interpretation holds for the actual energy flow in respect of its pertinent direct component as of its double-frequency alternating component.

Performing the *decomposition of power* in view of the *two forms* being *significant* of linear *energetic systems*, e.g., electric circuits (ac-power systems) as well as (linearized) mechanical systems in vibrations (vibration machines), it becomes obvious that the superposition principle remains valid for summing up partial products of the composite quadratic term of power. The prior condition of vector addition is preserved, thus the vector representation of power is true, [108].

Spectral Decomposition of Power into Average and Alternating Components

Taking up the total phasor of the instantaneous power $\underline{S}(t)$, Eq. (5.53c), the decomposition of this complex quantity into the phasor components

$$\begin{aligned}\underline{S}(t) &= \underline{S} + \underline{S_\omega}(t)\\ &= \underline{\tilde{F}}\,\underline{\tilde{\upsilon}}^* + \underline{\tilde{F}}\,\underline{\tilde{\upsilon}}\,\mathrm{e}^{\mathrm{j}2\omega_f t}\end{aligned} \qquad (5.54a)$$

corresponds to a decomposition of the instantaneous power expressed in terms of the real parts

$$\begin{aligned} S(t) &= S_{-} + S_{\omega}(t) \\ &= \tilde{F}\tilde{\upsilon}\,\mathrm{Re}\left[e^{j\varphi_1}\right] + \tilde{F}\tilde{\upsilon}\,\mathrm{Re}\left[e^{j(2\omega_f t+\varphi_\upsilon+\varphi_F)}\right] \\ &= \underbrace{\underbrace{\tilde{F}\tilde{\upsilon}}_{=S}\cos\varphi_1}_{=S_{-}=P} + \underbrace{\underbrace{\tilde{F}\tilde{\upsilon}}_{=S}\cos(2\omega_f+\varphi_\upsilon+\varphi_F)}_{=S_{\omega}(t)} \end{aligned} \qquad (5.54b).$$

Accordingly, the pulsating quantity of actual power flow is to be interpreted as the resultant of two power components

- the *average power* being the direct component S_{-} of the instantaneous power $S(t)$;
- the *wattless power* being the double-frequency alternating component $S_{\omega}(t)$ of the instantaneous power $S(t)$.

The direct component of Eq. (5.54b) is defined as the *mean value of the instantaneous power* $\overline{S(t)}$

$$S_{-} = P = \frac{1}{T}\int_{t_1}^{t_1+T} F_{\omega}(t)\upsilon_{\omega}(t)\,dt = \overline{F_{\omega}(t)\cdot\upsilon_{\omega}(t)} = \overline{S(t)} \qquad (5.55a)$$

taken over one period of the pulsating quantity $S(t)$ at a port, which is identical to the active power P.

Average Power S_{-}. The power component S_{-}, Eq. (5.54b), corresponds to the *dissipative flow of energy* which is absorbed by the real one-ported network owing to *losses of effort and flow*.

The consumption of energy being concentrated on the resistance element (damper) presupposes an *effective energy transaction* across the transmission link of the network being considered as a generalized impedance. The energy flow entering the port must be an equivalent on the average to the dissipative energy flow, and is delivered by a power supply (mechanical source) to the network (mechanical circuit).

Wattless Power $S_{\omega}(t)$. This designation applied to the power component $S_{\omega}(t)$, Eq. (5.54b), is appropriate for turning out the *idle flow of energy* which is swapped back and forth by the real one-ported network on account of *dynamic effects*.

The alternating component of Eq. (5.54b) thus reflects the total swappage of energy including the *energy storage interaction* between the capacitance and the inertance element (spring and mass) as well as the *compensative energy transaction* across the boundaries of the linear one-port impedance (mechanical source and circuit). The energy continuously interchanged with the power supply originates in the dynamic effects of energy storage (mechanical circuit store) being alternately governed by the capacitance or the inertance. Thus, periodic fluctuations in energy storage capacity occur in the network.

The peak value $\hat{S}$ of the alternating component $S_{\omega}(t)$, called the *apparent power S*

$$\hat{S} = \tilde{F}\tilde{\upsilon} = \left(\hat{F}\hat{\upsilon}/2\right) = S \qquad (5.55b),$$

is defined as the product of the r.m.s. values of the sinusoidal quantities force $\tilde{F}$ and velocity $\tilde{\upsilon}$ at a port.

The extent to which the circuit absorbs power from the source is characteristically indicated by $\cos\varphi_1$, Eqs. (5.54b), (5.55). This characteristic magnitude, termed the *power factor* λ

$$\lambda = P/S = \cos\varphi_1 \qquad (5.56a),$$

is defined as the ratio of the active power *P* to the apparent power *S* at a port.

The actual energy being a pulsating quantity flows from the circuit back into the source as well as from the source into the circuit, in general, energy flows in both directions.

Assuming a resistance element as existent, then more energy on the average flows into the network than is returned to the source. That is to say, for a passive network the average power *P* is always less or at most equal to the apparent power *S*:

$$P \leq S \quad ; \quad \lambda \leq 1 \qquad (5.56b).$$

The value of the power factor λ is unity for $\varphi_1 = 0$ and zero for $\varphi_1 = \pm\pi/2$. The first case occurs if the impedance contains only a resistance (dc-case), e.g., at resonance, whereas the latter case results from the absence of any resistance (lossless circuit).

Vector Representation of Power Decomposition (polar form). The representation of the power flow in the complex plane bases on the complex notation of instantaneous power, $\underline{S}(t)$, in the decomposition of the one form referred to Eq. (5.54a).

The summing up of power components of the pulsating scalar quantity $S(t)$, Eq. (5.54b), corresponds to the vector addition of a *constant (resting) phasor*, and a *time-varying phasor* which rotates at double the frequency upon the tip of the resting phasor. The latter one denoting the *complex power* $\underline{S}$, thus termed *phasor power*, represents the direct component $S_{-} = P$, whereas the rotating phasor $\underline{S}_{t\omega}$ localized outside the origin represents the superimposed alternating component $S_{t\omega}$.

Taking the projections on the real axis, both of the actual power components, S_{-} and $S_{t\omega}$, are then reproduced corresponding to Eq. (5.54b), thus the *actual energy flow* S_t can be completely traced out in real rectangular coordinates, [120].

Spectral Decomposition of Power into Active and Reactive Components

An other, most convenient decomposition of power follows from eliminating one of the two initial phases assigned to effort and flow with respect to the second term of the actual power flow $S(t)$, Eq. (5.54b). Related to the phase of force φ_f, or to that one of velocity φ_υ, this will be carried out with more efficiency by taking up the *phasor of the instantaneous power* $\underline{S}(t)$ and applying operations of conversion to the complex representation of actual power, Eq. (5.54a).

Adjusting to Rectangular Form (elimination of phase of force). The other decomposition of the complex quantity $\underline{S}(t)$ follows from adjusting the phasor components to only in-phase P and quadratic terms Q. For this reason an unit phasor is inserted as an adjusting multiplier to the effect that the phase of force φ_f is eliminated. Thus, the rotating phasor component of Eq. (5.54a) may be reduced to

the form of a *scalar triple product*

$$\begin{aligned}\underline{S}(t) &= \underline{P}(t) + \underline{Q}_\omega(t) = \underline{S}_- + \underline{P}_\omega(t) + \underline{Q}_\omega(t)\\ &= \underline{\tilde{F}}\,\underline{\tilde{\upsilon}}^* + \underline{\tilde{F}}\,\underline{\tilde{\upsilon}}^* \frac{\underline{\tilde{\upsilon}}\,\underline{\tilde{\upsilon}}}{\underline{\tilde{\upsilon}}^*\underline{\tilde{\upsilon}}} e^{j2\omega_f t}\end{aligned} \qquad (5.57a).$$

This dot product contains the resting phasor component $\underline{\tilde{F}}\,\underline{\tilde{\upsilon}}^*$ as the first factor. The adjusting *unit phasor*

$$\frac{\underline{\tilde{\upsilon}}\,\underline{\tilde{\upsilon}}}{\underline{\tilde{\upsilon}}^*\underline{\tilde{\upsilon}}} = \frac{\underline{\tilde{\upsilon}}\,\underline{\tilde{\upsilon}}}{|\underline{\tilde{\upsilon}}|^2} = e^{j2\varphi_\upsilon} \qquad (5.58a)$$

is a constant phasor which only involves a rotation through twice the phase of velocity $2\varphi_\upsilon$ in the counterclockwise direction.

This rotation is done (once) about the origin of the phasor which represents the first factor of the triple product.

Specifying this first factor by the rectangular form of the resting phasor, where P and Q are the real and imaginary parts, the complex quantity $\underline{S}(t)$ will be rewritten

$$\underline{S}(t) = (P + jQ) + (P + jQ)e^{j2(\omega_f t + \varphi_\upsilon)} \qquad (5.57b),$$

and the actual scalar quantity $S(t)$ can be expressed in terms of the real part of the complex notation

$$\begin{aligned}S(t) &= P(t) + Q_\omega(t) = P_- + P_\omega(t) + Q_\omega(t)\\ &= P + \mathrm{Re}\left[(P + jQ)e^{j2(\omega_f t + \varphi_\upsilon)}\right]\\ &= \tilde{F}\tilde{\upsilon}\,\mathrm{Re}\left[e^{j\varphi_1}\right] + \tilde{F}\tilde{\upsilon}\,\mathrm{Re}\left[e^{j\varphi_1}\, e^{j2(\omega_f t + \varphi_\upsilon)}\right]\end{aligned} \qquad (5.57c).$$

Applying the rule of multiplying complex numbers to the second term written in trigonometric form

$$\mathrm{Re}[\underline{z}\,\underline{r}] = (\underline{z}\,\underline{r})' = z'r' - z''r'' \qquad (5.59)$$

the power state is described in an alternative form by the instantaneous power in real notation

$$S(t) = \tilde{F}\,\tilde{\upsilon}\cos\varphi_1 + \tilde{F}\,\tilde{\upsilon}\left[\cos\varphi_1 \cos 2(\omega_f t + \varphi_\upsilon) - \sin\varphi_1 \sin 2(\omega_f t + \varphi_\upsilon)\right]$$

$$= \underbrace{\underbrace{\tilde{F}\,\tilde{\upsilon}\cos\varphi_1}_{=P_- = P}[1 + \underbrace{\cos 2(\omega_f t + \varphi_\upsilon)}_{=P_\omega(t)}]}_{=P(t)} \underbrace{- \underbrace{\tilde{F}\,\tilde{\upsilon}\sin\varphi_1}_{=Q} \sin 2(\omega_f t + \varphi_\upsilon)}_{=Q_\omega(t)} \qquad (5.57\text{db.})$$

The operation on complex numbers results in the known addition formula in trigonometry which involves the geometric interpretation of vector representation in the complex plane.

Accordingly, the pulsating quantity of actual power flow is to be interpreted as the resultant of two power components

- the *actual active power* turning out to be by itsself a periodic quantity $P(t)$ of non-zero mean value;
- the *actual reactive power* being a double-frequency (symmetrical) alternating component $Q_\omega(t)$.

The direct component of the first power component of Eq. (5.57d) is defined as the *mean value of the actual active power* $\overline{P(t)}$

$$P_- = \overline{P(t)} = \tilde{F}\,\tilde{\upsilon}\cos\varphi_1 = \left(\hat{F}\hat{\upsilon}/2\right)\cos\varphi_1 = S_- = P \qquad (5.60\text{a})$$

taken over one period of the pulsating quantity $P(t)$ at a port, which is identical to the average power S_-, Eqs. (5.54b), (5.55a), thus being equivalent to the active power P.

Active Power P. The power characteristic P of Eq. (5.57d) equals the mean value P_-, Eq. (5.60a), thus corresponding to the *dissipative flow of energy* owing to *losses of effort and flow*.

The alternating component of the first power component of Eq. (5.57d), called the *alternating active power* $P_\omega(t)$, is associated with the *idle flow of energy* on account of *dynamic effects*.

Contrary to the wattless power $S_\omega(t)$, Eq. (5.54b), only the *energy storage interaction* between the capacitance and the inertance element (spring and mass) is included in $P_\omega(t)$.

The *peak value* $\hat{P}$ of the alternating active power $P_\omega(t)$

$$\hat{P} = \tilde{F}\tilde{\upsilon}\cos\varphi_1 = P_- = P \qquad (5.60\text{b})$$

is equal to the direct component P_-, Eq. (5.60a), thus being equivalent to the active power P.

Reactive Power Q. The power characteristic Q of Eq. (5.57d) equals the peak value $\hat{Q}$ of the alternating component $Q_\omega(t)$

$$\hat{Q} = \tilde{F}\tilde{\upsilon}\sin\varphi_1 = Q \qquad (5.61).$$

The second power component of Eq. (5.57d), called the *alternating reactive power* $Q_\omega(t)$, is likewise associated with the *idle flow of energy* on account of *dynamic effects.*

Contrary to the wattless power $S_\omega(t)$, Eq. (5.54b), only the *compensative energy transaction* between the power supply and energy storage (mechanical source and circuit) is included in $Q_\omega(t)$.

Vector Representation of Power Decomposition (rectangular form). The representation of the power flow in the complex plane bases on the complex notation of instantaneous power, $\underline{S}(t)$, Eq. (5.54a), however in the decomposition of the other form referred to Eqs. (5.57a,b).

The summing up of actual power components being adjusted to only in-phase P and quadrature terms Q, Eq. (5.57c), corresponds to the vector addition of phasor components, Fig. 5.2.

The *constant (resting) phasor components,* P and jQ, form the resultant constant vector, and the *time-varying phasor components* result in a phasor which rotates at double the frequency upon the tip of the resting vector. The latter one denoting the *complex power* $\underline{S}$, thus termed *phasor power,* represents the direct component $P_{-} = P$ of the actual active power P_t, whereas the rotating phasor components $\underline{P}_{\mathrm{t}\omega}$ and $\underline{Q}_{\mathrm{t}\omega}$, localized outside the origin, represent the alternating components as $P_{\mathrm{t}\omega}$ of the actual active power P_t, and $Q_{\mathrm{t}\omega}$ denoting the actual reactive power.

The families of resultant and orthogonal component phasors, $\underline{S}$, P, jQ, as $\underline{S}_{\mathrm{t}\omega}$, $\underline{P}_{\mathrm{t}\omega}$, $\underline{Q}_{\mathrm{t}\omega}$, span two right *triangles of power* being plane geometric figures distinguished by *congruence*, Fig. 5.2a, shaded areas.

The triangle outside the origin rotates about the top vertex of the triangle fixed by one of the base vertices in the origin. At the instant specified by $t = -\varphi_\upsilon / \omega_\mathrm{f}$ the shaded geometric figures are in parallel position, because there is no difference in direction between the resultant constant vector, defined as the *phasor power* $\underline{S}$, and the resultant *time-varying phasor* $\underline{S}_{\mathrm{t}\omega}$. Only at this definite point of time the dot-dashed phasor of the instantaneous power $\underline{S}(t)$ just equals the double phasor power $\underline{S}$, thus $\underline{S}(t) = 2\underline{S}$, as traced out.

Finally, by locating the quadrature phasor component $\underline{Q}_{\mathrm{t}\omega}$ on the origin the decomposition of power is fitted to be performed as customary by taking the projections of three position vectors on the real axis. Those vectors located in distinct vertices are the phasor power $\underline{S}$, and the *two rotating phasors* $\underline{P}_{\mathrm{t}\omega}$, $\underline{Q}_{\mathrm{t}\omega}$. The actual power components implying the pulsating active term P_t, as well as the alternating reactive term $Q_{\mathrm{t}\omega}$, are then reproduced taking the projections on the real axis in respect of Eq. (5.57d), thus the *actual energy flow* S_t can be completely traced out in real rectangular coordinates, Fig. 5.2b, dot-dashed curve.

Adjusting to Rectangular Form (elimination of phase of velocity). A complementary form of decomposition into active and reactive components is also convenient and efficiently performed with reference to the complex representation of

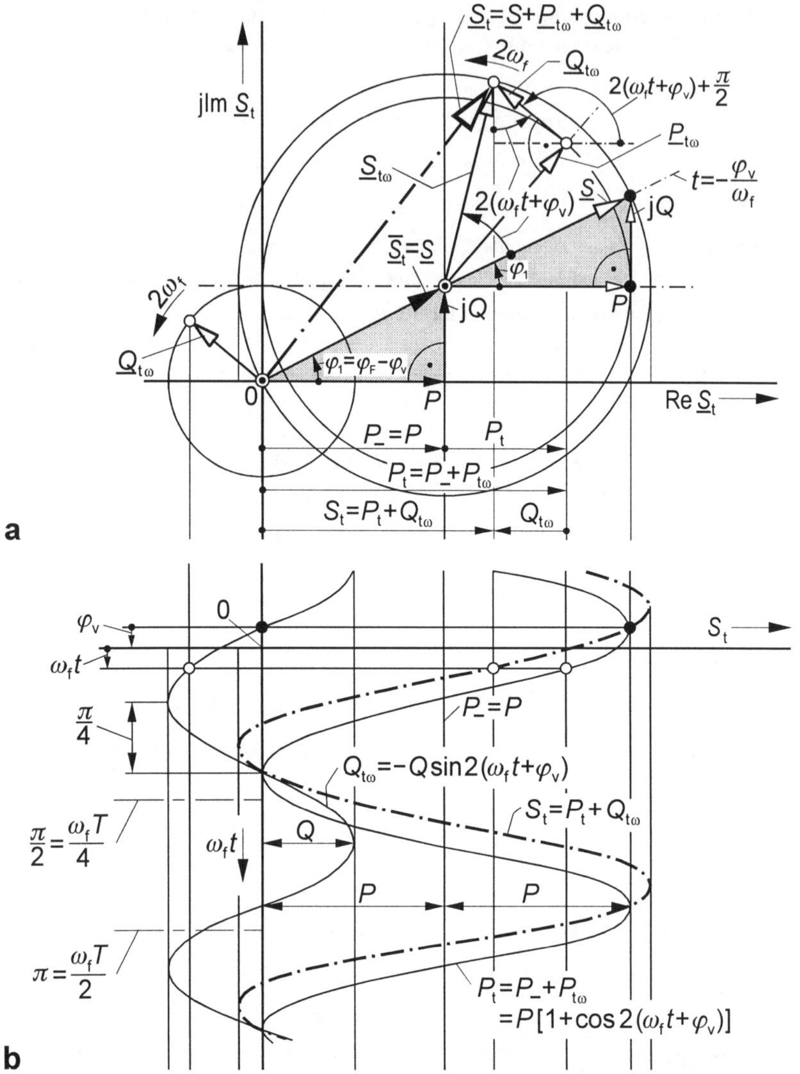

Fig. 5.2. Complex representation of instantaneous power implying the decomposition into active and reactive components and their projections on the real axis. **a** Phasor diagram of power; **b** time history of power

actual power, Eq. (5.54a). Again an unit phasor is inserted as an adjusting multiplier but to the effect that the phase of velocity φ_υ is eliminated.

Thus, the rotating phasor component of Eq. (5.54a) may be reduced to the form of a *scalar triple product*

$$\underline{S}(t)=\underline{\tilde{F}}\,\underline{\tilde{\upsilon}}^{*}+\underline{\tilde{F}}^{*}\,\underline{\tilde{\upsilon}}\,\frac{\underline{\tilde{F}}\,\underline{\tilde{F}}}{\underline{\tilde{F}}^{*}\,\underline{\tilde{F}}}\,\mathrm{e}^{\mathrm{j}2\omega_f t} \qquad (5.62\mathrm{a}),$$

which contains the conjugate of the resting phasor component $\underline{\tilde{F}}\underline{\tilde{\upsilon}}^*$ as the first factor. The adjusting *unit phasor*

$$\frac{\underline{\tilde{F}}\,\underline{\tilde{F}}}{\underline{\tilde{F}}^*\underline{\tilde{F}}} = \frac{\underline{\tilde{F}}\,\underline{\tilde{F}}}{\left|\underline{\tilde{F}}\right|^2} = \mathrm{e}^{\mathrm{j}2\varphi_\mathrm{F}} \tag{5.58b}$$

is a constant phasor involving a rotation through twice the phase of force $2\varphi_\mathrm{F}$ in the counterclockwise direction.

Following Eq. (5.57b) the complex quantity $\underline{S}(t)$ will be rewritten

$$\underline{S}(t) = (P + \mathrm{j}Q) + (P - \mathrm{j}Q)\mathrm{e}^{\mathrm{j}2(\omega_\mathrm{f}t+\varphi_\mathrm{F})} \tag{5.62b),$$

and the actual scalar quantity S_t can be expressed in terms of the real part of the complex notation

$$\begin{aligned} S_\mathrm{t} &= P + \mathrm{Re}(P - \mathrm{j}Q)\mathrm{e}^{\mathrm{j}2(\omega_\mathrm{f}t+\varphi_\mathrm{F})} \\ &= \tilde{F}\tilde{\upsilon}\,\mathrm{Re}\left[\mathrm{e}^{\mathrm{j}\varphi_1}\right] + \tilde{F}\tilde{\upsilon}\,\mathrm{Re}\left[\mathrm{e}^{-\mathrm{j}\varphi_1}\,\mathrm{e}^{\mathrm{j}2(\omega_\mathrm{f}t+\varphi_\mathrm{F})}\right] \qquad (5.62\mathrm{c}). \\ &= \underbrace{\underbrace{\tilde{F}\tilde{\upsilon}\cos\varphi_1}_{=P_-=P}\,[1 + \underbrace{\cos 2(\omega_\mathrm{f}t + \varphi_\mathrm{F})}_{=P_\omega(t)}\,]}_{=P(t)} + \underbrace{\underbrace{\tilde{F}\tilde{\upsilon}\sin\varphi_1}_{=Q}\sin 2(\omega_\mathrm{f}t + \varphi_\mathrm{F})}_{=Q_\omega(t)} \end{aligned}$$

The 1-port Power Transmission in Summary

Concept of Phasor Power. A most considerable result of spectral decomposition of power is the specification of actual energy flow by use of the phasor method, in particular by the reduction of the time-varying complex quantity $\underline{S}(t)$ to a complex parameter given by the phasor product $\underline{\tilde{F}}\underline{\tilde{\upsilon}}^*$ being invariable with changing the form of decomposition of power. Whichever of the spectral components are favoured, representing either the average and wattless power, Eq. (5.54a), or the active and reactive power, Eqs. (5.57a), (5.62a), in any case the first term of the complex quantity $\underline{S}(t)$ coincides with the phasor product mentioned above. This phasor connective, identified as the *complex power* $\underline{S}$, constitutes the concept of phasor power as pointed out in the following.

Vector Representation of Complex Power (phasor power). Returning to the vector representation of sinusoids modified to power relations, 5.2.2, the scalar quantity $S(t)$ in the steady state (instantaneous power) can be gathered solely from the one complex parameter, the complex power $\underline{S}$, Eq. (5.63a), as demonstrated by use of modified phasor method.

For characterizing the sinusoidal steady-state behaviour not the pulsating quantity of actual power flow is in one's particular interest. The steady-state scalar quantities being significant of actual power are called the characteristic parameters of power. Those parameters, graphically interpreted as resultant and orthogonal component phasors, make the superposable sides of two right *triangles of power*,

Fig. 5.2a, shaded areas. Hence, owing to congruence of the power triangles in diagram representation solely the *triangle fixed in the origin* must be delineated. It is illustrated by the shaded plane geometric figures that for both of the decomposition forms of power the *significant power parameters* are associated with the complex power $\underline{S}$, likewise with its conjugate $\underline{S}^*$.

Thus, in diagram representation the steady-state power relation is figured in terms of constant (resting) phasors, corresponding to the power variables, which form the resultant vector defined as
the *phasor power* $\underline{S}(\mathrm{j}\omega_\mathrm{f})$

$$\begin{aligned}\underline{S}(\mathrm{j}\omega_\mathrm{f}) &= \underline{\widetilde{F}}\,\underline{\widetilde{\upsilon}}^* = \frac{1}{2}\underline{\hat{F}}\,\underline{\hat{\upsilon}}^* \\ &= \widetilde{F}\,\mathrm{e}^{\mathrm{j}\varphi_\mathrm{F}}\,\widetilde{\upsilon}^{-\mathrm{j}\varphi_\upsilon} = \widetilde{F}\,\widetilde{\upsilon}\,\mathrm{e}^{\mathrm{j}(\varphi_\mathrm{F}-\varphi_\upsilon)} = \widetilde{F}\,\widetilde{\upsilon}\,\mathrm{e}^{\mathrm{j}\varphi_1} \\ &= \widetilde{F}\,\widetilde{\upsilon}\cos\varphi_1 + \mathrm{j}\widetilde{F}\,\widetilde{\upsilon}\sin\varphi_1 = P + \mathrm{j}Q\end{aligned} \tag{5.63a}$$

for a 1-port component.

To each forcing (angular) frequency ω_f corresponds a single (inner or dot) *product* of the *(r.m.s.) force phasor* $\underline{\widetilde{F}}$, Eq. (5.50a), and the *conjugate of the (r.m.s.) velocity phasor* $\underline{\widetilde{\upsilon}}^*$, Eq. (5.50d), being assigned to a particular value of the *complex power*: $\underline{S}(\mathrm{j}\omega)$.

It should be pointed out that the product of force phasor and velocity phasor depends on the angular frequency ω, thus, the resultant vector changes both, its length and its direction, as ω varies from 0 to ∞. For each frequency variable assigned to a particular forcing frequency $\omega = \omega_\mathrm{f}$ the corresponding phasor power may be plotted as a single point in the complex plane, called, in this case, the *complex power plane*, Fig. 5.5.

Thus, a transformation of the time-dependent scalar quantity $S(t)$ into the frequency domain is performed.

Review of Phasor Power Concept. Historically, *E.A. Guillemin*, [16], thought it logical to regard the resultant vector of average and reactive power component as a complex quantity denoted "vector power" which proves useful in electric circuits to form energy functions for the sinusoidal steady state.

The term "phasor power" being proposed by electric power engineers has been combined with a concept which simplifies to define the operating condition of an electric power system. This has been outlined by *W.A. Blackwell*, [52]. Applying the concept of phasor power to more complex networks, e.g., unbalanced polyphase alternating-current systems, *V.N. Nedelcu*, [108], defined the energy function "complex power" as

the *mean value of the complex quantity* $\overline{\underline{S}(t)}$

$$\begin{aligned}\underline{S}_{-} = \underline{S}(\mathrm{j}\omega_\mathrm{f}) &= \frac{1}{T}\int_{t_1}^{t_1+T}\frac{1}{2}\underline{F_\omega}(t)\left[\underline{\upsilon_\omega}(t) + \underline{\upsilon_\omega}^*(t)\right]\mathrm{d}t \\ &= \overline{\frac{1}{2}\underline{F_\omega}(t)\left[\underline{\upsilon_\omega}(t) + \underline{\upsilon_\omega}^*(t)\right]} = \overline{\underline{S}(t)}\end{aligned} \tag{5.64a),}$$

where the complex quantity of power, $\underline{S}(t)$, is expressed by complex constituents and their conjugates for sinusoidal quantities, Eq. (5.53b), in the suited (shortened) form

$$\begin{aligned}\underline{S}_{\mathrm{t}} &= \frac{1}{2}\underline{F}_{\omega}(t)\left[\underline{\upsilon}_{\omega}(t)+\underline{\upsilon}_{\omega}^{*}(t)\right] = \underline{\tilde{F}}\,\mathrm{e}^{\mathrm{j}\omega_{\mathrm{f}}t}\left[\underline{\tilde{\upsilon}}\,\mathrm{e}^{\mathrm{j}\omega_{\mathrm{f}}t}+\underline{\tilde{\upsilon}}^{*}\mathrm{e}^{-\mathrm{j}\omega_{\mathrm{f}}t}\right] \\ \underline{S}_{\mathrm{t}}^{*} &= \frac{1}{2}\underline{F}_{\omega}^{*}(t)\left[\underline{\upsilon}_{\omega}(t)+\underline{\upsilon}_{\omega}^{*}(t)\right] = \underline{\tilde{F}}^{*}\mathrm{e}^{-\mathrm{j}\omega_{\mathrm{f}}t}\left[\underline{\tilde{\upsilon}}\,\mathrm{e}^{\mathrm{j}\omega_{\mathrm{f}}t}+\underline{\tilde{\upsilon}}^{*}\mathrm{e}^{-\mathrm{j}\omega_{\mathrm{f}}t}\right]\end{aligned} \qquad (5.65).$$

For the power product in complex notation sometimes the conjugate form is preferred resulting in

the *phasor power conjugate* $\underline{S}^{*}(\mathrm{j}\omega_{\mathrm{f}})$

$$\begin{aligned}\underline{S}^{*}(\mathrm{j}\omega_{\mathrm{f}}) &= \underline{\tilde{F}}^{*}\,\underline{\tilde{\upsilon}} = \frac{1}{2}\underline{\hat{F}}^{*}\,\underline{\hat{\upsilon}} \\ &= \tilde{F}\mathrm{e}^{-\mathrm{j}\varphi_{\mathrm{F}}}\tilde{\upsilon}\,\mathrm{e}^{\mathrm{j}\varphi_{\upsilon}} = \tilde{F}\tilde{\upsilon}\,\mathrm{e}^{-\mathrm{j}(\varphi_{\mathrm{F}}-\varphi_{\upsilon})} = \tilde{F}\tilde{\upsilon}\,\mathrm{e}^{-\mathrm{j}\varphi_{1}} \\ &= \tilde{F}\tilde{\upsilon}\cos\varphi_{1} - \mathrm{j}\tilde{F}\tilde{\upsilon}\sin\varphi_{1} = P-\mathrm{j}Q\end{aligned} \qquad (5.63b)$$

for a 1-port component to be defined, respectively, as

the *mean value of the conjugate of the complex quantity* $\overline{\underline{S}^{*}(t)}$

$$\begin{aligned}\underline{S}_{-}^{*} &= \underline{S}^{*}(\mathrm{j}\omega_{\mathrm{f}}) = \frac{1}{T}\int_{t_1}^{t_1+T}\frac{1}{2}\underline{F}_{\omega}^{*}(t)\left[\underline{\upsilon}_{\omega}(t)+\underline{\upsilon}_{\omega}^{*}(t)\right]\mathrm{d}t \\ &= \overline{\frac{1}{2}\underline{F}_{\omega}^{*}(t)\left[\underline{\upsilon}_{\omega}(t)+\underline{\upsilon}_{\omega}^{*}(t)\right]} = \overline{\underline{S}^{*}(t)}\end{aligned} \qquad (5.64b).$$

In vibrations power transactions have been treated to a certain extent though belonging to a subsidiary matter as outlined in 5. However, for applying the vibration method to *working processes* the attention of mechanical engineers turned to the pulsating quantity of power flow regarding *economical aspects of power supply* as well as to the alternating quantity of force transmission respecting the *design of structural members* for vibration machines.

C.L.P. Fleck, [110], improved the directly drived vibration generator system by additionally mounted springs tuning them to resonance to the effect that both the reactive component in power transaction as the force in the driving mechanism of the crank connecting rod-type generator have been minimized.

Various mechanical vibration generator systems, or mechanical vibrators for short, have been developed the main types of which are classified as *direct drive* and *reactive-type vibrators*. The latter class generates an excitation force by means of a rotating unbalance or a reciprocating mass.

H. Schieferstein, [111], has been engaged in a comparative analysis of direct-drive and reaction type vibrators in view of mechanical energetic systems getting benefit from the resonant effect. Thus, the effective power or the economy of operation became a subject to investigations.

E. Lehr, [63], provided a comprehensive survey of mechanical, likewise of electric vibrators. Varying fundamental configurations of structures and applying harmonic excitation to different points on a structure the steady state behaviour of *basic vibration systems* is analysed. To prove their usefulness for system design several *guiding performance criteria* have been specified. The balancing of forces is graphically interpreted by use of vector representation as customary in electric engineering. Yet, the alternating components of power interaction are illustrated as the sides of a *rotating right triangle*, thus performing a graphical interpretation of *rectangular power decomposition* basing on *real diagram representation*.

Apart from the requirements of vibration machines the development of mechanical vibrators has progressed due to exacting specifications arising from *vibration tests* as pointed out next to E. Lehr, [63], by *W. Späth*, [112], and *S. Berg*, [113].

In recent years, different wave forms of vibrations have been generated for environmental testing procedures, [114], likewise for methods of fatigue testing, [115], [116]. As regards *vibration and shock generating equipments*, [117], [118], as to *fatigue testing machines*, [119], mechanical vibrators have been widely replaced by electrodynamic and hydraulic vibration generator systems. Their advantages in the main equipment characteristics concerning waveform types and waveform distortion, as parameter controlling and accuracy controlling, brought into focus vibration control methods. Since system approaches peculiar to controls, respective to vibrations, are combined the basis is formed to understand the transformation and flow of energy for *mixed physical domains* (mixed-domain systems). Contrary to measuring systems or system analysis the energetic system design requires a representation of driving power flow in the frequency domain.

D. Findeisen, [120], [121], proposed by the two attempts of energetic system approaches, 5.1 and 5.2, to use the efficient relationship of *complex power* for expressing the energy flow characteristic in terms of an algebraic function in the frequency domain (ω-domain). In contrast to the customary procedure still prevalent in vibrations treating the scalar quantity of actual power only in real notation the realized explicit form of power state bases on the complex notation. To gain acceptance in vibrations the abstract term "complex power" has been graphically interpreted by *vector representation modified to power relations*, 5.2.2. As demonstrated, the significant power parameters are uniquely associated with a resultant vector called *phasor power*. This graphic form of power state hitherto reserved to electric power engineering also may have become accessible to vibrations. By this, vibration theory gains an effective tool of system design in the steady state by *combining* the concepts of *dynamic compliance* and *phasor power*. Thus, by completing functional relationships the energetic system approach is applied to basic vibration systems representing fatigue testing machines. Using system phasor models (ω-domain models) both, the dynamic characteristics, and the energy flow characteristics are specified by polar plots referred to load or displacement control (force or stroke).

Specifying Significant Power Parameters. Considering the *polar form* of Eq. (5.63a), the phasor power is specified by its modulus denoting
the *apparent power S*, Eq. (5.55b),

$$\left|\underline{S}(\mathrm{j}\omega_\mathrm{f})\right| = S = \widetilde{F}\,\widetilde{\upsilon} \qquad (5.65a);$$

and by its argument designated as
the *impedance angle* φ_1

$$\operatorname{arc}\underline{S}(\mathrm{j}\omega_\mathrm{f}) = \varphi_1 = \varphi_\mathrm{F} - \varphi_\upsilon \qquad (5.65b).$$

Using the phasor representation of the individual conjugate variables, force and velocity, the apparent power is the *product of the magnitudes* of force and velocity phasors (r.m.s. values product); whereas the impedance angle is the angle of the force phasor with respect to the velocity phasor, in short the *relative phase angle* (phase difference).

With regard to the *rectangular form* of Eq. (5.63a), the phasor power is specified by its real and imaginary parts which represent transacting components being significant for the insight into stationary flow of energy through a 1-port component. The real part of phasor power describing the active (or in-phase) component is called
the *active power P*, Eqs. (5.55a), (5.60a),

$$\mathrm{Re}\left[\underline{S}(\mathrm{j}\omega_\mathrm{f})\right] = P = S\cos\varphi_1 = \widetilde{F}\,\widetilde{\upsilon}\cos\varphi_1 \qquad (5.66a);$$

and the imaginary part of phasor power which represents the reactive (or quadrature) component is called
the *reactive power Q*, Eq. (5.61)

$$\mathrm{Im}\left[\underline{S}(\mathrm{j}\omega_{\mathrm{f}})\right] = Q = S\sin\varphi_1 = \widetilde{F}\,\widetilde{\upsilon}\sin\varphi_1 \qquad (5.66b).$$

Using individual phasor variables the active power is geometrically interpreted as the *product of the magnitude of velocity phasor* and that component of *force phasor being in phase* with velocity phasor; whereas the reactive power results from the *product of the magnitude of the velocity phasor* and that component of *force phasor being in quadrature* to the velocity phasor. The first product yields the average power component, the second one a wattless or quadrature component.

The apparent power S as its in-phase and quadrature components P, Q are specifying the resultant phasor power $\underline{S}$. The family of real characteristic parameters S, P, Q span a (resting) right triangle of power in the complex plane, Fig. 5.2a. Thus, the phasor diagram of power yields with ease the impedance angle φ_1. This is the fourth parameter being significant for deriving a characteristic magnitude in power engineering, the *power factor* λ, Eq. (5.56a).

To the specification of significant power parameters it is irrelevant if the phasor product $\underline{\widetilde{F}}\,\underline{\widetilde{\upsilon}}^*$ or its conjugate $\underline{\widetilde{F}}^*\,\underline{\widetilde{\upsilon}}$, i.e., whether $\underline{S}(\mathrm{j}\omega_{\mathrm{f}})$, Eq. (5.63a), or $\underline{S}^*(\mathrm{j}\omega_{\mathrm{f}})$, Eq. (5.63b), is defined as the resultant phasor appropriate to applications of phasor power concept in vibrations. The transacting components of stationary flow of energy are equal except for the sign of the reactive (or quadrature) component.

The *sign of reactive power* is an arbitrary choice which associates

$$Q < 0 \quad \text{with} \quad \varphi_1 < 0 \quad \text{(velocity lagging force);}$$
$$Q > 0 \quad \text{with} \quad \varphi_1 > 0 \quad \text{(velocity leading force).}$$

Assuming for causal 1-port components a phase of responding velocity which lags the exciting force the *conjugate form of complex power*, Eq. (5.63b), may be taken as a basis for the following phasor power calculations.

5.2.3
Connections with Frequency-response Characteristics. *Combining Dynamic Compliance and Phasor Power Concepts*

By transforming time-dependent power transactions into the frequency domain the spectral decomposition of power has been reduced to decomposing a resultant vector, termed the *phasor power*, which permits the insight into the steady-state behaviour of power and energy flow through linear 1-port components.

The general effort-flow dynamic functional relationship is now expressed in terms of the complex ratio of the excitation force phasor to the velocity response phasor. The corresponding phasor ratio (complexor), designated the *mechanical impedance* (phasor impedance), 4.2.1, will be fitted in the sequel to a force-to-displacement ratio representing the *dynamic stiffness* of the structure (phasor stiffness). This equivalent definition of a frequency-response function expressing motion response as displacement represents a dynamic characteristic related to phasor impedance but basing on static elastic analysis with force and displacement as the principle variables in mechanics, 5.2.1. Exemplified by the power flow of a basic vibration system the vector representation of phasor ratios (complexors), Fig. 3.20b, 3.2.10, will be completed by the vector representation of phasor power (complex power), Fig. 5.2a, 5.2.2, thus illustrating the usefulness of combining the concepts of dynamic compliance and phasor power for vibration design applications.

Driving Power of Vibrators (phasor power concept)

The model system of a vibration machine is reduced to a one-degree-of-freedom system undergoing a harmonic motion that provides excitation at the point *B*. the simplified configuration consists of a mechanical source acting by a displacement of the amplitude $\hat{u}$ on a mass m and a two-element subsystem k, c in parallel supported from a frame at *A*. The device is directly driven by the driving mechanism of crank connecting rod-type, Fig. 5.3.

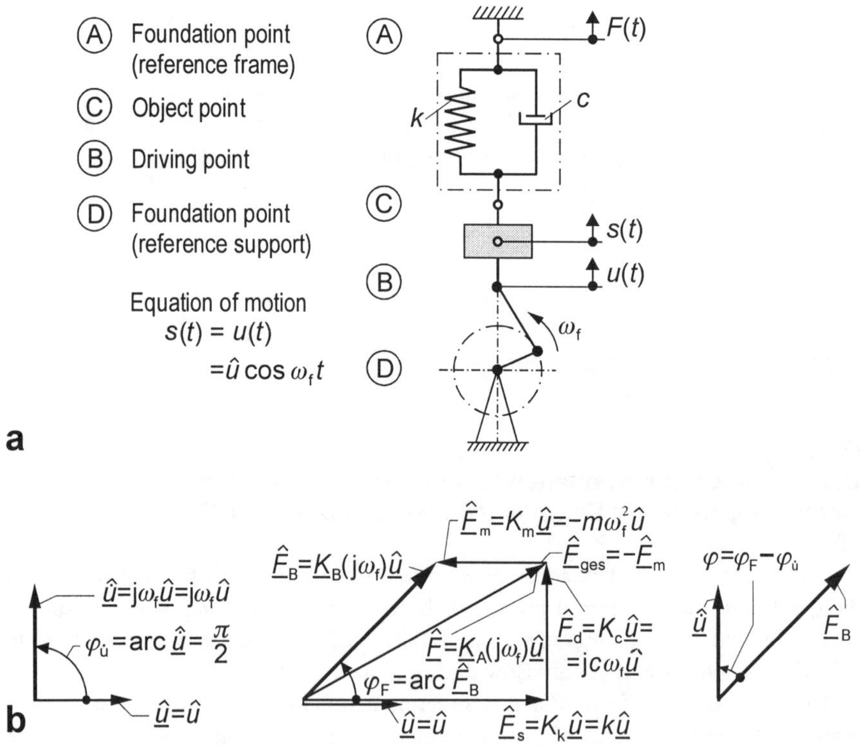

Fig. 5.3. Direct drive vibration machine. **a** Schematic diagram of mechanical vibration generator connected with fundamental configuration in parallel; **b** phasor diagrams of driving displacement and velocity, element forces and rod force

One-port Relation (phasor ratio). The overall behaviour of the one-ported network is defined by an impedance relation which will be fitted to principle variables in mechanics. Using a transform model (ω-domain model) the corresponding phasor ratio of the dynamic functional relationship between driving point B and foundation point A is directly driven by the procedure of network simplification (basic tool of network diagram reduction). Spring and damper forming a two-element section in parallel the dynamic stiffnesses of structural components incident to the connecting reference frame A are added, 4.2. Thus, the phasor relation between responding force and exciting displacement at different points is specified by a frequency-response characteristic designated

the *transfer dynamic stiffness* (phasor stiffness) $\underline{K}_{\mathrm{A}}(\mathrm{j}\omega_{\mathrm{f}})$

$$\underline{K}_{\mathrm{A}}(\mathrm{j}\omega_{\mathrm{f}}) = \frac{\hat{\underline{F}}}{\hat{\underline{u}}} = \underline{K}_{\mathrm{k}} + \underline{K}_{\mathrm{c}} = k + \mathrm{j}c\omega_{\mathrm{f}} = \underline{C}_{\mathrm{A}}^{-1}(\mathrm{j}\omega_{\mathrm{f}}) \qquad (5.67)$$

of the parallel-connected structure, or the inverse of the transfer dynamic compliance $\underline{C}_{\mathrm{A}}(\mathrm{j}\omega_{\mathrm{f}})$.

Power State at the Single Port of Entry. For the representation of driving power flow an explicit form of power state will be expressed in terms of the pertinent effort-flow product at the entering port in complex notation.

Individual Variables (effort-flow couple). Approximating the translational driving motion and the force at the driving point B by simple harmonics the individual conjugate variables *velocity* and *force* have to be specified at first.
The actual exciting displacement

$$u_{\omega}(t) = \hat{u}\cos(\omega_{\mathrm{f}} t + \varphi_{\mathrm{u}}) = \frac{\hat{\underline{u}}\,\mathrm{e}^{\mathrm{j}\omega_{\mathrm{f}} t} + \hat{\underline{u}}^{*}\,\mathrm{e}^{-\mathrm{j}\omega_{\mathrm{f}} t}}{2} \qquad (5.68\mathrm{a})$$

is related to its time rate of change or power variable, the actual velocity

$$\dot{u}_{\omega}(t) = -\omega_{\mathrm{f}}\hat{u}\sin(\omega_{\mathrm{f}} t + \varphi_{\mathrm{u}}) = -\frac{\omega_{\mathrm{f}}\,\hat{\underline{u}}\,\mathrm{e}^{\mathrm{j}\omega_{\mathrm{f}} t} - \omega_{\mathrm{f}}\,\hat{\underline{u}}^{*}\,\mathrm{e}^{-\mathrm{j}\omega_{\mathrm{f}} t}}{2\mathrm{j}}$$

$$= \underbrace{\omega_{\mathrm{f}}\hat{u}}_{=\hat{\dot{u}}}\cos\Big[\,\omega_{\mathrm{f}} t + \underbrace{\Big(\varphi_{\mathrm{u}} + \frac{\pi}{2}\Big)}_{=\varphi_{\dot{\mathrm{u}}}}\Big] = \frac{\hat{\dot{\underline{u}}}\,\mathrm{e}^{\mathrm{j}\omega_{\mathrm{f}} t} + \hat{\dot{\underline{u}}}^{*}\,\mathrm{e}^{-\mathrm{j}\omega_{\mathrm{f}} t}}{2} \qquad (5.69\mathrm{a}).$$

The sinusoidal motion variables are equivalently expressed in terms of paired off complex quantities thus serving as the correct substitute for the actual excitation and its first derivative with respect to time.

The complex constituents are determined by
the *phasor of the exciting displacement*

$$\hat{\underline{u}} = |\hat{\underline{u}}|\,\mathrm{e}^{\mathrm{j}\,\mathrm{arc}\,\hat{\underline{u}}} = \hat{u}\,\mathrm{e}^{\mathrm{j}\varphi_{\mathrm{u}}} \qquad (5.68\mathrm{b}),$$

and the *phasor of the exciting velocity*

$$\hat{\dot{\underline{u}}} = \mathrm{j}\omega_{\mathrm{f}}\,\hat{\underline{u}} = |\hat{\dot{\underline{u}}}|\,\mathrm{e}^{\mathrm{j}\,\mathrm{arc}\,\hat{\dot{\underline{u}}}} = \omega_{\mathrm{f}}\,|\hat{\underline{u}}|\,\mathrm{e}^{\mathrm{j}(\mathrm{arc}\,\hat{\underline{u}} + \frac{\pi}{2})} = \underbrace{\omega_{\mathrm{f}}\hat{u}}_{=\hat{\dot{u}}}\,\mathrm{e}^{\overbrace{\mathrm{j}(\varphi_{\mathrm{u}} + \frac{\pi}{2})}^{=\varphi_{\dot{\mathrm{u}}}}} \qquad (5.69\mathrm{b}).$$

The second power variable complementary to velocity is the force at the branch rod linked up with the connecting-rod head and driven by the crank, for short, the actual rod force

$$F_{\mathrm{B}_\omega}(t) = \hat{F}_{\mathrm{B}} \cos(\omega_{\mathrm{f}} t + \varphi_{\mathrm{F}}) = \frac{\underline{\hat{F}_{\mathrm{B}}}\, \mathrm{e}^{\mathrm{j}\omega_{\mathrm{f}} t} + \underline{\hat{F}^*_{\mathrm{B}}}\, \mathrm{e}^{-\mathrm{j}\omega_{\mathrm{f}} t}}{2} \quad (5.70\mathrm{a}).$$

The complex constituents are determined by the phasor equation expressing the *phasor of the rod force*

$$\underline{\hat{F}_{\mathrm{B}}} = \underline{K_{\mathrm{B}}}\, \underline{\hat{u}} = \left|\underline{\hat{F}_{\mathrm{B}}}\right| \mathrm{e}^{\mathrm{j}\,\mathrm{arc}\underline{\hat{F}_{\mathrm{B}}}} = \left|\underline{K_{\mathrm{B}}}\right| \left|\underline{\hat{u}}\right| \mathrm{e}^{\mathrm{j}\,\mathrm{arc}(\underline{\hat{F}_{\mathrm{B}}} + \underline{\hat{u}})} \quad (5.70\mathrm{b}).$$

Contrary to the presupposed velocity $\dot{u}$ the second power variable F_{B} is a quantity indefinite in magnitude and phase.

The determination of the associated force phasor will be readily performed by use of one-port relations (impedance relations), 5.2.1. Thus, the overall behaviour of the configuration at the single port of entry is a functional relationship specified by the corresponding phasor ratio, 4.2.3.

Applying the approved procedure of network simplification to the three-element section in parallel the dynamic stiffnesses of the components incident to the junction (node or mechanical vertex) B are added to form the overall frequency-response characteristic $\underline{K}(\mathrm{j}\omega_{\mathrm{f}})$, Eq. (4.25a), 4.2.3. Thus, the phasor relation between responding rod force and exciting displacement at the same point is specified by a dynamic characteristic designated the *direct dynamic stiffness (phasor stiffness)* $\underline{K_{\mathrm{B}}}(\mathrm{j}\omega_{\mathrm{f}})$

$$\underline{K_{\mathrm{B}}}(\mathrm{j}\omega_{\mathrm{f}}) = \frac{\underline{\hat{F}_{\mathrm{B}}}}{\underline{\hat{u}}} = \underline{K}(\mathrm{j}\omega_{\mathrm{f}}) = K_{\mathrm{k}} + K_{\mathrm{c}} + K_{\mathrm{m}} = \left(k - m\omega_{\mathrm{f}}^2\right) + \mathrm{j}c\omega_{\mathrm{f}} \quad (5.71\mathrm{a})$$

of the parallel-connected structure.

The phasor relations between the element forces $\hat{F}_{\mathrm{i}}$ and the exciting displacement $\hat{u}$

$$\text{elastic (or spring) force} \quad \underline{\hat{F}_{\mathrm{s}}} = K_{\mathrm{k}}\, \underline{\hat{u}} = k\, \underline{\hat{u}} \quad (5.72\mathrm{a})$$

$$\text{damping (or damper) force} \quad \underline{\hat{F}_{\mathrm{d}}} = K_{\mathrm{c}}\, \underline{\hat{u}} = \mathrm{j}c\omega_{\mathrm{f}}\, \underline{\hat{u}} \quad (5.72\mathrm{b})$$

$$\text{inertial (or mass) force} \quad \underline{\hat{F}_{\mathrm{m}}} = K_{\mathrm{m}}\, \underline{\hat{u}} = -m\omega_{\mathrm{f}}^2\, \underline{\hat{u}} \quad (5.72\mathrm{c})$$

are characterized by the phasor stiffnesses of the elementary (1-port) components, Eqs. (4.58a,b.c), 4.2.5. The balancing of element forces and rod force can be illustrated by the force phasor polygon, Fig. 5.3b.

The transformation in variable and scale, outlined in 3.2.11, results in the adjusted (data-reduced) form of a normalized response ratio of phasors designated the *normalized direct dynamic stiffness* $\underline{X}_{\mathrm{k}}(\mathrm{j}\eta_1)$, Eq. (4.62), being characterized by the normalized element stiffnesses, Eqs. (4.64a,b,c), 4.2.5.

For predicting *performance criteria*, such as the rod force amplitude $\hat{F}_{\mathrm{B}}$, the phasor ratio polygon proves useful, Fig. 5.4.

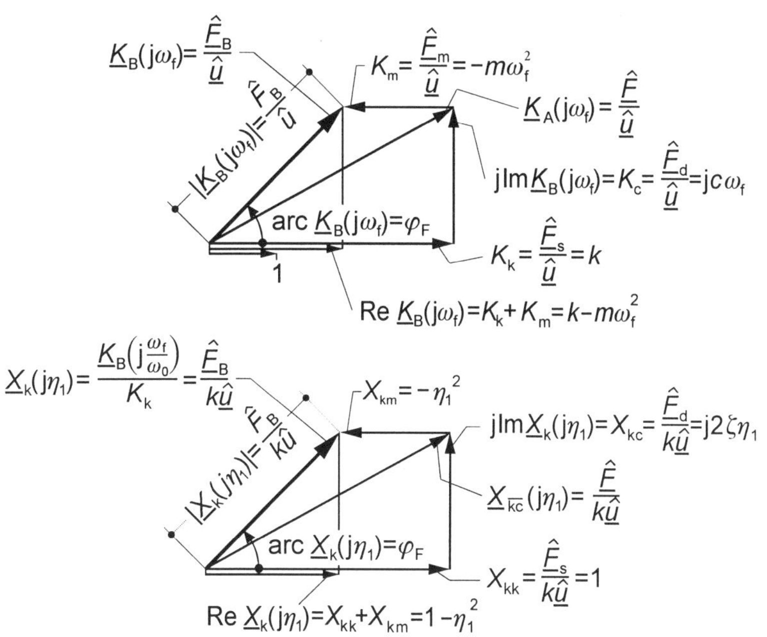

Fig. 5.4. Phasor diagrams for parallel-connected basic elements in the dynamic stiffness plane, and data-reduced in the normalized stiffness plane

The rod force, to be transmitted by the vibration generator for applying a presupposed constant-amplitude motion $\hat{u}$ or $\hat{\dot{u}}$, at the driving point B is related to individual phasor rations, the contributions of which result in the composite characteristic of dynamic stiffness.

With reference to phasor power calculation, Eq. (5.63b), 5.2.2, the conjugate of the frequency-response characteristic may be defined as

the *(direct) dynamic stiffness conjugate* $\underline{K}_B^*(\mathrm{j}\omega_f)$

$$\underline{K}_B^*(\mathrm{j}\omega_f) = \frac{\underline{\hat{F}}_B^*}{\underline{\hat{u}}^*} = \underline{K}^*(\mathrm{j}\omega_f) = K_k - K_c + K_m = \left(k - m\omega_f^2\right) - \mathrm{j}c\omega_f \qquad (5.71b),$$

normalized form

$$\frac{\underline{K}_B^*(\mathrm{j}\frac{\omega_f}{\omega_0})}{K_k} = \frac{\underline{\hat{F}}_B^*}{\underline{\hat{F}}_{stat}} = \underline{X}_k^*(\mathrm{j}\eta_1) = X_{kk} - X_{kc} + X_{km} = \left(1 - \eta_1^2\right) - \mathrm{j}2\zeta\eta_1 \quad (5.71c).$$

Writing the conjugate of the response ratio of phasors in rectangular form the composite characteristic is specified by performance criteria as

the *modulus of dynamic stiffness*

$$\begin{aligned} \left|\underline{K}_B^*(\mathrm{j}\omega_f)\right| &= \left|\sqrt{\mathrm{Re}^2\left[\underline{K}_B^*(\mathrm{j}\omega_f)\right] + \mathrm{Im}^2\left[\underline{K}_B^*(\mathrm{j}\omega_f)\right]}\right| \\ &= \left|\sqrt{(k - m\omega_f^2)^2 + (c\omega_f)^2}\right| = k\left|\sqrt{(1-\eta_1^2)^2 + (2\zeta\eta_1)^2}\right| \end{aligned} \qquad (5.73a),$$

and the *argument of the dynamic stiffness conjugate*

$$\operatorname{arc} \underline{K}_{\mathrm{B}}^{*}(\mathrm{j}\omega_{\mathrm{f}}) = \arctan \frac{\operatorname{Im}\left[\underline{K}_{\mathrm{B}}^{*}(\mathrm{j}\omega_{\mathrm{f}})\right]}{\operatorname{Re}\left[\underline{K}_{\mathrm{B}}(\mathrm{j}\omega_{\mathrm{f}})\right]} = -\arctan \frac{c\omega_{\mathrm{f}}}{k - m\omega_{\mathrm{f}}^{2}} = -\arctan \frac{2\zeta\eta_{1}}{1-\eta_{1}^{2}} \quad (5.73\mathrm{b}).$$

Actual Energy Flow (effort-flow product). To realize the energetic system approach there is no need for expressing the driving power state in terms of the actual energy flow. Although it adds some illustrative interest to power calculations by applying the modified phasor method.

By use of the phasor of the *instantaneous power* at the driving point *B* the actual scalar quantity can be expressed in terms of the real part of the complex notation

$$\begin{aligned} S_{\mathrm{B}}(t) = F_{\mathrm{B}_{\omega}}(t)\cdot \dot{u}_{\omega}(t) &= \operatorname{Re}\left[\frac{1}{2}\underline{\hat{F}_{\mathrm{B}}}\,\underline{\hat{\dot{u}}}^{*} + \frac{1}{2}\underline{\hat{F}_{\mathrm{B}}}\,\underline{\hat{\dot{u}}}\,\mathrm{e}^{\mathrm{j}2\omega_{\mathrm{f}}t}\right] = \operatorname{Re}\left[\underline{S(t)}\right] \\ &= \operatorname{Re}\left[\frac{1}{2}\underline{\hat{F}_{\mathrm{B}}^{*}}\,\underline{\hat{\dot{u}}} + \frac{1}{2}\underline{\hat{F}_{\mathrm{B}}^{*}}\,\underline{\hat{\dot{u}}}^{*}\,\mathrm{e}^{-\mathrm{j}2\omega_{\mathrm{f}}t}\right] = \operatorname{Re}\left[\underline{S^{*}(t)}\right] \\ &= \operatorname{Re}\left[\frac{1}{2}\underline{\hat{F}_{\mathrm{B}}^{*}}\,\underline{\hat{\dot{u}}} + \frac{1}{2}\underline{\hat{F}_{\mathrm{B}}^{*}}\,\underline{\hat{\dot{u}}}\,\frac{\underline{\hat{\dot{u}}}^{*}\,\underline{\hat{\dot{u}}}^{*}}{\underline{\hat{\dot{u}}}\,\underline{\hat{\dot{u}}}^{*}}\,\mathrm{e}^{-\mathrm{j}2\omega_{\mathrm{f}}t}\right] \\ &= \underbrace{\frac{1}{2}\hat{F}_{\mathrm{B}}\,\hat{\dot{u}}}_{=S}\operatorname{Re}\left[\mathrm{e}^{-\mathrm{j}\varphi_{1}}\right] + \frac{1}{2}\hat{F}_{\mathrm{B}}\,\hat{\dot{u}}\operatorname{Re}\left[\mathrm{e}^{-\mathrm{j}\varphi_{1}}\,\mathrm{e}^{-\mathrm{j}2(\omega_{\mathrm{f}}t+\varphi_{\dot{u}})}\right] \end{aligned} \quad (5.74\mathrm{a}),$$

finally, in real notation

$$S_{\mathrm{B}}(t) = \underbrace{\frac{1}{2}\hat{F}_{\mathrm{B}}\,\hat{\dot{u}}\cos\varphi_{1}}_{=P}\left[1+\cos 2(\omega_{\mathrm{f}}t+\varphi_{\dot{u}})\right] - \underbrace{\frac{1}{2}\hat{F}_{\mathrm{B}}\,\hat{\dot{u}}\,\sin\varphi_{1}}_{=Q}\sin 2(\omega_{\mathrm{f}}t+\varphi_{\dot{u}}) \quad (5.74\mathrm{b}).$$

The decomposition of the conjugate of the complex quantity $\underline{S}_{\mathrm{t}}^{*}$ into phasor components corresponds to the decomposition into active and reactive components, Eq. (5.62c).

Significant power parameters are specified by the angle of the rod force phasor with respect to the exciting velocity phasor, called
the *impedance angle* φ_1, Eq. (5.65b),

$$\varphi_{1} = \varphi_{\mathrm{F}} - \varphi_{\dot{u}} = \varphi_{\mathrm{F}} - \frac{\pi}{2};\ \text{provided that}\ \varphi_{\mathrm{u}} = 0 \quad (5.75\mathrm{a}),$$

the ratio of active power to apparent power, termed
the *power factor* λ, Eq. (5.56a),

$$\frac{P}{S} = \cos\varphi_{1} = \cos\left(\varphi_{\mathrm{F}} - \frac{\pi}{2}\right) = \sin\varphi_{\mathrm{F}} \quad (5.75\mathrm{b}),$$

and its orthogonal characteristic magnitude, the ratio of reactive power to apparent power

$$\frac{Q}{S} = \sin\varphi_{1} = \sin\left(\varphi_{\mathrm{F}} - \frac{\pi}{2}\right) = -\cos\varphi_{\mathrm{F}} \quad (5.75\mathrm{c}).$$

The characteristic parameter *velocity phase angle* (or initial phase) $\varphi_{\dot{u}}$ reduces to

$$\varphi_{\dot{u}} = \varphi_{\mathrm{u}} + \frac{\pi}{2} = \frac{\pi}{2};\ \text{provided that}\ \varphi_{\mathrm{u}} = 0 \quad (5.76\mathrm{a})$$

in the case of usually chosen zero-crossing excitation, thus omitting the initial phase of excitation displacement φ_{u}.

By simplifying the sinusoids of the components

$$\cos 2(\omega_f t + \varphi_{\dot{u}}) = \cos 2\left(\omega_f t + \frac{\pi}{2}\right) = -\cos 2\omega_f t$$
$$\sin 2(\omega_f t + \varphi_{\dot{u}}) = \sin 2\left(\omega_f t + \frac{\pi}{2}\right) = -\sin 2\omega_f t \tag{5.76b}$$

the actual scalar quantity S_t reduces to

$$S_B(t) = \underbrace{\frac{1}{2}\hat{F}_B\,\hat{\dot{u}}\sin\varphi_F}_{=P}\left(1-\cos 2\omega_f t\right) - \underbrace{\frac{1}{2}\hat{F}_B\,\hat{\dot{u}}\cos\varphi_F}_{=Q}\sin 2\omega_f t \quad (5.74c).$$

An other characteristic parameter, the *rod force phase angle* (or initial phase) φ_F, also can be determined by directly reading off the phasor relations between the element forces, Fig. 5.3b, or the real and imaginary parts of the (direct) stiffness, Fig. 5.4,

$$\varphi_F = \arctan\frac{\hat{F}_d}{\hat{F}_s - \hat{F}_m} = \arctan\frac{\mathrm{Im}\left[\underline{K}_B(\mathrm{j}\omega_f)\right]}{\mathrm{Re}\left[\underline{K}_B(\mathrm{j}\omega_f)\right]} = \arctan\frac{c\omega_f}{k - m\omega_f^2} \quad (5.77a),$$

$$\varphi_F = \arccos\frac{\hat{F}_s - \hat{F}_m}{\hat{F}_B} = \arccos\frac{\mathrm{Re}\left[\underline{K}_B(\mathrm{j}\omega_f)\right]}{\left|\underline{K}_B(\mathrm{j}\omega_f)\right|} = \arccos\frac{k - m\omega_f^2}{\left|\sqrt{(k-m\omega_f^2)^2 + (c\omega_f)^2}\right|} \quad (5.77b),$$

$$\varphi_F = \arcsin\frac{\hat{F}_d}{\hat{F}_B} = \arcsin\frac{\mathrm{Im}\left[\underline{K}_B(\mathrm{j}\omega_f)\right]}{\left|\underline{K}_B(\mathrm{j}\omega_f)\right|} = \arcsin\frac{c\omega_f}{\left|\sqrt{(k-m\omega_f^2)^2 + (c\omega_f)^2}\right|} \quad (5.77c).$$

Expressing now the actual active and reactive power components in a more explicit form the pulsating quantity of actual power flow may be rewritten

$$S_B(t) = \frac{1}{2}\hat{F}_B\,\hat{\dot{u}}\frac{c\omega_f}{\left|\sqrt{(k-m\omega_f^2)^2 + (c\omega_f)^2}\right|}(1-\cos 2\omega_f t)$$
$$-\frac{1}{2}\hat{F}_B\,\hat{\dot{u}}\frac{c - m\omega_f^2}{\left|\sqrt{(k-m\omega_f^2)^2 + (c\omega_f)^2}\right|}\sin 2\omega_f t \tag{5.74d}$$

Basing on phasor magnitude relations between

rod force $\quad \hat{F}_B = \left|\underline{K}_B\right|\hat{u} = \left|\sqrt{(k-m\omega_f^2)^2 + (c\omega_f)^2}\right| \quad$ (5.78a),

and the element forces

spring force $\quad \hat{F}_s = K_k \cdot \hat{u} = k\cdot\hat{u}$

damper force $\quad \hat{F}_d = K_c \cdot \hat{u} = c\omega_f \cdot \hat{u} \quad$ (5.78b)

mass force $\quad \hat{F}_m = K_m \cdot \hat{u} = -m\omega_f^2 \cdot \hat{u}$

by the stiffnesses of the elementary components K_i an alternative form for the actual scalar quantity S_t is given by

$$S_B(t) = \underbrace{\underbrace{\frac{1}{2}\hat{F}_d\,\hat{\dot{u}}}_{=P}(1-\cos 2\omega_f t)}_{=P(t)=P_- + P_\omega(t)} \underbrace{-\underbrace{\frac{1}{2}\left(\hat{F}_s+\hat{F}_m\right)\hat{\dot{u}}}_{=Q}\sin 2\omega_f t}_{=Q_\omega(t)} \tag{5.74e}$$

expressing significant power parameters in terms of element force amplitudes related to the phasor stiffnesses of the components,
the *active power P*

$$P = \frac{1}{2}\hat{F}_B\,\hat{\dot{u}}\cos\varphi_1 = \frac{1}{2}\hat{F}_B\hat{\dot{u}}\sin\varphi_F = \frac{1}{2}\hat{F}_d\hat{\dot{u}} = \frac{1}{2}\hat{u}^2\omega_f K_c = \frac{1}{2}\left(\hat{u}\omega_f\right)^2 c \tag{5.79a}$$

and the *reactive power Q*

$$Q = \frac{1}{2}\hat{F}_B\,\hat{\dot{u}}\sin\varphi_1 = \frac{1}{2}\hat{F}_B\hat{\dot{u}}\cos\varphi_F = \frac{1}{2}\left(\hat{F}_s+\hat{F}_m\right)\hat{\dot{u}} = \frac{1}{2}\hat{u}^2\omega_f\left(K_k+K_m\right) = \frac{1}{2}\hat{u}^2\omega_f\left(k - m\omega_f^2\right) \tag{5.79b}$$

Complex Input Power (driving phasor power). The energetic system approach is completely performed by determining the significant power parameters being associated with the complex power $\underline{S}$. Following Eq. (5.64b), 5.2.2, this energy function is defined as
the *mean value of the conjugate of the complex quantity*

$$\begin{aligned}\underline{S}_B^*(j\omega_f) &= \frac{1}{T}\int_{t_1}^{t_1+T}\frac{1}{2}\underline{F}_{B_\omega}^*(t)\left[\underline{\dot{u}}_\omega(t)+\underline{\dot{u}}_\omega^*(t)\right]dt \\ &= \overline{\frac{1}{2}\underline{F}_{B_\omega}^*(t)\left[\underline{\dot{u}}_\omega(t)+\underline{\dot{u}}_\omega^*(t)\right]} = \underline{S}_t^*\end{aligned} \tag{5.80}$$

The steady-state power relation is readily expressed in explicit form by multiplying together the pertinent phasor quantities assigned to the same point B, Eq. (5.63a). Thus, the power product written in the preferred conjugate form of complex notation, Eq. (5.63b), results in
the *driving phasor power conjugate*

$$\begin{aligned}\overline{\underline{S}^*(t)} = \underline{S}_B^*(j\omega_f) &= \underline{\tilde{F}}_B^*\,\underline{\tilde{\dot{u}}} = \underline{\tilde{F}}_B^*\,j\omega_f\,\underline{\tilde{u}} \\ &= \frac{1}{2}\underline{\hat{F}}_B^*\,\underline{\hat{\dot{u}}} = \frac{1}{2}\underline{\hat{F}}_B^*\,j\omega_f\,\underline{\hat{u}}\end{aligned} \tag{5.81a}$$

being the product of the conjugate of the (r.m.s.) rod force phasor $\underline{\tilde{F}}_B^*$, Eq. (5.50a), and the (r.m.s.) exciting velocity phasor $\underline{\tilde{\dot{u}}}$, Eq. (5.50d).

Using the phasor relation between responding rod force and exciting displacement at the same point, Eq. (5.70b)

$$\underline{\hat{F}}_B = \underline{K}_B(j\omega_f)\,\underline{\hat{u}} \tag{5.82}$$

together with the phasor of the exciting velocity, Eq. (5.69b)

$$\underline{\hat{\dot{u}}} = j\omega_f\,\underline{\hat{u}} \tag{5.83a}$$

the energy function can be rewritten

$$\underline{S}_{\mathrm{B}}^{*}(\mathrm{j}\omega_{\mathrm{f}}) = \frac{1}{2}\underline{K}_{\mathrm{B}}^{*}(\mathrm{j}\omega_{\mathrm{f}})\,\underline{\hat{u}}^{*}\,\mathrm{j}\omega_{\mathrm{f}}\,\underline{\hat{u}} = \frac{1}{2}\underline{\hat{u}}^{*}\,\underline{\hat{u}}\,\mathrm{j}\omega_{\mathrm{f}}\,\underline{K}_{\mathrm{B}}^{*}(\mathrm{j}\omega_{\mathrm{f}}) \qquad (5.81\mathrm{b}).$$

Combining the displacement phasors to the square of displacement magnitude

$$\underline{\hat{u}}^{*}\,\underline{\hat{u}} = |\,\underline{\hat{u}}\,|^{2} = \hat{u}^{2} \qquad (5.83\mathrm{b})$$

and multiplying the frequency-response characteristic, Eq. (5.71a), by $\mathrm{j}\omega_{\mathrm{f}}$ the (direct) dynamic stiffness $\underline{K}_{\mathrm{B}}^{*}(\mathrm{j}\omega_{\mathrm{f}})$ (phasor stiffness) is fitted for the phasor product of power (resultant vector) in direction. As a result
the *converted (direct) dynamic stiffness*

$$\underline{K}_{\mathrm{B}}^{*\prime}(\mathrm{j}\omega_{\mathrm{f}}) = \mathrm{j}\omega_{\mathrm{f}}\,\underline{K}_{\mathrm{B}}^{*}(\mathrm{j}\omega_{\mathrm{f}}) = \mathrm{j}\omega_{\mathrm{f}}\left[\left(k - m\omega_{\mathrm{f}}^{2}\right) - \mathrm{j}\left(c\omega_{\mathrm{f}}\right)\right] = \omega_{\mathrm{f}}\left[c\omega_{\mathrm{f}} + \mathrm{j}\left(k - m\omega_{\mathrm{f}}^{2}\right)\right] \qquad (5.84)$$

is defined being abbreviated to $\underline{K}_{\mathrm{B}}^{*\prime}(\mathrm{j}\omega_{\mathrm{f}})$. Using this related dynamic characteristic the steady-state power relation can be calculated very conveniently expressing the driving phasor power conjugate, Eq. (5.81b), in
the *explicit form*

$$\underline{S}_{\mathrm{B}}^{*}(\mathrm{j}\omega_{\mathrm{f}}) = \frac{1}{2}\hat{u}^{2}\,\underline{K}_{\mathrm{B}}^{*\prime}(\mathrm{j}\omega_{\mathrm{f}}) = \frac{1}{2}\left[c\omega_{\mathrm{f}} + \mathrm{j}\left(k - m\omega_{\mathrm{f}}^{2}\right)\right]\hat{u}^{2}\cdot\omega_{\mathrm{f}} \qquad (5.81\mathrm{c}).$$

The *significant power parameters* being associated with the conjugate of the complex input power can be gathered from this complex parameter by decomposition of power. Preferring the rectangular form the insight of stationary flow of driving energy is represented by the real and imaginary parts of the phasor power conjugate. The first power component, termed
the *active power P*

$$P = \frac{1}{2}F_{\mathrm{B}\hat{\dot{u}}}\,\underline{\hat{\dot{u}}} = \frac{1}{2}\left[\underline{S}_{\mathrm{B}}^{*}(\mathrm{j}\omega_{\mathrm{f}}) + \underline{S}_{\mathrm{B}}(\mathrm{j}\omega_{\mathrm{f}})\right] = \mathrm{Re}\left[\underline{S}_{\mathrm{B}}^{*}(\mathrm{j}\omega_{\mathrm{f}})\right]$$

$$\mathrm{Re}\left[\underline{K}_{\mathrm{B}}^{*\prime}(\mathrm{j}\omega_{\mathrm{f}})\right] = c\omega_{\mathrm{f}}\cdot\omega_{\mathrm{f}} \qquad (5.85\mathrm{a}),$$

$$P = \frac{1}{2}\hat{u}^{2}\,\mathrm{Re}\left[\underline{K}_{\mathrm{B}}^{*\prime}(\mathrm{j}\omega_{\mathrm{f}})\right] = \frac{1}{2}c\omega_{\mathrm{f}}\hat{u}^{2}\cdot\omega_{\mathrm{f}} = \frac{1}{2}(\hat{u}\omega_{\mathrm{f}})^{2}c$$

is the product of the magnitude $\hat{\dot{u}}$ of velocity phasor $\underline{\hat{\dot{u}}}$ and the component $F_{\mathrm{B}\hat{\dot{u}}}$ of rod force phasor $\underline{\hat{F}}_{\mathrm{B}}^{*}$ being in phase with the velocity phasor. The second power component, termed
the *reactive power Q*

$$Q = \frac{1}{2}F_{\mathrm{B}\,\mathrm{j}\hat{\dot{u}}}\,\underline{\hat{\dot{u}}} = \frac{1}{2\mathrm{j}}\left[\underline{S}_{\mathrm{B}}^{*}(\mathrm{j}\omega_{\mathrm{f}}) - \underline{S}_{\mathrm{B}}(\mathrm{j}\omega_{\mathrm{f}})\right] = \mathrm{Im}\left[\underline{S}_{\mathrm{B}}^{*}(\mathrm{j}\omega_{\mathrm{f}})\right]$$

$$\mathrm{Im}\left[\underline{K}_{\mathrm{B}}^{*\prime}(\mathrm{j}\omega_{\mathrm{f}})\right] = \left(k - m\omega_{\mathrm{f}}^{2}\right)\cdot\omega_{\mathrm{f}} \qquad (5.85\mathrm{b})$$

$$Q = \frac{1}{2}\hat{u}^{2}\,\mathrm{Im}\left[\underline{K}_{\mathrm{B}}^{*\prime}(\mathrm{j}\omega_{\mathrm{f}})\right] = \frac{1}{2}\left(k - m\omega_{\mathrm{f}}^{2}\right)\hat{u}^{2}\cdot\omega_{\mathrm{f}}$$

is the product of the magnitude $\hat{\dot{u}}$ of velocity phasor $\hat{\underline{\dot{u}}}$ and the component $F_{\mathrm{Bj}\hat{\underline{\dot{u}}}}$ of rod force phasor $\hat{\underline{F}}_{\mathrm{B}}^{*}$ being in quadrature to the velocity phasor.

Vector Representation of Complex Input Power (driving phasor power). Taking up the vector representation modified to power relations, 5.2.2, the explicit form of power state at the single port of entry can be specified in aid of graphical interpretation, Fig. 5.5.

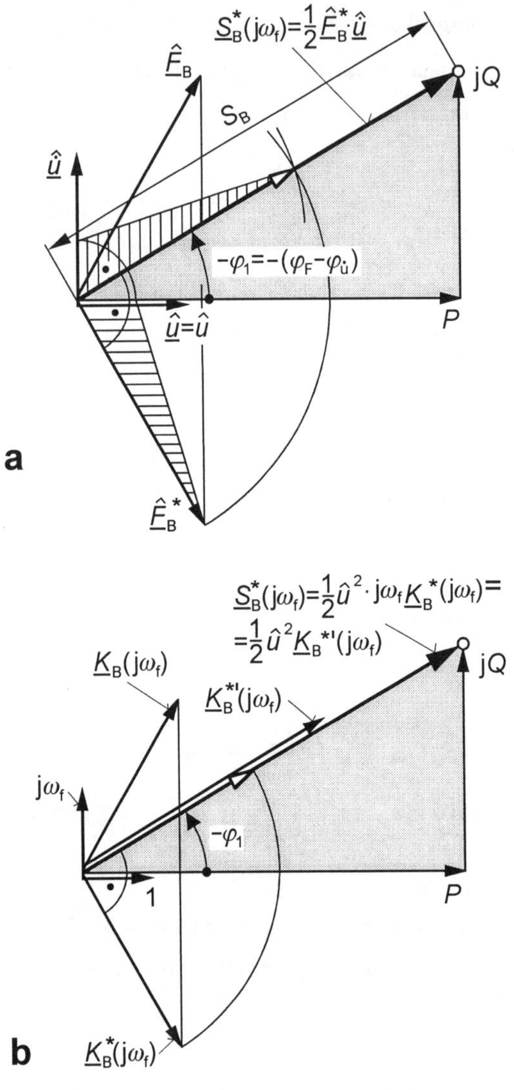

Fig. 5.5. Phasor diagrams of power geometrically interpreted by vector rotation and stretching applied to the conjugate of: **a** Rod force phasor $\hat{\underline{F}}_{\mathrm{B}}^{*}$; **b** phasor stiffness $\underline{K}_{\mathrm{B}}^{*}(\mathrm{j}\omega_{\mathrm{f}})$

For visualizing the phasor product $\underline{\hat{F}}_B^* \underline{\hat{\dot{u}}}$ in diagram representation the corresponding *vector calculus of multiplying*, Eq. (5.81a), starts from the *pair of individual phasor variables* being figured in terms of the rod force phasor $\underline{\hat{F}}_B$, Eq. (5.70b), respectively its conjugate, $\underline{\hat{F}}_B^*$, and the exciting velocity phasor $\underline{\hat{\dot{u}}}$, Eq. (5.69b). The two (resting) phasors of the effort-flow couple represent the corresponding sinusoids, Eqs. (5.70a), (5.69a). The operation of multiplying is performed by *rotation* around the angle $+\pi/2\,\text{rad}$ and *stretching* about the velocity magnitude $\hat{\dot{u}} = \omega_f \hat{u}$ being applied to the rod force phasor conjugate $\underline{\hat{F}}_B^*$, Fig. 5.5a. As a result the conjugate form of the complex input power can be figured in terms of the driving phasor power conjugate $\underline{S}_B^*(j\omega_f)$, Eq. (5.81a). This resultant vector of the effort-flow product represents the corresponding stationary flow of driving energy.

To impart the insight of energy interactions the vector representation takes advantage of involving impedance relations (frequency-response characteristics) for power calculation. For that reason the individual phasor variables are replaced by the *response phasor ratio* being figured in terms of the (direct) dynamic stiffness $\underline{K}_B(j\omega_f)$, Eq. (5.71a), respectively its conjugate $\underline{K}_B^*(j\omega_f)$, Eq. (5.71b), and the *factor* $j\omega_f$ representing the first derivative of phasors, Fig. 5.5b. Then the *vector calculus of multiplying*, Eq. (5.81b) may be interpreted as a *differentiating operation* applied to the conjugate $\underline{K}_B^*$ of the response phasor ratio $\underline{K}_B$. By rotation (around $\pi/2\,\text{rad}$) and stretching (about ω_f) the frequency-response characteristic $\underline{K}_B^*(j\omega_f)$ results in the *converted dynamic stiffness* $\underline{K}_B^{*\prime}(j\omega_f)$, Eq. (5.84), the phasor of which is *perpendicular to phasor stiffness* $\underline{K}_B^*(j\omega_f)$. Thus, being fitted for the phasor product of power in direction the required driving phasor power conjugate is obtained by an additional stretch about a constant multiplier amounting to half the square of exciting displacement $(1/2)\hat{u}^2$.

The *significant power parameters* are specified either by modulus and phase of the resultant constant phasor $\underline{S}_B^*(j\omega_f)$, designated as the apparent power S_B and the negative impedance angle $-\varphi_1$, Eq. (5.75a), or, more conveniently, by the vector components, called the active power P and the reactive power Q, Eqs. (5.85a,b). The real characteristic parameters S_B, P, Q span the (resting) *right triangle of driving power* in the complex power plane, Fig. 5.5a,b, shaded areas.

Based on developing the force-to-displacement ratio (phasor stiffness) into the force-and-velocity product (phasor power) the geometric interpretation demonstrates how to *combine* the concepts of *dynamic compliance* and *phasor power* thus delivering an ω-domain energetic approach for the design of a mechanical vibration generator.

Complex Input Power and Frequency Normalization. In the operation of *data processing* for obtaining frequency-response characteristics the *normalization* is a tool of vibration data analysis being of considerable usefulness. This also holds for the energy function in its explicit form $\underline{S}_B^*(j\omega_f)$ turning out to be a convenient data reduction method suited to graphical interpretation of phasor products in the frequency domain. For that, the *complex power* (phasor power) is *adjusted in variable and scale* as shown in 3.2.11 and 4.2.5.

The constant quantity S_0 at $\omega = \omega_0$ defines the *static power factor*. This real particular parameter is identical with
the *phasor power of the spring at resonance*

$$S_0 = \frac{1}{2\mathrm{j}}\underline{\hat{F}}_0^* \underline{\hat{u}} = \frac{1}{2\mathrm{j}}\underline{\hat{F}}_0^* \mathrm{j}\omega_0 \underline{\hat{u}} = \frac{1}{2}K_\mathrm{k} \underline{\hat{u}}^* \omega_0 \underline{\hat{u}} = \frac{1}{2}k\underline{\hat{u}}^* \omega_0 \underline{\hat{u}} = \frac{1}{2}k\omega_0 \hat{u}^2 = \frac{1}{2}k\sqrt{k/m}\,\hat{u}^2 \tag{5.86}$$

If the complex parameter $\underline{S}_\mathrm{B}^*(\mathrm{j}\omega/\omega_0)$ will be related to the multiplying constant S_0 the transformation results in
the *normalized phasor product of power* (normalized resultant phasor)

$$\frac{\underline{\underline{S}}_\mathrm{B}^*(\mathrm{j}\omega_\mathrm{f}/\omega_0)}{S_0} = \frac{\underline{\hat{F}}_\mathrm{B}^*}{\underline{\hat{F}}_0^*}\mathrm{j}\frac{\underline{\hat{u}}}{\underline{\hat{u}}_0} = \frac{\underline{\hat{F}}_\mathrm{B}^*}{k\underline{\hat{u}}^*}\frac{\underline{\hat{u}}}{\omega_0\underline{\hat{u}}} = \mathrm{j}\frac{\omega_\mathrm{f}}{\omega_0}\frac{\underline{\hat{F}}_\mathrm{B}^*}{\underline{\hat{F}}_0^*} = \mathrm{j}\frac{\omega_\mathrm{f}}{\omega_0}\frac{\underline{\underline{K}}_\mathrm{B}^*(\mathrm{j}\omega_\mathrm{f}/\omega_0)}{K_\mathrm{k}} = \frac{\underline{\underline{K}}_\mathrm{B}^{*\prime}(\mathrm{j}\omega_\mathrm{f}/\omega_0)}{K_\mathrm{k}} \tag{5.87a}$$

or, as a product of the nondimensional response ratio of phasors (phasor stiffness) $\underline{X}_\mathrm{k}^*(\mathrm{j}\eta_1)$ and the normalized differential factor $\mathrm{j}\eta_1$
the *normalized driving phasor power conjugate*

$$\begin{aligned}\frac{\underline{\underline{S}}_\mathrm{B}^*(\mathrm{j}\omega_\mathrm{f}/\omega_0)}{S_0} = \frac{\underline{\underline{K}}_\mathrm{B}^{*\prime}(\mathrm{j}\omega_\mathrm{f}/\omega_0)}{K_\mathrm{k}} &= \underline{X}_\mathrm{k}^{*\prime}(\mathrm{j}\eta_1) = \mathrm{j}\eta_1\underline{X}_\mathrm{k}^*(\mathrm{j}\eta_1) = \mathrm{j}\eta_1\left[(1-\eta_1^2) - \mathrm{j}2\zeta\eta_1\right] \\ &= \eta_1\left[2\zeta\eta_1 + \mathrm{j}(1-\eta_1^2)\right]\end{aligned} \tag{5.87b}$$

is defined being abbreviated to $\underline{X}_\mathrm{k}^{*\prime}(\mathrm{j}\eta_1)$.

The spectral decomposition into the rectangular power components is termed the *normalized active power*

$$\frac{P(\omega_\mathrm{f}/\omega_0)}{S_0} = \frac{F_{\mathrm{B}\hat{\underline{u}}}}{k\underline{\hat{u}}^*}\frac{\hat{u}}{\omega_0\underline{\hat{u}}} = \frac{1}{2}\left[\underline{X}_\mathrm{k}^{*\prime}(\mathrm{j}\eta_1) + \underline{X}_\mathrm{k}'(\mathrm{j}\eta_1)\right] = \mathrm{Re}\left[\underline{X}_\mathrm{k}^{*\prime}(\mathrm{j}\eta_1)\right] = 2\zeta\eta_1\cdot\eta_1 = 2\zeta\eta_1^2 \tag{5.88a}$$

and the *normalized reactive power*

$$\frac{Q(\omega_\mathrm{f}/\omega_0)}{S_0} = \frac{F_{\mathrm{B}\,\mathrm{j}\hat{\underline{u}}}}{k\underline{\hat{u}}^*}\frac{\hat{u}}{\omega_0\underline{\hat{u}}} = \frac{1}{2\mathrm{j}}\left[\underline{X}_\mathrm{k}^{*\prime}(\mathrm{j}\eta_1) - \underline{X}_\mathrm{k}'(\mathrm{j}\eta_1)\right] = \mathrm{Im}\left[\underline{X}_\mathrm{k}^{*\prime}(\mathrm{j}\eta_1)\right] = \left(1-\eta_1^2\right)\cdot\eta_1 \tag{5.88b}$$

Polar Representation of the Complex Power (Nyquist plot)

Using the vector representation modified to power relations the phasor product of power (resultant vector) $\underline{S}$, likewise its conjugate $\underline{S}^*$, may be plotted for each frequency variable assigned to a particular forcing frequency $\omega = \omega_\mathrm{f}$ as a single point in the complex power plane.

The corresponding *phasor power* (conjugate), Eq. (5.63b), is assigned to a particular value of (the conjugate form of) the *complex power* $\underline{S}^*(\mathrm{j}\omega) = \underline{S}^*(\mathrm{j}\omega_\mathrm{f})$. The power frequency plot can be traced out in terms of polar coordinates by the locus of the tip of the resultant vector, as the driving frequency ω_f (henceforth equivalent to ω) varies from 0 to ∞.Thus, a transformation of the actual energy flow $S(t)$ into the frequency domain is performed by sweeping out the total frequency range. The frequency locus plot connecting all the points represents the totality of power states (stationary flows of energy) for varying forcing frequency $\omega = \omega_\mathrm{f}$. As a result the required *energy flow characteristic* in the ω-domain is figured, called the *polar frequency response locus of the complex power*, in short the *polar plot of power*, Fig. 5.6.

The graphical representation of power interactions referred to linear one-port networks (general impedance elements interconnected) by a polar plot in the *complex power plane* presupposes *specifications* of

- the pair of *dynamic variable quantities* (effort-flow couple) defined or measured as the individual conjugate variables;
- the *single point in space* referred to the power state (effort-flow product) at an energy port (input or output port) with terminal impedance specified.

The considered translational mechanical system, Fig. 5.3, is represented by a steady-state power relation expressed in terms of the phasor product of *force* (conjugate) $\hat{\underline{F}}^*$ and *velocity* $\hat{\underline{\upsilon}}^*$. This relationship expressed in terms of an alge-

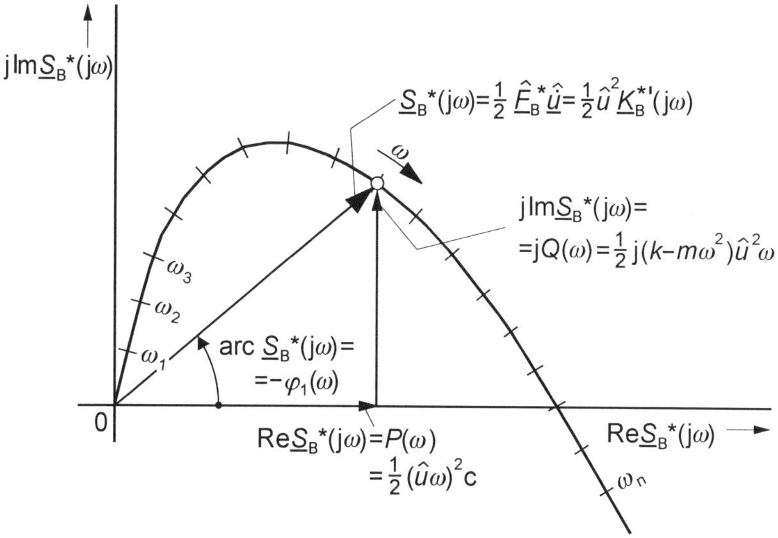

Fig. 5.6. Phasor diagram and polar plot of power at the driving point (complex input power)

braic function is taken at the *single port of entry* marked by the point B, where the terminal dynamic characteristic is specified by (the conjugate form of) the *direct (or driving-point) dynamic stiffness* (phasor stiffness) $\underline{K}_B^*(j\omega_f)$, Eqs. (3.188b), (4.14), (5.71b).

In such a way uniquely defined the power frequency plot represents graphically the phasor products of responding rod force (conjugate) $\underline{\hat{F}}_B^*$, and exciting velocity $\underline{\hat{u}}$, or specified in a more explicit form by (the conjugate form of) related characteristics, the converted phasor stiffnesses $\underline{K}_B^{*\prime}(j\omega_f)$, Eq. (5.84), multiplied by $(1/2)\hat{u}^2$, Fig. 5.5a,b.

Thus, by connecting all the points for varying angular frequency $\omega_f = \omega$ the totality of driving phasor power (conjugates) $\underline{S}_B^*(j\omega_f)$ is gained, termed the polar plot of the *complex input power* $\underline{S}_B^*(j\omega)$.

Vibratory Specifications of Power by Varying Forcing Frequency

Basing on graphical methods mechanical system design derives benefit from polar diagrams for judging the vibratory effects on machine structures in operation. It has been shown that the polar representation of frequency-response characteristics, in particular of dynamic stiffness, proves useful for selecting vibratory specifications, 4.2.5.

In like manner the *polar representation of energy flow characteristics* is of benefit to design, e.g., for proportioning of the entering energy flow to the flow of energy being absorbed and stored in machinery. Considering the steady state of energy systems the polar diagram of complex input power is appropriate for extracting *performance criteria* which are fundamental to energetic system approach, in that case to rating and optimizing the power supply required for vibratory drives. Performance criteria in terms of calculated or estimated parameter values related to energy functions may be called *characteristic parameters of power*.

Specifying Significant Power Parameters. With regard to the rectangular form of spectral decomposition the transacting components of stationary flow of energy are specified by the real and imaginary parts of driving phasor power (conjugate) $\underline{S}_B^*(j\omega_f)$, Eq. (5.63b).

The phasor product of power (resultant vector) is associated to a single forcing frequency, and thus assigned to one point of frequency locus plot. The spectral components of the power product being identical with the components of the resultant vector are related to the individual phasor variables by multiplying the velocity phasor magnitude $\hat{u}$ by the in-phase component $F_{B\underline{\hat{u}}}$, respectively the quadrature component $F_{Bj\underline{\hat{u}}}$ of the constant rod force phasor. Thereupon, the components of energy flow entering the system follow from the *active power* P,

Eqs. (5.66a), (5.85a), and the *reactive power Q*, Eqs. (5.66b), (5.85b), being either the in-phase or, respectively, the quadrature component of driving phasor power (conjugate) $\underline{S}_\mathrm{B}^*(\mathrm{j}\omega_\mathrm{f})$.

If forcing frequency is allowed to vary within the frequency range of interest the change of the *significant power parameters, P* and *Q*, can be gathered by the frequency locus plot following the change of the components of the resultant vector. The assembly of vector components appertaining to the totality of driving phasor power (conjugates) is equal to the assembly of rectangular coordinates (with respect to the complex power plane).

One problem of system design in the frequency domain arises from predicting *performance criteria for a vibration generator* involving not only vibrations but also the transformation and flow of energy in system-to-system interactions. Supposing a simple harmonic vibration the phasor relations, Eq. (5.85a,b), permit to determine "dependent on frequency" the *dissipative and the idle flow of energy*, which is to be delivered, respectively to be compensated by the power supply for generating or maintaining a presupposed constant operating motion $\hat{s}$, or $\hat{\upsilon}$ (constant-amplitude motion).

Specifying Significant System Parameters. For illustrating frequency effects on energy functions the *normalizing transformation of complex power* (phasor power) is used being a convenient data reduction method suited to graphical interpretation in the frequency domain. For that, the phasor product of power is adjusted in variable and scale as already demonstrated, Eqs. (5.86), (5.87a,b).

The *normalized driving phasor power* (conjugate) is figured in terms of an adjusted resultant vector defined by the "nondimensional phasor product" of power $\underline{X}_\mathrm{k}^{*\prime}(\mathrm{j}\eta_1)$, Fig. 5.7a.

The number of parameters is diminished by combining element parameters to system parameters, so that the new function, Eq. (5.87b), is expressed in terms of only two quantities, the frequency ratio η and the damping ratio ζ. Thus, the effect of the *performance criterion for related energy dissipation* ζ on the energy flow characteristic will be represented graphically in the normalized form by an *adjusted (data-reduced) polar plot* for a fixed amount of damping as η varies over some driving frequency range, Fig. 5.7b.

Response Data Plotting. The polar plots of *normalized complex power* are shown in Fig. C.1 of the Appendix C.

The new coordinates being in general polar coordinates of magnitude and phase (Nyquist plots) are replaced by *rectangular (Cartesian) coordinates* assigned to the real and imaginary parts of the complex energy function, Eq. (5.87b). By this, the graphical representation in the complex power plane gives the preference to spectral decomposition of power in the convenient rectangular form. The coordinates of (vector) components are uniformly scaled in absolute numbers making it possible to estimate at each *frequency ratio* η the significant power parameters, designated as normalized active and reactive power, Eqs. (5.88a,b).

The effect of the system parameter *damping ratio* ζ on the *energy flow characteristic* (complex power at the driving point or complex input power) $\underline{S}_\mathrm{B}^*(\mathrm{j}\omega)$ is represented graphically in

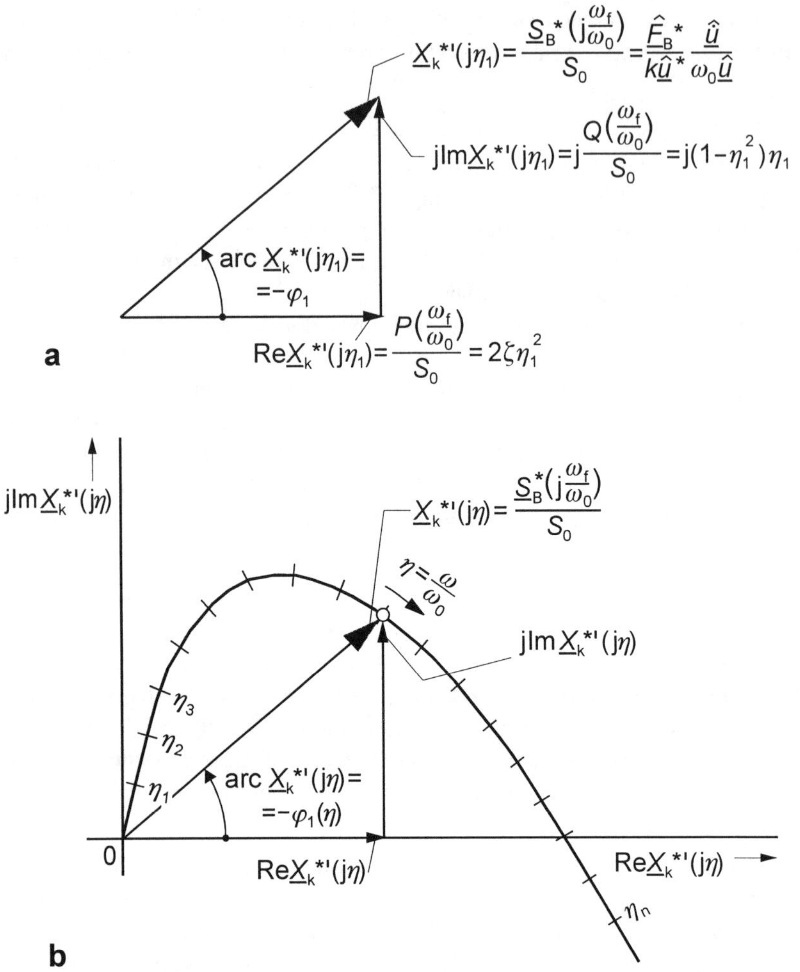

Fig. 5.7. Adjusted diagrams of power at the driving point. **a** Phasor diagram of normalized driving phasor power; **b** polar diagram of normalized complex input power

the normalized form $\underline{X}_k^{*\prime}(j\eta)$ by a set (or family) of polar plots for various amounts of damping. The damping effect is indicated by the *change in normalized active power component*, i.e., by the change in the dissipative flow of energy.

Specifications of Polar Plots of Power. A polar plot (Nyquist plot) is specified by the curve shape, the frequency scale, and the shape signature at high and low frequencies as treated in 3.2.11.

Respecting the polar representation of a set of normalized frequency characteristics of power the adjusted *curve shapes* are defined by the loci (hereat higher parabolic plane curves) for selected damping ratio values $\zeta = const$, and the *frequency scales* are outlined as intersecting lines running parallel with the abscissa for selected frequency ratio values $\eta = const$.

The *shape signature* is portrayed by a locus starting perpendicularly to the real axis from zero at $\eta = 0$. Assuming a resistance or lossy element (damper) as existent the dissipative flow of energy delivered by the power supply (source) is absorbed by the network (generalized impedance).

Thus, the *real parts of the loci*, representing the normalized active or average power, only can be positively valued for a *passive system*. On the average, more energy flows into the network than is returned to the source. That is to say, the *magnitudes of the loci* (lengths of adjusted phasor powers), representing the normalized apparent power, are *greater than or equal* to their real parts for all frequency ratios. This performance criterion is according to the condition of power factor $\lambda \leq 1$, Eq. (5.56b).

The *imaginary parts of the loci*, representing the normalized reactive power, indicate the extent to which the source participates in energy storage interaction between the capacitance and inertance element (spring and mass). Depending on frequency ratio this idle flow of energy can be *positively or negatively valued*. The latter case makes it physically possible for a passive network to supply the source. Under condition of minor resonant vibration $\eta < 1$ the average energy stored in the capacitance (spring) exceeds that stored in the inertance (mass), thus the sign of reactive power is positive $Q > 0$. Under condition of major resonant vibration $\eta > 1$ the exceeding of energy storage, respectively the sign of reactive power are vice versa $Q < 0$.

Power Component Frequency Plots

A common alternative to the frequency locus in the complex $\underline{S}(\mathrm{j}\omega)$-plane (complex power plane) results from the graphical representation of rectangular frequency responses of power, i.e., by portraying the complex power, Eq. (5.81c), separately by the power components versus frequency. The pair of real-valued frequency response curves of power is called the *power component frequency plots*.

Response Data Plotting. The power component frequency plots of the *normalized active and reactive power* $\mathrm{Re}\left[\underline{X}_{\mathrm{k}}^{*\,\prime}(\mathrm{j}\eta)\right]$, respectively $\mathrm{Im}\left[\underline{X}_{\mathrm{k}}^{*\,\prime}(\mathrm{j}\eta)\right]$, are shown in Fig. C.2 and C.3 of the Appendix C.

The new coordinates being rectangular axes of either real part or imaginary part versus frequency have a natural scale on both axes. The rectangular coordinates thus generate coordinate frames for a pair of real-valued frequency plots of power as *natural curve charts*.

Respecting the rectangular representation of a set of normalized power components the *pair of adjusted curves* is defined by the normalized active and reactive power, both for selected damping ratio values $\zeta = const$ and for a natural *range of frequency ratio*, the "nondimensional" bandwidth starting from zero $\Delta\eta = \eta_2 - \eta_1 = 2{,}5 - 0 = 2{,}5$.

Significant Frequencies of Resonance. The compensation of energy transactions across the boundaries of the system (mechanical circuit) depending on frequency is to be performed by the power supply (mechanical source). The geometric locus of the maxima of idle energy flow (largest swappage of energy between source and circuit) according to greatest values of the imaginary parts for all amounts of damping (horizontal top line) indicates the *distinct frequency ratio*

$$\eta_{\mathrm{max,\,Im}} = 0{,}5774 \tag{5.86}$$

referred to the *maximum of normalized reactive power* $Q = Q_{\mathrm{max}}$.

At the singular value of forcing frequency $\omega_f = \omega_0$, defined in 3.1.2 as the *condition of resonance*, the magnitudes of the loci (adjusted phasor powers) are equal to their real parts (normalized active power) for all amounts of the damping. This is a consequence of the fact that, on the average, the amounts of energy stored in capacitance and inertance elements (spring and mass) are equal. Then *resonance* occurs characterized by the phenomenon that the imaginary parts of the loci vanish, and the performance criterion of power factor equals unity $\lambda = 1$, Eq. (5.56b). The *resonance frequency ratio* $\eta_0 = 1$, Eq. (3.46a), referred to *zero reactive power* $Q = 0$, is associated with the horizontal direction of phasor powers. The geometric locus of normalized complex input power in resonance (no compensative energy transaction) coincides with the (positive) real axis.

Q factor and Low-loss Network. Since capacitance and inertance elements merely swap energy back and forth between them the source is not called upon to participate in the idle flow of energy. This energy state condition specifying resonance relieves the source from compensative energy transactions, and is referred in electric power circles as *power-factor correction.*

Besides resonance the energy flow behaviour in the vicinity of resonance is of special interest. Considering the bandwidth or half-power method, 3.2.11, the *Q factor* (quality factor) has already been defined as one-half the reciprocal of the damping ratio, Eq. (3.39c). This performance criterion commonly used with reference to a lightly damped system can be approximately defined as 2π times the stored total energy to the energy dissipation per cycle. Expressed in terms of

$$Q = 2\pi \frac{\hat{E}_p}{\Delta \hat{E}_d} \tag{3.87}$$

the Q factor can be interpreted on an energy basis relating 2π times the maximum of circuit energy storage

$$\hat{E}_p = \frac{1}{2} k \hat{u}^2 = \frac{1}{2} m \omega_0^2 \hat{u}^2 = \frac{1}{2} m \hat{\dot{u}}^2 = \hat{E}_k \tag{3.88a}$$

and the energy dissipated per cycle

$$\Delta E_d = \frac{1}{2} c \omega_f \hat{u}^2 = P \cdot (T/2\pi) \tag{3.88b}.$$

The energetically based definition, Eq. (3.87), not only provides an independent approach to the estimation of efficiency and economy of operation (useful in situations where parameter calculations are difficult or not feasible) but also illustrates the meaning of *low-loss or high-Q networks*, as called in electric circuitry, [16]. Namely, it is one in which the loss per cycle is small compared with the peak value of the total stored energy. In order to obtain a circuit with a sharp resonance curve, one must strive to obtain as large an energy storage as possible relative to the associated loss per cycle. This is utilized for mechanical vibrators taking benefit from the resonant effect.

Driving Power of other Types of Vibrators

The calculation of power flow using the phasor power concept has been exemplified by a *directly driven vibration generator system.* The usefulness of the energy approach basing on the explicit form of complex power proves for any type of vibration generator. So the *reactive-type vibrators* getting benefit from the resonant effect are energy systems the steady state of which is illustrated in a distinct form by use of phasor power calculation. As a result the energy flow characteristics are available which may be graphically interpreted by the polar representation of the complex power (Nyquist plots), [120], [121].

Response Data Plotting. The polar plots of *normalized complex power* are shown in Fig. C.4 and C.5 of the Appendix C.

The effect of the system parameter *damping ratio* ζ on the *energy flow characteristic* is represented graphically by the normalized complex power at the driving point by a set (or family) of polar plots for a fixed value of damping ratio ζ.

Beyond the two quantities η and ζ a third system parameter must be introduced specifying the *stiffness ratio* κ of the reaction type vibrator with an additional spring k_2, respectively the mass ratio μ in the case of adding an unbalanced rotating mass m_2 to the vibrator.

Appendix A

Time-history Curves of the Displacement Response and of the Force Excitation

Free Vibration Histories
Complete Response Histories
Fourier Spectra
Time-response Characteristics
Shock Response Spectrum
Random Data Plots

Performance Criteria

Steady-state response specifications
Transient-response specifications
Statistical response specifications

Displacement Time History

$s(\tau)\,/\,s_0 - \tau$-curve

Initial-condition free vibration (transient motion)
of the underdamped oscillator, Eq. (3.23b),
starting from a specific initial value ratio $\dot{s}_0/(\omega_0 s_0) = s_0'/s_0$

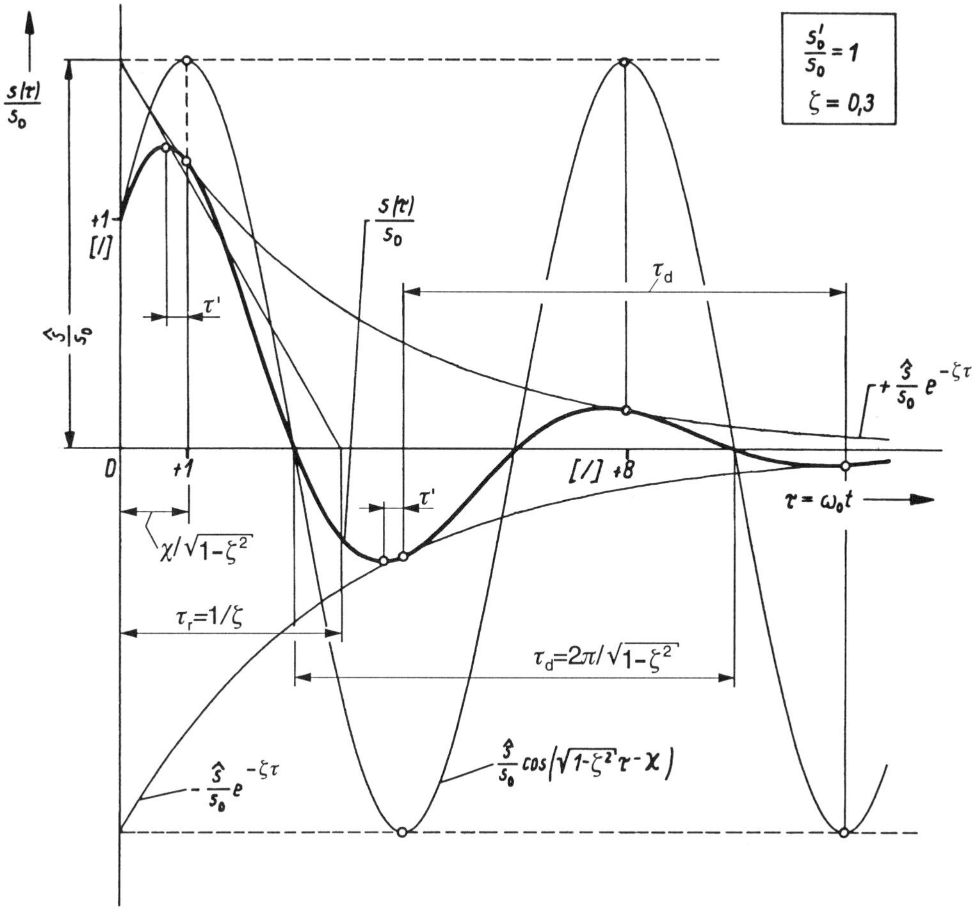

Transient-response specifications

Normalized

time constant (relaxation time) $\tau_r = T_r\omega_0 = 3{,}33$

damped natural period $\tau_d = T_d\omega_0 = 6{,}59$

Logarithmic decrement $\Lambda = \tau_d/\tau_r = T_d/T_r = 1{,}98$

Fig. A.1. Free vibration (transient motion), underdamped oscillator

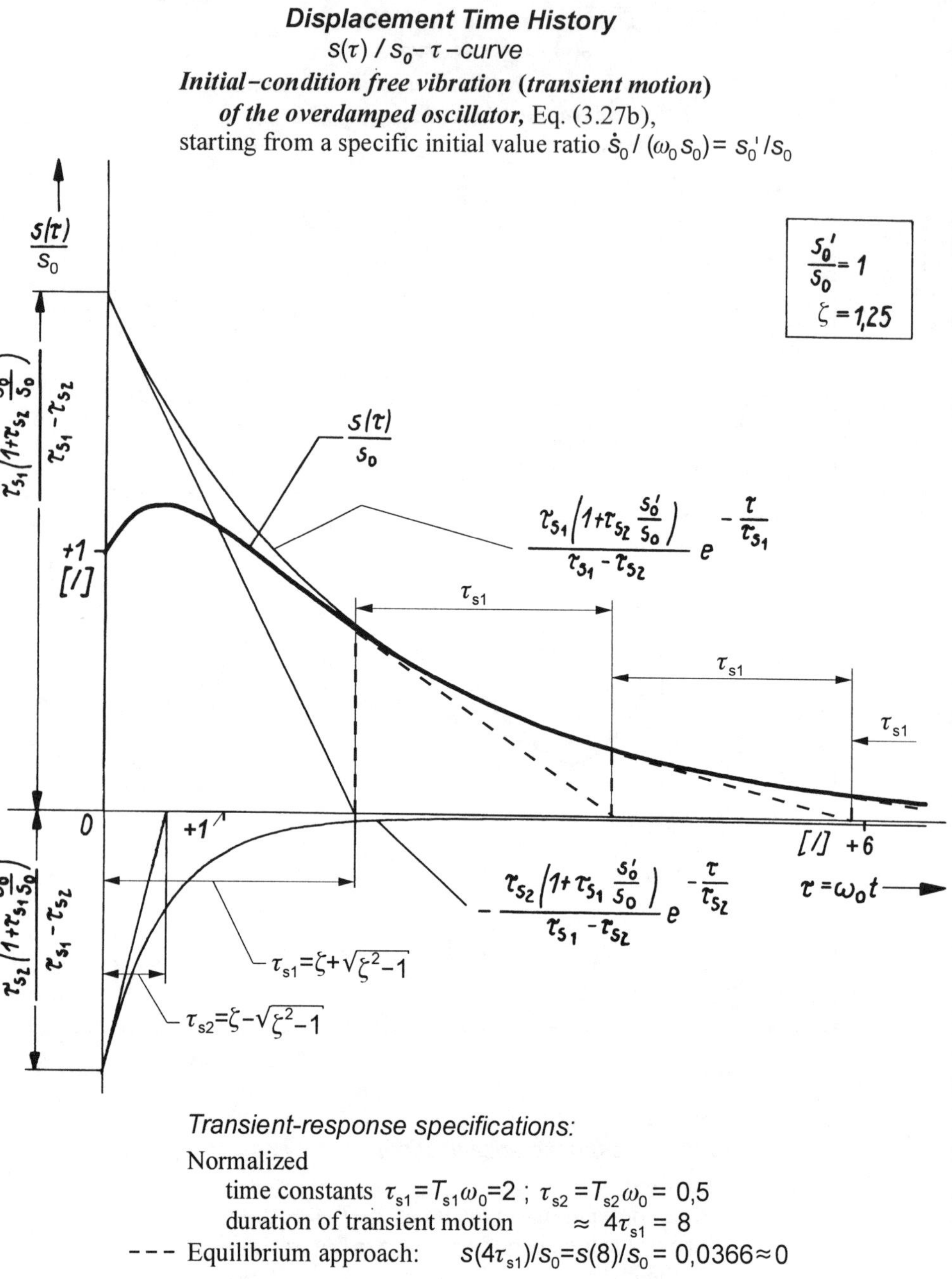

Fig. A.2. Free vibration (transient motion), overdamped oscillator

Displacement Time History

$s(\tau) / s_0 - \tau$-curves

Initial-condition free vibration (transient motion)
of the overdamped and the critically damped oscillator,
Eqs. (3.27b), (3.29b)
Effect of initial value ratio $\dot{s}_0/(\omega_0 s_0) = s_0'/s_0$ on time-history curve

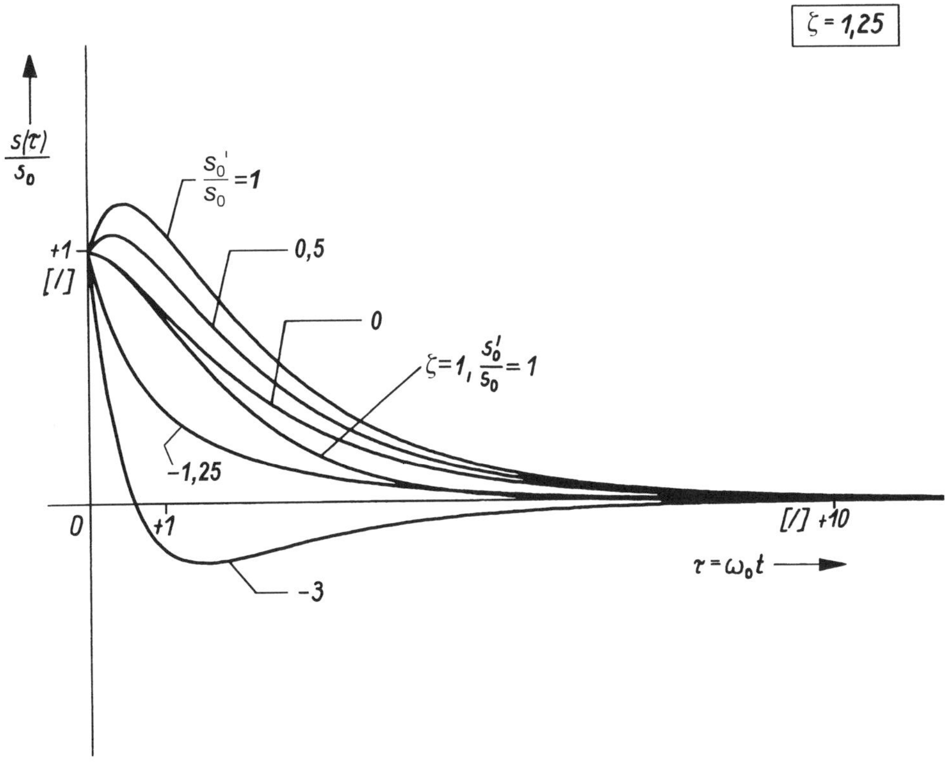

Fig. A.3. Free vibration (transient motion), over- and critically damped oscillator

Response Time History

$s(\tau)\,/\,s_{stat} - \tau$ -curve

Complete displacement response (combined motion) of the underdamped oscillator, Eq. (3.44b), caused by a transient harmonic excitation $F(\tau)/\hat{F}$, Eq. (3.35a), (dash-and-dot line) beneath the resonance condition $\omega_f/\omega_0=\eta_1=1$, i.e., for $\eta_1<|\sqrt{1-\zeta^2}|$ or $\eta_1<1$ due to $|\sqrt{1-\zeta^2}|\approx 1$ for $\zeta\ll 1$

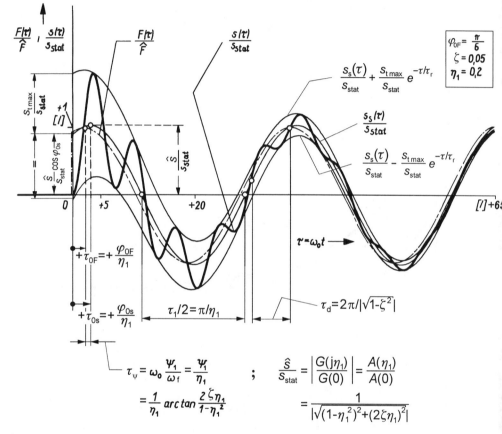

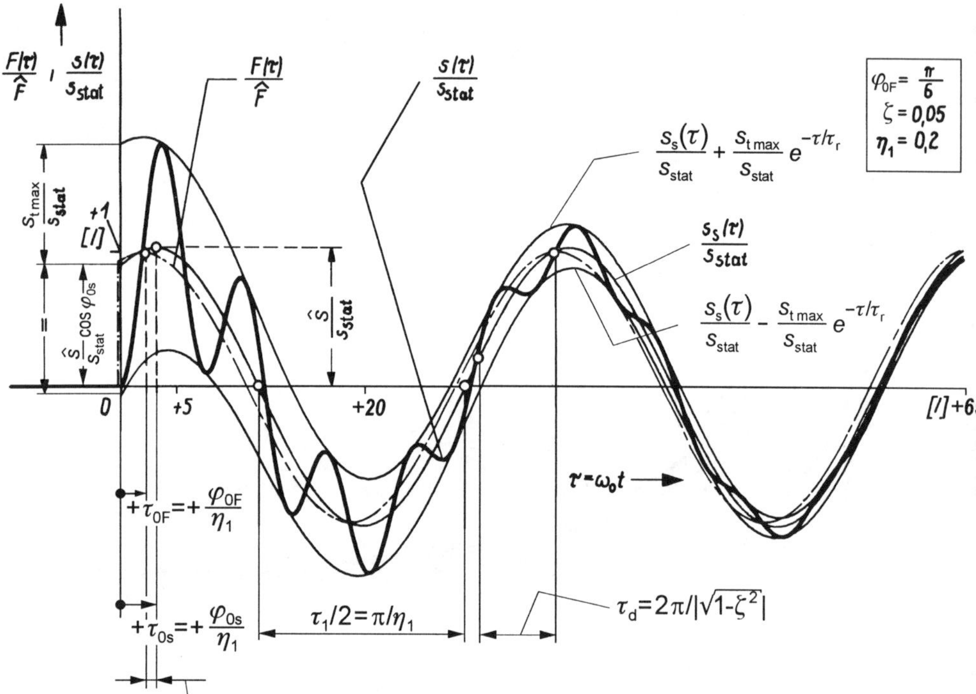

$$\tau_\psi = \omega_0 \frac{\psi_1}{\omega_f} = \frac{\psi_1}{\eta_1} = \frac{1}{\eta_1} \arctan \frac{2\zeta\eta_1}{1-\eta_1^2} \quad ; \quad \frac{\hat{s}}{s_{stat}} = \left|\frac{G(j\eta_1)}{G(0)}\right| = \frac{A(\eta_1)}{A(0)} = \frac{1}{|\sqrt{(1-\eta_1^2)^2+(2\zeta\eta_1)^2}|}$$

Transient-response specifications:

$\tau_r = T_r\,\omega_0 = 20$; $\tau_d = T_d\,\omega_0 = 6{,}29$, see Fig. A.1

Steady-state response specifications:

Magnification factor $\hat{s}/s_{stat} = k\hat{s}/\hat{F} = 1{,}04$

Normalized time lag (phase shift) $\tau_\psi = T_\psi\,\omega_0 = 0{,}104$

Fig. A.4. Complete response (combined motion), underdamped oscillator

Response Time History

$s(\tau)\,/\,s_{stat}$ – τ – curve

Complete displacement response (combined motion) of the overdamped oscillator, Eq. (3.45b),
caused by a transient harmonic excitation $F(\tau)/\hat{F}$, Eq. (3.35a),
(dash-and-dot line)
beneath the resonance condition $\omega_f/\omega_0=\eta=1$, i.e.,
for $\eta_1<1$

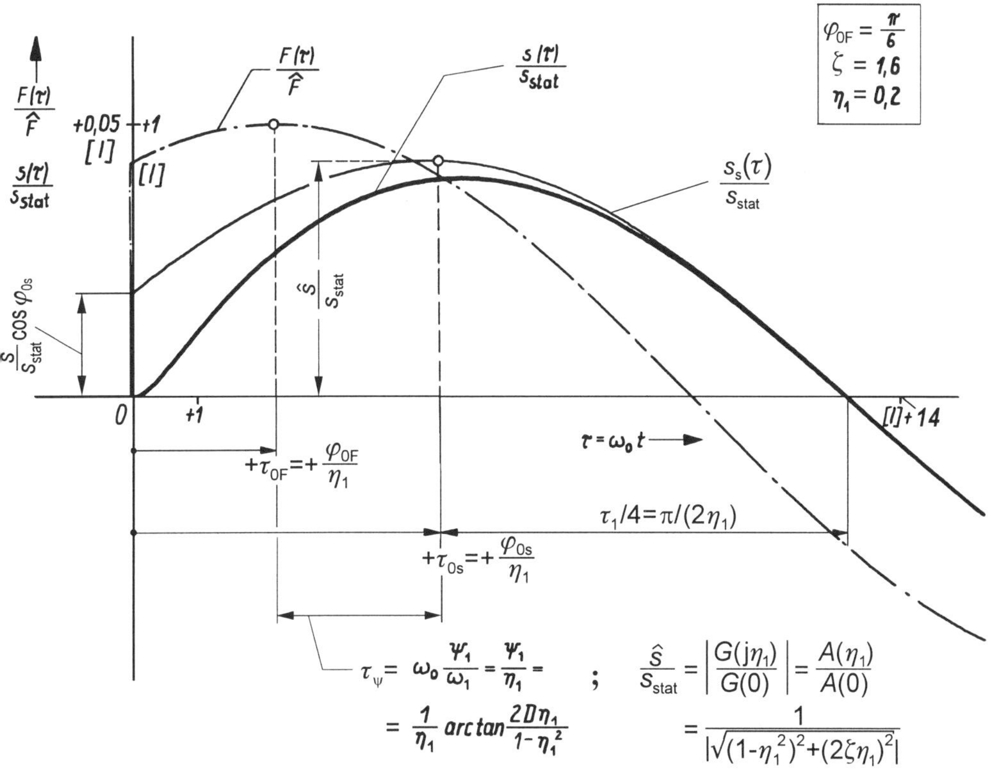

Transient-response specifications:

$\tau_{s1}=T_{s1}\,\omega_0=2{,}85$; $\tau_{s2}=T_{s2}\,\omega_0=0{,}35$, see Fig.A.2

Steady-state response specifications:

Magnification factor $\hat{s}/s_{stat}=k\hat{s}/\hat{F} = 0{,}87$
Normalized time lag (phase shift) $\tau_\psi=T_\psi\omega_0 = 2{,}94$

Fig. A.5. Complete response (combined motion), overdamped oscillator

Response Time History
$s(\tau) / s_{stat} - \tau$ -curve
Complete displacement response
of the undamped oscillator, Eq. (3.46c),
caused by a transient harmonic excitation $F(\tau)/\hat{F}$, Eq. (3.35a),
at frequency in resonance $\eta_1 = \omega_f / \omega_0 = 1$

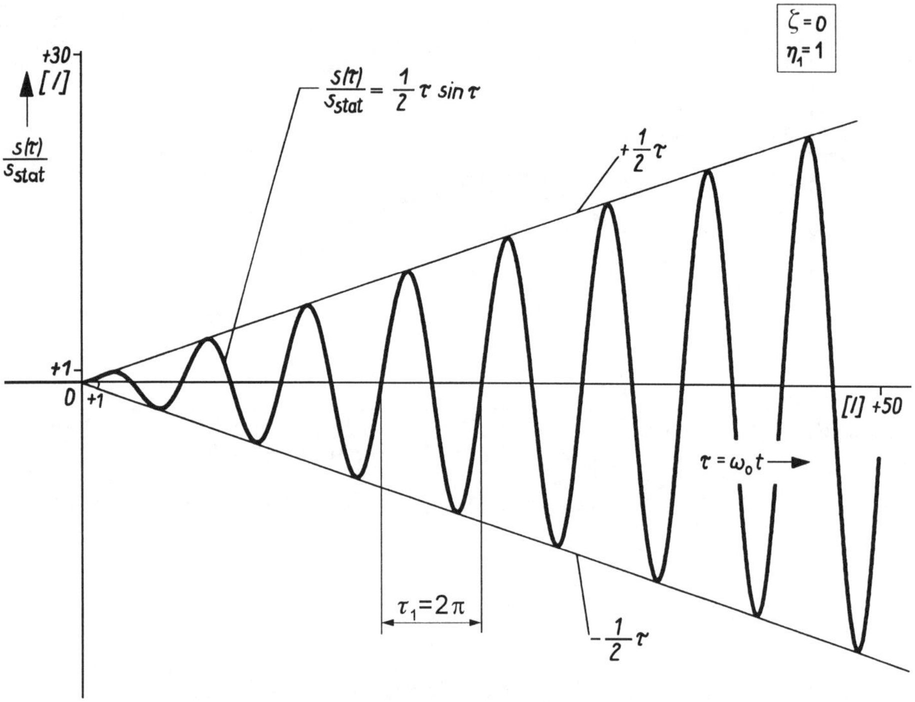

Resonance Phenomenon
(unstable forced motion)
Normalized
forcing period $\tau_1 = T_f \omega_0 = 6{,}28$

Fig. A.6. Complete response , undamped oscillator

Response Time History

$s(\tau) / s_{stat} - \tau$ *-curve*

Complete displacement response

of the undamped oscillator, Eq. (3.48b),

caused by a transient harmonic excitation $F(\tau)/\hat{F}$, Eq. (3.35a),

at a frequency slightly different to resonance $\eta_1 = \omega_f / \omega_0 \approx 1$

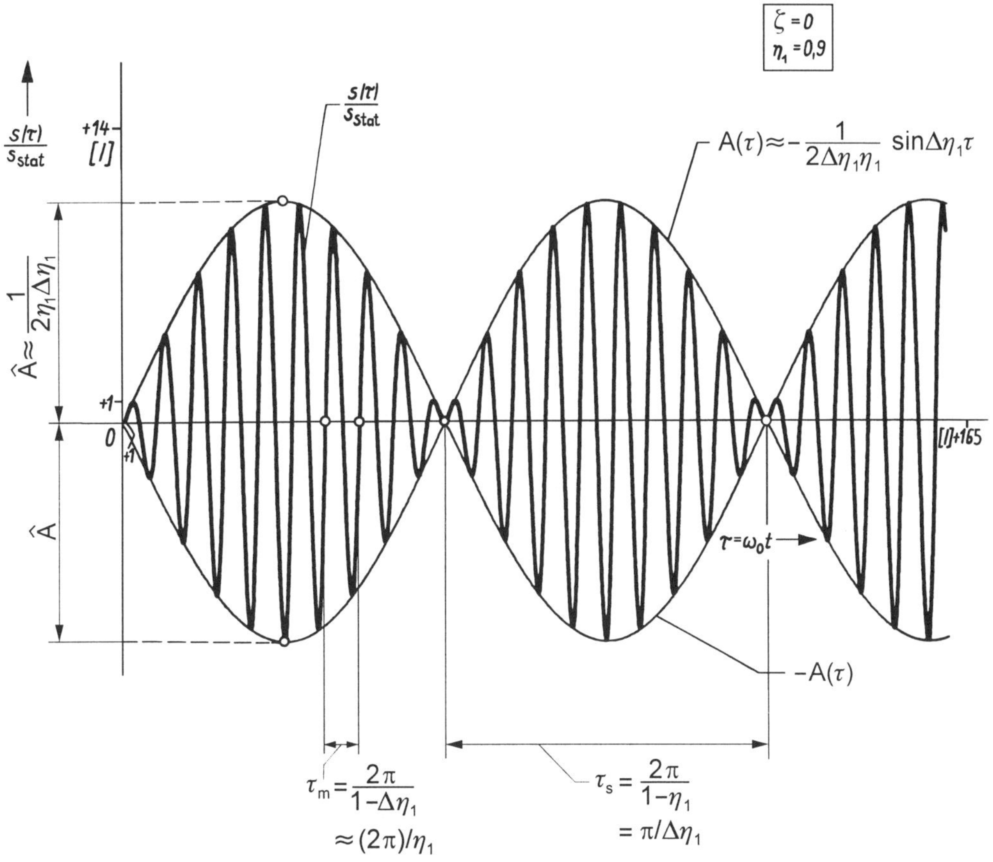

Beating Phenomenon

Normalized

"period" of beat $\tau_s = T_s \omega_0 = 62{,}83$

"period" of motion $\tau_m = T_m \omega_0 = 6{,}98$

Fig. A.7. Complete response , undamped oscillator

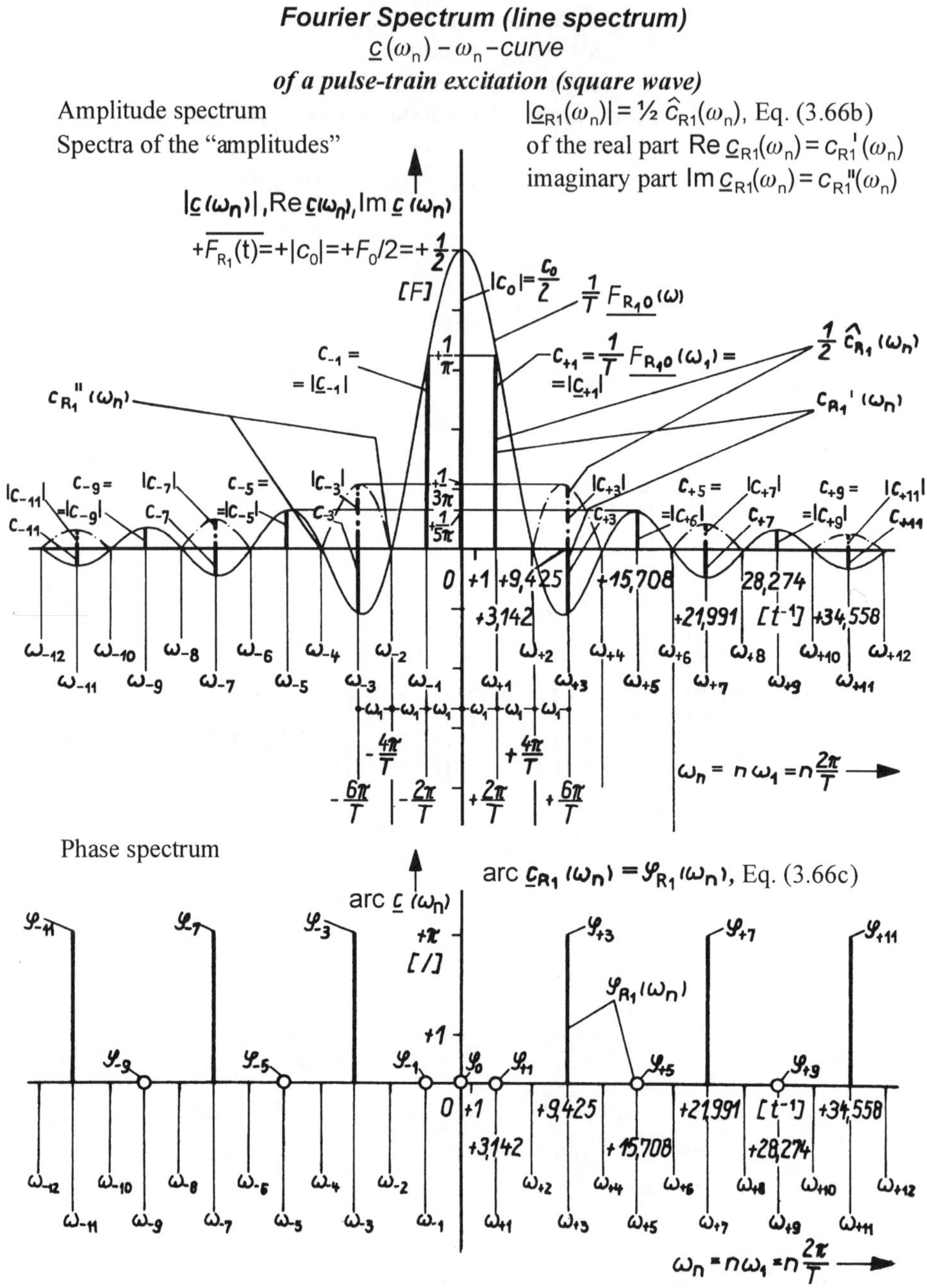

Fig. A.8. Discrete frequency spectrum of a periodic excitation force

Fourier Spectrum (line spectrum)

$|\underline{c}(\omega_n/\omega_1)|/|c_0|-(\omega_n/\omega_1)$ *-curve*

of a pulse-train excitation (square wave)

Amplitude function $|\underline{c}_{R1n}|/|c_0| = \hat{c}_{R1n}/c_0$, Eq. (3.67b)

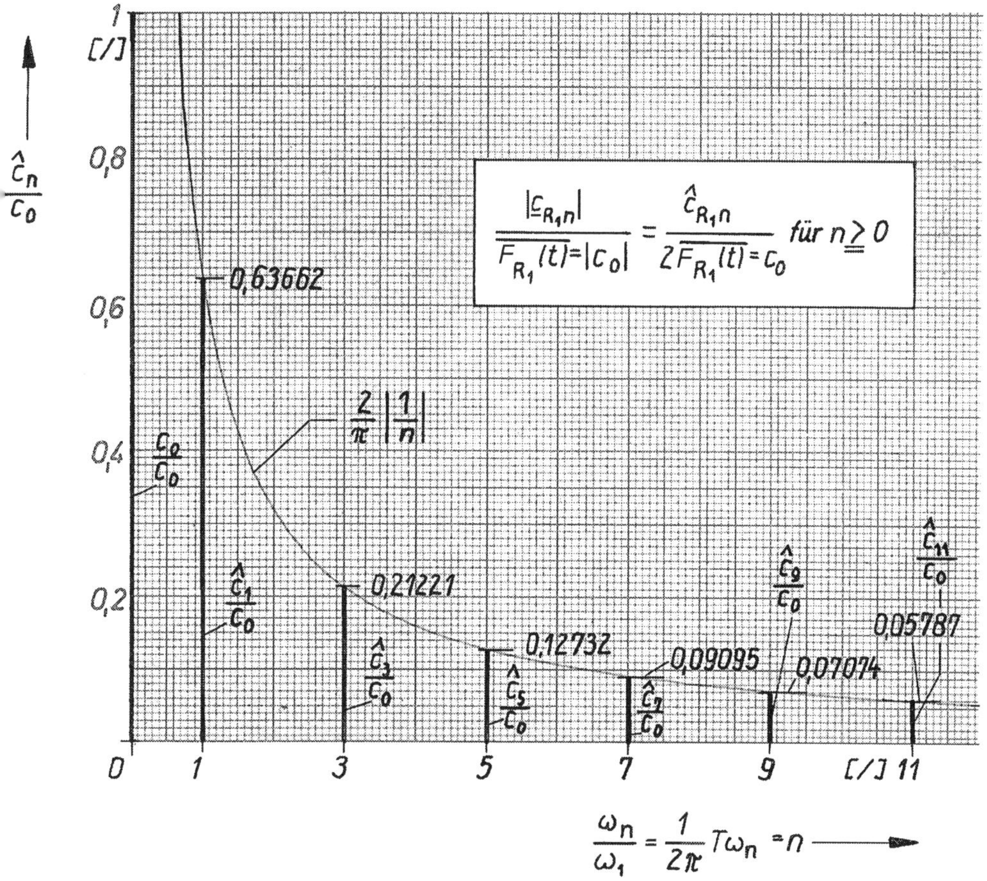

Fig. A.9. Discrete magnitude ratio spectrum

Fourier Spectrum (line spectrum)

of a pulse-train excitation (rectangular pulses)

Effect of duty cycle $\vartheta = \tau_0/T$ on amplitude spectrum $|\underline{c}_{R_1}(n)|$ for equal mean value (average) $\overline{F_R(t)} = F_0\vartheta = 1\,[F]$

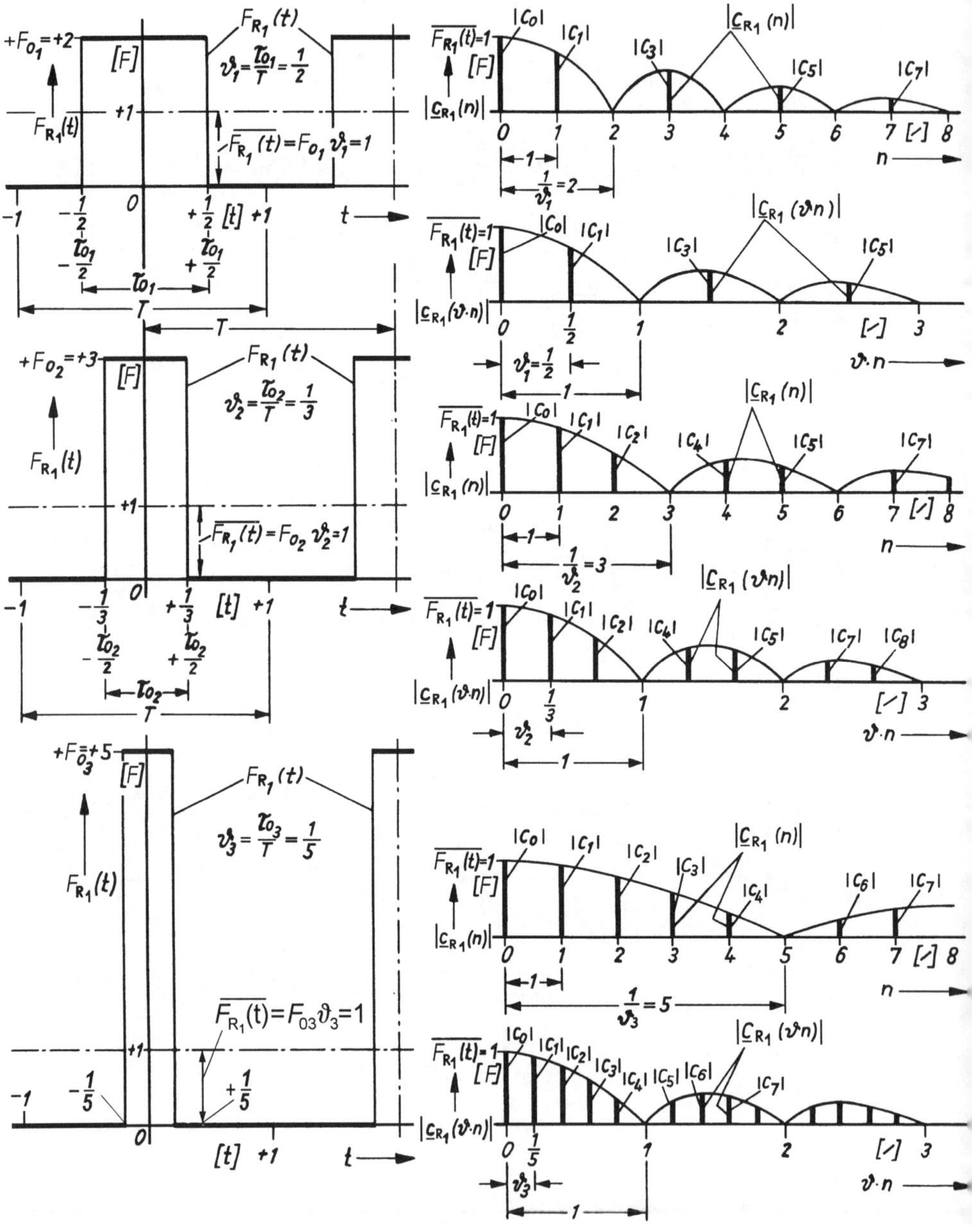

Fig. A.10. Change of duty cycle (pulse control factor) on a periodic excitation force

Fourier Spectrum (continuous spectrum)

$\underline{F}(\omega) - \omega$-curve

of a single-pulse excitation (rectangular pulse)

Amplitude density function $|\underline{F}_{R10}(\omega)| = A_{R10}(\omega)$, Eq. (3.72a),

Spectra of the "amplitude densities" of the real part $\mathrm{Re}\,\underline{F}_{R10}(\omega) = R_{R10}(\omega)$

imaginary part $\mathrm{Im}\,\underline{F}_{R10}(\omega) = X_{R10}(\omega)$

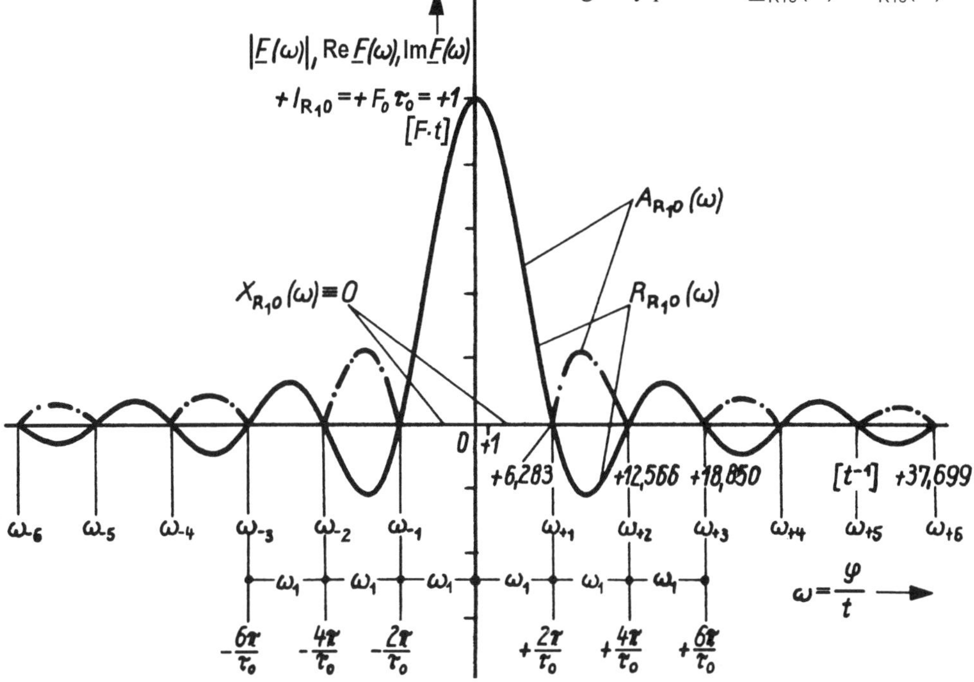

Phase spectrum arc $\underline{F}_{R10}(\omega) = \phi_{R10}(\omega)$, Eq.(3.72b)

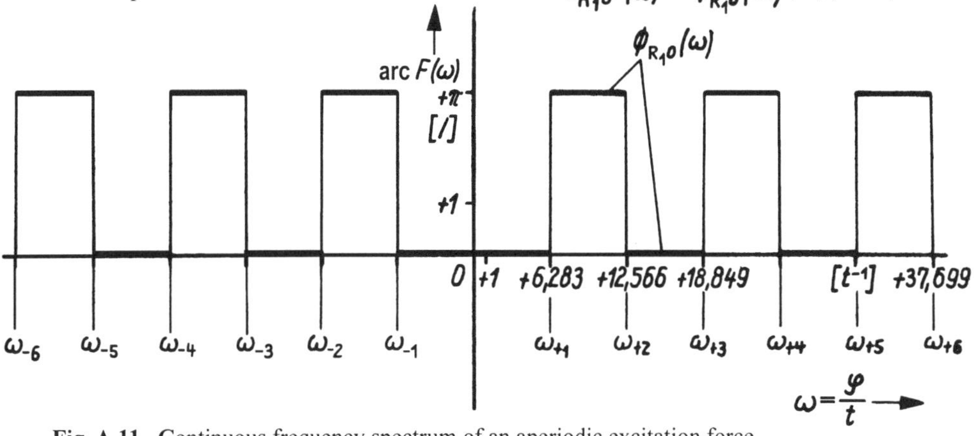

Fig. A.11. Continuous frequency spectrum of an aperiodic excitation force

Fourier Spectrum (continuous spectrum)
$|\underline{F}(\omega/\omega_1)|/F_0 - \omega/\omega_1$-*curve*
of a single-pulse excitation (rectangular pulse)

Amplitude density function $A_{R10}(\frac{\omega}{\omega_1})/I_{R10}$, Eq. (3.73)

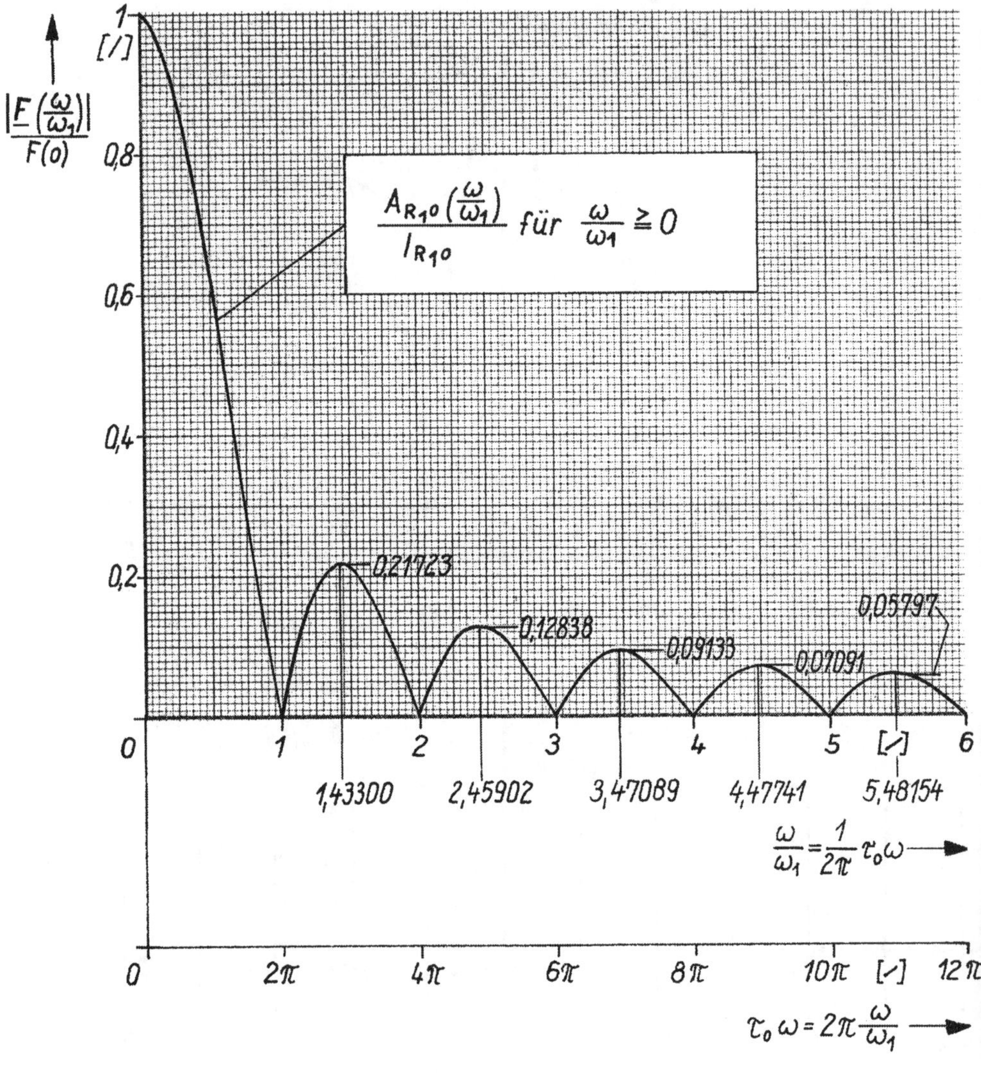

Fig. A.12. Continuous magnitude ratio spectrum

Fourier Spectrum (continuous spectrum)

of a single-pulse excitation (rectangular pulse)

Effect of pulse duration τ_0 on amplitude density function $|\underline{F}_{R10}(\omega)|$

for equal pulse area (impulse) $I_{R10}=F_0\tau_0=1$ $[F\cdot t]$

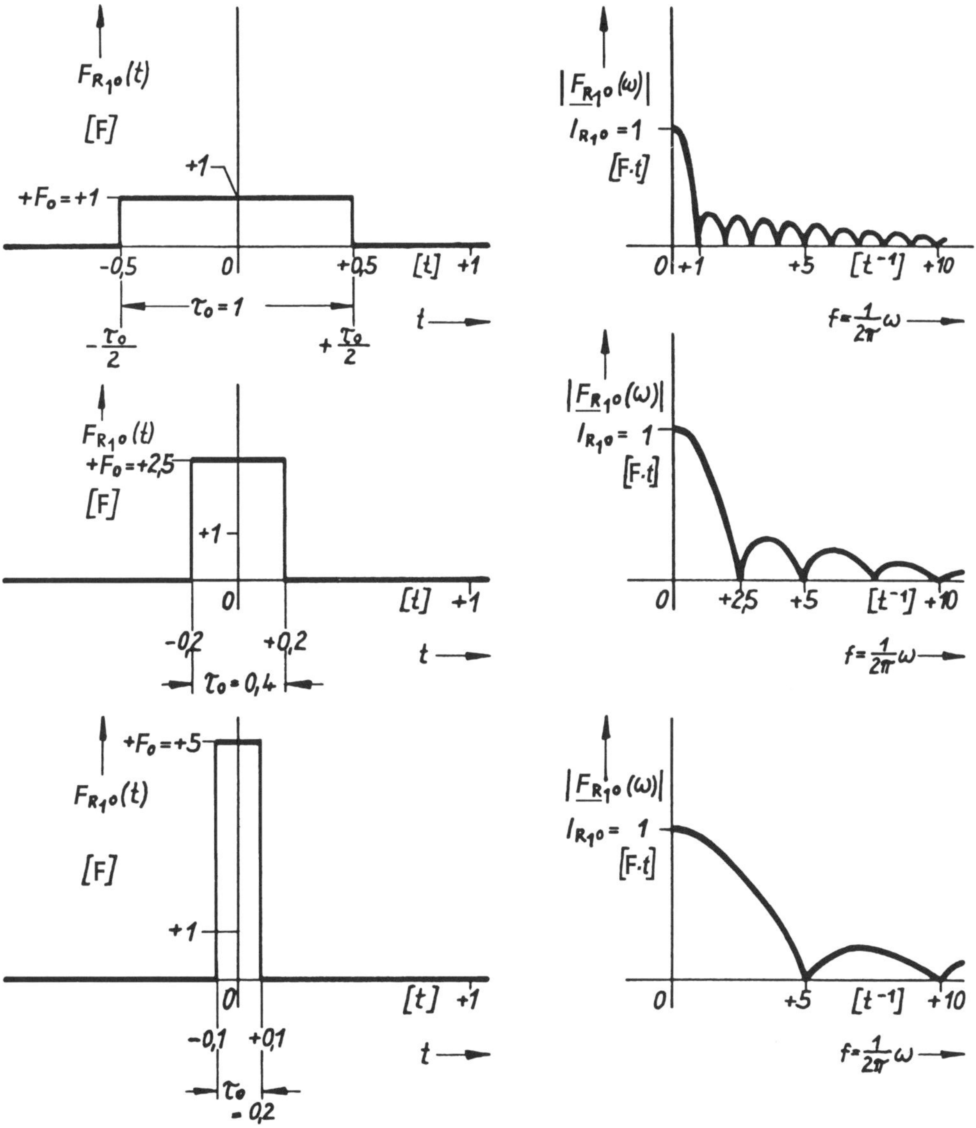

Fig. A.13. Change of pulse duration on an aperiodic excitation force

Graphical Interpretation of the Convolution Integral,

Eq. (3.92b), exemplified for
a lightly damped system: "weighting" function $g(\tau)$, Eq. (3.83b),
and a constant-slope rising excitation: ramp force function $F(\tau) = F_1 u_1(\tau)$
$F(\tau)=F_1 u_1(\tau)=(F_0/\tau_1)u_1(\tau)=(F_0/\tau_1)\tau$ for $\tau \geqq 0$

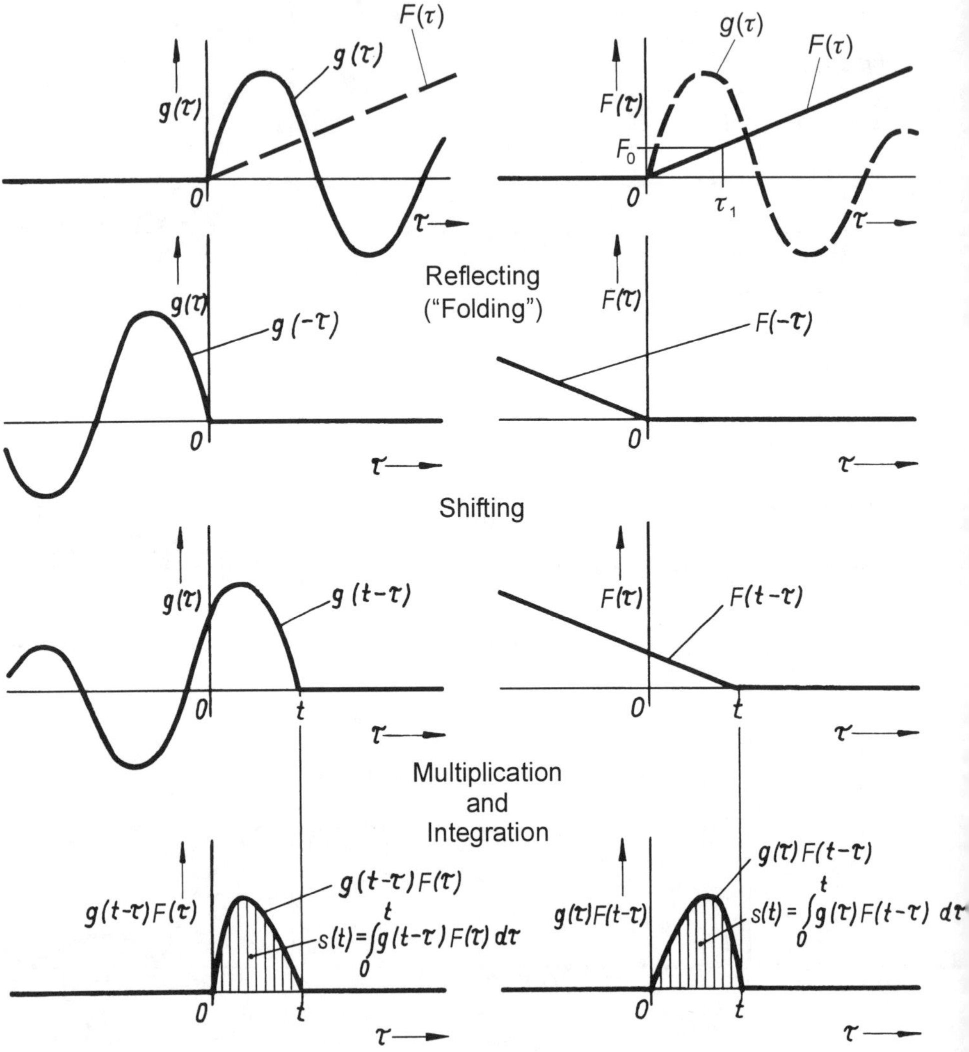

Symmetry of integral transformation in $F(t)$ *and* $g(t)$

Fig. A.14. Convolution integral or superposition integral

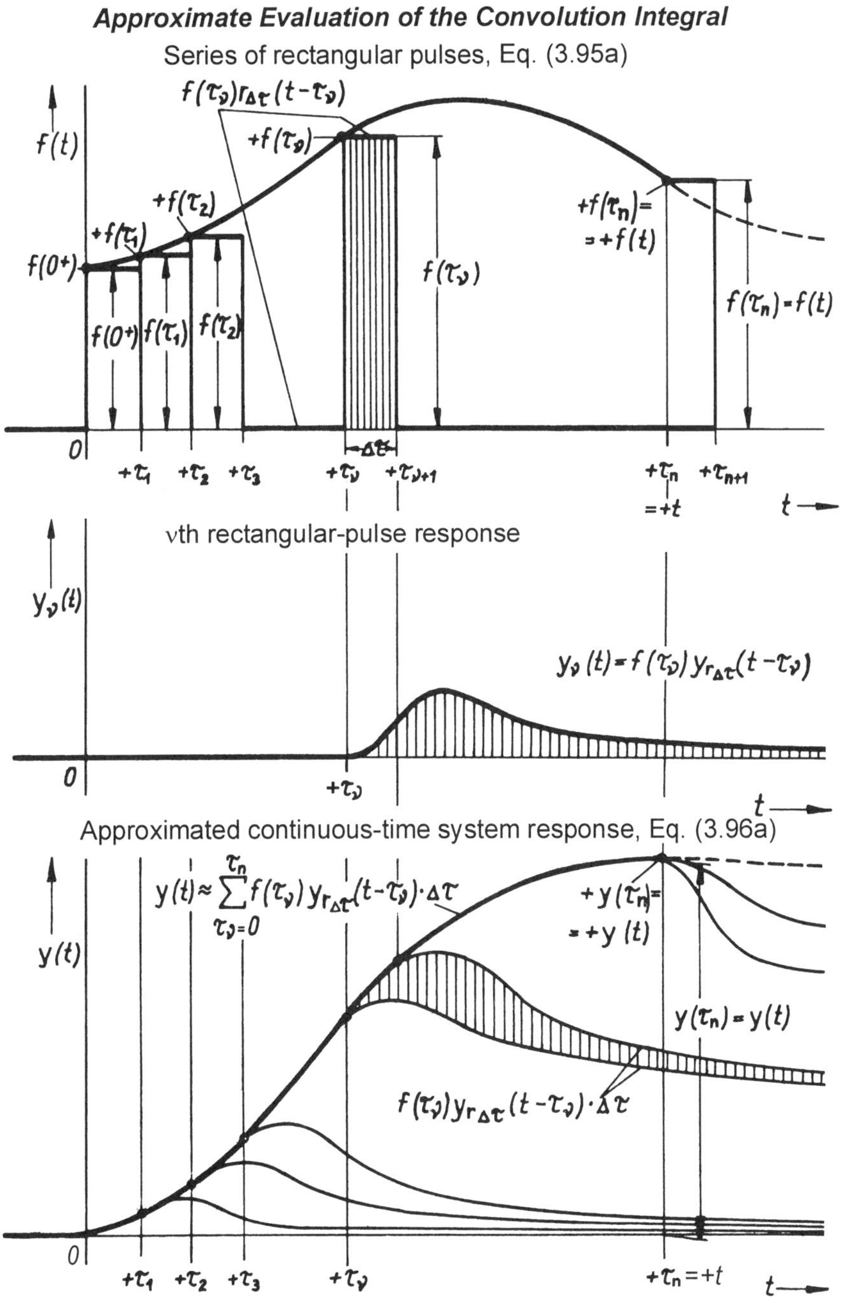

Fig. A.15. Pulse approximation of an arbitrary excitation

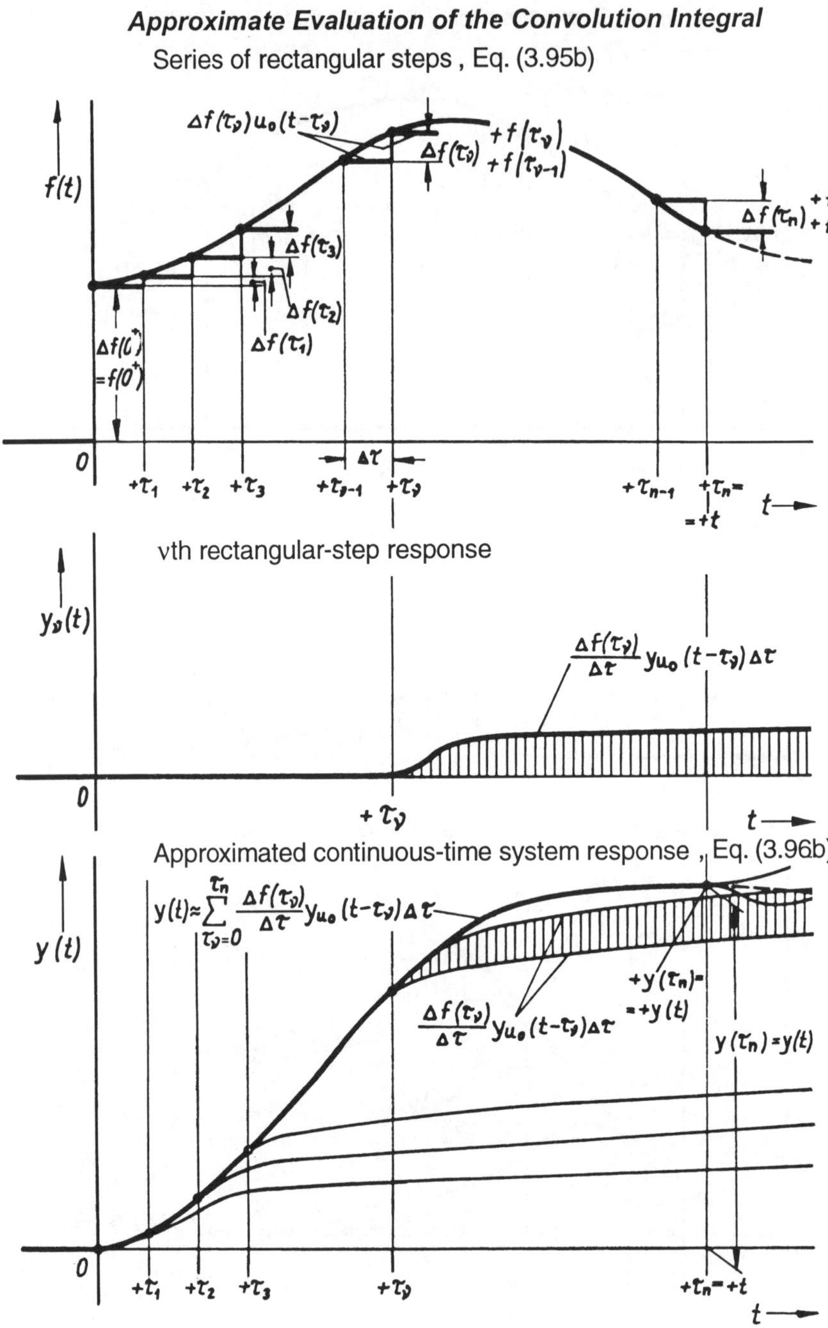

Fig. A.16. Step approximation of an arbitrary excitation

Time-response Characteristic

$s(\tau)\,/\,(F_{-1}/\sqrt{mk})$ – τ – curves

Unit pulse response (weighting function) of the damped oscillator (second-order-system),

Eqs. (3.83b),(3.84b);

caused by the unit pulse excitation $\delta(\tau)$

(dash-and-dot line)

Effect of damping ratio $\zeta = c/c_c$ on time-response curve

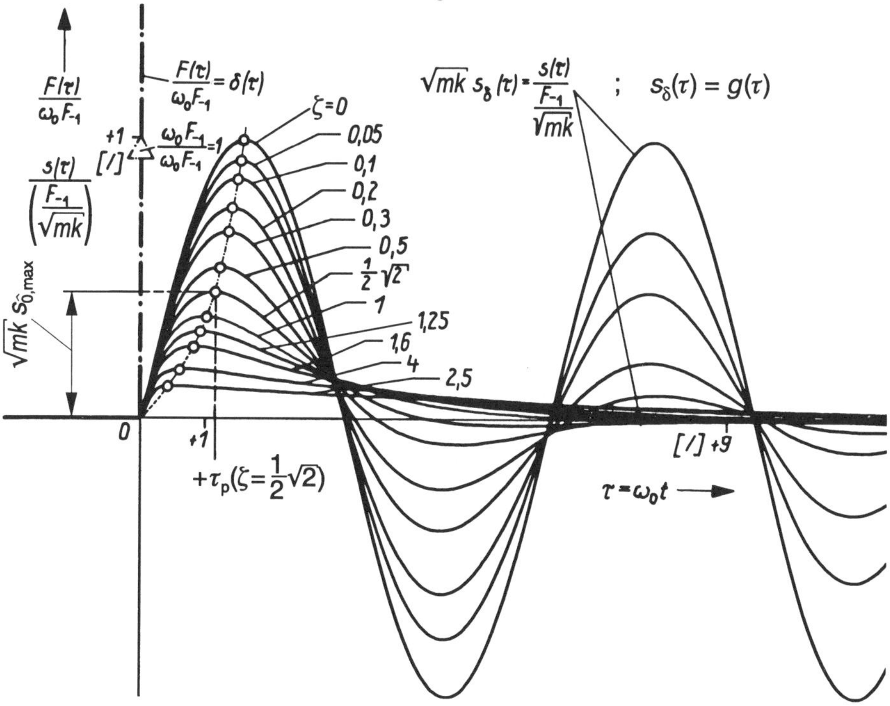

Transient-response specifications:

Normalized peak time: $\tau_p = \omega_0 t_p \approx \dfrac{\pi}{2|\sqrt{1-\zeta^2}|}$;

Relative maximum: $\sqrt{mk}\, s_{\delta,\max} = s_{\max} / (F_{-1}/\sqrt{mk}) = \dfrac{1}{|\sqrt{1-\zeta^2}|}\, e^{-\pi\zeta/(2|\sqrt{1-\zeta^2}|)}$

Fig. A.17. Idealized force pulse-motion history

Time-response Characteristic
$s(\tau)\,/\,s_{\mathrm{stat}} - \tau$ *-curves*
Unit step response
of the damped oscillator (second-order-system),

Eqs. (3.104b,c) for $\tau_0 = 0$

caused by the unit step excitation $u_0(\tau)$

(dash-and-dot line)

Effect of damping ratio $\zeta = c/c_c$ on time-response curve

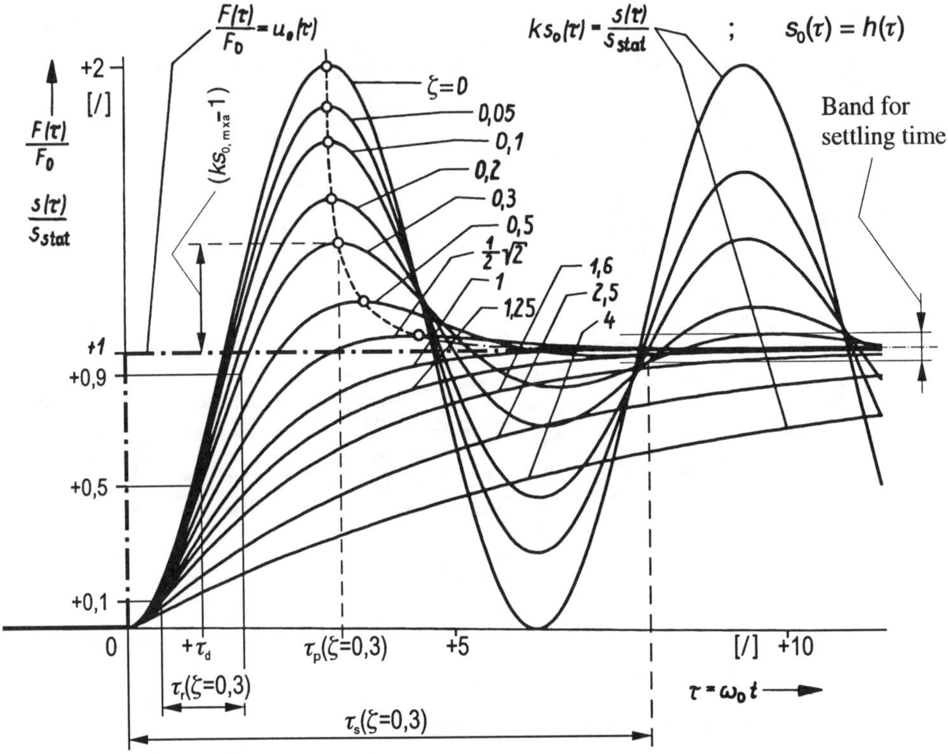

Time-domain (transient-response) specifications:

Normalized

peak time : $\tau_p = \omega_0 t_p \approx \dfrac{\pi}{|\sqrt{1-\zeta^2}|}$;

delay time : $\tau_d = \omega_0 t_d \approx (1+0{,}7\zeta)$;

rise time : $\tau_r = \omega_0 t_r \approx (0{,}8+2{,}5\zeta)$;

the latter two for $0 \leqq \zeta \leqq 1$;

2% settling time : $\tau_s = \omega_0 t_s \approx (4/\zeta) = 4\tau_{s1}$ for $\zeta < 0{,}7$;

Relative

maximum overshoot :

$$ks_{0,\max} - 1 = (s_{\max}/s_{\mathrm{stat}}) - 1 = e^{-\pi\zeta/|\sqrt{1-\zeta^2}|}$$

Fig. A.18. Response to an ideal shock excitation

Response Time History
$s(t/\tau_0)\,/\,s_{stat}$ - t/τ_0 - curves
Complete displacement response (shock motion)
of the underdamped (and critically damped) oscillator,
Eqs. (3.101c),(3.104b),
caused by the rectangular pulse excitation $F(t/\tau_0)/F_0$
of short duration (small period ratio τ_0/T_0)
(dash-and-dot line)
Effect of damping ratio $\zeta = c/c_c$ on time-history curve

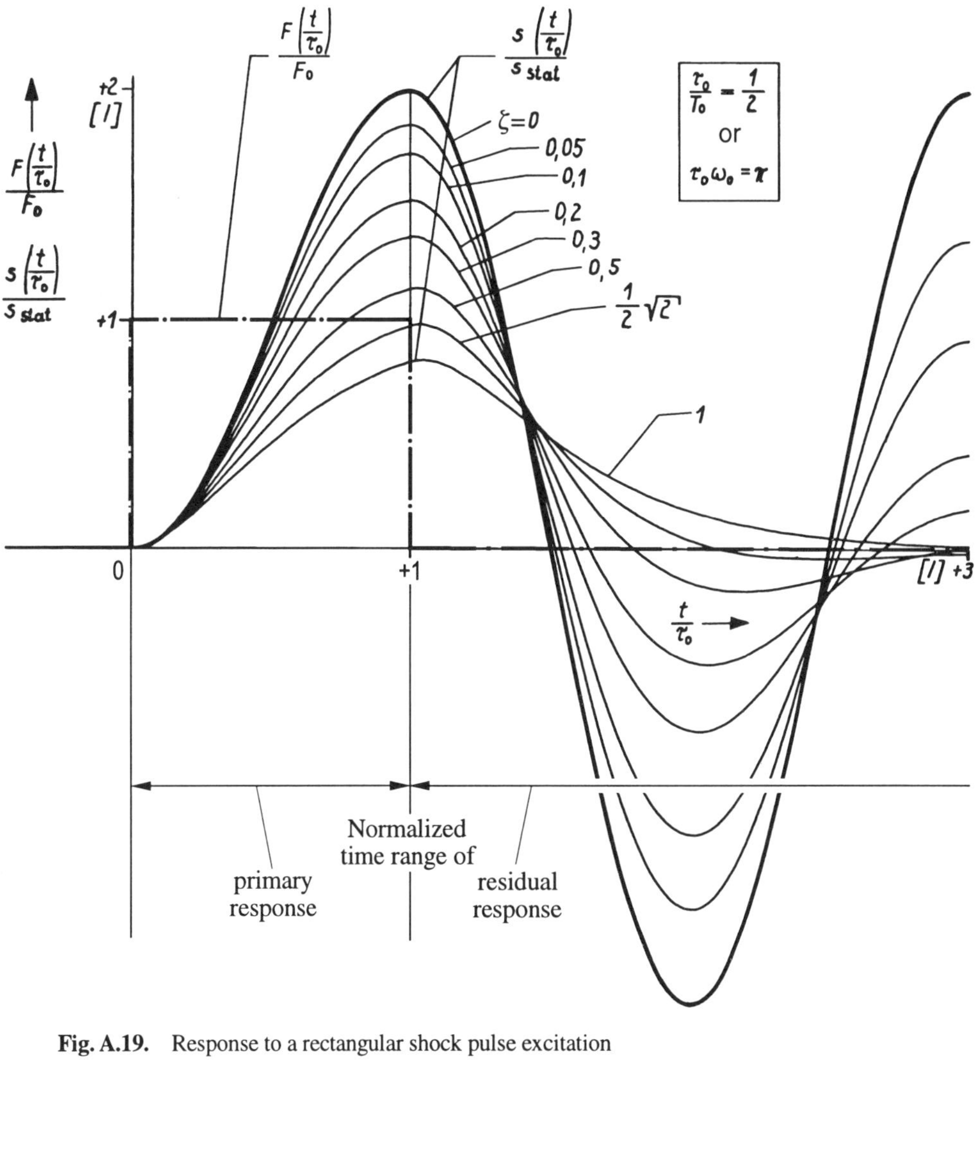

Fig. A.19. Response to a rectangular shock pulse excitation

Response Time History

$s(t/\tau_0)/s_{stat} - t/\tau_0$*-curves*

Complete displacement response (shock motion)

of the underdamped (and critically damped) oscillator,

Eqs. (3.101c),(3.104b),

caused by a rectangular pulse excitation $F(t/\tau_0)/F_0$

of long duration (large period ratio τ_0/T_0)

(dash-and-dot line)

Effect of damping ratio $\zeta = c/c_c$ on time-history curve

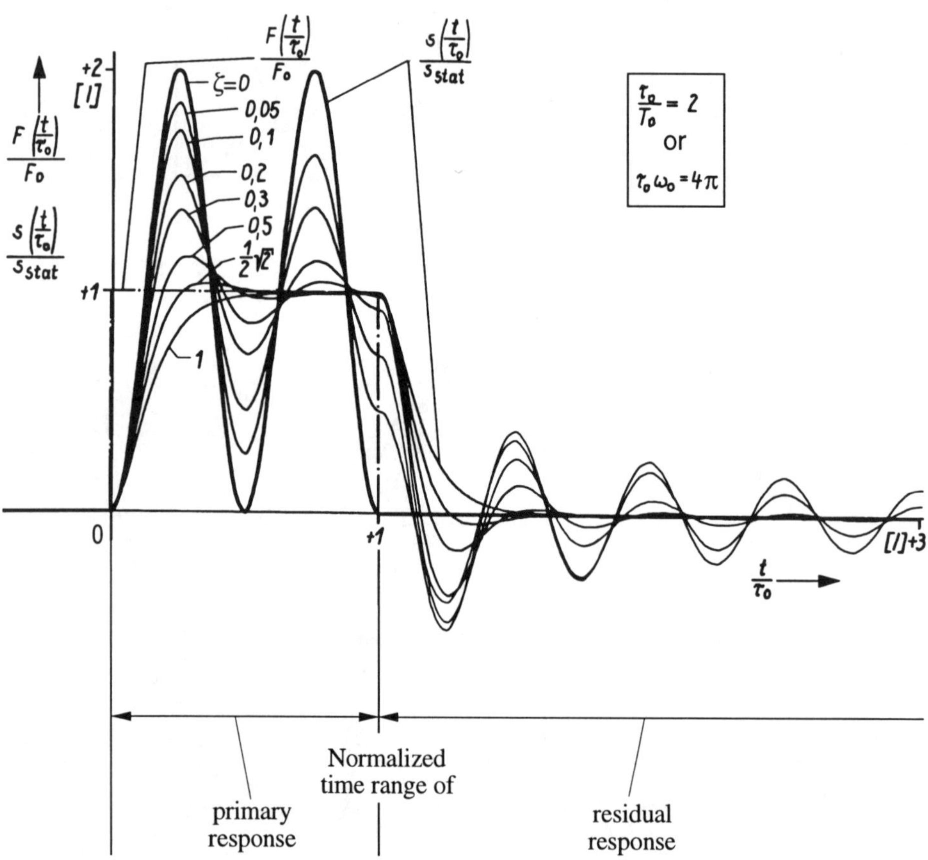

Fig. A.20. Response to a rectangular shock pulse excitation

Shock Response Spectrum

$s_{max}(\tau_0/T_0)/s_{stat} - \tau_0/T_0$ *-curves*

Response maxima (shock motion peaks)

of the undamped oscillator

caused by a rectangular pulse excitation $F(t/\tau_0)/F_0$

—— Primary shock spectrum $s_{max\,prim}/s_{stat}$, Eq.(3.102b)

—— Residual shock spectrum $s_{stat\,res}/s_{stat}$, Eq.(3.102a)

–·–·– Maximax shock spectrum s_{max}/s_{stat} (dash-and-dot-line)

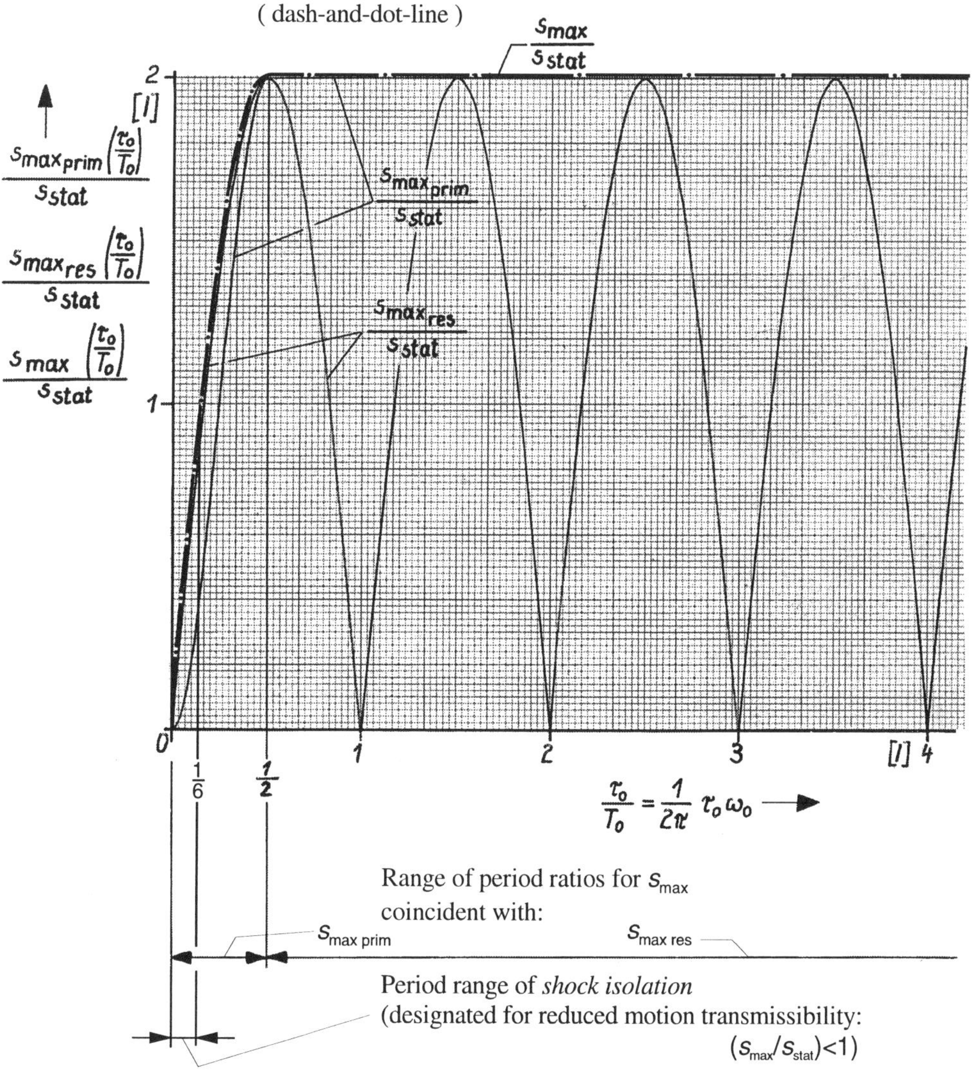

smax / sstat
2
[1]
smaxprim(τ0/T0) / sstat
smaxres(τ0/T0) / sstat
smax(τ0/T0) / sstat
1
0
smaxprim / sstat
smaxres / sstat
1/6
1/2
1
2
3
[1] 4
τ0/T0 = 1/2π τ0ω0
Range of period ratios for smax
coincident with:
smax prim
smax res
Period range of shock isolation
(designated for reduced motion transmissibility:
(smax/sstat)<1)

Fig. A.21. Displacement Shock Response Spectrum to a rectangular shock pulse excitation

Autocorrelation Curves

$\varphi_{ss}(\omega_0|\tau|)$ - $\omega_0|\tau|$-curves

Response autocorrelation function at displacement of the underdamped oscillator, Eq.(3.130b)

caused by a white noise excitation process

Effect of damping ratio $\zeta = c/c_c$ on AC-curve

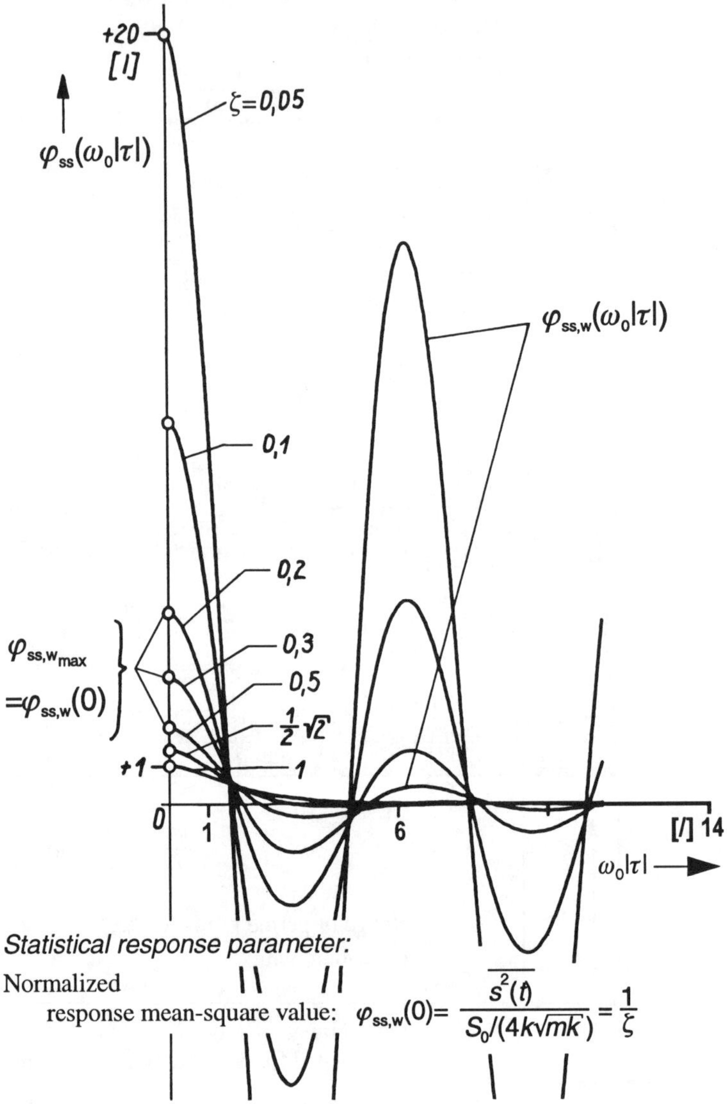

Statistical response parameter:

Normalized response mean-square value: $\varphi_{ss,w}(0) = \dfrac{\overline{s^2(t)}}{S_0/(4k\sqrt{mk})} = \dfrac{1}{\zeta}$

Fig. A.22. White random vibration and its autocorrelation function

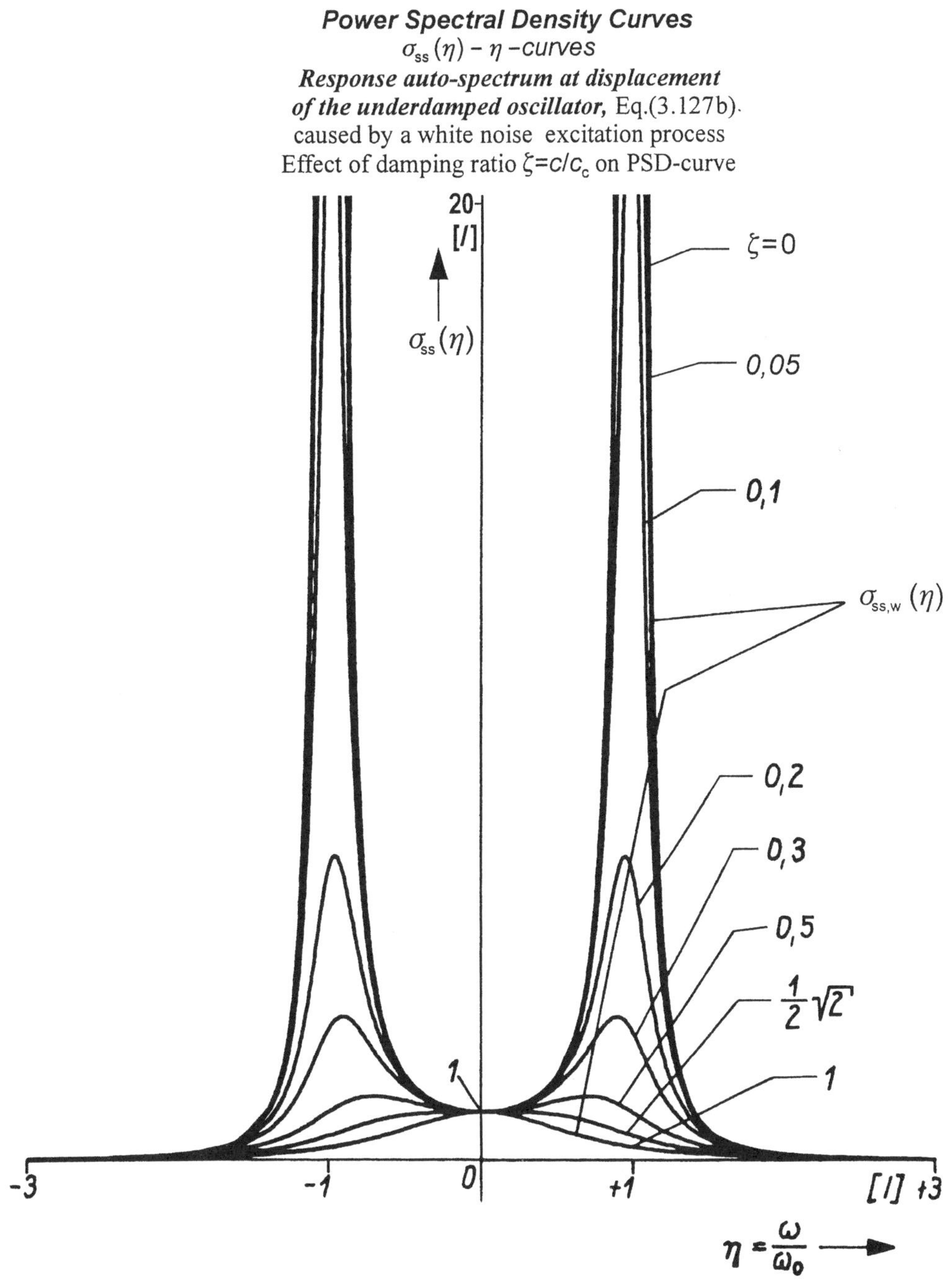

Fig. A.23. White random vibration and its auto-spectral density

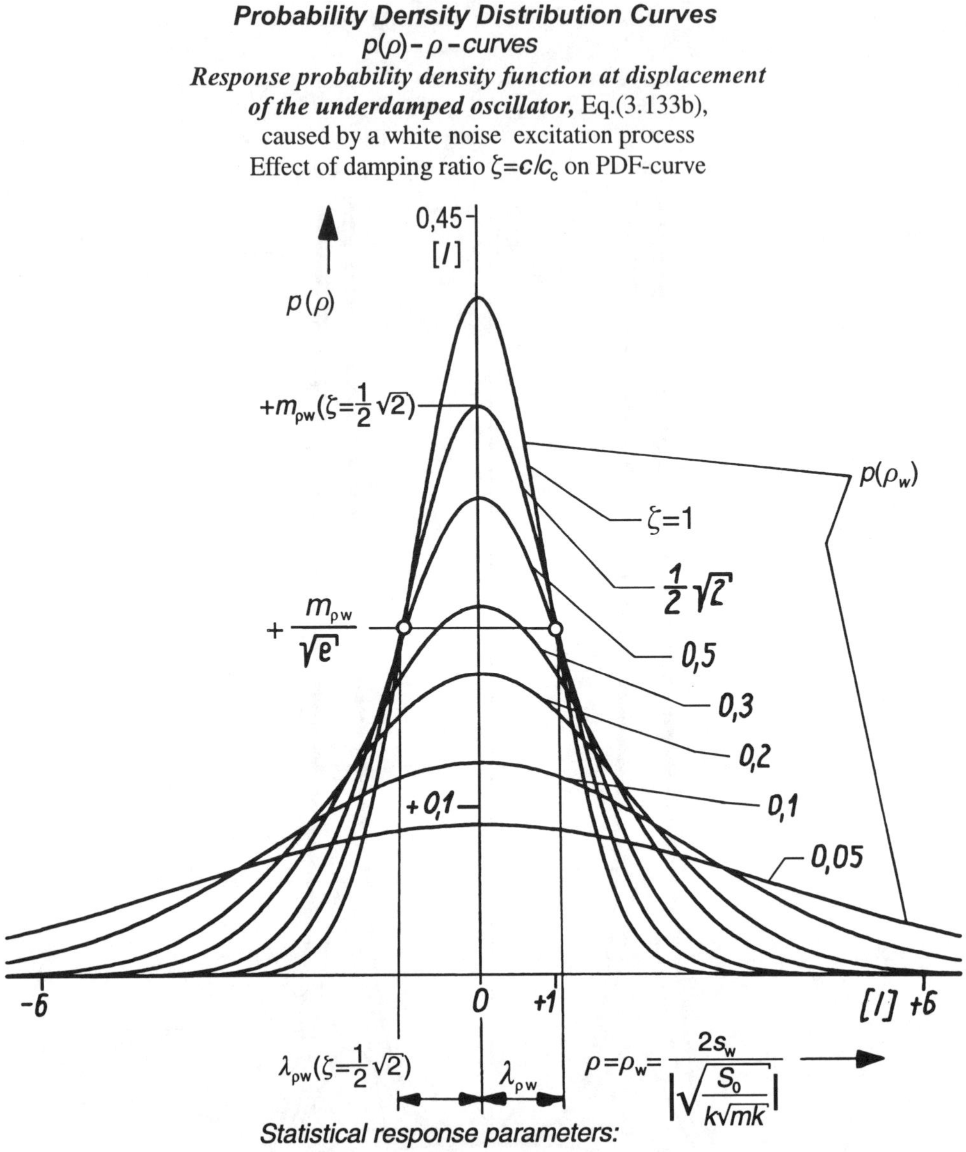

Statistical response parameters:

Normalized

response standard devation (r.m.s.value)

$$\lambda_{\rho w} = |\sqrt{\rho_w^2(t)}| = \rho_{w,eff} = |\sqrt{\zeta}|/\zeta \quad \text{, Eq. (3.134b)}$$

response median

$$m_{\rho w} = p(\rho_w = 0) = |\sqrt{\zeta}|/|\sqrt{2\pi}| \quad \text{, Eq. (3.134c)}$$

Fig. A.24. White random vibration and its probability density function

Appendix B

Frequency Response Plots

Dynamic Compliance Plots
(Nyquist, Nichols, Bode plots)
Dynamic Stiffness Plots
Mobility Plots
Mechanical Impedance Plots

Performance Criteria

Frequency-response specifications

Frequency Response Plots

$\mathrm{Im}[\underline{Y}_k(j\eta)] - \mathrm{Re}[\underline{Y}_k(j\eta)]$ *-curves*

Polar frequency-response loci (Nyquist plots)
of the normalized dynamic compliance of a structure

$\underline{Y}_k(j\eta)=\underline{C}_{ii}(j\omega/\omega_0)/C_k$; Eqs.(3.186a,b),(3.198c)

within the range of damping ratio $(D=)\zeta \leqq 1$

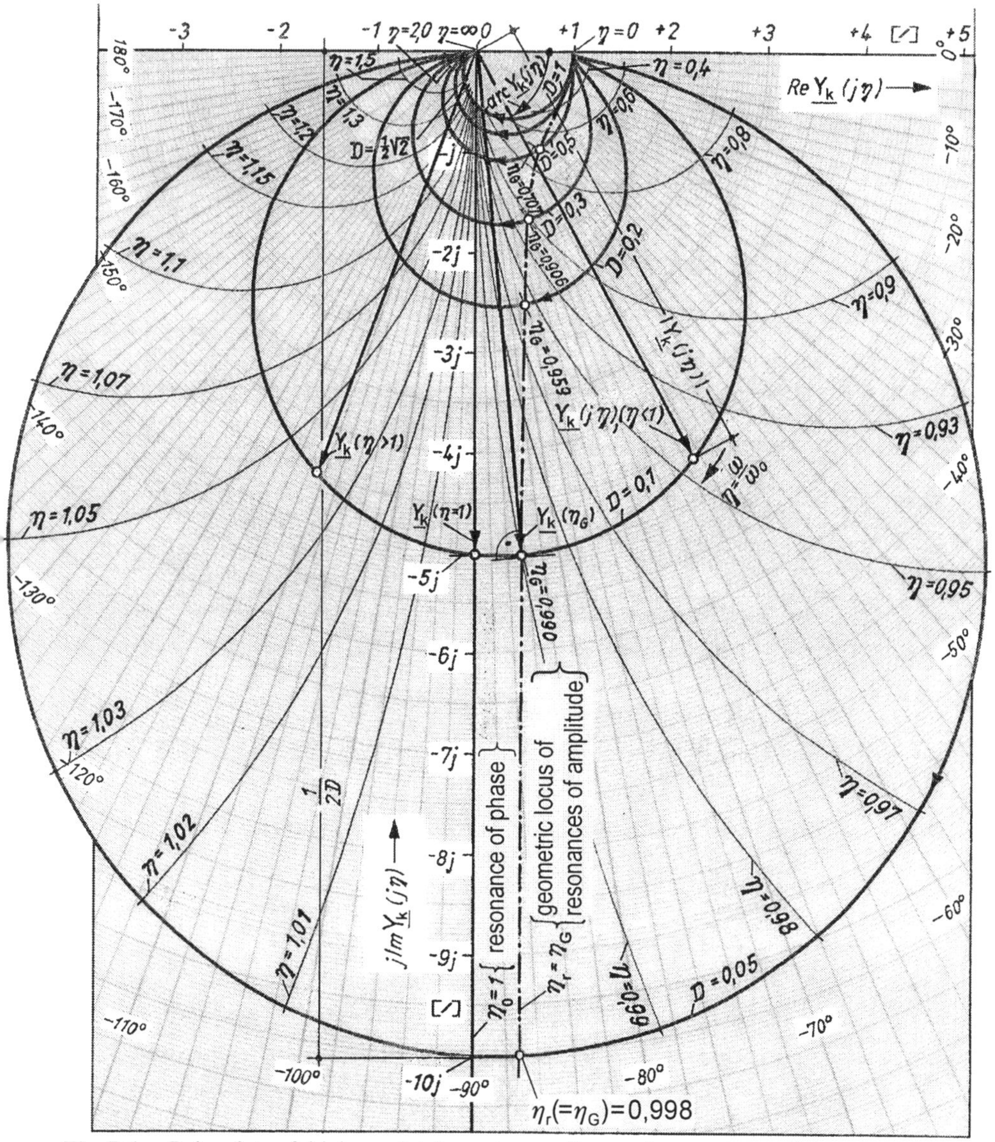

Fig. B.1. Polar plots of driving-point dynamic compliance, elements in parallel, excitation force external or via spring

Frequency Response Plots

$\lg[\underline{Y}_k(j\eta)] - \mathrm{arc}[\underline{Y}_k(j\eta)]$ *-curves*

Logarithmic frequency-response loci (Nichols plots) of the normalized dynamic compliance of a structure

$\lg[\underline{Y}_k(j\eta)] = \lg[\underline{C}_{ii}(j\omega/\omega_0)/C_k]$; Eq.(3.205)

within the range of damping ratio $(D=)\zeta \leqq 1$

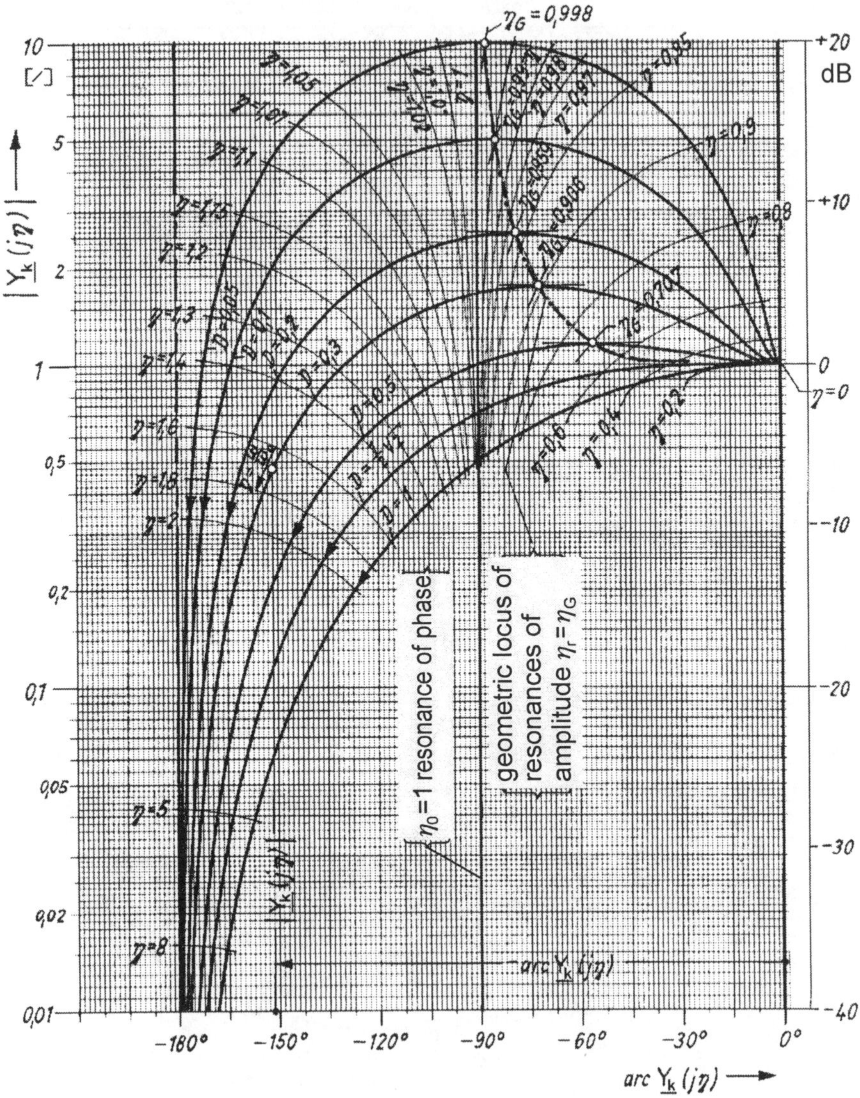

Fig. B.2. Log magnitude-phase diagrams of driving-point dynamic compliance, elements in parallel, excitation force external or via spring

Frequency Response Plots

$\lg|\underline{Y}_k(j\eta)| - \lg\eta$ - and $\mathrm{arc}[\underline{Y}_k(j\eta)] - \lg\eta$ -*curves*

Logarithmic frequency plots (Bode plots)
of the normalized dynamic compliance of a structure

$\lg[\underline{Y}_k(j\eta)] = \lg[\underline{C}_{ii}(j\omega/\omega_0)/C_k]$; Eq.(3.205)

within the range of damping ratio $(D=)\,\zeta \lesseqgtr 1$

Logarithmic gain (magnitude ratio in decibel*)*

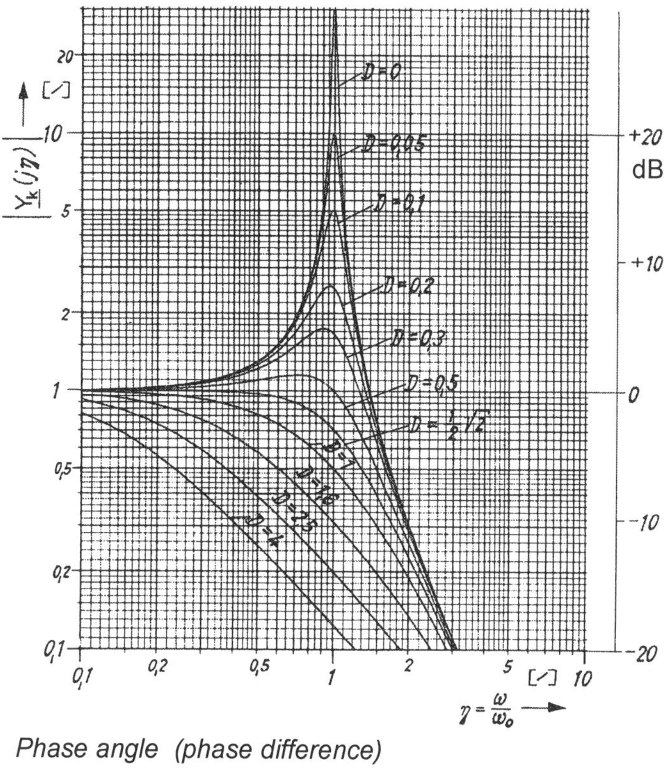

Phase angle (phase difference)

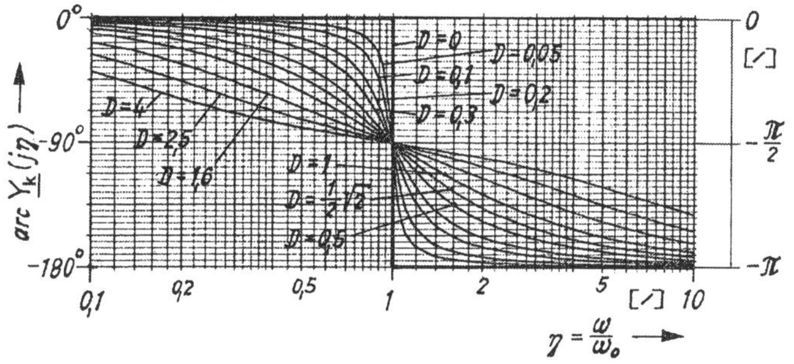

Fig. B.3. Magnitude and phase plots of driving-point dynamic compliance

Frequency Response Plot

$\lg[\underline{Y}_k(j\eta)] - \lg \eta$*-curve*

Logarithmic gain (magnitude ratio in* decibel*)

of the normalized dynamic compliance of a structure

$\lg[\underline{Y}_k(j\eta)] = \lg[\underline{C}_{ii}(j\omega/\omega_0)/C_k]$; Eqs.(3.214a,b),(3.215a)

for the value of damping ratio $\zeta = 0{,}2$

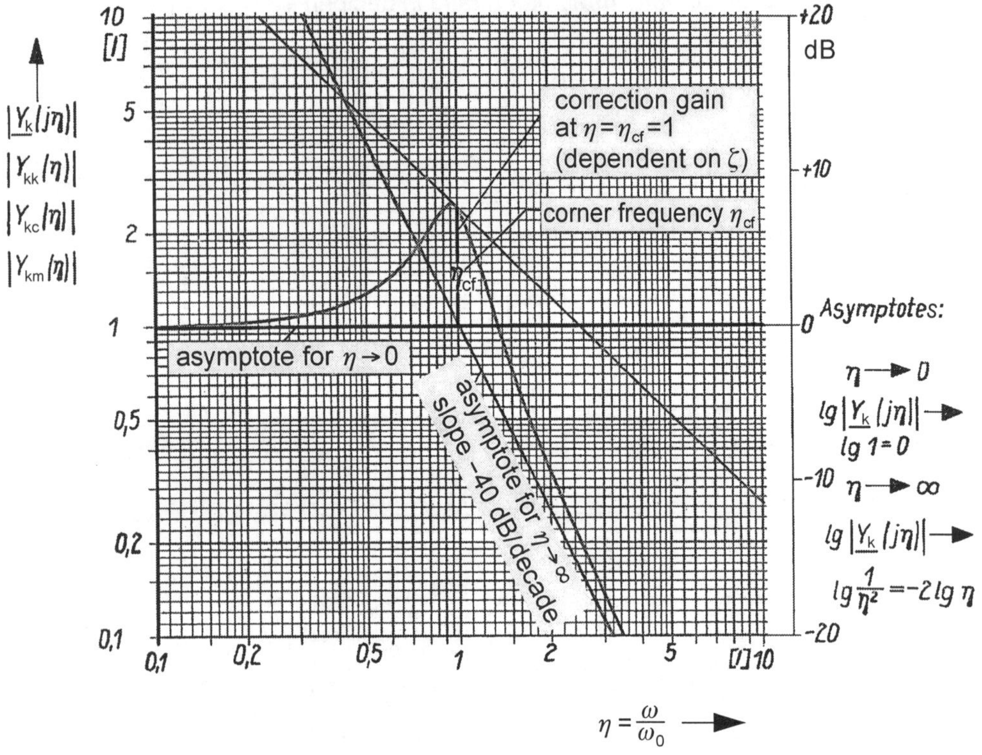

Frequency response specifications:

Low-frequency asymptote of slope 0

High-frequency asymptote of slope −2 or −40 dB/decade

Corner frequency ratio $\eta_{cf} = 1$

Correction gain $\lg|Y_{kc}|_{dB} = 7{,}96$ dB; Eq.(3.215c)

Fig. B.4. Magnitude plot and straight-line approximation of driving-point dynamic compliance

Frequency Response Plots

$\mathrm{Im}[\underline{X}_k(j\eta)] - \mathrm{Re}[\underline{X}_k(j\eta)]$ *-curves*

Polar frequency-response loci (Nyquist plots)
of the normalized dynamic stiffness of a structure

$\underline{X}_k(j\eta) = \underline{K}_{ii}(j\omega/\omega_0)/K_k$; Eqs.(3.188a,b),(3.216)
within the range of damping ratio $(D=)\zeta \leqq 1$

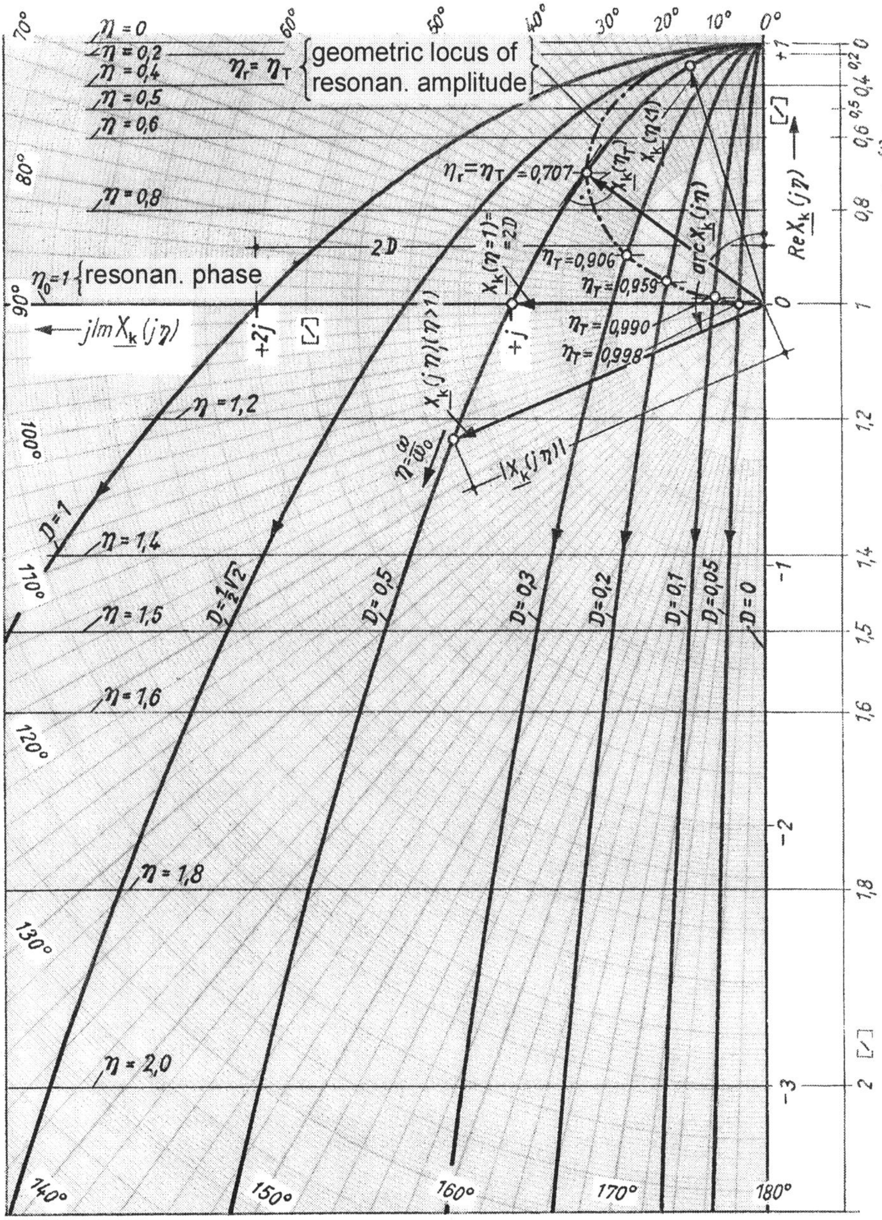

Fig. B.5. Polar plots of driving-point dynamic stiffness

Frequency Response Plots

$\lg[\underline{X}_k(j\eta)]$ – $\mathrm{arc}[\underline{X}_k(j\eta)]$ –*curves*

Logarithmic frequency-response loci (Nichols plots) of the normalized dynamic stiffness of a structure

$\lg[\underline{X}_k(j\eta)] = \lg[\underline{K}_{ii}(j\omega/\omega_0)/K_k]$; Eq.(3.216)

within the range of damping ratio $(D=)\zeta \leqq 1$

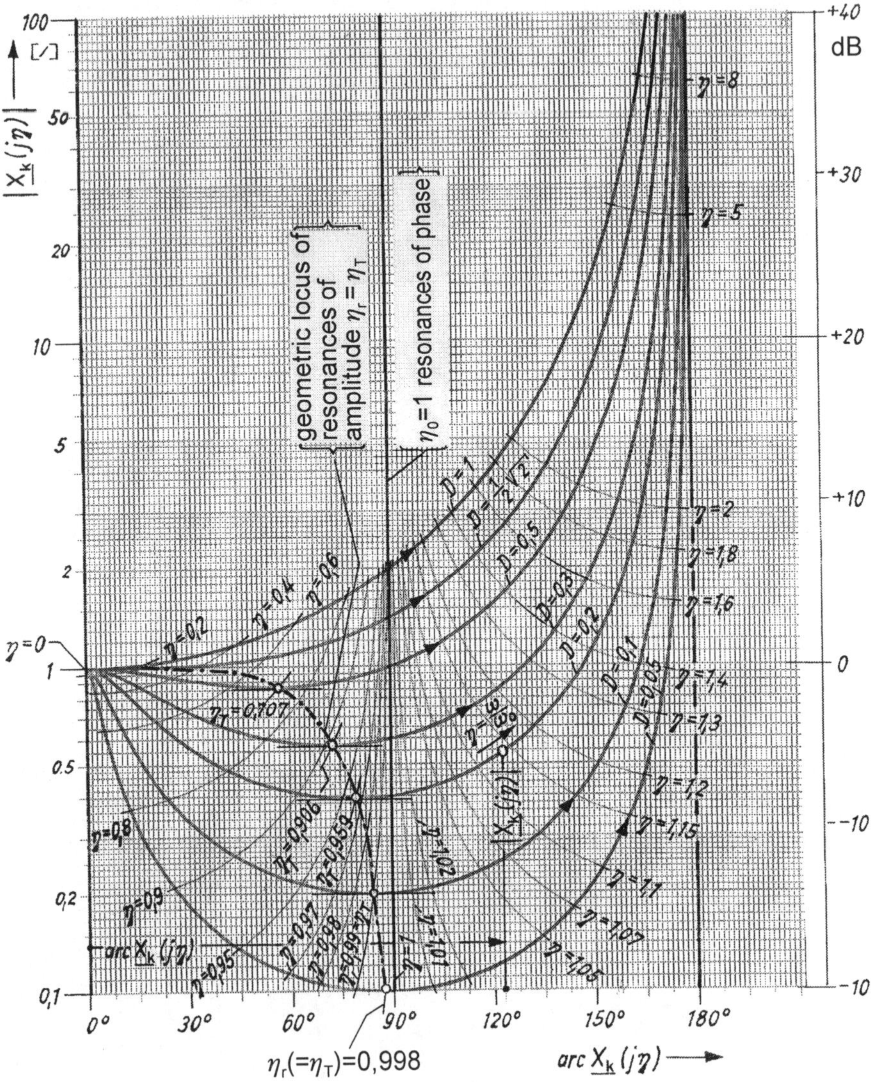

Fig. B.6. Log magnitude-phase diagrams of driving-point dynamic stiffness, elements in parallel, excitation force external or via spring

Frequency Response Plots

$\lg|\underline{X}_k(j\eta)| - \lg\eta$ – and $\text{arc}[\underline{X}_k(j\eta)] - \lg\eta$ –*curves*

Logarithmic frequency plots (Bode plots)

of the normalized dynamic stiffness of a structure

$\lg[\underline{X}_k(j\eta)] = \lg[\underline{K}_{ii}(j\omega/\omega_0)/K_k]$; Eq.(3.216)

within the range of damping ratio $(D=)\ \zeta \lesseqgtr 1$

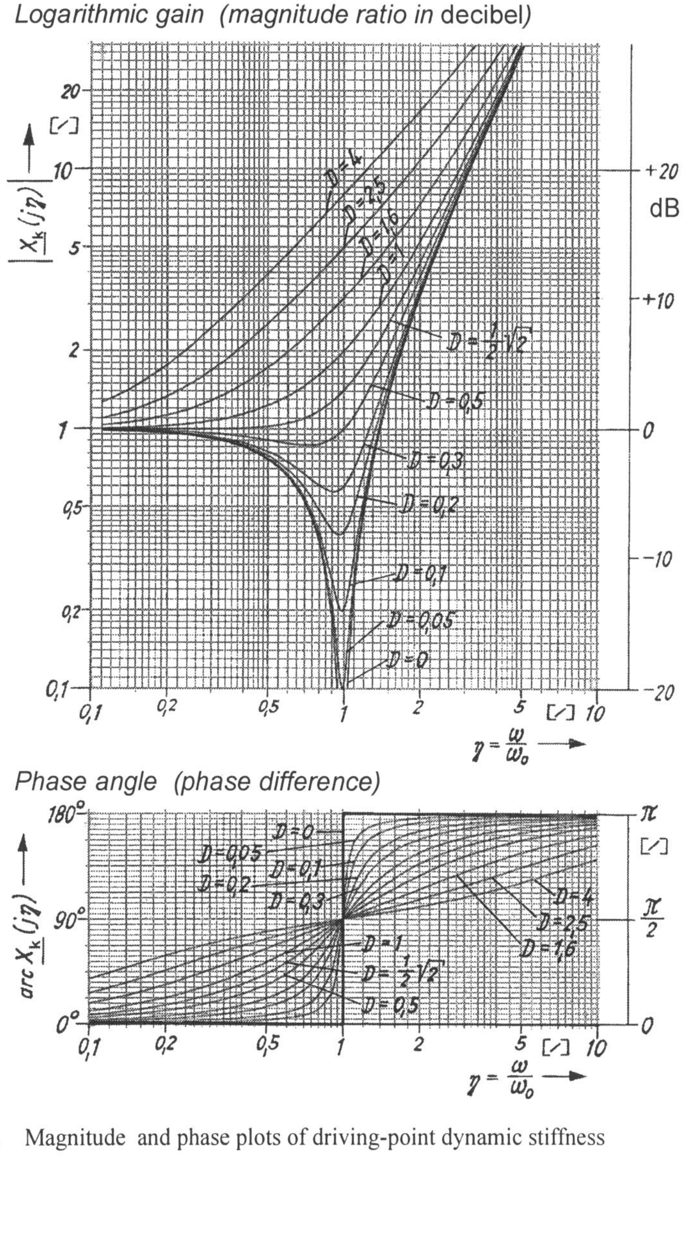

Fig. B.7. Magnitude and phase plots of driving-point dynamic stiffness

Frequency Response Plot

$\lg|\underline{X}_k(j\eta)| - \lg\eta$*-curve*

Logarithmic gain (magnitude ratio in* decibel*)
of the normalized dynamic stiffness of a structure

$\lg[\underline{X}_k(j\eta)] = \lg[\underline{K}_{ii}(j\omega/\omega_0)/K_k]$; Eqs.(3.216a-c)
for the value of damping ratio $\zeta = 0{,}2$

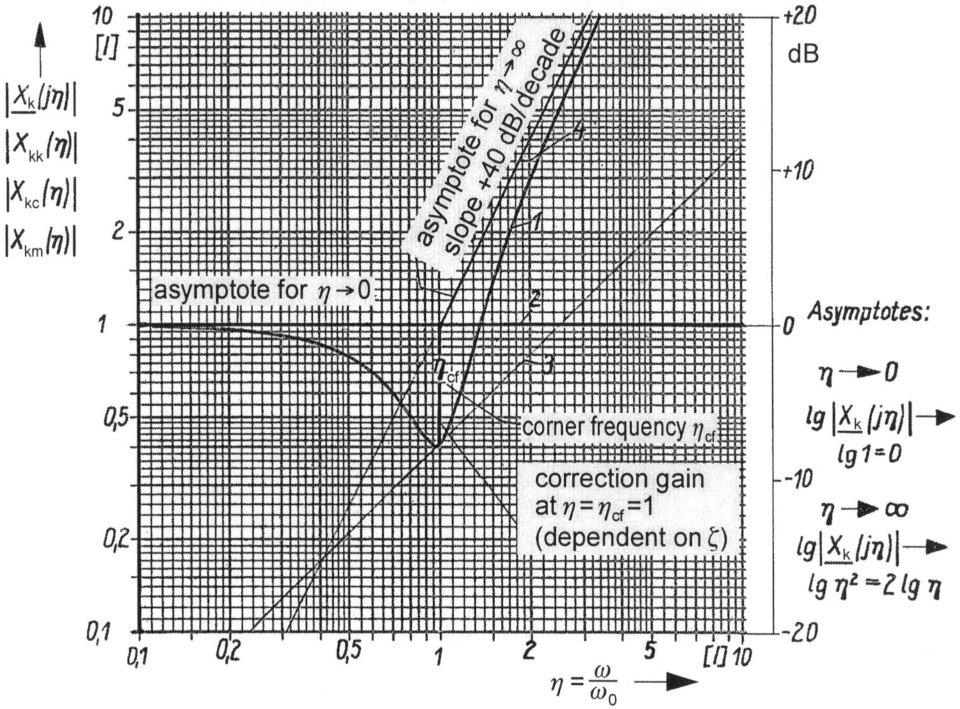

Frequency response specifications:

Low-frequency asymptote of slope 0
High-frequency asymptote of slope +2 or +40 dB/decade
Corner frequency ratio $\eta_{cf} = 1$
Correction gain $\lg|X_{kc}|_{dB} = -7{,}96$ dB; Eq.(3.215c)

Fig. B.8. Magnitude plot and straight-line approximation of driving-point dynamic stiffness

Frequency Response Plots

$\lg[\underline{Y}'_k(j\eta)] - \mathrm{arc}[\underline{Y}'_k(j\eta)]$ -*curves*

Logarithmic frequency-response loci (Nichols plots) of the normalized (mechanical) mobility of a structure

$\lg[\underline{Y}'_k(j\eta)] = \lg[j\eta\underline{Y}_k(j\eta)] = \lg[\underline{Y}_{ii}(j\omega/\omega_0)/Y_k]$; Eq.(3.190)

within the range of damping ratio $(D=)\zeta \leqq 1$

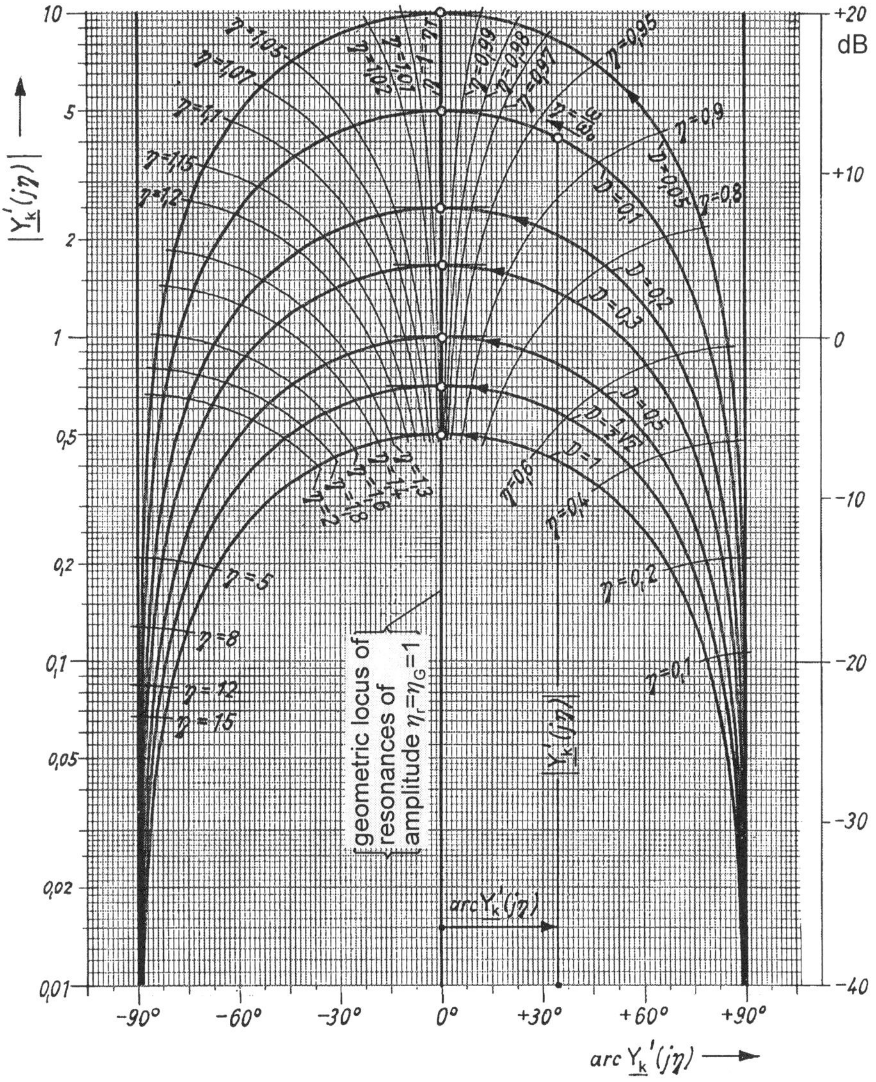

Fig. B.9. Log magnitude-phase diagrams of driving-point mobility, elements in parallel,excitation force external or via spring

Frequency Response Plots

$\lg|\underline{Y}_k'(j\eta)| - \lg\eta -$ and $\mathrm{arc}[\underline{Y}_k'(j\eta)] - \lg\eta$ *-curves*

Logarithmic frequency plots (Bode plots)

of the normalized (mechanical) mobility of a structure

$\lg[\underline{Y}_k'(j\eta)] = \lg[j\eta\underline{Y}_k(j\eta)] = \lg[\underline{Y}_{ii}(j\omega/\omega_0)/Y_0]$; Eq.(3.190)

within the range of damping ratio $(D=)\,\zeta \lesseqgtr 1$

Logarithmic gain (magnitude ratio in decibel)

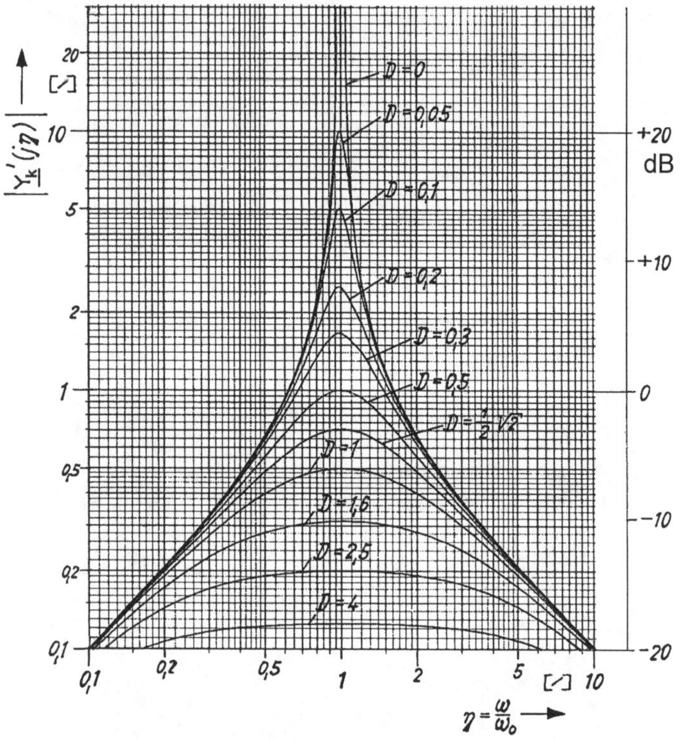

Phase angle (phase difference)

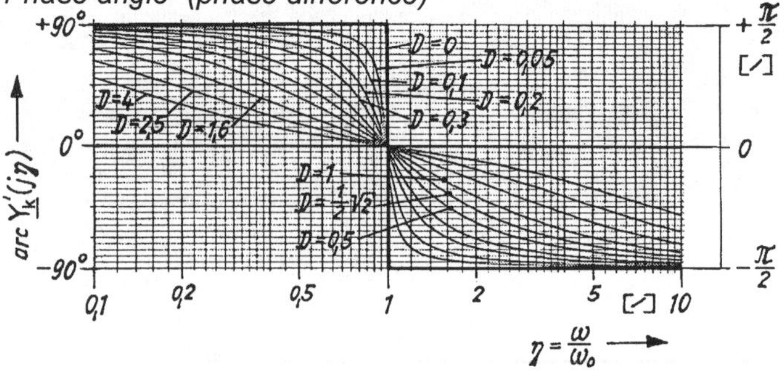

Fig. B.10. Magnitude and phase plots of driving-point mobility

Frequency Response Plots

lg[$\tilde{\underline{X}}_k(j\eta)$] - arc[$\tilde{\underline{X}}_k(j\eta)$] -*curves*

Logarithmic frequency-response loci (Nichols plots) of the normalized mechanical impedance of a structure

$\lg[\tilde{\underline{X}}_k(j\eta)]=\lg[(-j/\eta)\underline{X}_k(j\eta)=\lg[\underline{Z}_{ii}(j\omega/\omega_0)/Z_k]$; Eq.(3.191)

within the range of damping ratio $(D=)\zeta \leqq 1$

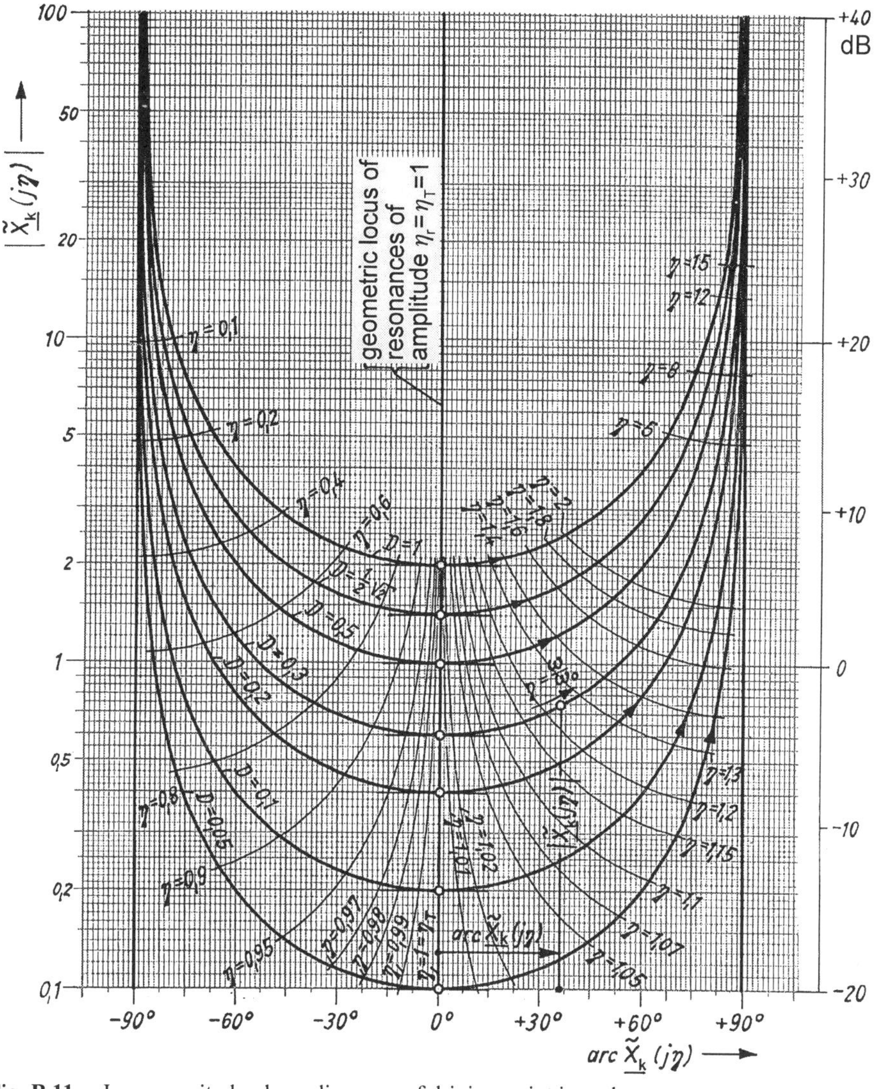

Fig. B.11. Log magnitude-phase diagrams of driving-point impedance, elements in parallel, excitation force external or via spring

Frequency Response Plots

$\lg|\underset{\sim}{\tilde{X}}_k(j\eta)| - \lg\eta$ – and $\mathrm{arc}[\underset{\sim}{\tilde{X}}_k(j\eta)] - \lg\eta$ *-curves*

Logarithmic frequency plots (Bode plots)

of the normalized mechanical impedance of a structure

$\lg[\underset{\sim}{\tilde{X}}_k(j\eta)] = \lg[-j/\eta\,\underset{\sim}{X}_k(j\eta)] = \lg[\underset{\sim}{Z}_{ii}(j\omega/\omega_0)/Z_0]$; Eq.(3.192)

within the range of damping ratio $(D=)\,\zeta \lesseqgtr 1$

Logarithmic gain (magnitude ratio in decibel)

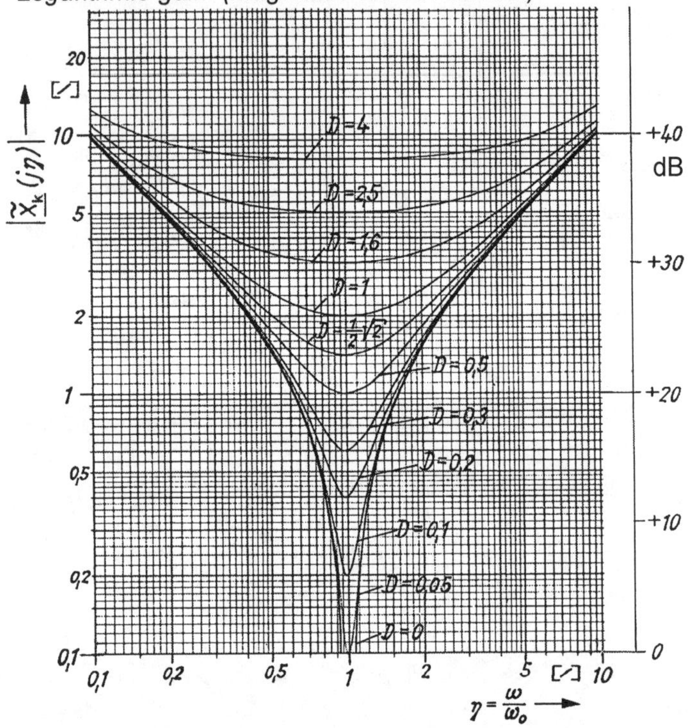

Phase angle (phase difference)

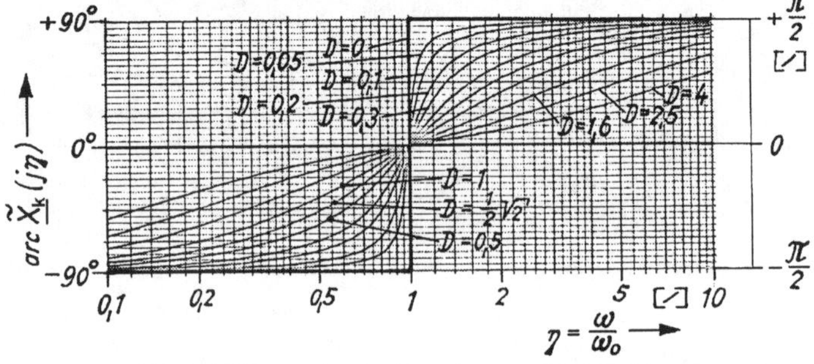

Fig. B.12. Magnitude and phase plots of driving-point impedance

Frequency Response Plots

$\mathrm{Im}[-\underline{Y}_m(j\eta)] - \mathrm{Re}[-\underline{Y}_m(j\eta)]$ *-curves*

Polar frequency-response loci (Nyquist plots)
of the normalized dynamic compliance of a structure

$-\underline{Y}_m(j\eta) = -\underline{C}_{ii}(j\omega/\omega_0)/C_m$;

within the range of damping ratio $(D=)\zeta \leqq 1$

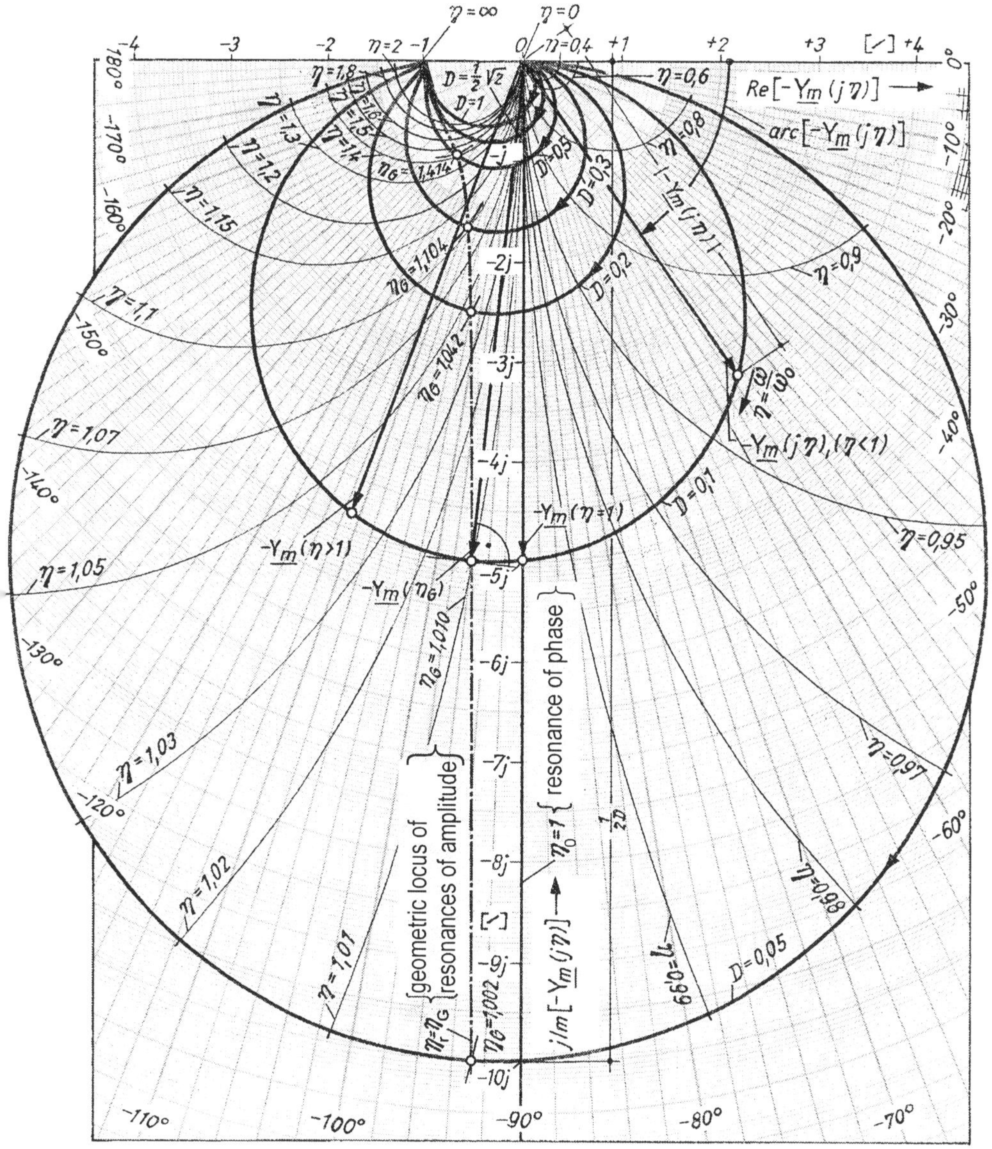

Fig. B.13. Polar plots of driving-point dynamic compliance, elements in parallel, excitation force via unbalanced rotating mass

Frequency Response Plots

$\lg|-\underline{Y}_m(j\eta)| - \lg\eta$ - and $\operatorname{arc}[-\underline{Y}_m(j\eta)] - \lg\eta$ -*curves*

Logarithmic frequency plots (Bode plots)
of the normalized dynamic compliance of a structure

$\lg[-\underline{Y}_m(j\eta)] = \lg[\underline{C}_{ii}(j\omega/\omega_0)/C_k]$;

within the range of damping ratio $(D=)\,\zeta \lesseqgtr 1$

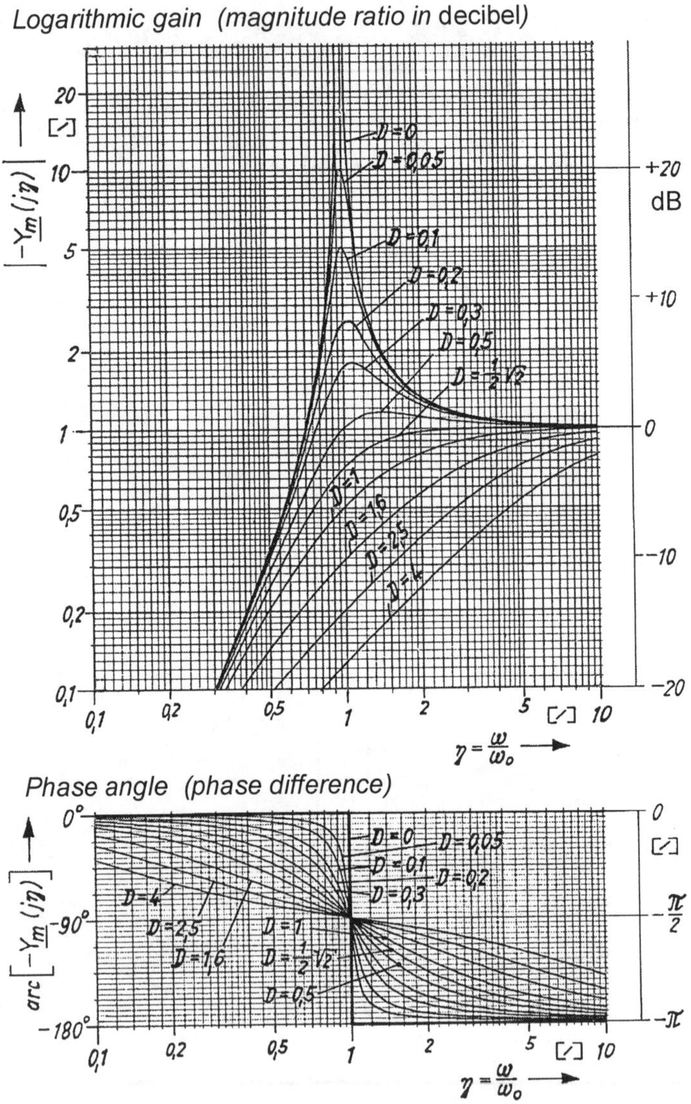

Fig. B.14. Magnitude and phase plots of driving-point dynamic compliance

Frequency Response Plots

Im $[-\underline{X}_m(j\eta)]$ - Re $[-\underline{X}_m(j\eta)]$ *-curves*

Polar frequency-response loci (Nyquist plots)
of the normalized dynamic stiffness of a structure

$$-\underline{X}_m(j\eta) = -\underline{K}_{ii}(j\omega/\omega_0)/K_k \;;$$

within the range of damping ratio $(D=)\zeta \leqq 1$

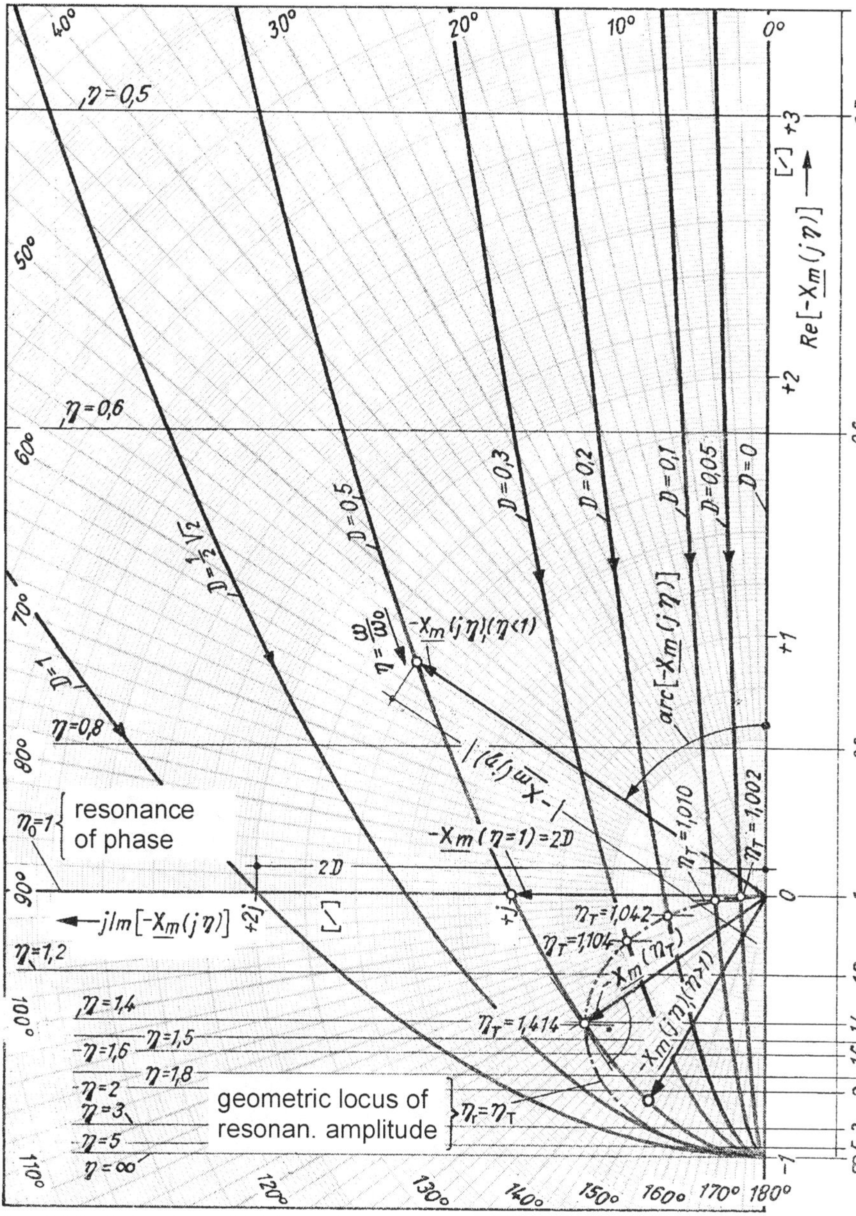

Fig. B.15. Polar plots of driving-point dynamic stiffness

Frequency Response Plots

$\lg|-\underline{X}_m(j\eta)| - \lg\eta$ – and $\text{arc}[-\underline{X}_m(j\eta)] - \lg\eta$ –*curves*

Logarithmic frequency plots (Bode plots) of the normalized dynamic stiffness of a structure

$$\lg[-\underline{X}_m(j\eta)] = \lg[-\underline{K}_{ii}(j\omega/\omega_0)/K_k] \;;$$

within the range of damping ratio $(D=)\ \zeta \lesseqgtr 1$

Logarithmic gain (magnitude ratio in decibel*)*

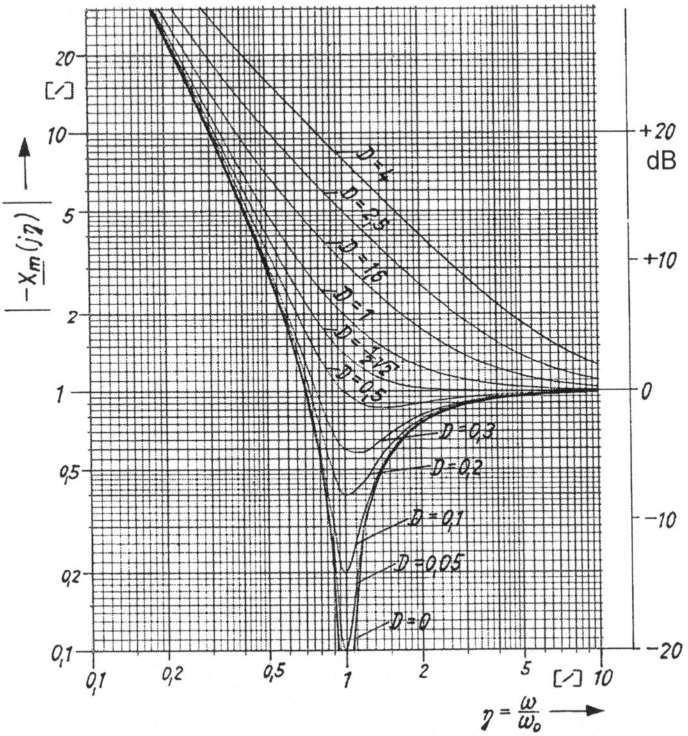

Phase angle (phase difference)

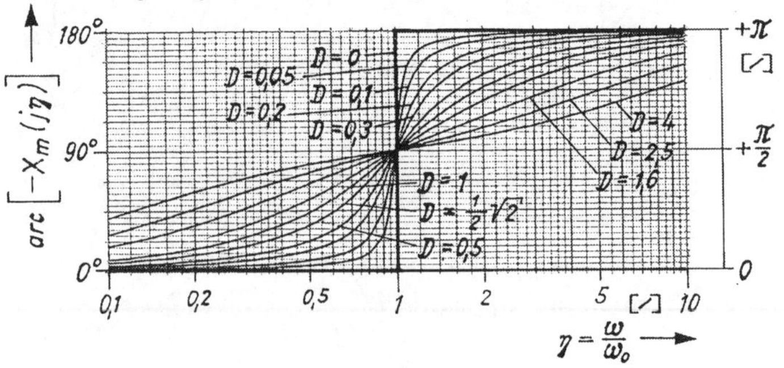

Fig. B.16. Magnitude and phase plots of driving-point dynamic stiffness

Appendix C

Frequency Response Plots of Power

Complex Power Plots
(Nyquist plots, power component plots)

Performance Criteria

Frequency-response specifications of power

Frequency Response Plots

$\mathrm{Im}[\underline{X}_k^{*\prime}(j\eta)]$ - $\mathrm{Re}[\underline{X}_k^{*\prime}(j\eta)]$ -*curves*

Polar frequency-response loci (Nyquist plots) of the normalized complex power at the driving point

$\underline{X}_k^{*\prime}(j\eta)=j\eta\underline{X}_k^{*}(j\eta)=\underline{S}_B^{*}(j\omega/\omega_0)/S_0$; Eq.(5.87b)

within the range of damping ratio $(D=)\zeta \leqq 1$

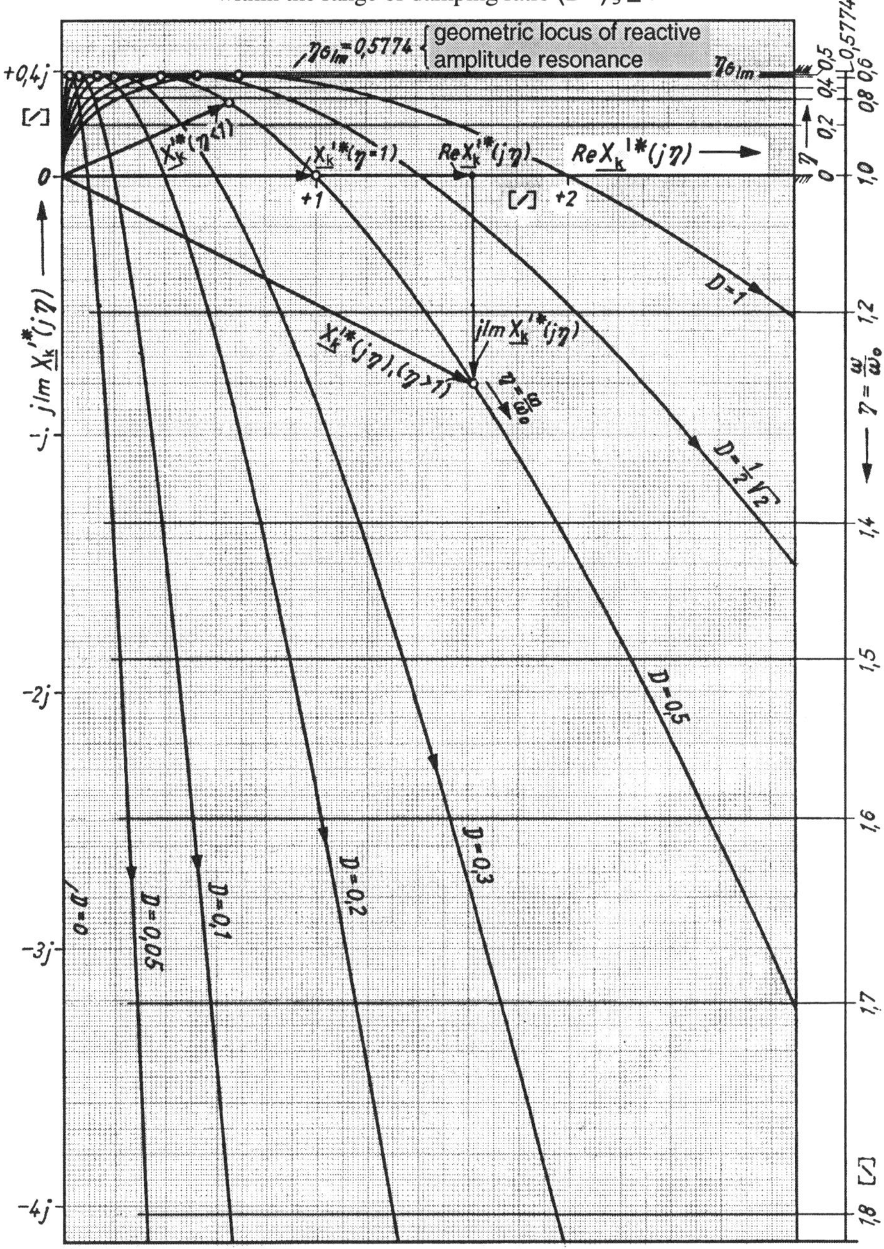

Fig. C.1. Polar plots of driving-point complex power, elements in parallel, exciting motion by a displacement

Power Frequency Plots

Re[$\underline{X}_k^{*\prime}(j\eta)$] - η-*curves*

Frequency-response plots of the normalized active power at the driving point

Re[$\underline{X}_k^{*\prime}(j\eta)$]=$P(\omega/\omega_0)/S_0$; Eq.(5.88a)

within the range of damping ratio (D=)$\zeta \leqq 4$

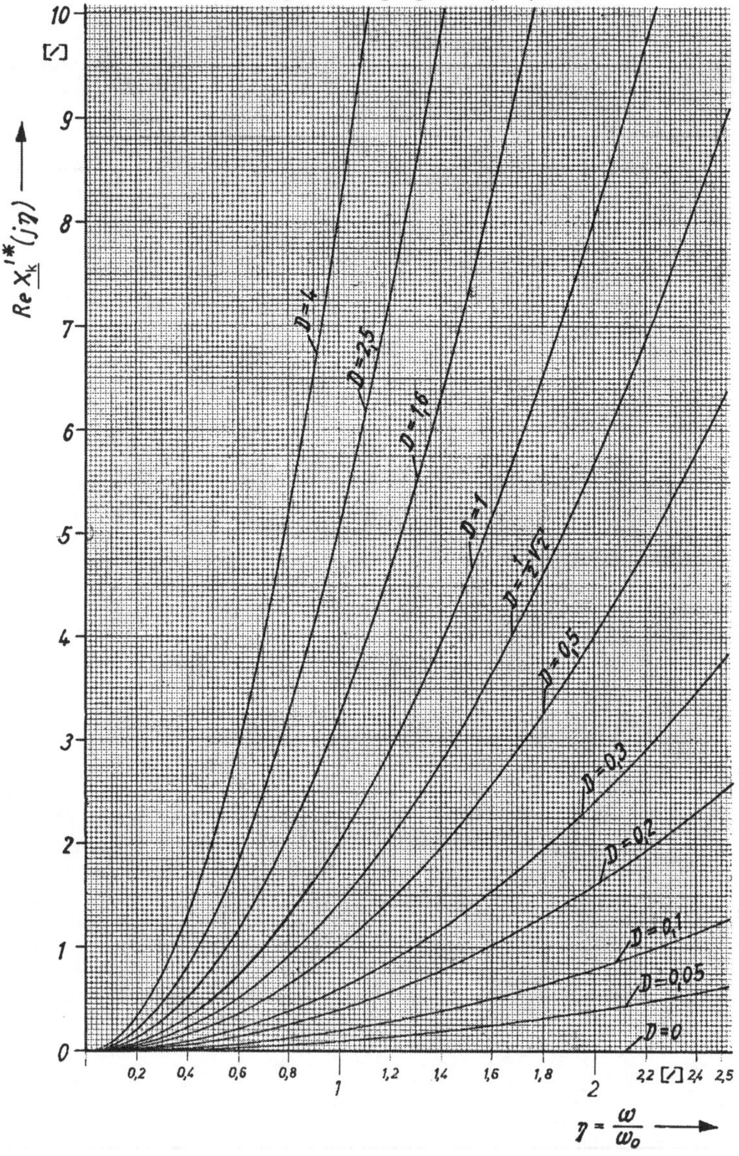

Fig. C.2. Frequency plots of driving-point active power

Power Frequency Plot

$\mathrm{Im}[\underline{X}_k^{*\prime}(j\eta)]-\eta-$ *curve*

Frequency-response plot
of the normalized reactive power at the driving point

$\mathrm{Im}[\underline{X}_k^{*\prime}(j\eta)]=Q(\omega/\omega_0)/S_0$; Eq.(5.88b)

for all values of damping ratio ζ

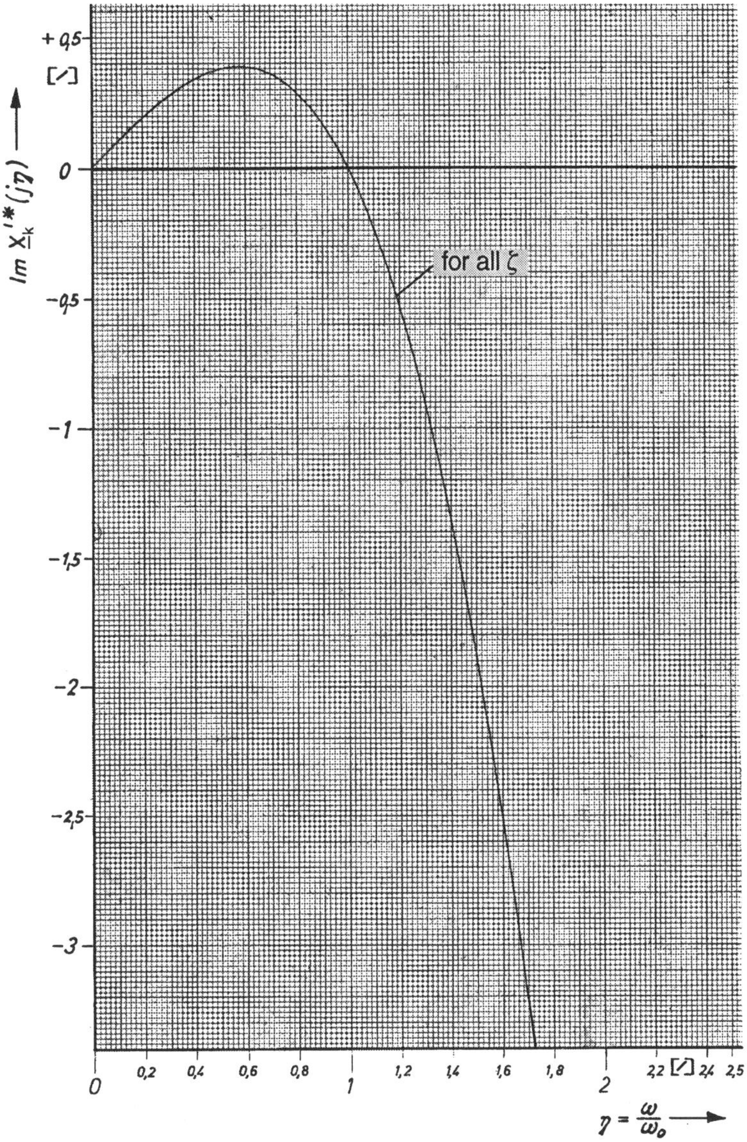

Fig. C.3. Frequency plot of driving-point reactive power

Power Frequency Plots

Polar frequency-response loci (Nyquist plots)
of the normalized complex power at the driving point

$\kappa j\eta\,[1-\kappa \underline{Y}_k^*(j\eta)] = \underline{S}_B^*(j\omega/\omega_0)/S_0$; [120], Eq.(213)

for the value of the damping ratio ζ=0,05

within the range of stiffness ratio $0{,}04 \leqq \kappa = k_2/(k_1+k_2) \leqq 0{,}1$

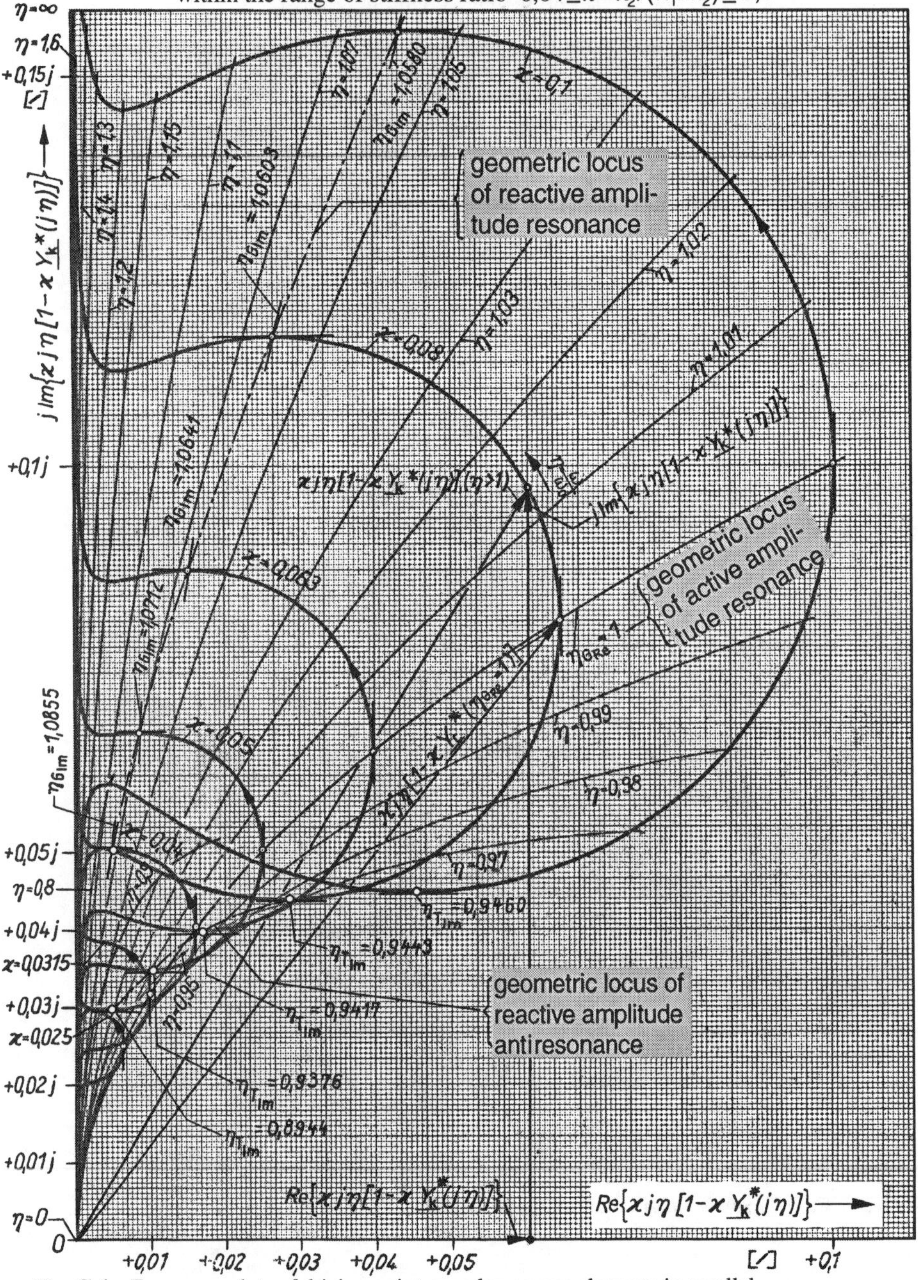

Fig. C.4. Frequency plots of driving-point complex power, elements in parallel, exciting force via additional spring

Power Frequency Plots

$\mathrm{Im}\{\,\mu \underline{X}_{km} j\eta[1-\mu \underline{Y}_m^*(j\eta)]\,\}$ - Re{...} - *curves*

Polar frequency-response loci (Nyquist plots)
of the normalized complex power at the driving point

$\mu X_{km} j\eta[1-\mu \underline{Y}_m^*(j\eta)]=\underline{S}_B^*(j\omega/\omega_0)/S_0$;[120], Eq.(217)

for the value of the damping ratio $\zeta=0{,}05$

within the range of mass ratio $0{,}04 \leqq \mu = m_2/(m_1+m_2) \leqq 0{,}1$

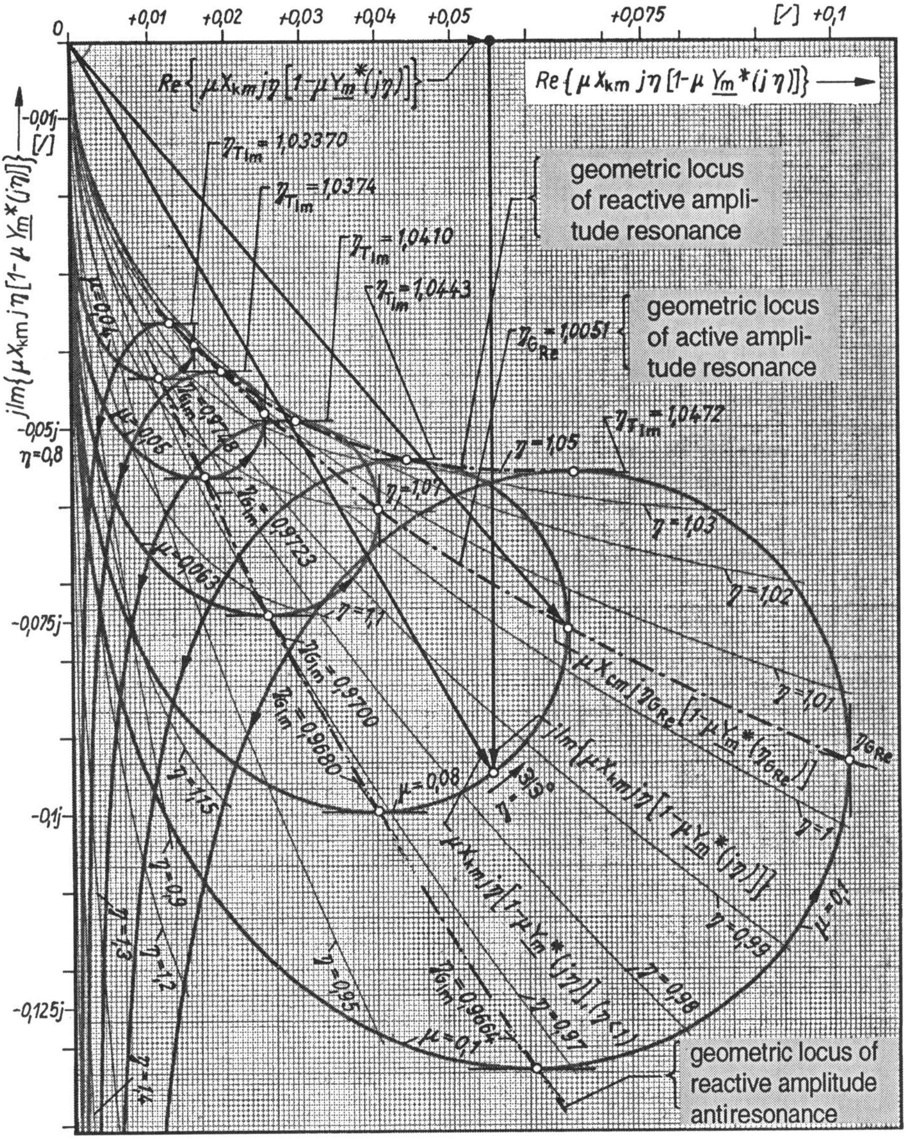

Fig. C.5. Frequency plots of driving-point complex power, elements in parallel, excitation force via unbalanced rotating mass

References

1. Küpfmüller, K.: Die Systemtheorie der elektrischen Nachrichtenübertragung. 3. Aufl. Stuttgart: S. Hirzel 1968
2. Meirovitch, L.: Introduction to Dynamics and Control. New York: J. Wiley 1985
3. Palm, W.J.: Control Systems Engineering. New York: J. Wiley 1986
4. Grantham, W.; Vincent, T.L.: Modern Control Systems Analysis and Design. New York: J. Wiley 1993
5. IEC 60 027-2: 1972 Letter symbols to be used in electrical technology. Part 2: Telecommunications and electronics
6. IEC 60 027-2A: 1975 First supplement. Chapter XI. Automatic control science and technology
7. IEV: 1991 International Electrotechnical Vocabulary. Chapter 351
8. Sommerfeld, A.: Lectures on theoretical physics. Vol. 3: Electrodynamics. New York: Academy Print 1961
9. MacFarlane, A.G.J.: Engineering Systems Analysis. London: G.G. Harrap 1964
10. Feldtkeller, R.: Einführung in die Vierpoltheorie der elektrischen Nachrichtentechnik. Stuttgart: S. Hirzel 1963
11. Molloy, C.T.: Four Pole Parameters in Vibration Analysis. In: Colloquium on Mechanical Impedance Methods for Mechanical Vibrations. ASME: New York 1958, 43–68
12. Hähnle, W.: Die Darstellung elektromechanischer Gebilde durch rein elektrische Schaltbilder. Wiss. Veröffentl. Siemens-Konz. 11 (1932) 1-23
13. Firestone, F.A.: New Analogy between Mechanical and Electrical Systems. J. Acoust. Soc. Amer. 4 (1933) 49–267
14. Trent, H.M.: Isomorphisms between oriented linear graphs and lumped physical systems. J. Acoust. Soc. Amer. 27 (1955) 500–527
15. MacFarlane, A.G.J.: Dynamical System Models. London: G.G. Harrap 1970
16. Guillemin, E.A.: Introducing Circuit Theory. New York: J. Wiley 1953
17. Crandall, S.H.: Impedance and Mobility Analysis of lumped Parameter Systems. In: Colloquium on Mechanical Impedance Methods for Mechanical Vibrations. ASME: New York 1958, 5–18
18. ISO 7626-1: 1986(E) Vibration and shock-Experimental determination of mechanical mobility. Part 1: Basic definitions and transducers
19. Blackburn, J.F.; Reethof, G.; Shearer, J.C.: Fluid Power Control. New York: J. Wiley 1960
20. McCloy, D.; Martin, H.R.: Control of Fluid Power. Analysis and Design. 2nd Ed. New York: J. Wiley 1980
21. Feldmann, D.G.: Untersuchung des dynamischen Verhaltens hydrostatischer Antriebe. Konstruktion 23 (1971) 420–428

22. Paynter, H.M.: Analysis and Design of Engineering Systems. Cambridge, Mass.: The M.I.T. Press 1961
23. Karnopp, D.; Rosenberg, C.: Analysis and Simulation of Multiport Systems. The Bond Graph Approach to Physical System Dynamics. Cambridge, Mass.: The M.I.T. Press 1968
24. Thoma, J.U.: Simulation by Bondgraphs. Introduction to a Graphical Method. Berlin: Springer 1990
25. Magnus, K.: Schwingungen. Eine Einführung in die theoretische Behandlung von Schwingungsproblemen. 2. Aufl. Stuttgart: Teubner 1969
26. Lehr, E.: Schwingungstechnik. Bd. 1. Berlin: Springer 1930
27. Truxal, J.: Introductory Systems Engineering. New York: McGraw-Hill 1971
28. Perkins, W.R.; Cruz, J.B.: Engineering of Dynamic Systems. New York: J. Wiley 1969
29. Skelton, R.E.: Dynamic Systems Control. Linear Systems Analysis and Synthesis. New York: J. Wiley 1988
30. Meirovitch, L.: Dynamics and Control of Structures. New York: J. Wiley 1990
31. Beards, C.F.: Vibrations and Control Systems. E. Horwood Ltd. New York: J. Wiley 1988
32. Blake, R.E.: Basic Vibration Theory. Chapt. 2 in: Shock and Vibration Handbook. 4th Ed. New York: McGraw-Hill 1995
33. Shabana, A.A.: Theory of Vibration. An Introduction. 2nd Ed. New York: Springer 1996
34. Steidel, R.F.: An Introduction to Mechanical Vibrations. 3rd Ed. New York: J. Wiley 1989
35. Fertis, D.G.: Mechanical and Structural Vibrations. New York: J. Wiley 1995
36. Papoulis, A.: The Fourier Integral and its Applications. New York: McGraw-Hill 1962
37. Hildebrand, F.B.: Advanced Calculus for Applications. Prentice-Hall. Englewood Cliffs, New Jersey: 1962
38. Schwartz, L.: Théorie des distributions. 2. Ed. T.I. Paris: Herman 1957
39. Zemanian, A.H.: Generalized Integral Transformations. New York: J. Wiley 1968
40. Meirovitch, L.: Analytical Methods in Vibrations. New York: The McMillan Comp. 1967
41. Vernon, J.B.: Linear Vibration Theory. Generalized Properties and Numerical Methods. New York: J. Wiley 1967
42. Ayre, R.S.: Transient Response to Step and Pulse Functions. Chapt. 8 in: Shock and Vibration Handbook. 4th Ed. New York: McGraw-Hill 1995
43. De Jong, R.G.: Statistical Methods for Analysing Vibrating Systems. Chapt. 11 in: see 42
44. Kreyszig, E.: Advanced Engineering Mathematics. New York: J. Wiley 1972
45. Aseltine, J.A.: Transform Method in Linear System Analysis. New York: McGraw-Hill 1958

46. ISO 7626-5: 1994 (E) Vibration and shock – Experimental determination of mechanical mobility. Part 5: Measurements using impact excitation with an exciter which is not attached to the structure
47. Gardner, M.F.; Barnes, J.L.: Transients in Linear Systems. Studied by the Laplace Transformation. Vol. 1. Lumped-Constant Systems. New York: J. Wiley 1950
48. Thomson, W.T.: Laplace Transformation. Theory and Engineering Applications. London: Longmans 1957
49. IEC 60027-3: 1989 Letter symbols to be used in electrical technology. Part 3: Logarithmic quantities and units
50. IEC 263:1982 Scales and sizes for plotting frequency characteristics and polar diagrams
51. Lynch, W.A.; Truxal, J.G.: Signals and Systems in Electrical Engineering. New York: Mc Graw-Hill 1962
52. Blackwell, W.A.: Mathematical Modeling of Physical Networks. New York: McMillan 1968
53. Klotter, K.: Technische Schwingungslehre. Bd. 2: Schwinger von mehreren Freiheitsgraden. 2. Aufl. Berlin: Springer 1960
54. Crandall, S.H.; Karnopp, D.C.; Kurtz, E. F.; Pridmore-Brown, D.C.: Dynamics of Mechanical and Electromechanical Systems. New York: McGraw-Hill 1968
55. Pestel, E.C.; Thomson, W.T.: Dynamics. New York: McGraw-Hill 1968
56. Thomson, W.T.: Theory of Vibration with Applications. London: George Allen & Unwin 1981
57. Prentis, J.M.; Leckie, F.A.: Mechanical Vibrations. An introduction to matrix methods. London: Longmans 1963
58. Pestel, E.C.; Leckie, F.A.: Matrix Methods in Elastomechanics, New York: McGraw-Hill 1963
59. Argyris, J.H.: Recent Advances in Matrix Methods of Structural Analysis. New York: Pergamon Press 1964
60. Desoer, Ch.A.; Kuh, E.S.: Basic Circuit Theory. New York: McGraw-Hill 1969
61. Firestone, F.A.; The Mobility Method of Computing the Vibration of Linear Mechanical Systems: Mechanical and Electrical Analogies. J. Appl. Physics 9 (1938) 128–134
62. Trent, H.M.: An Alternative Formulation of the Laws of Mechanics. J. Appl. Mech. 19 (1952) 147–150
63. Lehr, E.: Schwingungstechnik. Bd. 2. Berlin: Springer 1934
64. Lander, G.: Erweiterung und Präzisierung der Analogien zwischen elektrischen und mechanischen Schaltungen. Frequenz: Bd. 6, 1952, 235–246, 257–266
65. Federn, K.: Elektrisch-mechanische Analogien in erweiterter Form und ihre Anwendung auf elementare Schwingungsrechnungen. VDI-Berichte Bd. 35. Düsseldorf: VDI-Verlag 1959, 33–54
66. Cremer, L.; Klotter, K.: Neuer Blick auf die elektrisch-mechanischen Analogien. Ing.-Arch. 28 (1959) 27–38
67. O'Hara, G.J.: Mechanical Impedance and Mobility Concepts. J. Acoustic. Soc. Amer. 41 (1967) 1180–1183

68. Ewins, D.J.; Griffin, J.: A State-of-the-Art Assessment of Mobility Measurement Techniques-Results for the Mid-Range Structures (30 . . 3000 Hz). J. Sound and Vibration 78/2 (1981) 197–222
69. Lemon, J.R.; Sherlock, J.E.; Comstock, T.R.: Analyzing Dynamic Charakteristics of Machine Structures in Operation. ASME Design Engineering Conference Paper 67-DE-44, 1967
70. Klosterman, A.L.; Lemon, J.R.: Dynamic Design Analysis via the Building Block Approach. Shock & Vibration Bulletin 42 (1972) 1
71. Ewins, D.J.: Whys and Wherefores of Modal Testing. J. Soc. Environ. Eng. 18/3 (1979) 3–15
72. ISO 7626-7: 1990 (E) Vibration and shock – Experimental determination of mechanical mobility. Part 7: Modal parameter estimation
73. Plunkett, R.: Introduction to Mechanical Impedance Methods for Vibration Problems. In: see 17, 1–6
74. Duncan, W.J.; Biot, M.A.: Receptance replacing the term mechanical admittance. J. Roy. Aero. Soc. 58 (1954) 305
75. Bishop, R.E.D.; Johnson, D.C.: The Mechanics of Vibration. Cambridge: University Press 1960
76. Wundt, B.M.: Application of the Method of Dynamic Stiffness to the Evaluation of Transverse Frequencies of an Uniform Beam on Flexible Supports with Oil Film Considered. General Electric, Technical Information Series DF 56 TG 704, Fitchburg Mass. 1956
77. Caruso, W.J.: Prediction of Critical Speeds of Steam Turbines by Dynamic Stiffness Method. In: see 17, 137–145
78. Eisele, F.: Erregerquellen von Schwingungserscheinungen an Werkzeugmaschinen. VDI-Z 94 (1952) 843–848
79. Sadowy, M.: Zur rechnerisch-graphischen Bestimmung der dynamischen Steifigkeit und der dynamischen Nachgiebigkeit eines einfachen und eines gekoppelten Systems mit Dämpfung. Masch. Mkt. 63 (1957) WP 123–127, WP 141–146
80. Eisele, F.; Lysen, H. W.: Der Einsatz hochpolymerer Stoffe zur dynamischen Versteifung von Werkzeugmaschinen. Bericht 2. FoKoMa, S. 89–93, Würzburg: Vogel 1955
81. Eisele, F.; Korner, J.: Beeinflussung der dynamischen Steifigkeit von Werkzeugmaschinen durch Aufstellung und Zusatzdämpfung. Masch. Mkt. 67 (1961) 19–25
82. Loewenfeld, K.: Zusatzdämpfung von Werkzeugmaschinen durch Lamellenpakete. Masch. Mkt. 65 (1959) 28–35
83. Corbach, K.: Die dynamische Steifigkeit ruhender und beweglicher Verbindungen an Werkzeugmaschinen. Masch. Mkt. 72 (1966) 2099–2109
84. Sadowy, M.; Armbruster, A.: Statische und dynamische Steifigkeit von Werkzeugmaschinen bei Verwendung von Dämpfungselementen im Kraftfluß der Maschinen. Konstruktion 25 (1973) 351–356
85. Stefaniak, H.St.; Tröster, R.: Dynamische Eigenschaften einer Dämpfungseinrichtung mit elastisch nachgiebigen Zylinderböden. Masch. Mkt. 72(1963) WP 162–166
86. Weck, M.; Teipel, K.: Das dynamische Verhalten spanender Werkzeugmaschinen. Ind. Anz. 100 (1978) 26–28

87. Ewins, D.J.: State-of-the-Art Assessment of Mobility Measurement Techniques. J. Soc. Environ. Eng. 20/1 (1981) 3–13
88. Federn, K.: Beherrschung und Ausnutzung von Schwingungen als Konstruktionsaufgabe. VDI-Z. 100 (1958) 1220–1232
89. Woernle, H.-Th.: Kinetische Nachgiebigkeit und Steifigkeit. ZAMM 42 (1962) 89–102
90. Peters, E.: Neue Versteifungsbauart für Werzeugmaschinenbetten. Werkstattstechnik 14 (1920) 441
91. Krug, C.: Die DISKUS-Stahlzellenbauweise. Werkstattstechnik 27 (1933) 15–17
92. Krug, C.: Der Stahlbau bei Werkzeugmaschinen. Werkstattstechnik 30 (1936) 201–205
93. Bobek, K.; Heiß, A.; Schmidt, F.: Stahlleichtbau von Maschinen. Konstruktionsbuch. Bd.1, 2. Aufl. Berlin: Springer 1955
94. Weck, M.; Petuelli, G.: Steifigkeits- und Dämpfungskennwerte verschraubter Fügestellen. Konstruktion 33 (1981) 241–245
95. Krug, H.: Leichtbau von Werkzeugmaschinen. VDI-Berichte, Bd. 28, 105–112, Düsseldorf: VDI-Verlag 1958
96. Koblischek, P.J.: Synthetic resin-bound concrete. Proc. 1st Internat. Congr. an Polymer Concretes, London: 1975. GB-Lancaster: The Contruction Press 1976, pp. 409–419
97. Stawowy, J.: Polymer concrete made on polyester resin basis with very low shrinkage and the possibilities of reinforcement resulting thereof. Proc. 4th Internat. Congr. on Polymers in Concrete (ICPIC), Darmstadt 1984. Marl: Hüls Publication 5227, 1985
98. Körner, J.: Beeinflussung der dynamischen Steifigkeit von Werkzeugmaschinen durch Zusatzdämpfung und dämpfende Aufstellung. Diss. TH München 1966
99. Saegert, G.: Die elastische Lagerung als Mittel zur Aktiv- und Passivisolierung von Maschinen. In: VDI-Berichte: Dynamik von Maschinen, Bauwerken und Strukturen. Nr. 978, S. 109–148. Düsseldorf: 1992
100. Reichert, G.: Helicopter Vibration Control-A Survery. 6th European Rotorcraft and Powered Lift Aircraft Forum. GB University Bristol: 1980, 124–136
101. Meller, T.; Frühauf, F.: Variable Damping-philosophy and experiences of a prefered system. Proceedings of the IMech E. London 1988, 113–118
102. Hagemeister, W.: Hochdynamische hydraulische Aktoren zur aktiven Lagerung von Fräsmaschinenspindeln. In: Tagungsband 12. Aachener Fluidtechnisches Kolloquium, Aachen 1996, 383–400
103. Pfeiffer, F.: Präzision durch Echtzeit. Aktive Schwingungsdämpfung erhöht Effizienz von Robotern. KEM 25 (1988) 120
104. Scott, R.E.: Linear Circuits. Part 1. Reading, Mass.: Addison-Wesley 1960
105. DIN EN ISO 7500-1:1999 Metallic materials – Verification of static uniaxial testing machines. Part 1: Tension/compression testing machines –Verification and calibration of the force measuring system
106. ISO 7626-2. 1990(E) Vibration and shock – Experimental determination of mechanical mobility. Part 2: Measurements using single-point translation excitation with an attached vibration exciter

107. VDI 205-210: 1956. Messung mechanischer Schwingungen
108. Sawla, A.: Measurement of dynamic forces and compensation of errors in fatigue testing. Proceedings of the 12th IMEKO World Congress "Measurement and Progress". China, Beijing: 1991
109. Nedelcu, V.N.: Einheitliche Leistungstheorie der Mehrphasensysteme. ETZ-A 84 (1963) 153–157
110. Fleck, C.L.P.: Einrichtung zur Ausgleichung von Beschleunigungsdrucken durch Spannwerke. DRP 50945. Berlin: 1889
111. Schieferstein, H.: Wirtschaftlichkeit der in resonanzbetriebenen Kraft- und Arbeitsmaschinen. VDI-Z 77 (1933) 69–73
112. Späth, W.: Theorie und Praxis der Schwingungsprüfmaschinen. Berlin: Springer 1934
113. Berg, S.: Gestaltfestigkeit. Versuche mit Schwingern. Düsseldorf: Deutscher Ingenieur Verlag 1952
114. IEC 68-2-27:1987 Basic environmental testing procedures. Tests: Shock
115. ISO 1099:1975 Metals-Axial load fatigue testing
116. ISO/TC 164/SC 5-1 Methods of fatigue testing. Part 1: Guide to general principles
117. ISO 5344: 1980 (E) Electrodynamic test equipment for generating vibration: Methods of describing equipment characteristics
118. ISO/TC 108/SC 6 Vibration and shock generating equipment: Vibration generating equipment guidance
119. DIN EN ISO 7500-1 Supplement 3: 1999 Metallic materials – Verification of static uniaxial testing machines. Part 1: Tension/compression machines-Verification and calibration of the force measuring system – General for requirements, verification and calibration of fatigue testing machines
120. Findeisen, D.: Dynamisches System Schwingprüfmaschine. Fortschr.-Ber. VDI-Z., Reihe 11, Nr. 18. Düsseldorf: VDI-Verlag 1974 (Diss. TU Berlin 1974)
121. Findeisen, D.: Rechnerische Aussagen in der Konzeptphase über das dynamische Verhalten am Anwendungsbeispiel Schwingprüfmaschine. Konstruktion Bd. 27 (1975) 223–232
122. Findeisen, D.: Durchführung des Prüfverfahrens Schwingfestigkeitsversuch. Anforderungen an die Schwingprüfmaschine aus systemtheoretischer Sicht. Materialprüfung 17 (1975) 32–41 und 96–104
123. Findeisen, D.: Berechnung transienter Schwingungen bei stoßartiger Einwirkung unter Anwendung linearer Integraloperatoren. Forsch. Ing.-Wes. 45 (1979) 137–151
124. Findeisen, D.: Gerätetechnische Verwirklichung von Schwingprüfmaschinen. Zwanglaufantriebe. Fortschr.-Ber. VDI-Z., Reihe 1, Nr. 116. Düsseldorf: VDI-Verlag 1984 (Habil. TU Berlin 1984)
125. Findeisen, D.: Ungleichförmig übersetzende Getriebe durch fluidtechnische Energieübertragung. Anwendungsbeispiel Schwingprüfmaschine. Konstruktion 37 (1985) 133–140

Index

2-port parameters 21, 263

across variable 4
– energy 7
– power (effort variable) 7
active power 269, 273, 290
– actual 290
– parameter 296, 305
actuator 42
admittance matrix 24
amplifier 33
amplitude ratio 173, 238
amplitude spectrum 100, 106, 327
analysis (relationship) 202, 236
– circuit transform (loop velocities) 237
– general lumped-system 66, 203, 236, 249
– vertex transform (node forces) 237
analysis, (deterministic) vibration data
– Fourier series (harmonic) 96, 100, 106
– frequency-response (spectral) 105, 106, 108, 167, 174, 180, 198, 202, 207, 212, 217
– phasor-response 89, 90, 93
– time-response (real-time) 76, 108, 144
– transfer-function 143, 145, 167
analysis random vibration data 127, 129, 136
– correlation 127, 135, 174
– spectral density 128, 134, 174
analytic function 117, 156
angular displacement 219, 253
apparent power 288, 297
average power 269, 273, 287

back effect 41, 72
band-pass filter 137, 190
bandwidth method 183
beating 88
bilateral signal flow 39, 72, 258
block diagram 9, 71
Bode plot 186, 347, 351
bond graph 63, 72
– acausal 65
– augmented 66
– system 68, 69
branch 58
breakpoint (corner) frequency 187

cascade connection 11, 22, 31, 46, 280
causal description 258
causal matrices 262
causality principle 114
chain (transmission) matrix 21, 22, 48
chain (transmission) parameters 21
characteristic equation 78, 148
characteristic parameter (of a sinusoid) 93
– magnitude 84, 92
– phase angle 84, 92
characteristic values (eigenvalues) 119, 148
circuit (loop) 58
circuit diagram 48
closed-loop system 12
combined-flow diagram 58, 61
combined motion 87, 162, 322, 323
comparator 13
compensative energy transaction 291
complete (total) response (combined motion) 86, 142, 144, 162
complex angular frequency 139
complex power
– input 266, 268
– output 266, 271
– transmission factor 270, 275 (Rayleigh quotient)
– transmission ratio 265 (of phasor products)
complexor (phasor ratio) 92, 172, 207
component (element) dynamic characteristics
– dynamic compliances 212, 225, 228, 242
– dynamic stiffnesses 214, 220, 238, 300
– mechanical impedances 221, 238
– mobilities 212, 225, 229
component relationship 28, 50, 65, 202
– constitutive (material) 204, 205

– dynamic (temporal) 204, 205
composite 2-port 22
composite mechanical system 217
computerized model 3
concept of supernodes 249
condition of beating 89
condition of resonance 88, 182
conformal mapping 166
conservative system 51
constant-force generator 37
continuous system 47, 281
contour integral
– corresponding 118, 156
– intended 121
contral law 13
control system 2
control system design 13, 247
control system structure 12
control theory 77
convergence criterion 103, 129, 141
convolution approach 112, 123, 131, 135, 153, 165
convolution integral 112, 154
– approximate evaluation 115, 333, 334
– graphical interpretation 115, 332
corner (breakpoint) frequency 187
corner frequency ratio 188
correction gain 191, 194
correlation function 127
– auto- 127, 131, 135, 340
– cross- 128, 130
– filter 132, 135
correspondence 103, 140, 154
coupler 33, 60
– converter 33, 47
– transformer 33
critical damping coefficient 76
cutset equations 249, 251, 254
cut-set
– of a network 58
– simplification 253

d'Alembert's principle 54, 199
damper 30, 51
damping coefficient
– attenuation 75
– (viscous) linear 74
damping ratio (fraction of critical damping) 76
data processing 106, 125, 180
delta function 109, 167
delta functional 135, 168
design specifications 237, 244, 316
detailed model 47
detuning (tuning off resonance) 246
diagram 9, 71
– schematic (visually descriptive) 9, 70
– systematic (interconnection) 9, 71
differential equation 55, 60, 75
direct drive 249
displacement (response) 74, 75
– amplitude 84
– phase angle 84
displacement response factor (amplitude ratio) 85, 92, 173
distribution (generalized function) 104, 194
domain
– original (time-; t-) 103
– subsidiary, corresponding (frequency; ω -; p-) 103, 140
driving power of vibrators 298
Duhamel's integral 112
duty cycle 99
dynamic characteristics (of a structure) 207, 231, 235
– component (element) 212, 214
– overall (composite) 208, 217
dynamic compliance (receptance) 174, 207
– concept of 207, 211, 297
– direct (driving-point) 177, 208, 223, 229, 253
– plane 174, 226, 230, 345, 357
– techniques (analysis) 249, 252
– transfer 177, 208
dynamic equilibrium statement 54
dynamic stiffness 178, 208
– concept of 211
– converted (direct) 305, 310
– direct (driving-point) 178, 208, 221, 300
– plane 222, 349, 359
– transfer 178, 208, 299
dynamic system 2
dynamics 199
– analytical 199
– synthetical 200

effort variable 7, 21, 26, 64, 258
effort-flow characteristic (frequency responses of individual components) 265
effort-flow couple (individual components) 20, 52, 64, 262, 265, 299, 309
effort-flow product (actual energy flow) 265, 274, 285, 302, 309

electromechanical analogy 26
– dual (force-to-voltage) 26, 31, 38
– true-connected (force-to-current) 26
element, general type of 5, 50
– P-storage 5, 30, 51
– T-storage 5, 28, 50
– dissipator 30, 51
element force 50, 94, 240, 243, 300
– elastic (spring) 50, 300
– inertial (mass) 51, 300
– damping (damper) 52, 300
element laws, fundamental (linear element characteristics) 28, 50, 213
element parameters 50, 74
– elastic (spring) constant (stiffness) 50, 74
– linear (viscous) damping coefficient 51, 74
– mass 51, 74
elementary transform pair 161, 164, 169, 170, 171
energetic system approach 261
energetic system 257, 283
energy continuity 67
energy dissipation 51, 65
energy flow characteristic (frequency response of power) 265, 286, 309, 311
energy transaction 257
– compensative 291
– dissipative 290
equation of motion 55, 60, 74
equivalent viscous damping coefficient 81
ergodic process 126
excitation (input) 75
– actual 90
– complex 90
– phasor 92
– transform (spectral density) 102, 104
excitation (function, types of)
– harmonic, simple (sinusoidal) 84
– non-periodic 98, 104, 109
– periodic 96
– pulse-train 98
– random 129
– single-pulse 104
– singularity functions 167
– transient (shock) 109, 116, 143, 154
excitation (quantity)
– displacement 178, 249, 298
– force 37, 52, 65, 74, 177
– torque 253
– velocity 37, 52 65, 178
expansion theorem 161
experimental modelling 1
fatigue testing machine 249, 296
feedback control system 12
filtering property (mechanical) 137, 189
finite element method 201
flat tuning 246
flexibility influence coefficients 200
flow (general)
– energy 58, 63
– signal 58, 63
flow of (power and) energy 257
– dissipative (average) 279, 287, 290, 311
– idle (wattless) 279, 287, 290, 311
flow variable 7, 21, 26, 64, 258
fluid capacitance 47
fluid conductor 47
fluid impedance 47
fluid inductance 47
fluid resistance 47
force excitation (external force) 75, 84, 98, 104, 141, 167
force transmissibility 28, 252
forced vibration 83, 85, 93, 322
forcing (driving) frequency
– angular 84, 181
forcing function (excitation)
– harmonic, simple (sinusoidal) 84
– non-periodic (transient) 98, 104, 116, 141, 167
– periodic 96
forcing period 84
Fourier integral 102
Fourier (series)
– amplitude 96
– coefficient, complex 96, 99
– expansion 97, 101
– phase angle 96
– series, complex 96, 97
Fourier spectrum 106
– continuous 105, 329
– line 100, 326
Fourier transform 102, 104
Fourier transform method 106, 107
Fourier transform pair 103
Fourier-transformable function 103
free vibration 79, 82, 83, 319
frequency
– angular (circular) 76, 102
– cyclic 76
frequency normalization 180, 307
frequency ratio 85, 180
frequency response 174, 180
frequency-response characteristic 109, 130, 146

– Fourier (ω-domain), frequency-response function 109, 130, 134
– Laplace (p-domain), transfer function 146, 166
frequency-response function
– of a structure (dynamic characteristic) 176, 198, 207, 283
– direct (driving-point) 177, 253, 265
– transfer 177, 251, 265
frequency response plot 176, 343
– logarithmic frequency plots (Bode) 186, 347, 351
– polar (Nyquist) 176, 345, 349
– rectangular, logarithmic (Nichols) 184, 346, 350
frequency response plot of power 308, 361
– polar power frequency plot 309, 312, 363
– power component frequency plot 313, 364, 365
functional block diagram 10
fundamental configurations 11, 22, 28, 67, 219, 232
– cascade (tandem) 11, 22, 31, 46, 280
– mixed (parallel-series) 24, 219, 232
– parallel 12, 24, 31, 220
– series 23, 227
fundamental frequency 97
fundamental sets of network equations 206

Gaussian distribution 126, 133, 136, 342
generalized function (distribution) 104, 113, 115, 135, 168, 194
generalized impedance 281
generalized Kirchhoff's laws 66, 203, 236
generalized transport process (2-port) 261, 262
generic system 68, 74, 187

half-power method 183, 314
Hamilton's principle 200
harmonic frequency 96
harmonic number 99
Heaviside function (unit step) 112, 169
Hermitian form 269
high-frequency asymptote 190, 193
high-Q network 314
hydraulic servomechanism 43
hydrostatic power transmission 47

impedance angle 296
impedance matrix 24
impulse (pulse area) 99, 104
indicial mechanical impedance 240
individual conjugate variables (pair of rate variables) 20, 52, 64, 262, 265, 299, 309
inertial frame 53
inertial reference system 53
instantaneous power 52, 258, 285
instantaneous value 94
integral theorem, Cauchy's 117, 156
integral transform method 106, 144, 194
integral-transformed model 95, 198, 212, 236
interacting loop (back effect) 41, 72
interconnection requirement 22, 54, 66, 202, 236
– force (equilibrium statement) 57, 204, 217
– motion (compatibility requirement) 57, 204, 218
interconnective (spatial) relationship (dual; pair of postulates) 22, 57, 66, 202
– boundary (circuit; across-measurement principle) 57, 203, 217
– incidence (vertex; through-measurement principle) 57, 203, 218
inverse Fourier transform 102
inverse Laplace transform 139
inversion formula approach 117, 134, 155
isolation amplifier 41

junction, ideal energy 67
– flow- (0-) 67, 68, 69
– effort- (1-) 67, 68, 69
junction structure 68

Lagrangian mechanics 200
Laplace integral 139
Laplace transform 139
Laplace transform method 143, 144
Laplace transform pair 140
Laplace-transformable function 141
linearity 13
logarithmic decrement 80
logarithmic gain 184, 187
– dropping 190
loop (circuit) 58
low-frequency asymptote 190, 193
low-loss network 314
low-pass filter 189, 191
lumped parameter model 74
lumped-parameter equivalent system 48
lumped-system analysis, general 202, 232, 236, 249

magnification factor 85, 189
major resonant vibration 94, 245
mass 30, 51
mass cancellation 252
mass matrix 56
mass-damper-spring system 74
mathematical model 2
mathematical modelling 1
matrix of the form 267
maximum height (pulse) 99, 104
mean value of the instantaneous power 287
mean-square value 127, 129, 132, 133, 136
mechanical 2-port 25
– composite (structure) 30
– elementary (degenerate) 28
– supplementary 32
mechanical admittance (mobility) 208
mechanical bond graph 68
mechanical circuit 53
mechanical circuit theorems 231
mechanical circuitry 202
mechanical duality 204
mechanical elements (1-ports)
– elementary (fundamental) 50, 57, 65, 74
– supplementary 50
mechanical generator (source)
– complete 36, 37
– ideal constant-force 35, 37
– ideal constant-velocity 35, 37
mechanical impedance 27, 178, 208
– concept of 209, 282
– direct (driving-point) 178, 208, 221
– plane 222
– transfer 28, 178, 208
mechanical multiport 68
mechanical network analysis 202
mechanical plant 77
mechanical system design 237, 310
mechanical two port 25, 28, 30, 32
mechanics
– analytical (Lagrangian) 200
– vectorial (Newtonian) 199
method of influence coefficients 200
method of matrix iteration 201
method of residues 119, 159
minor resonant vibration 94, 244
mixed connection (2-port configuration) 24
mixed domain dynamic system 58
mixed physical domain 63
mobility measurements 207, 212
mobility (mechanical)
– concept of 207, 209
– direct (driving-point) 178, 208, 224, 229
– plane 226, 230
– transfer 28, 178, 208
modal analysis 210
model system 2
model validation 3
model verification 3
modified Hermitian form 267
modified quadratic form 272
modified Rayleigh quotient 274
modulator 41
modulus (magnitude) of
– dynamic compliance 243
– dynamic stiffness 239, 244, 301
– mechanical impedance 239
– mobility (mechanical) 243
moment of inertia 253
multiple-order pole 163
multiport 63
– component (basic) 64, 65
– interconnected (bonded) 64, 66
– field 69
multivariable system 39

narrow-band filter 137, 190
natural frequency
– damped 79
– undamped 76
natural period 76, 79
network diagram 48, 53, 72
Nichols plot 184, 346, 350
node (junction point; vertex) 54, 58, 300
noncausal description 257
non-conservative system 51
non-dimensional (natural) time 76
nonloading element 41, 72
normal distribution 126
normalized dynamic characteristic (element; overall)
– dynamic compliance 181, 189, 242
– dynamic stiffness 192, 240, 241, 301
– mechanical impedance 240, 241
– mobility (mechanical) 243
Norton's theorem 235
Nyquist plot 176, 345, 349

one port 65
open-loop system 12
orientation conventions, general 19, 20, 54, 57, 66, 202
oriented line segment 56

oriented linear graph 56
original variable (Fourier, Laplace)
– time 103, 139
overall (composite) dynamic characteristic
– dynamic compliance 217, 228
– dynamic stiffness 220
– mechanical impedance 221
– mobility (mechanical) 217, 229
overdamped system 81

parallel connection 12, 24, 31, 220
parameter estimation
– element 243
– system 248, 311
partial fraction 158, 164
partial noninteraction 45
path (branch) 58
path of integration 118, 156
performance criteria (control system design)
– robustness and linearity 13
– speed and accuracy 13, 17, 18
– stability characteristics 13, 18
performance criteria (mechanical system design; structural design)
– driving (rod) force amplitude 239, 301
modulus of dynamic stiffness 244, 246
– operating motion amplitude 243
– related energy dissipation (damping effect) 240
– stability criteria (of motion) 148
performance criteria (stationary energy flow; driving power)
– high-Q (low-loss network) 314
– power efficiency 278, 280
– power factor 297, 313, 314
– power parameters, significant 294, 296, 305
– power transmission factor 270, 276, 277
– transmission ratio (of power products) 265
– zero reactive power 314
performance equation (force-motion) 55
phase difference 85, 173, 189
phase lag (phase shift) 85, 189
phasor 27, 90
– configuration (polygon) 94, 222, 225, 230, 300
– constant (resting) 93, 286
– diagram 95, 216, 217, 223, 226, 231, 298, 301
– equation 91, 180
– method 91
– time-varying (rotating) 93, 286
phasor (of physical quantities) 90, 92
– acceleration 299
– displacement 90, 92
– dynamic characteristic 179, 180, 208, 221, 230, 299, 300
– force 92, 94, 300
– phasor ratio (complexor) 27, 172, 177, 179, 280, 207, 214, 236
phasor power (modified phasor method)
– concept of 261, 285, 293, 294, 297
– diagram 306
– driving (input) 304
– representation (vector) 286, 288, 291, 293, 306
physical system 2
pole (singularity)
– simple (first-order) 118, 159, 163
– multi-order 163
position control 15, 60
position control system 16, 61
potentiometer system 42, 60
power (stationary energy flow)
– actual (instantaneous) 285, 302
– complex 266, 293, 304, 307
– parameters (significant) 269, 288, 290, 294, 296, 305
– plane, complex 294, 308, 310, 363
power components
– active 269, 273, 290
– average 287
– reactive 270, 274, 290
– wattless 287
power decomposition, spectral
– polar form 286, 288
– rectangular form 269, 289, 291
power spectral density
– auto- 128, 132, 134, 341
– cross- 130
power state (transformation of)
– causal (vector) 258
– acausal (scalar) 257, 266
– steady-state 260
power transactions
– of energetic systems 257
– in vibrations 261
power transmission factor (Rayleigh quotient)
– active 270
– reactive 277
power-factor correction 314
probability density function 127, 136, 342

problem (dynamic system)
– analysis 1, 2, 3, 70, 198, 207, 210, 231, 252, 261
– instrumentation (measurement; calibration) 251, 261
– synthesis (design) 18, 70, 237, 244, 261, 298, 310
problem (of a 2-port network)
– insertion 263
– transfer 263, 265, 279
– transmission 263, 265, 279
processing of transducer signals 252
proportional control 16
pulsatance (angular frequency)
– real 102
– complex 139
pulse control factor 99
pulse duration 99
pulse-type function (transient excitation) 109, 116
P- (through) variable
– rate (flow variable) 7, 20, 50
– state 7

Q factor 182, 314
quadratic form 272
quotient of Fourier transforms 178

random process 126
rational function (complex variable function)
– factored form 119, 158, 163
– partial fraction (expansion) 158, 161, 164
– ratio of polynomials 155
reactance theorem 234
reactive power 270, 274, 290
– actual 290
– parameter 297, 305
receptance (dynamic compliance) 174, 208
reciprocity theorem 233
reduction of diagram (reduction rules)
– block 18
– bond graph 69
– network 53, 219, 232
– signal-flow graph 19
relationship (topological)
– component (constitutive; dynamic) 28, 65, 202
– interconnective (boundary; incidence) 22, 57, 66, 202
repeated structures (parallel-series configurations) 232, 249
residue theorem, Cauchy's 117, 157
resonance
– condition of 88
– of amplitude 182, 186, 192
– of phase 182, 186, 193
resonance frequency (displacement) 182
resonance frequency ratio 182
response (output), displacement
– actual 75, 93, 283
– complex 93, 284
– phasor 92
– transform 108, 117, 155, 236
response calculation 76, 111, 153
– classical approach 77, 83, 86, 89, 107
– convolution approach 112, 123, 131, 135, 153, 165
– inversion formula approach 117, 134, 155
response characteristic
– frequency- 109, 130, 146, 166
– time- 109, 112, 129, 147
response data plotting
– frequency response plot 176, 343
– power frequency plot 308, 361
– random data plot 137, 340
– shock response spectrum 124, 339
– time-history curve 162, 165, 317
review of dynamic compliance concept 211
review of dynamic stiffness concept 211
review of energy concept 257
review of mechanical impedance concept 209, 282
review of mobility concept 209
review of phasor method 284
review of phasor power concept 294
review of vibration-reduced design 246
root location 150
root-mean-square (r.m.s.) value (standard deviation) 137, 138
root-locus method 150

scalar triple product 289
selecting vibratory specifications 237
series connection 23, 227, 228
set of network equations (generalized Kirchhoff) 204, 236
– circuit (loop) 204, 237
– vertex (node) 204, 236
shock excitation (applied shock) 103, 106
– mechanical shock, ideal (constant; simple step) 112, 123, 167, 169, 336
– shock pulse, ideal (simple time function) 109, 335

– shock pulse, rectangular 104, 109, 331, 337, 338
– shock pulsating 98
shock motion (history) 116, 337, 338
– primary response 121, 123
– residual response 118, 123
shock response spectrum, displacement 124, 339
si-function 99
signal 4-pole 39, 263
signal amplifier 41
signal flow graph 18, 71
signal-flow-diagram 9
significant frequency 182, 313
simplification of networks 232, 249
– cut-set (analysis) 249, 253
– repeated configurations (parallel-series connections) 219, 220, 232, 249, 253
simulation block diagram 10
single degree-of-freedom system (mass-damper-spring system) 74, 179, 298, 314
single-variable system 9
singularity functions (typical wave forms) 167
– unit pulse 109, 167
– unit ramp 169
– unit step 112, 169
singularity function responses (time characteristics)
– unit pulse response 110, 147, 169, 335
– unit ramp response 170
– unit step response 113, 169, 336
solution 77, 83
– original (t-domain) 108, 143
– transformed (ω-domain) 108; (p-domain) 144
source (generalized)
– P-variable (flow variable) 34
– T-variable (effort variable) 34
source, electrical
– ideal current 35
– ideal voltage 35
– independent current 36
– independent voltage 36
source, mechanical (generator)
– complete 37
– ideal constant-force 37, 52, 65
– ideal constant-velocity 37, 52, 65
source, controlled 59
specifications (system parameters)
– control 77
– vibratory 75, 85, 180, 189
specifications, response
– frequency 106, 189, 238, 343
– frequency, of power 312, 361
– statistical 189, 317
– steady-state (frequency-response) 88, 317, 322, 323
– transient (time-response) 80, 88, 110, 113, 148, 317, 319, 320
specifications of polar plots 182
specifications of polar plots of power 312
spectral density 102, 106, 128
spring 28, 50
square wave (pulsating) 100
stable
– asymptotically 148, 149, 153
– neutrally 148, 153
– un- 148
standard deviation 136, 342
standard transform pair (of convolution) 115, 154
static equilibrium condition 55
static response factor 180, 187
stationary process 126
steady state of energetic systems 260
step-type function (transient excitation) 143, 154
stiffness (elastic constant) 50, 74
stiffness influence coefficients 201
stiffness matrix 56
stochastic process 126
straight-line approximation 186, 348, 352
subsidiary variable
– real (angular) frequency (Fourier) 102
– complex frequency (Laplace) 117, 139
summer 12
superposition (convolution) integral 112
superposition theorem 233
system parameters
– complex 92, 109, 146, 172, 207
– control 77, 82
– statistical 136
– vibratory 75, 85, 180, 189, 248, 311
systems modelling 1, 7, 9, 18, 48, 63, 69
– first step of 3

theory of analytical functions (complex variable functions) 117, 156
Thévenin's theorem 234
three-degree-of-freedom torsional system 253
three port 66
through variable 4
– energy 7
– power (flow variable) 7

time constant 76, 77
time-constant representation 77, 82, 187
time-(t-)domain 76, 103
time response (actual displacement)
- complete (total) 86, 87, 88, 89, 162, 165, 317, 322, 323
- steady-state (forced vibration) 85, 171, 317, 322, 323
- transient (free vibration) 79, 82, 166, 317, 319, 320
time-response characteristic 109, 112, 147, 150, 152
- unit pulse response (weighting function) 110, 129, 134, 147, 335
- unit step response (Heaviside response) 113, 122, 336
torsional stiffness 253
torsional vibration 252
total noninteraction 41
transducer 33
transfer function 146
transfer function block diagram 10
transfer function concept 10, 166
transfer matrices 201
transform analysis methods 198
transform methods (integral) 95, 194
transformed component relationships 214
transformer 33
transient excitations (shock excitations) 106
- pulse-type function 109, 116
- step-type function 147, 154
- transient harmonic 141, 142, 155
transmission loss 47
transmission matrix 22, 48, 262
transmission matrix of the ideal transformer 34
transmission matrix of the elements (fundamental mechanical) 29, 30
transmission matrix of the signal 4-pole 40
transmission parameters 21, 22, 263, 265
transmission system 9
true-connected analogy 26
tuning property (mechanical) 246
T- (across) variable 4
- rate (effort variable) 7, 20, 50
- state 7
two-degree-of-freedom (two-mass) system 49, 53, 61, 205, 250
two port, linear 66, 262
two-port network 20, 262
two-port diagram 20, 72
two-terminal-pair network 40

underdamped system 78

variable, dynamic system 4, 7, 74, 262
- P- (through; 1-point) 4
- T- (across; 2-point) 4
variable (transform method)
- original (time) 103, 139
- subsidiary (frequency) 103, 139
variance 136
vector representation (phasor method) 91
- of harmonics 97
- of phasor ratios (complexors) 172, 215, 216, 222, 225, 230, 301
- of sinusoids 93, 298
vector representation of power (modified phasor method) 285, 286, 292
- complex (input) power (input phasor power) 293, 306
- decomposition, polar form 288
- decomposition, rectangular form 291
- instantaneous power (total vector) 286
velocity transmissibility 28
vibration, types of
- deterministic 126
- (simple) harmonic (sinusoidal) 83, 85, 89, 93
- non-periodic (aperiodic) 101, 120, 122
- periodic (multi-sinusoidal) 96
- random 125, 126, 129
- steady-state (forced) 86, 93, 162
- transient (free) 79, 82, 83, 139, 162
vibration and shock generating equipment 296
vibration generator, mechanical
- direct-drive (crank connecting rod-; joke-; cam-type) 249, 295, 298, 314
- reaction type (rotating; reciprocating masses) 295, 314
vibration generator system
- mechanical 249, 296
- electrodynamic 249, 296
- hydraulic (servo-motor) 44, 249, 296
vibration machines (utility) 261, 286, 298
vibration-reduced structure
- by mechanical detuning (resilient heavy weight; stiff lightweight construction) 246
- by motion control 247, 248

white noise 133
white random vibration 133
Wiener-Khinchin equations 128

zero state 86
zero-driving (input) response 144
zero-state response 86, 144

Printing: Saladruck, Berlin
Binding: Buchbinderei Lüderitz & Bauer, Berlin